ISNM
International Series of Numerical Mathematics
Vol. 124

Variational Calculus, Optimal Control and Applications

International Conference in honour of L. Bittner and R. Klötzler
Trassenheide, Germany, September 23–27, 1996

Edited by

W.H. Schmidt
K. Heier
L. Bittner
R. Bulirsch

Springer Basel AG

Editors:

Werner H. Schmidt
Knut Heier
Leonhard Bittner
Ernst-Moritz-Arndt-Universität Greifswald
Institut für Mathematik und Informatik
Jahnstr. 15a
D-17487 Greifswald

Roland Bulirsch
Technische Universität München
Zentrum Mathematik
Arcisstr. 21
D-80290 München

1991 Mathematics Subject Classification 93-06, 49-06, 35-06

A CIP catalogue record for this book is available from the Library of Congress, Washington D.C., USA

Deutsche Bibliothek Cataloging-in-Publication Data
Variational calculus, optimal control and applications /
International Conference in Honour of L. Bittner and R. Klötzler,
Trassenheide, Germany, September 23–27, 1996. Ed. by W. H.
Schmidt ... – Basel ; Boston ; Berlin : Birkhäuser, 1998
(International series of numerical mathematics ; Vol. 124)
ISBN 978-3-0348-9780-8 ISBN 978-3-0348-8802-8 (eBook)
DOI 10.1007/978-3-0348-8802-8

Originally published by Birkhäuser Verlag, Basel, Switzerland in 1998
Softcover reprint of the hardcover 1st edition 1998
Printed on acid-free paper produced of chlorine-free pulp. TCF ∞
Cover design: Heinz Hiltbrunner, Basel

ISBN 978-3-0348-9780-8

9 8 7 6 5 4 3 2 1

Contents

2. Analysis and Synthesis of Control Systems and Dynamic Programming

3. Numerical Methods and their Application to Flight Path Optimization and Fluid Dynamics

4. Applications to Mechanical and Aerospace Systems

Preface

The 12th conference on "Variational Calculus, Optimal Control and Applications" took place September 23–27, 1996, in Trassenheide on the Baltic Sea island of Usedom. Seventy mathematicians from ten countries participated. The preceding eleven conferences, too, were held in places of natural beauty throughout West Pomerania; the first time, in 1972, in Zinnowitz, which is in the immediate area of Trassenheide.
The conferences were founded, and led ten times, by Professor Bittner (Greifswald) and Professor Klötzler (Leipzig), who both celebrated their 65th birthdays in 1996. The 12th conference in Trassenheide, was, therefore, also dedicated to L. Bittner and R. Klötzler. Both scientists made a lasting impression on control theory in the former GDR.
Originally, the conferences served to promote the exchange of research results. In the first years, most of the lectures were theoretical, but in the last few conferences practical applications have been given more attention. Besides their pioneering theoretical works, both honorees have also always dealt with applications problems. L. Bittner has, for example, examined optimal control of nuclear reactors and associated safety aspects. Since 1992 he has been working on applications in optimal control in flight dynamics. R. Klötzler recently applied his results on optimal autobahn planning to the south tangent in Leipzig.
The contributions published in these proceedings reflect the trend to practical problems; starting points are often questions from flight dynamics.
The first chapter deals with existence theory and optimality conditions. Relaxed problems are considered from various standpoints, from the classical point of view as well as that of L. C. Young. Suboptimality conditions are obtained by using the variational principle of Ekeland. An overview on the theory of extremal problems closes this chapter.
The heading of the second chapter reads "Analysis and Synthesis of Control Systems and Dynamic Programming." A modern interpretation of a solution of the Hamilton-Jacobi-Bellman equation is given. This is closely connected to the question of real-time or feedback control. New results for controlling under uncertainty are given and the problems of sensitivity and observability are investigated.
Recent advances in the field of "Numerical Methods and Their Application to Flight Path Optimization and Fluid Dynamics" are represented in the third chapter. Among these the reader will find nonlinear programming methods, accelerated multiple shooting and homotopy methods. The overview on SQP methods should be pointed out.
The fourth chapter is clearly named and deals with problems from a special area of research: "Applications to Mechanical and Aerospace Systems." Among the space

flight problems one can find solutions for many other complex practical optimal control problems: mobil robot controlling, geometrical extremal problems, fluid transport, fluid waves and human sciences.
These conferences on the Baltic coast were scientific, but also human meeting places for mathematicians from Eastern Europe (including the GDR) and Asia with colleagues from Western Europe (above all the Federal Republic of Germany). The 12th conference marked the first time that mathematicians from the USA took part. In the future, too, we want to bring together specialists from East and West!
In all twelve conferences the scientific program has been supplemented by cultural events and walks. In 1996 the participants visited the island of Ruden, which, before the political changes, was a military prohibited area and is now a wildlife refuge. The Baltic Sea resort Trassenheide and the idyllic island Ruden lie directly next to Peenemünde, where, in 1942, the space age began with the V-series of rocket experiments!
For the financial support of the 12th Conference we would like to sincerely thank the Deutsche Forschungsgemeinschaft, the Cultural Ministry of Mecklenburg-Vorpommern and the University of Greifswald.
For support in planning and conducting the conference and the development of these proceedings, we thank the employees of the Greifswald Research Group in Optimal Control. We are especially indebted to Frau Sabine Dempwolf for her thorough and careful work in creating the manuscript.
We thank Birkhäuser Verlag for its willingness to publish this volume of proceedings and for their cooperation.

ROLAND BULIRSCH
KNUT HEIER
WERNER SCHMIDT
Greifswald, München, Autumn 1997

Prof. L. Bittner

Leonhard Bittner was born on September 9, 1931, in Koeben on the Oder, that is, in Silesia. He attended school in the district of Breslau. The war and post-war chaos took the Bittner family on a one-and-a-half-year long odyssey to the Altmark. From 1950 to 1955, L. Bittner studied mathematics in Halle after he had worked as a soil-tester. Halle must have had good soil for mathematicians; many of his fellow students became well-known mathematicians. His mathematical interest was directed to applied mathematics by Professor Schubert. In 1955, L. Bittner became Professor Willer's assistant in Dresden; later he worked at the Institute for Applied Mathematics under the direction of Professor Heinrich. Here he concerned himself with numerical mathematics, worked with the D2 developed by N. Lehmann, studied optimization and algebraic topology. In 1960 he obtained a doctorate. In the same year he finished supplementary studies under Gawurin in Leningrad. The famous book by Pontrjagin, Boltjanski, Gamkrelidse, and Mischenko stimulated him to deal with control theory. In 1965, L. Bittner habilitated and became a university lecturer in Dresden and held guest lectures in Ilmenau. In 1967 he became professor for Numerical Mathematics in Greifswald. At that time Greifswald was strongly theoretically influenced and it was not easy for L. Bittner to assert himself as representative of the applied mathematics among "cybernetic greats in topological spaces." This all the more because L. Bittner does not like non-scientific arguments at all, doesn't like to come to decisions in personnel matters and puts off (sometimes necessary) administrative work. So he was almost lost to Greifswald in 1970 when the Academy in Berlin tried to lure him there.

L. Bittner has published more than 60 works; plus pre-prints, popular articles and articles about the history of mathematics. He edited the translations of four Russian books and is co-editor of the Lexicon Optimization. Among his works are 2 in Russian, produced during a stay as guest in Lithuania. Only very few of us will know that the article "On Error Bounds of Approximate Solutions by Means of Variational Methods and Duality Theory", published in "Treatise of the Mathematicial Seminar of the University in Hamburg" in 1971 was written by L. Bittner. He had sent the work, written in honour of Lothar Collatz, to Hamburg without the special permissions that were required at that time in the GDR. After a routine talk with the director of the department, L. Bittner decided to publish the article under the pseudonym H. L. Port.

L. Bittner has always been interested in many different mathematical questions. In the years in Dresden (1959–1967) pure numerical works, mainly for the solution of equations, were predominant. In the following fifteen years, works about optimal control, especially necessary conditions of optimality for an abstract model, about a discrete maximum principle, as well as estimations for methods of variation with the help of duality theorems dominated. The next ten years are mostly characterized by applications (such as control of atomic reactors, research of the Boltzmann equation and application to neutron transport, ARMA-models as means of safety research for nuclear reactors). Since 1992 he has been working together with his research group on existence theorems and conditions of optimization of higher order in the special field of research "Transatmospheric Flight Systems". Sometimes the weak relaxations he introduced can help to prove the existence of optimal solutions.

L. Bittner contributed greatly to the introduction and propagation of computer engineering in Greifswald in the seventies and eighties. For many years he has advised students writing diploma theses on computer engineering and programming. L. Bittner was and is a committed lecturer who had always much time for his students. Never did he strive for offices; he collaborated with the editorial staff of the ZAMM and is member of the Leopoldina. Always diligent, with a huge amount of knowledge, surprising ideas, critical towards results by himself or others, he devotes his life to mathematics and his family.

W. H. Schmidt

Prof. R. Klötzler[1]

On January 11, 1996, Professor Klötzler celebrated his 65th birthday. He can proudly look back on forty successful years as a university teacher and mathematician. After positions at the Academy of Science, the Hochschule für Bauwesen in Leipzig and the Martin–Luther–University in Halle, where he was head of the Mathematical Institute from 1965 to 1971, he returned to the University of Leipzig in 1972 as a full professor. He had been a student there from 1949 to 1953 and attended lectures by Ernst Hölder and Herbert Beckert. His scientific treatises are extensive and have made great contributions to many branches of calculus of variations, optimal control and mathematical programming. The common thread running through his work is the solution of Hilbert's problems concerning the calculus of variations and its development, control theory. His scientific interest first focussed on questions concerning extension of field theory of calculus of variations and existence theory of global geodesic fields, from which optimality criteria could be derived. In addition, he was able to establish eigenvalue criteria for weak optimality of extremals of regular variational problems with multiple integrals.

In the seventies, Rolf Klötzler successfully devoted himself to the theory of optimal control, a field which gained enormous practical significance through the development of space flight. He made a substantial contribution to its theoretical penetration when he developed a duality theory, based essentially on field theory of calculus of variations thereby giving new foundations for the computation of optimal solutions of one– and multidimensional control problems. By creating this duality theory he was able to give his answer to Hilbert's 23rd problem.

Connecting theoretical findings with the solution of practical problems has always been a basic concern of his. Rolf Klötzler has been particularly fascinated by geometrical optimization problems. As a result of his many years of research contact to partners in industry, optimization problems arose for symmetrical n–orbiforms which he treated from the point of view of control theory. Having found a new methodical approach to compute volume–optimal packings he was able to completely solve this problem both analytically and numerically.

The research group "Optimization" founded by Joachim Focke and later headed by Rolf Klötzler, provided an excellent working environment. R. Klötzler frequently produced more ideas than he had time to publish. Fundamental results on rough convexity and Favard's problem exist only as unpublished manuscripts.

Since the beginning of the nineties he has once again worked intensively on the establishment of a maximum principle for control problems with multiple integrals. In particular, he succeeded in revealing the nature of a suitable dual problem for this control problem – he identified it as a flow problem – and established connections with continuous and discrete flow optmization.

Rolf Klötzler remains a very committed university teacher. Now, as much as ever, he is in demand as a thesis tutor. He has guided countless students through their dissertations and many have gone on to complete their habilitation. Together with his colleague J. Focke he developed and published a basic concept for dynamic program-

[1]Summary of a honour to Rolf Klötzler published in Optimization, 1996, Vol. 38, 305–307

ming which decisively influenced corresponding syllabi, and incorporated this into his teaching.
He is an elected member of the German Academy of scientists "Leopoldina". Since 1968 he has worked as a co–editor of the journals "Beiträge zur Analysis" and "Optimization". He put a lot of effort into the foundation of the "Journal of Analysis and Its Applications". From 1982 to 1992 Rolf Klötzler acted as its editor–in–chief. In 1990, as chairman of the Mathematical Society of the GDR, he brought together the two German Mathematical Societies at the annual meeting of the German Mathematicians' Association in Bremen.
For his friends, colleagues and students it is unthinkable that such an engaged mathematician, who continues to produce so many interesting ideas and results and has stayed so young, will retire. We wish him good health and hope he remains active so that he can go on inspiring us for a long time to come.

S. Pickenhain
A. Kripfganz

Participants

H. ABESSER, Technische Universität Ilmenau, Ilmenau, Germany
B. ANAN'EV, IMM Uro RAN, Ekaterinburg, Russia
E. ANDREEVA, Tver State University, Tver, Russia
D. AUGUSTIN, Universität Münster, Münster, Germany
H. BEHNCKE, Universität Osnabrück, Osnabrück, Germany
U. BESSELT, Universität Greifswald, Greifswald, Germany
L. BITTNER, Universität Greifswald, Greifswald, Germany
CH. BÜSKENS, Universität, Münster, Münster, Germany
S. BUTZEK, Universität Greifswald, Greifswald, Germany
I. CAPUZZO DOLCETTA, Università di Roma, Roma, Italia
A. CEGIELSKI, Wyzsza Szkola Inzynierska, Zielona Gora, Poland
K. CHUDEJ, Technische Universität München, München, Germany
S. DEMPWOLF, Universität Greifswald, Greifswald, Germany
S. DIETZE, Technische Universität Dresden, Dresden, Germany
B. DITTMAR, Universität Halle-Wittenberg, Halle, Germany
Z. EMIRSAJLOW, Technical University of Szczecin, Szczecin, Poland
U. FELGENHAUER, Technische Universität Cottbus, Cottbus, Germany
P. E. GILL, University of California San Diego, La Jolla, California, USA
M. GOEBEL, Universität Halle-Wittenberg, Halle, Germany
H. GOLDBERG, Technische Universität Chemnitz-Zwickau, Chemnitz, Germany
E. GRIGAT, Technische Universität München, München, Germany
A. HAMEL, Universität Halle-Wittenberg, Halle, Germany
K. HEIER, Universität Greifswald, Greifswald, Germany
A. HOFFMANN, Technische Universität Ilmenau, Ilmenau, Germany
H. IRRGANG, Universität Greifswald, Greifswald, Germany
W. KAMPOWSKY, Fachhochschule Stralsund, Stralsund, Germany
M. KATZSCHMANN, Technische Universität Ilmenau, Ilmenau, Germany
P. KENDEROV, Bulgarian Academy of Sciences, Sofia, Bulgaria
M. KIEHL, Technische Universität München, München, Germany
U. KLEMT, Technische Universität Cottbus, Cottbus, Germany
R. KLÖTZLER, Universität Leipzig, Leipzig, Germany
V. I. KOROBOW, Kharkov University, Kharkow, Ukraine
O. KOSTJUKOWA, Academy of Sciences, Minsk, Belarus
D. KRAFT, Fachhochschule München, München, Germany
W. KRATZ, Universität Ulm, Ulm, Germany
A. KRIPFGANZ, Universität Leipzig, Leipzig, Germany

B. KUGELMANN, Technische Universität München, München, Germany
A. B. KURZHANSKIJ, Moscow State University, Moscow, Russia
R. LACHNER, Technische Universität Clausthal, Clausthal, Germany
G. LEITMANN, University of California, Berkeley, California, USA
F. LEMPIO, Universität Bayreuth, Bayreuth, Germany
H. LEONPACHER, Fachhochschule München, München, Germany
D. LIEBSCHER, Universität Ulm, Ulm, Germany
A. LOZOWICKI, Technical University of Szczecin, Szczecin, Poland
H. MAURER, Universität Münster, Münster, Germany
R. MEHLHORN, Technische Universität München, München, Germany
S. MIGORSKI, Jagiellonian University, Krakow, Poland
A. NOWAKOWSKI, University of Lodz, Lodz, Poland
H. J. OBERLE, Universität Hamburg, Hamburg, Germany
D. OESTREICH, Hochschule für Technik und Wirtschaft Dresden, Dresden, Germany
G. OPFER, Universität Hamburg, Hamburg, Germany
A. PAWELL, Technische Universität Cottbus, Cottbus, Germany
H. J. PESCH, Technische Universität Clausthal, Clausthal, Germany
H. X. PHU, Academy of Sciences, Hanoi, Vietnam
S. PICKENHAIN, Technische Universität Cottbus, Cottbus, Germany
J. PRESTIN, Universität Rostock, Rostock, Germany
U. RAITUMS, University of Riga, Riga, Latvia
W. RÖMISCH, Universität Berlin, Berlin, Germany
T. ROUBIČEK, Charles University, Prague, Czech Republic
H. SCHMELING, Universität Greifswald, Greifswald, Germany
W. H. SCHMIDT, Universität Greifswald, Greifswald, Germany
O. VON STRYK, Technische Universität München, München, Germany
CH. TAMMER, Universität Halle, Halle, Germany
V. M. TIKHOMIROV, Moscow State University, Moscow, Russia
F. TRÖLTZSCH, Technische Universität Chemnitz-Zwickau, Chemnitz, Germany
D. TSCHARNUTER, Technische Universität München, München, Germany
F. UNGER, Technische Universität Freiberg, Freiberg, Germany
G.-W. WEBER, Technische Hochschule Darmstadt, Darmstadt, Germany
H. WERNER, Universität Greifswald, Greifswald, Germany
Z. ZWIERZEWICZ, Maritime Academy Szczecin, Szczecin, Poland

Presentations not Contained in this Book

The following list consists of those presentations which were given at the conference but are not included in this book.

B. Anan'ev
On the Minimax State Estimation for Statistically Uncertain Hereditary Systems

A. Cegielski
Projection Methods in Convex Minimization Problems

S. Dietze
A Survey on Some Direct Methods for Solving Optimal Control Problems

B. Dittmar
A Stekloff Eigenvalue Problem in Ring Domains

H. Goldberg
A Multigrid Approach for Solving Optimal Control Problems with Parabolic State Equations

A. Hoffmann
Approximation von Punkt-Menge-Abbildungen durch stetige Funktionen im Sinne von Chebyshev

P. Kenderov
Fragmentability of Banach Spaces and Well-posedness of Optimization Problems

V. I. Korobow
Min-problem Moments in the Time Optimal Control

F. Lempio
Approximating Reachable Sets by Set-valued Interpolation

A. Lozowicki
On Optimization and Uniform Optimization of Feedback Control Systems for Any Class of Input Signals

D. Oestreich
Mathematische Modellierung des Einsatzes von Gießpfannen in der Sekundärmetallurgie

H. J. Pesch
Applications of Direct Methods of Optimal Control to Differential-algebraic Systems

J. Prestin
Basis Properties of Algebraic Polynomial Wavelets

W. RÖMISCH
Real-time Power Dispatch via Stochastic Programming

O. VON STRYK
Efficient Robot Trajectory Optimization by Utilizing Dynamical Structure

F. TRÖLTZSCH
On a Lagrange-Newton Method for Parabolic Control Problems

G.-W. WEBER
On Aspects of Structure and Stability in Optimal Control

Z. ZWIERZEWICZ
On a Computer Oriented Analytical Method for Solving Optimal Control Problems via Maximum Principle – Zermelo Navigational Problem

Existence Theory and Optimality Conditions

International Series of Numerical Mathematics
Vol. 124, © 1998 Birkhäuser Verlag, Basel

On the Convexification of Optimal Control Problems of Flight Dynamics

Leonhard Bittner *

Abstract. The control structure of the differential equations of a typical flight dynamical control problem is studied and methods for defining a relaxed problem with the aid of a least number of additional control parameters are presented. The usefulness of the relaxed problems for proving the existence of optimal solutions and calculating approximately optimal solutions is explained in detail.

1. The aim of this paper

All theorems assuring the existence of an optimal solution for a dynamical control problem (cf. [8]) require some kind of convexity for the sets of allowed velocities (for the hodographs) which, as a rule, is not satisfied. Hence it is recommEnded to define a convexified, a relaxed problem and to try to calculate approximately optimal solutions of the original problem by means of an optimal solution of the convexified problem. From the vast variety of problems of flight dynamics let us pick out some special problems with which we got familiar by papers of Sachs, Mehlhorn [9] and Bulirsch, Chudej [4], because they are in some sense typical for such problems and well suited to illustrate the convexification. For the sake of simplicity let us mainly confine ourselves to the simpler setting of [9] as we understand it: A vehicle, driven by a rocket engine, is moving in the atmosphere around the rotating earth in an equatorial plane. The vehicle (cf. Fig. 1) is considered to be a mass point and subject to the usual forces, namely thrust $\vec{T}$, lift $\vec{L}$, drag $\vec{D}$ and gravity $\vec{G}$. The movement is controlled by a control station (tower) situated at a fixed point of the equator and equipped with a coordinate system, the horizontal y-axis being tangential and the vertical h-axis being normal to the equator. The earth (equator) of radius R is assumed to rotate counterclockwise with constant velocity ω. The vehicle has mass m, coordinates y and h, velocity $\vec{v}$ and a flight path angle γ, measured as angle between $\vec{v}$ and the normal to the stretch connecting the midpoint of the earth with the vehicle. Let v, D, L, T be the scalar values of $\vec{v}, \vec{D}, \vec{L}, \vec{T}$ and α_T be the angle between $\vec{v}$ and $\vec{T}$. g denotes the gravitational acceleration on the equator, σ_0 and σ are given functions of v and h. S is the distance between the vehicle and the midpoint of the earth, i.e.

$$S = ((R+h)^2 + y^2)^{\frac{1}{2}} \tag{1.1}$$

Newton's laws, applied to the mass point m in a fixed inertial system with its origin in the midpoint of the earth, afford the following dynamical system after a transformation

*Ernst-Moritz-Arndt-Universität Greifswald, Institut für Mathematik und Informatik Jahnstr. 15a, D-17487 Greifswald, e-mail: bittner@rz.uni-greifswald.de

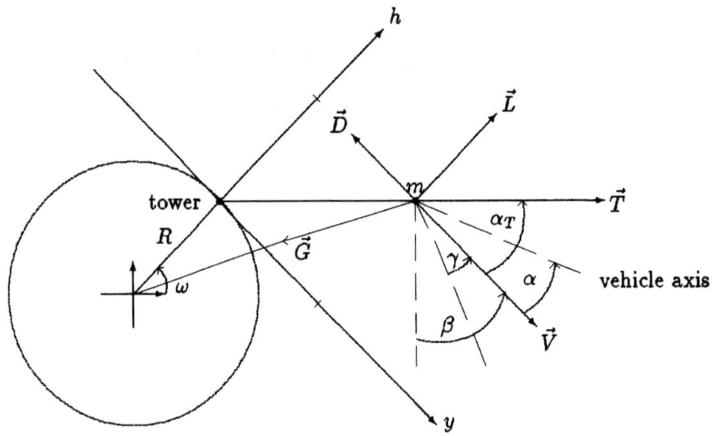

Figure 1

to the moving (y, h)-system

$$
\begin{aligned}
\dot{x}_1 &= \dot{v} = f_1(t,x,u) = \tfrac{1}{m}[T\cos\alpha_T - D] + [\omega^2 S - \tfrac{gR^2}{S^2}]\sin\gamma,\\
\dot{x}_2 &= \dot{\gamma} = f_2(t,x,u) = \tfrac{1}{mv}[T\sin\alpha_T + L] + \left[\omega^2 S - \tfrac{gR^2}{S^2}\right]\tfrac{\cos\gamma}{v} + \tfrac{v\cos\gamma}{S} - 2\omega,\\
\dot{x}_3 &= \dot{m} = f_3(t,x,u) = \sigma_0 - \sigma T, \qquad (1.2)\\
\dot{x}_4 &= \dot{y} = f_4(t,x,u) = [(R+h)\cos\gamma + y\sin\gamma]\tfrac{v}{S},\\
\dot{x}_5 &= \dot{h} = f_5(t,x,u) = [(R+h)\sin\gamma - y\cos\gamma]\tfrac{v}{S}.
\end{aligned}
$$

There is another possibility to define a flight path angle, namely as angle β between $\vec{v}$ and the normal to the stretch connecting the tower with the vehicle. β is more convenient for practical measurements, but the differential equations become "singular" for $y = h = 0$. Put

$$r = (y^2 + h^2)^{\frac{1}{2}}. \qquad (1.3)$$

Then the relation between γ and β reads

$$
\begin{aligned}
\cos\beta &= \frac{1}{S}\left(r + \frac{Rh}{r}\right)\cos\gamma - \frac{1}{S}\frac{Ry}{r}\sin\gamma, \qquad (1.4)\\
\sin\beta &= \frac{1}{S}\frac{Ry}{r}\cos\gamma + \frac{1}{S}\left(r + \frac{Rh}{r}\right)\sin\gamma.
\end{aligned}
$$

The control u consists of three components: the throttle coefficient δ, the thrust angle α_T and the angle of attack α (angle between $\vec{v}$ and the symmetry axis of the vehicle) or the lift coefficient C_L. We prefer C_L so that

$$u = (u_1, u_2, u_3) = (\delta, \alpha_T, C_L). \qquad (1.5)$$

The state x consists of five components

$$x = (x_1, \dots, x_5)^T = (v, \gamma, m, y, h)^T. \tag{1.6}$$

Put

$$f(t, x, u) = \Big(f_1(t, x, u), \dots, f_5(t, x, u)\Big)^T.$$

We adopt the relations

$$\begin{aligned} T &= \delta T_{\max} = \delta \cdot T_{\max}(v, h, y), \\ L &= q(v, h) C_L, \quad q(v, h) = \frac{1}{2}\varrho(h) v^2 F, \\ D &= q(v, h)\Big(k_0(v, h) + k_1(v, h) C_L^2\Big) \end{aligned} \tag{1.7}$$

from [4], where $T_{\max}, \varrho, k_0, k_1$ are given smooth functions, $F, \varrho, k_1, T_{\max}$ being positive.

2. Optimal trajectory problem

Given an initial point $x = (v_0, \gamma_0, m_0, y_0, h_0)^T$ with $y_0 = h_0 = 0, v_0 > 0, m_0 > 0$, given a terminal functional $\varphi(x(t_f))$, e.g.

$$\varphi(x(t_f)) = -\frac{y(t_f)}{m_0 - m(t_f)} \quad \text{or} \quad -v(t_f),$$

find a measurable control $u(t) = \Big(\delta(t), \alpha_T(t), C_L(t)\Big), \quad 0 \le t \le t_f$, with values

$$\begin{aligned} u(t) \in U &= \{u = (\delta, \alpha_T, C_L) \mid \underline{\delta} \le \delta \le 1, \ \underline{\alpha}_T \le \alpha_T \le \overline{\alpha}_T, \ |C_L| \le \overline{C}_L\} \\ &= \{u = (u_1, u_2, u_{,3}) \mid \underline{u}_i \le u_i \le \overline{u}_i \ (i = 1, 2, 3)\} \end{aligned} \tag{2.1}$$

so that the corresponding phase, i.e. the solution $x(\cdot)$ of the differential equations (1.2) with initial value $x(0) = x^0$, satisfies certain pure phase constraints of inequality type like

$$g(x(t)) \ge 0, \tag{2.2}$$

e.g. $g(x) = -q(v, h) + q_{\max}$, and, maybe, some terminal conditions of inequality type like

$$b(x(t_f)) \ge 0, \tag{2.3}$$

e.g. $b(x) = h_f - h$, and minimizes $\varphi(x(t_f))$.

3. The sets of allowed velocities

i.e. the sets

$$V(t, x) = \Big\{f(t, x, u) \mid u \in U\Big\}, \tag{3.1}$$

are not convex for all (t, x) as a rule. The convential convexified system is defined by [7] as

$$\dot{x}(t) = \sum_{i=0}^{n} \lambda_i(t) \ f\Big(t, x(t), u^i(t)\Big), \tag{3.2}$$

where the new control vector $v = (\lambda_0, \dots, \lambda_n, u^0, \dots, u^n)$ is restricted to $\lambda_i \geq 0, \sum \lambda_i = 1, u^i \in U$. v has a much greater number of components so that the convexified problem cannot be handled. It is the goal of this paper to provide a convexification with the least number of additional components.

4. Convexification in a special case

At first let us consider the case where the lift coefficient C_L is either constant or does not have a significant impact on the functions D or L. Now the control vector u consists only of two components $u_1 = \delta, u_2 = \alpha_T$ and the control region U is the rectangle $U = \{u = (u_1, u_2) \mid \underline{u}_i \leq u_i \leq \overline{u}_i \ (i = 1, 2)\}$. The system of differential equations (1.2) turns over to a system of the following "control structure"

$$\begin{aligned} \dot{x}_1 &= T_1 u_1 \cos u_2 + G_1, \\ \dot{x}_2 &= T_2 u_1 \sin u_2 + G_2, \\ \dot{x}_3 &= -T_3 u_1 + G_3, \\ \dot{x}_4 &= G_4, \\ \dot{x}_5 &= G_5, \end{aligned} \tag{4.1}$$

where the T_i and G_i are known smooth functions of t and x in an open set containing $(t_0, x^0) = (0, x^0)$ and the T_i are positive.

The sets of allowed velocities, truncated to the first three components, are

$$V_{tr}(t,x) = \Big\{ z = (z_1, z_2, z_3) \mid z_1 = T_1 u_1 \cos u_2 + G_1, \ z_2 = T_2 u_1 \sin u_2 + G_2, \\ z_3 = -T_3 u_1 + G_3, \ \underline{u}_i \leq u_i \leq \overline{u}_i \ (i = 1, 2) \Big\}; \tag{4.2}$$

they are sections of elliptical cones with vertices (G_1, G_2, G_3) and axises parallel to the z_3-axis (cf. Fig. 2). A cross section in the level $z_3 = -T_3 u_1 + G_3$ perpendicular to the z_3-axis forms a piece of an ellipse parametrized by

$$z = z(u_2) = \Big\{ \Big(T_1 u_1 \cos u_2 + G_1, T_2 u_1 \sin u_2 + G_2, -T_3 u_1 + G_3 \Big) \mid \underline{u}_2 \leq u_2 \leq \overline{u}_2 \Big\} \tag{4.3}$$

Let

$$\underline{z} = z(\underline{u}_2), \quad \overline{z} = z(\overline{u}_2) \tag{4.4}$$

be the boundary points of this elliptical arc.

There are several ways to generate the convex hull of V_{tr}, hence *conv* $V(t,x)$ by adjoining the fourth and fifth component G_4, G_5.

(i) Connect the boundary point $\underline{z}$ with every current point $z = z(u_2)$ of the elliptical arc by a stretch (cf. Fig. 3). Intermediate points are

$$z_\lambda = (1 - \lambda)\underline{z} + \lambda z, \quad 0 \leq \lambda \leq 1.$$

As a result one obtains

$$\begin{aligned} conv V_{tr}(t,x) = \Big\{ \Big(& T_1 u_1 [(1-\lambda) \cos \underline{u}_2 + \lambda \cos u_2] + G_1, \\ & T_2 u_1 [(1-\lambda) \sin \underline{u}_2 + \lambda \sin u_2], -T_3 u_1 + G_3 \Big) \mid \\ & 0 \leq \lambda \leq 1, \ \underline{u}_i \leq u_i \leq \overline{u}_i \ (i = 1, 2) \Big\} \end{aligned} \tag{4.5}$$

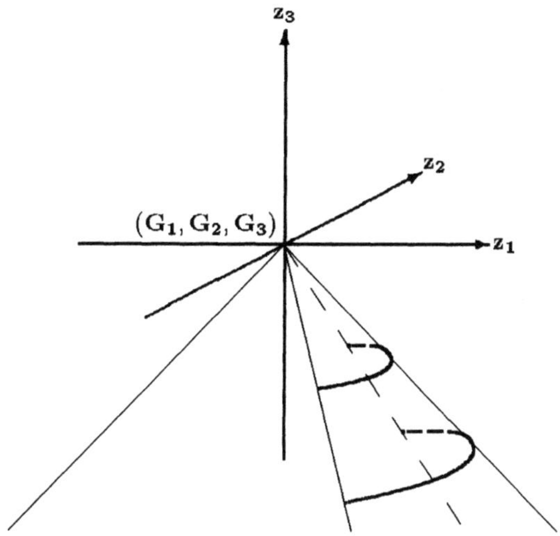

Figure 2

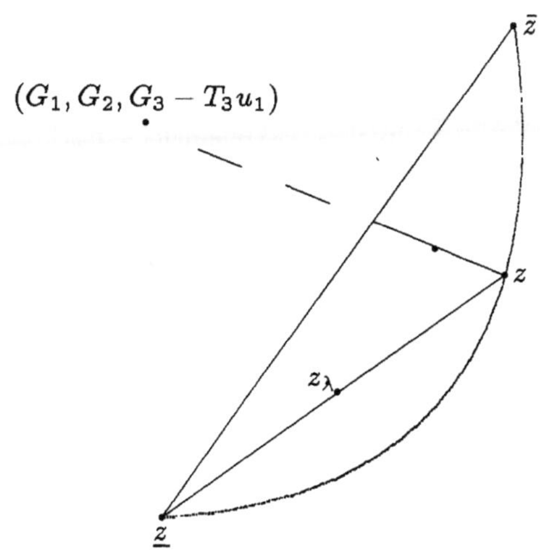

Figure 3

The corresponding convexified system of dynamical equations reads as follows

$$\dot{x} = (1-\lambda)f(t,x,\underline{u}) + \lambda f(t,x,u), \tag{4.6}$$

where

$$u = u(t) = \big(u_1(t), u_2(t)\big), \quad \lambda = \lambda(t),$$
$$\underline{u} = \underline{u}(t) = \big(u_1(t), \underline{u}_2\big), \quad x = x(t) = \big(x_1(t), \ldots, x_5(t)\big)^T$$

and $f(t,x,u)$ is defined by (1.2). The control functions $u_i(\cdot), \lambda(\cdot)$ have to be measurable and satisfy

$$\underline{u}_i \le u_i(t) \le \overline{u}_i \ (i = 1,2), \ 0 \le \lambda(t) \le 1 \tag{4.7}$$

(i′) Analoguously connect the boundary point $\overline{z}$ with every current point z of the elliptical arc by a stretch. The convexified system reads

$$\dot{x} = (1-\lambda)f(t,x,\overline{u}) + \lambda f(t,x,u), \tag{4.8}$$

where $\overline{u} = \overline{u}(t) \equiv (u_1(t), \overline{u}_2)$

(ii) Connect the point

$$z^0 = (G_1, G_2, G_3 - T_3 u_1), \tag{4.9}$$

lying on the axis of the elliptical cone, with every current point z of the elliptical arc by a stretch. The intermediate points

$$z_\lambda = (1-\lambda)z^0 + \lambda z$$

belong to the convex hull of this elliptical arc only for $\lambda^*(u) \le \lambda \le 1$, where

$$\lambda^*(u) = \frac{\sin(\overline{u}_2 - \underline{u}_2)}{\sin(\overline{u}_2 - u_2) + \sin(u_2 - \underline{u}_2)} \tag{4.10}$$

This way of creating $convV(t,x)$ yields the convexified system

$$\begin{aligned} \dot{x}_1 &= \lambda T_1 u_1 \cos u_2 + G_1, \\ \dot{x}_2 &= \lambda T_2 u_1 \sin u_2 + G_2, \\ \dot{x}_3 &= -T_3 u_1 + G_3, \\ \dot{x}_4 &= G_4, \\ \dot{x}_5 &= G_5 \end{aligned} \tag{4.11}$$

and the control restrictions

$$\underline{u}_i \le u_i = u_i(t) \le \overline{u}_i \ (i = 1,2), \ \lambda^*(u(t)) \le \lambda = \lambda(t) \le 1.$$

(iii) Connect the midpoint

$$z^m = \frac{1}{2}(\overline{z} + \underline{z})$$

of the chord $\underline{z}\overline{z}$ of the elliptical arc with every current point $z = z(u_2)$ by a stretch. The intermediate points

$$z_\lambda = (1-\lambda)z^m + \lambda z, \;\; 0 \le \lambda \le 1,$$

lead to the convexified differential system

$$\begin{aligned}
\dot{x}_1 &= T_1 u_1 \Big[\lambda \cos u_2 + \tfrac{1}{2}(1-\lambda)(\cos \underline{u}_2 + \cos \overline{u}_2)\Big] + G_1, \\
\dot{x}_2 &= T_2 u_1 \Big[\lambda \sin u_2 + \tfrac{1}{2}(1-\lambda)(\sin \underline{u}_2 + \sin \overline{u}_2)\Big] + G_2, \\
\dot{x}_3 &= -T_3 u_1 + G_3, \\
\dot{x}_4 &= G_4, \\
\dot{x}_5 &= G_5
\end{aligned} \tag{4.12}$$

and the control conditions

$$\underline{u}_i \le u_i = u_i(t) \le \overline{u}_i \;\; (i = 1, 2), \;\; 0 \le \lambda = \lambda(t) \le 1.$$

It can immediately be verified that the convexified systems have convex sets of allowed velocities. Note that the convexification is accomplished by introducing only one additional control component λ.

5. The advantage of the convexification

Each convexification assures that the crucial condition of existence theorems, the convexity of the sets of allowed velocities, is satisfied for the convexified system. If the other, the minor conditions are also satisfied, then the convexified control problem has an optimal solution. Now the question arises whether one can profit from this fact in order to solve the original problem (approximately).

What do we understand by a convexified optimal control problem?

Case a: There are neither phase constraints (2.2) nor terminal conditions (2.3) in the setting of the original problem.

Then the convexified control problem is the task to minimize the (same) terminal functional, but subject to the initial condition, the convexified differential equations and the corresponding control constraints for u_1, u_2, λ. Let the infimal value of the functional for the original and the convexified problem be I and I^C respectively. Evidently

$$I^C \le I, \tag{5.1}$$

since each feasible pair u_1, u_2 of the original problem yields the feasible triplet $u_1, u_2, 1$ for the convexified problem.

Case b: There are phase and terminal constraints in the setting of the original problem.

Then we choose an arbitrary (small) nonnegative number (vector) ε and define the convexified control problem as follows: Minimize the terminal functional subject to the initial condition, the convexified differential equations, the corresponding control restrictions and the ε-reinforced phase and terminal constraints

$$g\big(x(t)\big) - \varepsilon \geq 0 \quad \text{f. a. } t, \quad b\big(x(t_f)\big) - \varepsilon \geq 0. \tag{5.2}$$

Let I_ε^C be the infimal value of this convexified control problem. Analoguously let I_ε be the infimal value of the original problem with the ε-reinforced conditions (5.2) instead of the previous conditions (2.3), (2.4). Note that I_0 is the real infimum of the proper original problem. Evidently

$$I_0 \leq I_\varepsilon, \quad I_0^C \leq I_\varepsilon^C \leq I_\varepsilon \tag{5.3}$$

$I_\varepsilon, I_\varepsilon^C$ are nonincreasing for decreasing ε, hence

$$I_0^C \leq \lim_{\varepsilon \downarrow 0} I_\varepsilon^C \leq \lim_{\varepsilon \downarrow 0} I_\varepsilon, \quad I_0 \leq \lim_{\varepsilon \downarrow 0} I_\varepsilon. \tag{5.4}$$

Put

$$I_+ = \lim_{\varepsilon \downarrow 0} I_\varepsilon \tag{5.5}$$

I_+ can be interpreted as the infimum of the functional subject to the initial condition, the original differential equations, the original control constraints but subject to the strict phase and terminal constraints

$$g\big(x(t)\big) > 0 \ \forall t, \quad b\big(x(t_f)\big) > 0 \tag{5.6}$$

(g and b being continuous, may be, multidimensional). Suppose that the

Convexification (i) or (i′) is applied. The right-hand sides of the differential equations (4.6), (4.8) are convex linear combinations of $f(t, x, \cdot)$-vectors for different (feasible) controls $u, \underline{u}$ or $u, \overline{u}$. Therefore the well-known merging procedure can be used in order to construct an approximately optimal solution $u^\tau(\cdot), x^\tau(\cdot)$ for the original problem by means of an optimal solution $u(\cdot) = u^C(\cdot), \lambda(\cdot) = \lambda^C(\cdot), x(\cdot) = x^C(\cdot)$ of the convexified problem.

Assume a (sufficiently fine) subdivision

$$0 = \tau_0 < \tau_1 < \ldots < \tau_p = t_f \tag{5.7}$$

of the interval $[0, t_f]$ and put

$$\lambda_i = \frac{1}{\tau_{i+1} - \tau_i} \int_{\tau_i}^{\tau_{i+1}} \lambda(t)dt, \quad \mu_i = 1 - \lambda_i \tag{5.8}$$

$$u^\tau(t) = \begin{cases} u\big(\tau_i + \frac{1}{\lambda_i}(t - \tau_i)\big), & \tau_i \leq t < \tau_i + \lambda_i(\tau_{i+1} - \tau_i) \\ \underline{u}\big(\tau_i + \frac{1}{\mu_i}(t - \tau_i)\big), & \tau_i + \lambda_i(\tau_{i+1} - \tau_i) \leq t < \tau_{i+1} \end{cases} \tag{5.9}$$

or $\overline{u}(\cdot)$ instead of $\underline{u}(\cdot)$. Let $x^\tau(\cdot)$ be the solution of the initial conditions and the original differential equations corresponding to $u(\cdot) = u^\tau(\cdot)$. As it is known,

$$x^\tau(t) \longrightarrow x^C(t) = x(t) \ \text{uniformly, if} \ \max(\tau_{i+1} - \tau_i) \longrightarrow 0. \tag{5.10}$$

Case a: $u^\tau(\cdot), x^\tau(\cdot)$ is a feasible pair for the original problem and

$$g\big(x^\tau(t_f)\big) \longrightarrow I, \text{ if } \max_i(\tau_{i+1} - \tau_i) \longrightarrow 0. \tag{5.11}$$

Case b: The convexified problem has been defined with the aid of some parameter(vector) ε, therefore, being more precise, let us denote $u^\tau, x^\tau, x = x^C, u = u^C$ by $u_\varepsilon^\tau, x_\varepsilon^\tau, x_\varepsilon, u_\varepsilon$.

$u_\varepsilon^\tau(\cdot), x_\varepsilon^\tau(\cdot)$ is a feasible pair for the original problem, the strict conditions (5.6) are fulfilled and

$$g\big(x_\varepsilon^\tau(t_f)\big) \longrightarrow I_+, \text{ if at first } \max(\tau_{i+1} - \tau_i) \longrightarrow 0 \text{ and then } \varepsilon \longrightarrow 0. \tag{5.12}$$

This holds true due to (5.10)

$$I_+ \leq g\big(x_\varepsilon^\tau(t_f)\big) \longrightarrow g\big(x_\varepsilon(t_f)\big) = I_\varepsilon^C \leq I_\varepsilon \longrightarrow I_+$$

Hence $u^\tau(\cdot), x^\tau(\cdot)$ is an approximately optimal solution for the original problem, but with the strict conditions (5.6) instead of (2.3), (2.4), for a sufficiently fine subdivision τ and small positive ε.

Now let us suppose that the

Convexification (ii) is applied. We want to indicate its possible usage in confining us to the special case a. The Hamiltonian for the convexified problem reads as follows

$$H^C(x, u, \lambda, \psi) = \lambda u_1(\psi_1 T_1 \cos u_2 + \psi_2 T_2 \sin u_2) - \psi_3 T_3 u_1 + \sum_{k=1}^{5} \psi_k G_k.$$

The adjoined vector function $\psi(\cdot)$ for an optimal solution $u(\cdot), \lambda(\cdot), x(\cdot)$ of the convexified problem has to satisfy

$$\psi_k(t_f) = -\frac{\partial \varphi}{\partial x_k}(x(t_f)) \quad (k = 1, \ldots, 5).$$

Let us further suppose $\frac{\partial \varphi}{\partial x_1}(x) < 0$ for attainable x, $0 \leq t_f \ll 1$, $|\underline{u}_2|$ and $|\overline{u}_2| \ll 1$, $\underline{u}_1 > 0$ so that $\psi_1(t_f) > 0$, moreover $\psi_1(t) > 0$, $\cos u_2(t) \approx 1$, $\sin u_2(t) \approx 0$, $u_1(t) \geq \underline{u}_1 > 0$, $\lambda^*(u(t)) > 0$ and $u_1(t)\big[\psi_1(t)T_1 \cos u_2(t) + \psi_2(t) T_2 \sin u_2(t)\big] > 0$ f. a. a. t.. Remember $T_i = T_i(t, x(t)) > 0$ $(i = 1, 2)$. The maximum principle in particular yields

$$\max_{\lambda^*(u(t)) \leq \lambda \leq 1} H^C(x(t), u(t), \lambda, \psi(t)) = H^C(x(t), u(t), \lambda(t), \psi(t)),$$

in detail

$$\begin{aligned} &\max_\lambda \lambda u_1(t)\big[\psi_1(t)T_1 \cos u_2(t) + \psi_2(t) T_2 \sin u_2(t)\big] \\ &= \lambda(t) u_1(t)\big[\psi_1(t)T_1 \cos u_2(t) + \psi_2(t) T_2 \sin u_2(t)\big] \end{aligned}$$

f. a. a. t., which entails $\lambda(t) = 1$ and leads to the

Conclusion:

$$u(t) = (u_1(t), u_2(t)), \ 0 \leq t \leq t_f$$

is an optimal solution of the original problem.

In a similar, but rigorous way existence has been established for a reentry, an orbit and a constrained Brachistochrone problem in the reports [1], [2].

6. Taking account of the third control

As has been explained in section 1, the third control u_3 is the lift coefficient C_L, restricted by $|u_3| \leq \overline{u}_3 = \overline{C}_L$. The full control is $u = (u_1, u_2, u_3)$. The control structure of the dynamical system (1.2) now reads as follows

$$\begin{aligned} \dot{x}_1 &= f_1(t,x,u) = T_1 u_1 \cos u_2 - Q_1 u_3^2 + G_1, \\ \dot{x}_2 &= f_2(t,x,u) = T_2 u_1 \sin u_2 + Q_2 u_3 + G_2, \\ \dot{x}_3 &= f_3(t,x,u) = -T_3 u_1 + G_3, \\ \dot{x}_4 &= f_4(t,x,u) = G_4, \\ \dot{x}_5 &= f_5(t,x,u) = G_5, \end{aligned} \tag{6.1}$$

where the T_i, Q_i, G_i only depend smoothly on t, x and the T_i, Q_i are positive in a certain open set around $(0, x^0)$ in the (t,x)-space. Pose

$$b(t,x,u_3) = \Big(-Q_1 u_3^2, Q_2 u_3, 0, 0, 0 \Big)^T. \tag{6.2}$$

Then

$$f(t,x,u) = f(t,x,(u_1,u_2,0)) + b(t,x,u_3). \tag{6.3}$$

The truncated sets of allowed velocities are

$$V_{tr} = V_{tr}(t,x) = A + B, \tag{6.4}$$

where

$$A = \Big\{ (T_1 u_1 \cos u_2 + G_1, T_2 u_1 \sin u_2 + G_2, -T_3 u_1 + G_3)^T \mid \underline{u}_1 \leq u_1 \leq 1, \;\; \underline{u}_2 \leq u_2 \leq \overline{u}_2 \Big\}$$

$$B = \Big\{ (-Q_1 u_3^2, Q_2 u_3, 0)^T \mid |u_3| \leq \overline{u}_3 \Big\}.$$

Hence (cf. [6], chap. V.2)

$$conv V_{tr} = conv A + conv B \tag{6.5}$$

which implies a two-parameter convexification of the sets $V(t,x)$ of allowed velocities. In detail: *convA* can be established by a one parameter convexification due to method (i) or (i′) of the preceding section. The set B is a symmetric piece of a parabola in the (z_1, z_2)-plane, a parabola with the origin as vertex and the negative z_1-axis as axis (cf. Fig. 4). Evidently *convB* can be generated by all stretches connecting the point

$$\overline{\zeta} = (-Q_1 \overline{u}_3^2, Q_2 \overline{u}_3, 0)^T$$

with the current point

$$\zeta = (-Q_1 u_3^2, Q_2 u_3, 0)^T, \quad |u_3| \leq \overline{u}_3.$$

An intermediate point is

$$\zeta_\nu = (1-\nu)\overline{\zeta} + \nu\zeta = \overline{\zeta} + \nu(\zeta - \overline{\zeta}), \quad 0 \leq \nu \leq 1.$$

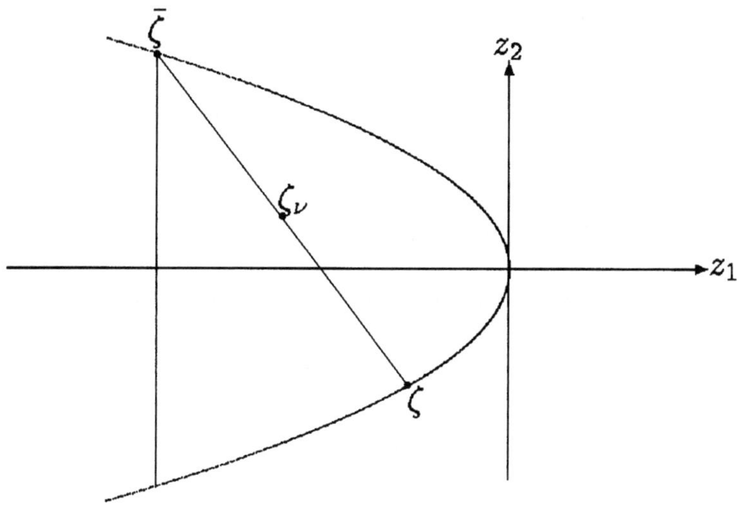

Figure 4

Finally this yields the convexified system

$$\dot{x} = (1-\lambda)f(t,x,(u_1,\underline{u}_2,0))+\lambda f(t,x,(u_1,u_2,0))+(1-\nu)b(t,x,\overline{u}_3)+\nu b(t,x,u_3), \quad (6.6)$$

where $x = x(t), u_i = u_i(t), \lambda = \lambda(t), \nu = \nu(t)$. The new control $(u_1, u_2, u_3, \lambda, \nu)$ has to satisfy $\underline{u}_i \leq u_i \leq \overline{u}_i,\ i = (1,2),\ |u_3| \leq \overline{u}_3,\ 0 \leq \lambda \leq 1,\ 0 \leq \nu \leq 1$.

All that has been said in the preceding section about the convexified problem, case a and b and ε-reinforcement remains valid after an appropriate modification even now.

If $u(\cdot), \lambda(\cdot), \nu(\cdot)$ is an optimal solution of the convexified problem, then an approximately optimal solution $u^\tau(\cdot)$ of the original problem is obtained by the merging procedure (5.9) applied to $(u_1(\cdot), u_2(\cdot))$ and λ in order to get $(u_1^\tau(\cdot), u_2^\tau(\cdot))$ and an analoguous merging procedure applied to $u_3(\cdot)$ and $\nu(\cdot)$ in order to get

$$\nu_i = \frac{1}{\tau_{i+1} - \tau_i} \int_{\tau_i}^{\tau_{i+1}} \nu(t)dt, \quad (6.7)$$

$$u_3^\tau(t) = \begin{cases} u_3(\tau_i + \frac{1}{\nu_i}(t - \tau_i)), & \tau_i \leq t < \tau_i + \nu_i(\tau_{i+1} - \tau_i) \\ \overline{u}_3, & \tau_i + \nu_i(\tau_{i+1} - \tau_i) \leq t < \tau_{i+1} \end{cases}$$

As to a justification refer to [3].

References

[1] Bittner, L.: *Zur Existenz von Optimallösungen für das Problem des Wiedereintritts in die Atmosphäre und das Problem des Einschwenkens in eine Umlaufbahn.* Rep. Nr. 20 des Sonderforschungsbereichs (SFB) 255 der DFG, 1994.

[2] Bittner, L.: *Zur Existenz einer optimalen Lösung des Brachistochroneproblems mit Tangentialwinkelbeschränkungen.* Rep. Nr. 29 des SFB 255 der DFG, 1995.

[3] Bittner, L.: *Zur Konvexifizierung von Optimalsteuerproblemen.* Rep. Nr. 33 des SFB 255 der DFG, 1996.

[4] Bulirsch, R.; Chudej, K.: *Combined optimization of trajectory and stage separation of a hypersonic two-stage space vehicle.* Z. Flugwiss. Weltraumforschung 19 (1995) 55-60.

[5] Butzek, S.; Schmidt, W.: *Konstruktion von Näherungslösungen für Steuerprobleme mit Hilfe von Lösungen relaxierter Probleme.* Rep. Nr. 26 des SFB 255 der DFG, 1996.

[6] Dunford, N.; Schwartz, J. T.: *Linear Operators* I. Interscience Publishers, New York, 1958.

[7] Lee, E. B.; Marcus, L.: *Foundations of Optimal Control Theory.* John Wiley Inc., New York, 1970.

[8] Macky, P.; Strauss, A.: *Introduction to Optimal Control Theory.* Springer, Berlin, 1988.

[9] Sachs, G.; Mehlhorn, R.; Dinkelmann, N.: *Efficient convexification of flight path optimization problems.* This volume.

International Series of Numerical Mathematics
Vol. 124,

Restricted Optimal Transportation Flows

Rolf Klötzler *

Abstract. This paper generalizes optimality criteria for free transportation flows to restricted flows.

1. Introduction

In 1994 the author introduced a new kind of optimization problem, so-called transportation flow problems (TFP), which we can interpret as a generalized version of problems of calculus of variations [2–3]. At once it is an improvement of L. C. Youngs concept of "generalized flows" from 1969 [4]. We characterize a (TFP) by the following problem.

Let Ω be a strongly Lipschitz domain of $\mathbb{E}^m$ and n the number of different transport products within Ω. L should be a given linear continuous functional on $C(\bar{\Omega})^n$ which describes the distribution of production and consumption of all considered n goods. By Riesz's representation theorem [1] $L(\sigma)$ with $\sigma = (\sigma_1, \ldots, \sigma_n)^T \in C(\bar{\Omega})^n$ can be represented uniquely by means of n regular Borel measures $\alpha_k (k > 1, \ldots, n)$ in the shape

$$L(\sigma) = \int_\Omega \sigma^T(x) d\alpha(x) \tag{1}$$

with $\alpha = (\alpha_1, \ldots, \alpha_n)^T$.

We require each α_k as equilized by the assumption

$$\int_\Omega d\alpha_k(x) = 0 \quad (k = 1, \ldots, n)\,. \tag{2}$$

That means, the kth product will be produced in the same quantity as the consumption behaves.

Further let us given a local cost rate r on $\Omega \times \mathbb{E}^{nm}$ with the following properties:

$$\left.\begin{array}{ll} r(\cdot, v) & \text{is summable on } \Omega \\ r(x, \cdot) & \text{is positive homogeneous of degree one and convex on } \mathbb{E}^{nm}\ \forall\, x \in \Omega \\ \multicolumn{2}{l}{\gamma_1 |v| \le r(x,v) \le \gamma_2 |v|\ \forall\, v \in \mathbb{E}^{nm},\ x \in \Omega \text{ for some constants } \gamma_1, \gamma_2 > 0.} \end{array}\right\} \tag{3}$$

* BTU Cottbus, Institut für Mathematik, Universitätsplatz 3-4, D - 03013 Cottbus

We describe admissible flows $\mu = (\mu_1, \dots, \mu_n)^T$ by elements $\mu \in L_\infty^{mn}(\Omega)^*$ which satisfy the *continuity equation*

$$\langle \nabla\sigma, \mu \rangle = L(\sigma) \quad \forall\, \sigma \in W_\infty^{1,nm}(\Omega). \tag{4}$$

The set of all admissible flows μ will be denoted by Y. In [3] the *cost functional*

$$\begin{aligned} K(\mu) \;:=\; & \int_\Omega r(x, d\mu(x)) := \sup\{\langle u, \mu\rangle \mid u \in L_\infty^{nm}(\Omega), u^T(x)v \le r(x,v) \\ & \forall v \in \mathbb{E}^{mn}, \text{a.e. } x \in \Omega\} \end{aligned} \tag{5}$$

and the corresponding *figuratrix set*

$$\mathcal{F}(x) := \{z \in \mathbb{E}^{nm} \mid ;\, z^T v \le r(x,v) \quad \forall\, v \in \mathbb{E}^{nm}, x \in \Omega\} \tag{6}$$

were studied as well as the (unrestricted) *transportation flow problem* (TFP)

$$K(\mu) \to \min_Y . \tag{7}$$

In comparison to that we shall extend this questions with respect to

$$K_V(\mu) \;:=\; \sup\{\langle u, \mu\rangle \mid u \in L_\infty^{nm}(\Omega), u^T(x)v \le r(x,v) \quad \forall\, v \in V(x), \text{a.e.} x \in \Omega\} \tag{8}$$

$$\mathcal{F}_V(x) \;:=\; \{z \in \mathbb{E}^{nm} \mid z^T v \le r(x,v) \quad \forall\, v \in V(x), x \in \Omega\} \tag{9}$$

and the *restricted transportation flow problem*

$$K_V(\mu) \to \min_Y . \tag{10}$$

Here $V(x)$ is a given continuous set valued map from Ω into $\mathbb{E}^{nm}$ with the property:

$$V(x) \text{ is a closed convex cone } \neq \{0\}\ \forall\, x \in \bar{\Omega}.$$

2. Some results on unrestricted problems

With respect to problem (7) we dispose under the assumptions (2) and (3) on a file of useful optimality statements according to [3]

Theorem 1: *A minimal solution μ_0 of problem (7) exists.*

Theorem 2: *The deposit problem*

$$L(S) \;\to\; \max_\Upsilon \tag{11}$$

$$\textit{with } \Upsilon := \{S = (S^1, \dots, S^n)^T \;\in\; W_\infty^{1,n}(\Omega) \mid \nabla S(x) \in \mathcal{F}(x) \quad \textit{a.e. on } \Omega\} \tag{12}$$

is a dual problem to (7), *strong duality is guaranted in the sense of*

$$\min_{Y} K = \max_{\Upsilon} L. \tag{13}$$

Theorem 3: *A flow* $\mu_0 \in Y$ *is optimal for* (7) *iff an element* $S^0 \in \Upsilon$ *exists which satisfies the generalized Pontryagin's maximum principle*

$$\langle \nabla S^0, \mu_0 \rangle \geq \langle \mu, \mu_0 \rangle \tag{14}$$

for every $u \in L_\infty^{mn}(\Omega)$ *with the property*

$$u(x) \in \mathcal{F}(x) \quad \text{a.e. on } \Omega. \tag{15}$$

Theorem 4: *A maximal solution* S^0 *of problem* (11) *exists and satisfies Pontryagin's maximum principle* (15) *with a certain* $\mu_0 \in Y$.

In the next chapter we shall generalize these results on problem (10).

3. Optimality criteria for restricted flows

We study the restricted (TFP) in the sense of (10) and its dual deposit problem

$$L(S) \quad \to \quad \max_{\Upsilon_V} \tag{16}$$

$$\text{with } \Upsilon_V \quad := \quad \{S \in W_\infty^{1,n}(\Omega) \mid \nabla S(x) \in \mathcal{F}_V(x) \text{ a.e. on } \Omega\}. \tag{17}$$

Indeed since formula (4) for all $S \in \Upsilon_V$, $\mu \in Y$

$$\begin{aligned} L(S) &= \langle \nabla S, \mu \rangle \\ &\leq \sup\{\langle u, \mu \rangle \mid u \in L_\infty^{nm}(\Omega), \quad u(x) \in \mathcal{F}_V(x) \text{ a.e. on } \Omega\} \\ &= K_V(\mu), \end{aligned}$$

such that the duality between (10) and (16) is proved.

We assume

$$\sup_{\Upsilon_V} L < \infty. \tag{18}$$

In general, now we cannot await the existence of an optimal S^0 of (16). Only in cases where $\mathcal{F}_V(x)$ is uniformly bounded on Ω – this means there is a ball $B(0, \varrho)$ with center 0 and radius ϱ such that $B(0, \varrho) \supset \mathcal{F}_V(x) \; \forall \; x \in \Omega$ – is the existence of such an S^0 guaranteed. This special situation is given for instance by $V(x) = \mathbb{E}^{nm} \; \forall \; x \in \Omega$ in consequence of (3). For this reason we introduce the following sequences of transportation flow problems:

$$K_V^k(\mu) \to \min_{Y} \tag{19}$$

where for $k \in \mathbb{N}$, $\mu \in Y$

$$K_V^k(\mu) := \sup\{\langle u, \mu\rangle \mid u \in L_\infty^{nm}(\Omega), \quad u(x) \in \mathcal{F}_V^k(x) \text{ a.e. on } \Omega\} \tag{20}$$

and

$$\mathcal{F}_V^k(x) := \mathcal{F}_V(x) \cap B(0,k). \tag{21}$$

By using the support function

$$r^k(x,v) := \sup\left\{(z^T v) \mid z \in \mathcal{F}_V^k(x)\right\} \tag{22}$$

of $\mathcal{F}_V^k(x)$ we can understand the cost functional $K_V^k(\mu)$ in the same sense as $K(\mu)$ according to (5) by the equivalent definition

$$K_V^k(\mu) := \sup\{\langle u, \mu\rangle \mid u \in L_\infty^{nm}(\Omega), \quad u^T(x)v \le r^k(x,v) \;\; \forall\, v \in \mathbb{E}^{nm}, \text{a.e. } x \in \Omega\}. \tag{20'}$$

Case I: Problem (16) has an optimal solution $S^0 \in \Upsilon_V$.

Then $\forall\, k > \|\nabla S^0\|$, $k \in \mathbb{N}$ the element S^0 is also an optimal solution of

$$L(S) \quad \to \quad \max_{\Upsilon_V^k} \tag{23}$$

$$\text{for } \Upsilon_V^k := \{S \in W_\infty^{1,n}(\Omega) \quad | \quad \nabla S(x) \in \mathcal{F}_V^k(x) \text{ a.e. on } \Omega\}. \tag{24}$$

With reference to Theorem 4 – applied for (23) – there is an element $\mu_0^k \in Y$ with the property

$$\begin{array}{rcl} \langle \nabla S^0, \mu_0^k\rangle & \ge & \langle u, \mu_0^k\rangle \quad \forall\, u \in L_\infty^{mn}(\Omega) \\ \text{with } u(x) & \in & \mathcal{F}_V^k(x) \text{ a.e. on } \Omega, \end{array} \tag{25}$$

and especially with $u(x) \in \mathcal{F}_V^l(x)$ a.e. on Ω for fixed $l < k$

$$\langle \nabla\sigma, \mu_0^k\rangle = L(\sigma) \quad \forall\, \sigma \in W_\infty^{1,n}(\Omega). \tag{26}$$

From (26) for $\sigma = S^0$ follows together with (25)

$$\begin{array}{rcl} L(S^0) & \ge & \langle u, \mu_0^k\rangle \quad \forall\, u \in L_\infty^{mn}(\Omega) \\ \text{with } u(x) & \in & \mathcal{F}_V^k(x) \text{ a.e. on } \Omega, \end{array}$$

respectively under consideration of (20') and (3)

$$L(S^0) \ge K_V^k(\mu_0^k) \ge \gamma_1 \|\mu_0^k\|. \tag{27}$$

Consequently, $\{\mu_0^k\}$ is a bounded sequence in $L_\infty^{mn}(\Omega)^*$ and by virtue of the Theorem of Alaoglu there is a weak-*-convergent subsequence $\{\mu_0^{k'}\}$ with $\mu_0^{k'} \overset{*}{\rightharpoonup} \mu_0 \in L_\infty^{mn}(\Omega)^*$.

Therefore from (25) and (26) we get especially for $k = k'$ and the limit $k' \to \infty$

$$\begin{array}{rcl} \langle \nabla S^0, \mu_0\rangle & \ge & \langle u, \mu_0\rangle \quad \forall\, u \in L_\infty^{mn}(\Omega) \\ \text{with } u(x) & \in & \mathcal{F}_V^l(x) \text{ a.e. on } \Omega, \end{array} \tag{28}$$

$$\langle \nabla\sigma, \mu_0\rangle = L(\sigma) \quad \forall\, \sigma \in W_\infty^{1,n}(\Omega). \tag{29}$$

In consequence of $\mathcal{F}_V(x) = \bigcup_{l\in\mathbb{N}} \mathcal{F}_V^l(x)$ (28) leads for $l \to \infty$ to

$$\begin{aligned} \langle \nabla S^0, \mu_0 \rangle &\geq \langle u, \mu_0 \rangle \quad \forall\, u \in L_\infty^{mn}(\Omega) \\ \text{with } u(x) &\in \mathcal{F}_V(x) \text{ a.e. on } \Omega. \end{aligned} \tag{30}$$

Formulas (29) and (30) show in our case I the validity of Theorem 4 for unrestricted (TFP), too. We get from (30) under consideration of (8) immediately

$$\max_{\Upsilon_V} L = L(S^0) = K_V(\mu_0) = \min_{Y} K_V(\mu). \tag{31}$$

Case II: (16) has no optimal solution.

Then we consider for each $k \in \mathbb{N}$ an optimal solution S_k^0 of (23) and its corresponding multiplicator μ_0^k which satisfies the formula

$$\begin{aligned} \langle \nabla S_k^0, \mu_0^k \rangle &\geq \langle u, \mu_0^k \rangle \quad \forall\, u \in L_\infty^{mn}(\Omega) \\ \text{with } u(x) &\in \mathcal{F}_V^k(x) \text{ a.e. on } \Omega \end{aligned} \tag{32}$$

and according to (20)

$$L(S_k^0) = K_V^k(\mu_0^k) \leq K_V^k(\mu) \quad \forall \mu \in Y. \tag{33}$$

On the other hand,

$$\infty > \sup_{\Upsilon_V} L \geq L(S_k^0) = K_V^k(\mu_0^k) \geq \gamma_1 \|\mu_0^k\| \tag{34}$$

follows, similarly to (27), from (33) since assumption (18) holds true. Therefore, $\{\mu_0^k\}$ is again a bounded sequence in $L_\infty^{mn}(\Omega)^*$ such that a subsequence $\{\mu_0^{k'}\}$ exists with $\mu_0^{k'} \overset{*}{\rightharpoonup} \mu_0 \in L_\infty^{mn}(\Omega)^*$.

Formula (32) leads with (26) to

$$\begin{aligned} L(S_{k'}^0) &\geq \langle u, \mu_0^{k'} \rangle \quad \forall\, u \in L_\infty^{mn}(\Omega) \\ \text{with } u(x) &\in \mathcal{F}_V^k(x) \text{ for fixed } l < k' \text{ a.e. on } \Omega. \end{aligned} \tag{35}$$

In virtue of (18) and the monotonicity of the sequence $\{L(S_{k'}^0)\}$ we get from (35)

$$\begin{aligned} \lim_{k'\to\infty} L(S_{k'}^0) &\geq \langle u, \mu_0 \rangle \quad \forall\, u \in L_\infty^{mn}(\Omega) \\ \text{with } u(x) &\in \mathcal{F}_V^l(x) \text{ a.e. on } \Omega. \end{aligned} \tag{36}$$

In consequence of $\mathcal{F}_V(x) = \bigcup_{l\in\mathbb{N}} \mathcal{F}_V^l(x)$ we obtain from (36)

$$\begin{aligned} \lim_{k'\to\infty} L(S_{k'}^0) &\geq \langle u, \mu_0 \rangle \quad \forall\, u \in L_\infty^{mn}(\Omega) \\ \text{with } u(x) &\in \mathcal{F}_V(x) \text{ a.e. on } \Omega \end{aligned} \tag{37}$$

and with reference to (10) and the proved duality between (10) and (16)

$$\lim_{k'\to\infty} L(S_{k'}^0) = K_V(\mu_0). \tag{38}$$

We summarize our main results of this chapter.

Theorem 5: *Under the assumptions* (2), (3) *and* (18) *between the restricted (TFP)*(10) *and its dual deposit problem* (16) *holds strong duality in the sense*

$$\sup_{\Upsilon_v} L(S) = \min_{Y} K_V(\mu) = K_V(\mu_0).$$

If $\max_{\Upsilon_V} L(S) = L(S^0)$ *exists, then Pontryagin's maximum principle* (30) *holds.*

References

[1] Kantorowitsch, L.W.; Akilow, A.P.: *Funktionalanalysis in normierten Räumen.* Berlin: Akademie-Verlag 1964.

[2] Klötzler, R.: *Flußoptimierung.* ZAMM **74**(1994)6, T591–T593.

[3] Klötzler, R.: *Optimal Transportation Flows.* Z. Anal. Anw. **14**(1995), 391–401.

[4] Young, L.C.: *Calculus of Variations and Optimal Control Theory.* Philadelphia: Saunders Comp. 1969.

International Series of Numerical Mathematics
Vol. 124,

Relaxation Gaps in Optimal Control Processes With State Constraints

Sandra Butzek* Werner H. Schmidt*

1. Introduction

There are a lot of very simple examples of control processes which have no optimal solution. A well-known fundamental existence theorem is that of ROXIN-FILLIPOV; unfortunately, the assumptions in this theorem are rather strong. Sometimes one can prove existence in absence of convexity by studying "bigger" problems obtained by relaxation, that means by convexification of the sets of speed vectors. Then we try to choose a special optimal solution of the relaxed problem and apply PONTRYAGIN'S maximum principle in order to discuss whether certain derived controls are optimal ones of the original problem (or not). BALDER [1] used a similar idea to prove new existence results for optimal control problems without convexity. He applies BAUER'S extremal principle instead of the maximum principle.
In some problems we are able to construct minimizing sequences for the original control problem starting with a solution of the relaxed problem. TIKHOMIROV [8] considered processes governed by ODE with objective functional and control restrictions, but without final conditions or state constraints. He proposed a "mixing procedure" to construct controls of the original problem such that the corresponding values of the cost functional converge to the optimal cost of the relaxed problems. The situation is more complicated when there are additional state constraints or isoperimetric conditions. In such cases it may happen that the minimal (optimal) costs of the relaxed problem are strictly less than the infimum of costs of the original problem: there is a relaxation gap! We will illustrate this in the paper by some concrete (academic) examples.

2. Original control problem and the corresponding relaxed problem

Let $[0, T]$ be a fixed time interval, $U \subseteq \mathbb{R}^r$ a compact control set and $x_0 \in \mathbb{R}^n$. f, g and h are assumed to be continuous mappings on $[0,T] \times \mathbb{R}^n \times U$ with values in $\mathbb{R}, \mathbb{R}^n$ and $\mathbb{R}^p$, respectively. Let $k : [0,T] \times \mathbb{R}^n \to \mathbb{R}^q$ and $\varphi : \mathbb{R}^n \to \mathbb{R}$, $K : \mathbb{R}^n \to \mathbb{R}^l$ be continuous.

The original problem (PO) is to minimize

$$J_0(x(\cdot), u(\cdot)) = \int_0^T f(t, x(t), u(t))dt + \varphi(x(T)) \tag{1}$$

* Ernst-Moritz-Arndt-Universität Greifswald, Institut für Mathematik und Informatik, Jahnstr. 15a, D-17487 Greifswald, email: wschmidt@rz.uni-greifswald.de

with respect to

$$\dot{x} = g(t, x, u(t)), \qquad x(0) = x_0, \tag{2}$$

$$u(\cdot) \in \mathcal{L}_\infty(\mathbb{R}^r), \quad u(t) \in U, \quad x(\cdot) \in W^1_\infty(\mathbb{R}^n) \tag{3}$$

and

$$\int_0^T h(t, x(t), u(t))dt \geq 0 \tag{4}$$

$$K(x(T)) \geq 0, \quad k(t, x(t)) \geq 0 \quad \text{for all} \quad 0 \leq t \leq T. \tag{5}$$

The essential assumption in the existence theorem of ROXIN-FILLIPOV, mentioned above, is the convexity of the sets

$$V(t,x) = \{(f(t,x,u), g(t,x,u), h(t,x,u)) | u \in U\} \qquad \text{for all} \qquad t, x,$$

which is not valid in many applications. Therefore we consider the following new problem which is called relaxed problem (PR): Minimize

$$J_R(x(\cdot), z(\cdot)) = \int_0^T F(t, x(t), z(t))dt + \varphi(x(T)) \tag{R 1}$$

subject to

$$\dot{x} = G(t, x(t), z(t)) \qquad x(0) = x_0 \tag{R 2}$$

$$z(\cdot) \in \mathcal{L}_\infty(\mathbb{R}^{m(r+1)}), \qquad z(t) \in Z, \qquad x(\cdot) \in W^1_\infty(\mathbb{R}^n) \tag{R 3}$$

and

$$\int_0^T H(t, x(t), z(t))dt \geq 0 \tag{R 4}$$

$$K(x(T)) \geq 0, \qquad k(t, x(t)) \geq 0, \qquad 0 \leq t \leq T, \tag{R 5}$$

where

$$z(\cdot) = (\alpha_1(\cdot), \ldots, \alpha_m(\cdot), \nu_1(\cdot), \ldots, \nu_m(\cdot))$$

$$Z = \{z \in \mathbb{R}^{m(r+1)} | z = (\alpha_1, \ldots, \alpha_m, \nu_1, \ldots, \nu_m), \sum_{i=1}^m \alpha_i = 1, \alpha_i \geq 0, \nu_i \in U \; \forall i\}$$

$$F(t,x,z) = \sum_{i=1}^m \alpha_i f(t,x,\nu_i), \quad G(t,x,z) = \sum_{i=1}^m \alpha_i g(t,x,\nu_i), \quad H(t,x,z) = \sum_{i=1}^m \alpha_i h(t,x,\nu_i).$$

Note that the sets $V_R(t,x) = \{(F(t,x,z), G(t,x,z), H(t,x,z)) | z \in Z\}$ are convex for all t, x if $m \geq n+p+2$. In optimal control problems this is also true for $m = n+p+1$. In this case (R1)–(R5) is called CARATHEODORY-relaxation of (1)–(5). Note that in concrete examples also $m < n+p+1$ may give convex sets $V_R(t,x)$.

3. Connections between (PO) and (PR)

Let us consider an original problem (1)–(3) without state constraints (5) and without isoperimetric conditions (4). The corresponding relaxed problem is given by (R1)–(R3). Then it is known [8]:

Theorem 1 *Assume (R2) has a solution for all admissible controls $u(\cdot)$, $V_R(t,x)$ are convex $\forall\, t, x$ and the set of all trajectories is uniformly bounded, then PR has an optimal solution and it is*

$$\min_{(R2)-(R3)} J_R(x(\cdot), z(\cdot)) = \inf_{(2)-(3)} J_0(x(\cdot), u(\cdot)).$$

If $x_R(\cdot), z_R(\cdot) = (\alpha_1(\cdot), \ldots, \nu_m(\cdot))$ is a solution of (R1)–(R3) then a minimizing sequence $x_s(\cdot), u_s(\cdot)$ for (1)–(3) can be constructed by the methods of TIKHOMIROV or LYAPUNOV (see [4] and [3]) by mixing the controls $\nu_1(\cdot), \ldots, \nu_m(\cdot)$. TIKHOMIROV's procedure is to divide the interval $[0,T]$ in s (equidistant) subintervals and then every small subinterval I^j once more into m intervals $I_1^j, \ldots, I_m^j$ of length $\frac{1}{s}\int_{I^j} \alpha_i(t)dt, i = 1, \ldots, m, j = 1, \ldots, s$. Define $u_s(t)$ to be $\nu_i(t)$ if $t \in I_i^j$.

The corresponding states $x_s(\cdot)$ converge to $x_R(\cdot)$ uniformly, provided g satisfies a Lipschitz-condition with respect to x.

Theorem 2 *Under the assumptions of theorem 1 it is*

$$\begin{aligned} J_R(x_R(\cdot), z_R(\cdot)) &= \min_{x(\cdot),z(\cdot)} J_R(x(\cdot), z(\cdot)) = \lim_{s\to\infty} J_0(x_R(\cdot), u_s(\cdot)) = \\ &= \lim_{s\to\infty} J_0(x_s(\cdot), u_s(\cdot)) = \inf_{x(\cdot),u(\cdot)} J_0(x(\cdot), u(\cdot)). \end{aligned} \tag{6}$$

Turn back to the problem (1)–(5) and (R1)–(R5), respectively. The controls $u_s(\cdot)$ and states $x_s(\cdot)$ defined as above by means of $x_R(\cdot), z_R(\cdot)$ usually are not admissible solutions of (1)–(5), even if $x_R(\cdot), z_R(\cdot)$ fulfil (R4), R(5). We are only able to prove

Theorem 3 *Let $x_R(\cdot), z_R(\cdot)$ be an admissible (or optimal) solution of (R1) - (R5) and let all assumptions of theorem 1 be fulfilled. Then for every $\varepsilon > 0$ there exist $N(\varepsilon) \in \mathbb{N}$ such that*

$$J_0(x_s(\cdot), u_s(\cdot)) \le J(x_R(\cdot), z_R(\cdot)) + \varepsilon$$

and

$$K(x_s(T)) \ge -\varepsilon, \quad \int_0^T h(t, x_s(t), u_s(t))dt \ge -\varepsilon, \quad k(t, x_s(t)) \ge -\varepsilon, \quad 0 \le t \le T,$$

for all $s \ge N(\varepsilon)$.

From theorem 3 it follows

Theorem 4 *Let $x_R(\cdot), z_R(\cdot)$ be an optimal pair of (R1)–(R5). Then*

$$J_R(x_R(\cdot), z_R(\cdot)) = \min_{(R2)-(R5)} J_R(x(\cdot), z(\cdot)) = \lim_{s\to 0} J_0(x_s(\cdot), u_s(\cdot))$$

and $$\min_{(R2)-(R5)} J_R(x(\cdot), z(\cdot)) \le \inf_{(2)-(5)} J_0(x(\cdot), u(\cdot))). \tag{7}$$

Definition 1 *The control problem (1)–(5) has a relaxation gap if*

$$\min_{(R2)-(R5)} J_R(x(\cdot), z(\cdot)) < \inf_{(2)-(5)} J_0(x(\cdot), u(\cdot)).$$

4. Examples

1. Find the minimum of

$$J_0(x(\cdot), u(\cdot)) = \int_0^1 [x(t) - 1]u^2(t)dt \tag{8}$$

subject to

$$\dot{x} = x^2 + \sin[x + \frac{\pi}{2}u(t)], \qquad x(0) = 0, \tag{9}$$

$$|u(t)| \leq 1, \qquad 0 \leq t \leq 1, \qquad \text{and} \qquad \int_0^1 x^2(t)dt \leq 0 \tag{10}$$

From the isoperimetric condition it follows $x(t) = 0$, therefore $\sin[\frac{\pi}{2}u(t)] = 0$ and $u(t) = 0$ for all t. There is only one admissible solution: $u(\cdot) = 0, x(\cdot) = 0$. The optimal control problem has a solution and it is $\min_{(9)-(10)} J_0(x(\cdot), u(\cdot)) = 0$.

The relaxed problem will be formulated with $m = 2$ because (10) does not depend on u:
Minimize

$$J_R(x(\cdot), z(\cdot)) = \int_0^1 [x(t) - 1][\alpha(t)\nu_1^2(t) + (1 - \alpha(t))\nu_2^2(t)]dt \tag{11}$$

where

$$\dot{x} = x^2 + \alpha(t)\sin[x + \frac{\pi}{2}\nu_1(t)] + (1 - \alpha(t))\sin[x + \frac{\pi}{2}\nu_2(t)] \tag{12}$$

$$x(0) = 0, \qquad |\nu_i(t)| \leq 1, \qquad i = 1, 2, \qquad 0 < \alpha(t) \leq 1, \qquad \int_0^1 x^2(t)dt \leq 0. \tag{13}$$

(PR) has the unique admissible trajectory $x(\cdot) = 0$, too. An admissible control is $\bar{z}(t) = (\frac{1}{2}, \frac{1}{2}, 1, -1)$ for all t and $\min_{(12)-(13)} J_R(x(\cdot), z(\cdot)) \leq J_R(0, \bar{z}(\cdot)) = -1 < 0 = \min_{(9)-(10)} J_0(x(\cdot), u(\cdot))$.

A simpler modification of this example is

2. Maximize

$$\int_0^1 u^2(t)dt \qquad \text{with respect to} \tag{14}$$

$$\dot{x} = u(t), \qquad u(t) \in \{-1, 0, 1\} \tag{15}$$

and

$$x^2(t) \leq 0, \qquad 0 \leq t \leq 1. \tag{16}$$

3. Maximize $\int_0^1 u^2(t)dt$ such that $\dot{x} = u$, $x^2(t) \leq 0$, $u(t) \in \{-1, 1\}$, $0 \leq t \leq 1$.
(PO) has no admissible solution, an optimal solution of (PR) is $\alpha_1(t) = \alpha_2(t) = \frac{1}{2}$, $\nu_1(t) = 1$, $\nu_2(t) = -1$ for all t.

4. A control problem which has infinite admissible solutions is:
Minimize

$$-\int_0^2 u^2(t)dt \qquad \text{subject to} \qquad \dot{x} = u(t), \qquad U = [-1, 1] \tag{17}$$

$$x(t) = 0 \qquad \text{if} \qquad 0 \leq t \leq 1 \tag{18}$$

$$x(t) \leq t - 1 \qquad \text{if} \qquad 1 < t \leq 2$$

Optimal controls of (PO) are given by $u(t) = 0$ if $0 \le t \le 1$ and $u^2 = 1$ if $1 \le t \le 2$. It follows $\min J_0(x(\cdot), u(\cdot)) = -1$. The relaxed problem (PR) is:
Minimize

$$-\int_0^2 [\alpha(t)\nu_1^2(t) + (1-\alpha(t))\nu_2^2(t)]dt \tag{19}$$

such that

$$\dot{x} = \alpha(t)\nu_1(t) + (1-\alpha(t))\nu_2(t) \tag{20}$$

$$\nu_1(t), \nu_2(t) \in U, \quad 0 \le \alpha(t) \le 1,$$

$$x(t) = 0, \quad 0 \le t \le 1 \qquad \text{and} \qquad x(t) \le t-1, \quad 1 \le t \le 2.$$

An optimal solution of (PR) is $\nu_1(t) = 1$, $\nu_2(t) = -1$ for $0 \le t \le 2$ and $\alpha(t) = \frac{1}{2}$ if $0 \le t \le 1$, $\alpha(t) = 0$ if $1 < t \le 2$. Then $\min J_R(x(\cdot), z(\cdot)) = -2$.

5. Find the infimum of

$$J_0(x(\cdot), u(\cdot)) = \int_0^1 [1-u(t)]^2 dt \tag{21}$$

with respect to

$$\dot{x} = u(t), \qquad x(0) = 0, \qquad u(t) \in \{0,1\} \tag{22}$$

$$[x(t) - \frac{1}{2}t][x(t) - \frac{2}{3}t]^2 \le 0, \qquad 0 \le t \le 1. \tag{23}$$

(PO) has admissible solutions, for instance, $u(t) = 0$, $x(t) = 0$ for all t. Minimizing controls are $u_s(t) = 0$ if $t_{2i} \le t < t_{2i+1}$ and $u_s(t) = 1$ if $t_{2i+1} \le t < t_{2i+2}$, where $t_{2i} = \frac{i}{s}$ and $t_{2i+1} = \frac{2i+1}{2s}$, $i = 0, \dots, s-1$; $s \in \mathbb{N}$. Then $\inf_{(22)-(23)} J_0(x(\cdot), u(\cdot)) = \frac{1}{2}$.

We put $m = 2$ to construct the relaxed problem:

$$J_R(x(\cdot), z(\cdot)) = \int_0^1 \{\alpha(t)[1-\nu_1(t)]^2 + [1-\alpha(t)][1-\nu_2(t)]^2\}dt$$

is to minimize subject to

$$\dot{x} = \alpha(t)\nu_1(t) + [1-\alpha(t)]\nu_2(t), \qquad x(0) = 0$$

and inequality (23), $0 \le \alpha(t) \le 1$; $\nu_1(t), \nu_2(t) \in \{0,1\}$.
The control set U has only the two values 0 and 1, respectively; therefore we can assume $\nu_1(t) = 1$ and $\nu_2(t) = 0$ for all t. Then (PR) can be transformed into:

$$J_R(\alpha(\cdot)) = \int_0^1 [1-\alpha(t)]dt \qquad \text{is to minimize} \tag{24}$$

under the conditions

$$\dot{x} = \alpha(t) \qquad x(0) = 0 \qquad 0 \le \alpha(t) \le 1 \tag{25}$$

$$[x(t) - \frac{1}{2}t][x(t) - \frac{2}{3}t]^2 \le 0. \tag{26}$$

Admissible trajectories are either given by $x(t) = \frac{2}{3}t$ or $x(t) \le \frac{1}{2}t$ for all $0 \le t \le 1$. It follows either $\alpha(t) = \frac{2}{3}$ for all $0 \le t \le 1$ or $1 - \int_0^1 \alpha(t)dt = 1 - x(1) \ge \frac{1}{2}$.

Therefore

$$\min_{(25)-(26)} J_R(x(\cdot), z(\cdot)) = \frac{1}{3} < \frac{1}{2} = \inf_{(22)-(23)} J_0(x(\cdot), u(\cdot)).$$

6. Minimize

$$\int_0^1 [1 - x_1^2(t)][1 - u^2(t)]dt \tag{27}$$

with respect to

$$\dot{x}_1 = u(t), \qquad x_1(0) = 0 \tag{28}$$

$$\dot{x}_2 = x_1^2, \qquad x_2(0) = 0, \qquad x_2(1) = 0 \tag{29}$$

$$-1 \leq u(t) \leq 1, \qquad 0 \leq t \leq 1. \tag{30}$$

It is $\dot{x}_2 \geq 0$ and therefore $x_2(\cdot) = 0$, $x_1(\cdot) = 0$, $u(\cdot) = 0$ and so $\min_{(28)-(30)} J_0(x(\cdot), u(\cdot)) = 1$. The essential part of $V(t, x)$ is $\{(1 - u^2, u, 0) | -1 \leq u \leq 1\}$, this is a concave curve and we can take $m = 2$. The relaxed problem is

$$\int_0^1 [1 - x_1^2(t)][1 - \alpha(t)\nu_1^2(t) - (1 - \alpha(t))\nu_2^2(t)]dt \tag{31}$$

is to minimize such that

$$\dot{x}_1 = \alpha(t)\nu_1(t) + (1 - \alpha(t))\nu_2(t), \qquad x_1(0) = 0 \tag{32}$$

$$\dot{x}_2 = x_1^2, \qquad x_2(0) = 0, \qquad x_2(1) = 0 \tag{33}$$

$$-1 \leq \nu_i(t) \leq 1, \qquad 0 \leq \alpha(t) \leq 1; \qquad i = 1, 2; \qquad 0 \leq t \leq 1. \tag{34}$$

It has the only trajectory $x_1(\cdot) = x_2(\cdot) = 0$. An admissible control is $\bar{z}(\cdot) = \left(\frac{1}{2}, \frac{1}{2}, 1, -1\right)$. It follows

$$\min_{(32)-(34)} J_R(x(\cdot), z(\cdot)) < J_R(0, \bar{z}(\cdot)) = 0 = \min_{(28)-(30)} J_0.$$

7. The following example (35)–(38) with terminal functional, state constraints and an isoperimetric inequality has no (optimal) solution and there is no relaxation gap. Consider at first

$$J_0(x(\cdot), u(\cdot)) = -x(1) \qquad \text{is to minimize} \tag{35}$$

subject to

$$\dot{x} = u \qquad x(0) = 0 \qquad u(t) \in \{0, 1\} \tag{36}$$

$$\frac{2}{3}t - x(t) \geq 0, \qquad 0 \leq t \leq 1. \tag{37}$$

This problem has optimal solutions, e.g. $u(t) = 0$ for $0 \leq t \leq \frac{1}{3}$ and $u(t) = 1$ for $\frac{1}{3} < t \leq 1$.
Now we add an isoperimetric condition more:

$$\int_0^1 [x(t) - tu(t)]dt \geq 0. \tag{38}$$

The control process (35)–(38) has admissible solutions, e.g. $u(\cdot) = 0$, $x(\cdot) = 0$, it follows $\inf_{(36)-(38)} J_0 \leq 0$.

To write down (PR) we can take $m = 2$, $\nu_1(t) = 1, \nu_2(t) = 0, 0 \leq t \leq 1$, because $U = \{0, 1\}$ has only two values.
Then minimize

$$J_R(x(\cdot), z(\cdot)) = -x(1) \tag{39}$$

with respect to

$$\dot{x} = \alpha(t), \qquad x(0) = 0 \tag{40}$$

$$\frac{2}{3}t - x(t) \geq 0, \qquad 0 \leq \alpha(t) \leq 1 \tag{41}$$

$$\int_0^1 [x(t) - t\alpha(t)]dt \geq 0. \tag{42}$$

In this example the relaxation is given by convexification of the control set U. A solution of (PR) is $\alpha(t) = \frac{2}{3}$, $x(t) = \frac{2}{3}t$, $0 \leq t \leq 1$, it follows $J_R(x(\cdot), z(\cdot)) = -\frac{2}{3} = \min J_R$.

By integration by parts (38) gives

$$0 \leq \int_0^1 [x(t) - tu(t)]dt = \int_0^1 x(t)dt - tx(t)\Big|_0^1 + \int_0^1 x(t)dt$$

hence

$$x(1) \leq 2\int_0^1 x(t)dt \leq 2\int_0^1 \frac{2}{3}tdt = \frac{2}{3}.$$

Suppose the original problem (35)–(38) has a solution $x(\cdot)$ with $x(1) = \frac{2}{3}$, then it follows $\int_0^1 x(t)dt = \frac{1}{3}$. Because of (37) it is $x(t) = \frac{2}{3}t$ and therefore $u(t) = \frac{2}{3}$ a.e. this is a contradiction to $U = \{0, 1\}$.
Consequently is either $\min J_0(x(\cdot), u(\cdot)) < \frac{2}{3}$ or (35)–(38) has no optimal solution. We show $\inf J_0(x(\cdot), u(\cdot)) = \frac{2}{3}$ by constructing a minimizing sequence of controls in a modified TIKHOMIROV-way:
Put $s \in \mathbb{N}$, $h = \frac{1}{s+1}$, $t_0 = 0$, $t_{2i+1} = t_{2i} + \frac{h}{3}$, $t_{2i+2} = t_{2i+1} + \frac{2h}{3}$, $i = 0, \ldots, s$ and define

$$u_s(t) = \begin{cases} 0 & if \quad t_{2i} < t \leq t_{2i+1} \quad \text{and} \quad t_{2s} < t \leq t_{2s+2} = 1, \\ 1 & if \quad t_{2i+1} < t \leq t_{2i+2}. \end{cases}$$

Note that

$$x_s(t) = \begin{cases} \frac{2}{3}t_{2i} & \text{if} \quad t_{2i} < t \leq t_{2i+1} \\ \frac{2}{3}t_{2s} & \text{if} \quad t_{2s} < t \leq t_{2s+2} \\ \frac{2}{3}t - \frac{1}{3}(t_{2i} + h) & \text{if} \quad t_{2i+1} < t \leq t_{2i+2} \quad i = 0, \ldots, s-1. \end{cases}$$

Then (37) is valid by construction and (38) follows by simple calculus: $\int_0^1 [x_s(t) - tu_s(t)]dt = \frac{7}{9}\frac{s}{(s+1)^2} > 0$. The considered $u_s(\cdot)$, $x_s(\cdot)$ are admissible. It is $-x_s(1) = -\frac{2}{3}\frac{s}{s+1}$ and therefore $J_0(x_s(\cdot), u_s(\cdot)) = -x_s(1)$ tends to $-\frac{2}{3} = \min J_R(x(\cdot), u(\cdot))$ if $s \to \infty$. There is no relaxation gap! The problem (35)–(38) has no optimal solution!

5. About existence of gaps

Theorem 5 *Let $x_R(\cdot)$, $z_R(\cdot)$ be an optimal solution of (5)–(8) which fulfils all inequalities strictly (in all components), that means*

$$K(x_R(T)) > 0, \qquad k(t, x_R(t)) > 0, \qquad 0 \leq t \leq T$$

and $\int_0^T H(t, x_R(t), z_R(t))dt > 0$. Then the problem has no relaxation gap.

Proof: Tikhomirov's minimizing sequence $x_s(\cdot)$, $u_s(\cdot)$ is admissible for the problem (1)–(4) if s is sufficiently large. It follows

$$\inf_{(1)-(4)} J_0(x(\cdot), u(\cdot)) \leq \lim_{s\to\infty} J_0(x_s(\cdot), u_s(\cdot)) = J_R(x_R(\cdot), z_R(\cdot)) \leq \inf_{(1)-(4)} J_0(x(\cdot), u(\cdot))$$

and we get equality everywhere. ■

Usually there are many possibilities to construct controls $u_s(\cdot)$ in Tikhomirov's procedure. If there is at least one sequence $u_s(\cdot)$, $x_s(\cdot)$ which is admissible for (PO) then there is no gap.

Example: Minimize $\int_0^1 x^2(t)dt$ subject to $\dot{x} = u(t)$, $x(t) \leq 0$, $u(t) \in \{-1, 1\}$. Optimal controls of (PR) are e.g. $z_1(t) = (\frac{1}{2}, \frac{1}{2}, 1, -1)$ and $z_2(t) = (\frac{1}{2}, \frac{1}{2}, -1, 1)$, $0 \leq t \leq 1$. The controls $u_s^1(\cdot)$ derived from $z_1(\cdot)$ are not admissible but those governed by $z_2(\cdot)$ fulfil the state constraints. There is no relaxation gap.

We consider now – without loss of generality – control problems with terminal functional and state constraints, that means $f = 0$, $h = 0$.
Define $\mathcal{E}_0$, $\mathcal{E}_R$, $\mathcal{K}$ to be the following subsets of $C(\mathbb{R}^n)$:

$$\mathcal{E}_0 = \{x(\cdot) |\, \exists\ u(\cdot) \in \mathcal{L}_\infty(U): \ \dot{x} = g(t, x, u(t)),\ x(0) = x_0,\ 0 \leq t \leq T\}$$

$$\mathcal{E}_R = \{x(\cdot) |\, \exists\ z(\cdot) \in \mathcal{L}_\infty(Z): \ \dot{x} = G(t, x, z(t)),\ x(0) = x_0,\ 0 \leq t \leq T\}$$

$$\mathcal{K} = \{x(\cdot) |\, K(x(T)) \geq 0,\ k(t, x(t)) \geq 0,\ 0 \leq t \leq T\}$$

$\mathcal{E}_0$ and $\mathcal{E}_R$ are the sets of reachable trajectories of the dynamical systems (2) and (R2), respectively. The closure of sets X in a normed space is denoted by cl(X).

Theorem 6 *Suppose the control problem has a relaxation gap, then*

$$\mathcal{E}_R \cap \mathcal{K} \supset cl(\mathcal{E}_0 \cap \mathcal{K}).$$

Consequently, there is no gap if

$$\mathcal{E}_R \cap \mathcal{K} = cl(\mathcal{E}_0 \cap \mathcal{K}).$$

Proof: Assume $\mathcal{E}_R \cap \mathcal{K} = \mathrm{cl}(\mathcal{E}_0 \cap \mathcal{K})$. Note that $\mathcal{E}_R = \mathrm{cl}(\mathcal{E}_0)$. Suppose $x_R(\cdot) \in \mathcal{E}_R \cap \mathcal{K}$ is an optimal trajectory of (PR) with corresponding control $z_R(\cdot)$. Then we find a sequence $x_s(\cdot) \in \mathcal{E}_0 \cap \mathcal{K}$ converging to $x_R(\cdot)$. From the continuity of φ and the natural inequality $\varphi(x_R(T)) \leq \inf_{(PO)} \varphi(x(T))$ we get $\lim_{s\to\infty} \varphi(x_s(T)) = \varphi(x_R(T))$ that means there is no relaxation gap. The first assertion is evident. ■

Theorem 7 *If $\mathcal{E}_R \cap \mathcal{K}$ is convex and $\mathcal{E}_R \cap int(\mathcal{K}) \neq \emptyset$ then $\mathcal{E}_R \cap \mathcal{K} = cl(\mathcal{E}_0 \cap \mathcal{K})$, consequently the control problem has no gap. More general: Suppose $\mathcal{E}_R \cap \mathcal{K} = \bigcup_{l=1}^{N}(\mathcal{E}^l \cap \mathcal{K})$ and all components $\mathcal{E}^l \cap \mathcal{K}$ are convex and fulfil $\mathcal{E}^l \cap int(\mathcal{K}) \neq \emptyset$ then there are no gaps.*

The theorem is proved in [5].

In all known examples the following criterion dealing with cuts of $\mathcal{E}_R \cap \mathcal{K}$ in $\mathbb{R}^n$ is valid, but it has not yet been proven!
Define $\mathcal{E}_0(t) = \{x | \ \exists \ x(\cdot) \in \mathcal{E}_0 : \ x = x(t)\}$ and $\mathcal{E}_R(t)$ analogously,

$$\mathcal{K}(t) = \{x | \ k(t,x) \geq 0\}, \ 0 \leq t < T \qquad \mathcal{K}(T) = \{x | \ k(T,x) \geq 0, \ K(x) \geq 0\}.$$

Thesis: Suppose the optimal control problem has a gap. Then it is

$$\mathcal{E}_R(t) \cap \mathcal{K} \supset \mathrm{cl}(\mathrm{int}(\mathcal{E}_0(t) \cap \mathcal{K}(t)))$$

for at least one $t \in]0,T]$. Hence, if equality holds for all $0 < t \leq T$ there is no relaxation gap!

References

[1] Balder, E.J.: *New existence results for optimal controls in the absence of convexity: The importance of extremality.* SIAM J. Control and Optimization, Vol. 33 (1994), No. 3, pp. 890–916.

[2] Bauer, H.: *Minimalstellen von Funktionen und Extremalpunkte.* Archiv d. Math. 9 (1958), 389–393, 11 (1960)), 200–205.

[3] Bittner, L.: *Zur Konvexifizierung von Optimalsteuerproblemen.* Report SFB 255, Nr.33, Greifswald und München, 1996.

[4] Butzek, S.; Schmidt, W.H.: *Konstruktion von Näherungslösungen für Steuerprobleme mit Hilfe von Lösungen relaxierter Probleme.* Report SFB 255, Nr. 26, Greifswald und München, 1996.

[5] Butzek, S.: *Aspekte des Zusammenhangs von Optimalsteuerproblemen und zugehörigen relaxierten Problemen.* PhD-thesis. Greifswald 1997.

[6] Roubiček, T.: *Relaxation in Optimization Theory and Variational Calculus.* W.De Gruyter-Verlag. Berlin 1997

[7] Schmidt, W.H.: *An existence theorem for a special control problem.* Proceedings of MMAR'96. Vol. I, 263–266. Szczecin, 1996.

[8] Tikhomirov, V.M.: *Grundprinzipien der Theorie der Extremalaufgaben.* Teubner-Texte zur Mathematik. Bd. 30. Leipzig, 1982

[9] Warga, J.: *Relaxed Variational Problems.* Journal of Math. Analysis and Appl. 4, 111–128, 1962.

International Series of Numerical Mathematics
Vol. 124, © 1998 Birkhäuser Verlag, Basel

Optimal Shape Design for Elliptic Hemivariational Inequalities in Nonlinear Elasticity

Zdzislaw Denkowski* Stanislaw Migórski*

Abstract. Optimal shape design problem for systems governed by elliptic hemivariational inequality which comes from nonlinear elasticity is considered. A general existence result for this problem is established by the mapping method.

1. Introduction

In this paper we consider optimal shape design problem for systems described by a hemivariational inequality (referred later to as HVI). The main goal is to provide an existence result for this problem by applying the mapping method. The latter supplies both a class of admissible shapes and a topology in this class of domains. The admissible shapes are obtained as the images of a fixed open bounded subset of $\mathbb{R}^N$ by regular (see Section 2 for details) bijections in $\mathbb{R}^N$. The shape design problem is formulated as a control one in which we minimize an appropriately defined cost functional, the HVI appears as a state equation and the role of controls is played by sets from a family of admissible shapes.

The HVI was introduced for the first time by Panagiotopoulos in 1981 in order to express the principle of virtual work in the inequality form for laws derived from nonsmooth and nonconvex superpotentials. The HVI is a generalization of the notion of variational inequality and it provides mathematical models for many problems in mechanics, economics and engineering (see [13], [10] and the references therein for more details concerning theory and applications).

Another approach to shape optimization is presented, for instance, in [14], [2] for problems with differential equations and in [7], [6] for variational inequalities.

The plan of the paper is as follows. In Section 2 we present a HVI modelling the nonlinear behavior of materials in elasticity and we recall the preliminaries about the mapping method. In Section 3 we consider a HVI which serves as a model for the problem formulated in Section 2. For this model we provide a result (cf. Proposition 1) on the closedness of the graph of the mapping which to every admissible shape assigns the solution set of HVI. It can be seen without difficulty that the main result of this paper (see Theorem 3) can be also obtained for systems governed by the class of HVIs given in Section 2.

*Jagiellonian University, Institute of Computer Science, ul. Nawojki 11, 30072 Cracow, Poland, denkowski@softlab.ii.uj.edu.pl, migorski@softlab.ii.uj.edu.pl

2. Motivation

The aim of this section is to give a short description of a type of HVIs which appear in nonlinear elasticity and which motivate the problem studied in the paper.

Let Ω be a bounded domain in $\mathbb{R}^3$ which is occupied by a deformable body in its undeformed state. In the framework of small deformations, we have the relations

$$\sigma_{ij,j} + f_i = 0 \quad \text{in} \quad \Omega \quad \text{(the equilibrium equation)}, \tag{1}$$

$$\varepsilon_{ij}(u) = \frac{1}{2}(u_{i,j} + u_{j,i}) \quad \text{(the strain -- displacement relation)}, \tag{2}$$

where $i, j = 1, 2, 3$ (summation convention), $\sigma = \{\sigma_{ij}\}$ (resp. $\varepsilon = \{\varepsilon_{ij}\}$) is the stress (resp. strain) tensor. We assume that the body obeys a subdifferential constitutive law of the form

$$\sigma_{ij} = [\nabla w(\varepsilon)]_{ij} \tag{3}$$

which is the nonlinear stress-strain relation and where $w\colon \mathbb{R}^6 \to \mathbb{R}$ is a convex Gateaux differentiable function (called the nonlinear elastic energy function). With an appropriate choice of w (see [8], [11], [12], [10]) the relation (3) appears as a constitutive law in several nonlinear elastic materials and it describes, for instance, the polygonal stress-strain law, the superlinear generalization of the deformation theory of plasticity and the generalization of rubber-like materials.

We assume further that the body forces f_i consist of two parts, $\overline{\overline{f_i}}$ which is the given external loading and $\overline{f_i}$ which is the reaction of the constraint introducing the skin effect (friction, adhesion, etc.). It is supposed that the reaction-displacement law is of nonconvex, nonlinear and possibly multivalued type:

$$-\overline{f} \in \partial j(u) \quad \text{on} \quad \Omega, \tag{4}$$

where $j\colon \mathbb{R}^3 \to \mathbb{R}$ is a locally Lipschitz superpotential and ∂j stands for the Clarke's generalized subdifferential of j. Moreover, we suppose for simplicity that on $\partial\Omega$ the homogeneous Dirichlet boundary condition

$$u = 0 \quad \text{on} \quad \partial\Omega \tag{5}$$

holds. By means of the Green-Gauss theorem, we may show (see e.g. [11] for details) that the BVP (1)–(5) leads to the following hemivariational inequality: find $u \in H_0^1(\Omega; \mathbb{R}^3)$ such that

$$W(\varepsilon(v)) - W(\varepsilon(u)) + \int_\Omega j^0(u; v - u)\, dx \geq \int_\Omega \overline{\overline{f_i}}\,(v_i - u_i)\, dx$$

for every $v = \{v_i\} \in H_0^1(\Omega; \mathbb{R}^3)$, where $W\colon L^2(\Omega; \mathbb{R}^6) \to \mathbb{R}$ is given by

$$W(\varepsilon) = \begin{cases} \displaystyle\int_\Omega w(\varepsilon)\, dx, & \text{if } w(\varepsilon) \in L^1(\Omega) \\ +\infty & \text{otherwise} \end{cases}$$

and j^0 denotes the directional derivative of j in the sense of Clarke (see Section 3).

We remark that if $w(\varepsilon) = \frac{1}{2}C_{ijhk}\varepsilon_{ij}\varepsilon_{hk}$, where the elasticity tensor $C = \{C_{ijhk}\}$ satisfies the well-known symmetry and ellipticity conditions, then (3) becomes the Hooke's law $\sigma_{ij} = C_{ijhk}\varepsilon_{hk}$ of the theory of linear elasticity. In this case the shape optimization problem was studied in [5] separately for two types of HVIs with one dimensional laws:

1) for HVI with a nonlinear law in a domain Ω in $\mathbb{R}^N$: find $u \in K(\Omega)$ such that

$$a(u, v-u) + \int_\Omega j^0(u, v-u)\, dx \geq \langle f, v-u\rangle_{V'\times V}\,, \quad \forall\, v \in K(\Omega) \tag{6}$$

and

2) for HVI with a nonlinear law on the boundary $\partial\Omega$ of Ω: find $u \in K(\Omega)$ such that

$$a(u, v-u) + \int_{\partial\Omega} j^0(u, v-u)\, d\sigma \geq \langle f, v-u\rangle_{V'\times V}\,, \quad \forall\, v \in K(\Omega). \tag{7}$$

Above $a(\cdot,\cdot)$ represents a bilinear form on $V = H^1(\Omega)$ (or $V = H_0^1(\Omega)$ resp.) and $K(\Omega)$ is a nonempty, closed, convex subset of V which in applications incorporates various unilateral conditions on Ω or on $\partial\Omega$. The subdifferential ∂j of a locally Lipschitz function $j: \mathbb{R} \to \mathbb{R}$ describes the nonmonotone and nonconvex law, respectively, in Ω (in problem (6)) and on $\partial\Omega$ (in problem (7)).

In the remaining part of this section we report on the mapping method which was introduced in [9] and consists of finding the optimal shapes in a class of admissible domains gained as images of a fixed set. More precisely, given a bounded open set C in $\mathbb{R}^N$ with its boundary ∂C of class $W^{i,\infty}$, $i \geq 1$ and such that $int\,\overline{C} = C$, we define for $k \geq 1$, the space $\mathcal{O}^{k,\infty}$ of bounded open sets of $\mathbb{R}^N$ which are isomorphic with C, i.e.

$$\mathcal{O}^{k,\infty} = \{\Omega \mid \Omega = T(C),\ T \in \mathcal{F}^{k,\infty}\},$$

where

$$\mathcal{F}^{k,\infty} = \{T: \mathbb{R}^N \to \mathbb{R}^N \mid T \text{ is bijective and } T, T^{-1} \in \mathcal{V}^{k,\infty}\},$$
$$\mathcal{V}^{k,\infty} = \{T: \mathbb{R}^N \to \mathbb{R}^N \mid T - I \in W^{k,\infty}(\mathbb{R}^N; \mathbb{R}^N)\}$$

and I denotes the identity in $\mathbb{R}^N$. We define on $\mathcal{O}^{k,\infty} \times \mathcal{O}^{k,\infty}$ a function

$$\delta_{k,\infty}(\Omega_1, \Omega_2) = \inf_{T\in\mathcal{F}^{k,\infty}, T(\Omega_1)=\Omega_2} \left(||T - I||_{k,\infty} + ||T^{-1} - I||_{k,\infty}\right),$$

where $||\cdot||_{k,\infty}$ denotes the usual norm in $W^{k,\infty}(\mathbb{R}^N; \mathbb{R}^N)$. The following result follows from Proposition 2.3, Theorem 2.2 and Theorem 2.4 of [9].

Theorem 1 *Let $k \geq 1$.*

(1) There exists a positive constant μ_k such that $d_{k,\infty}$ defined by $d_{k,\infty} = (\delta_{k,\infty} \wedge \mu_k)^{1/2}$ is a metric on $\mathcal{O}^{k,\infty}$.

(2) The space $\left(\mathcal{O}^{k,\infty}, d_{k,\infty}\right)$ is a complete metric space.

(3) If $k \geq 2$, then the embedding from $\mathcal{O}^{k,\infty}$ into $\mathcal{O}^{k-1,\infty}$ is compact. More precisely, if $k \geq 2$ and $\mathcal{B}$ is a bounded (in $\delta_{k,\infty}$), closed subset of $\mathcal{O}^{k,\infty}$, then for any sequence $\{\Omega_m\} \subset \mathcal{B}$, there exists a subsequence $\{\Omega_{m_\nu}\}$ of $\{\Omega_m\}$ and a set $\Omega \in \mathcal{B}$ such that $\Omega_{m_\nu} \to \Omega$ in $\mathcal{O}^{k-1,\infty}$.

In what follows we will need some facts on the mapping method.

Lemma 1
(a) If $T \in \mathcal{F}^{1,\infty}$, $\Omega = T(C)$, *then* $u \in L^2(\Omega)$ *iff* $u \circ T \in L^2(C)$; $u \in H^1(\Omega)$ *iff* $u \circ T \in H^1(C)$. *Moreover, if* $u_m \to u$ *in* $H^1(\Omega)$ *(or in* $H^1(C)$*) and* $T \in \mathcal{F}^{k,\infty}$ *with* $k \geq 1$, *then* $u_m \circ T \to u \circ T$ *in* $H^1(C)$ *(or* $u_m \circ T^{-1} \to u \circ T^{-1}$ *in* $H^1(\Omega)$*).*

(b) Let $u \in H^l(\mathbb{R}^N)$ *with* $l = 0$ *or* 1 *and* $k \geq 1$. *Then the mapping* $T \mapsto u \circ T$ *is continuous from* $\mathcal{V}^{k,\infty}$ *to* $H^l(\mathbb{R}^N)$ *at every point* $T \in \mathcal{F}^{k,\infty}$.

(c) Let $k \geq 1$. *The following mappings are continuous*

$$T \mapsto J_T^{-1} \text{ from } \mathcal{V}^{k,\infty} \text{ to } W^{k-1,\infty}(\mathbb{R}^N; \mathbb{R}^{2N}),$$

$$T \mapsto \det J_T \text{ from } \mathcal{V}^{k,\infty} \text{ to } W^{k-1,\infty}(\mathbb{R}^N; \mathbb{R})$$

at every point $T \in \mathcal{F}^{k,\infty}$ *(*J_T *denotes here the standard Jacobian matrix of* T*).*

(d) Let $\Omega_m = T_m(C)$, $\Omega_0 = T_0(C)$ *with* $T_m, T_0 \in \mathcal{F}^{k,\infty}$ *and* $\Omega_m \to \Omega_0$ *in* $\mathcal{O}^{k,\infty}$ *with* $k \geq 1$. *Then* $T_m \to T_0$, $T_m^{-1} \to T_0^{-1}$ *in* $W^{k,\infty}(\mathbb{R}^N; \mathbb{R}^N)$.

For the proof of (a)–(d) of Lemma 1, we refer, respectively to Lemma 4.1, Lemma 4.4 (i), Lemma 4.3 and 4.2, and Section 2.4 of [9].

For the relationships between the convergence in $\mathcal{O}^{k,\infty}$ and other types of convergence of sets, see [9], [14], [7], [5].

We close this section with the hypothesis which will be useful in the next sections.

$\underline{(H_0)}$: C is a bounded open set in $\mathbb{R}^N$ with boundary of class $W^{i,\infty}$, $i \geq 1$ such that $int\,\overline{C} = C$ and $\mathcal{B}$ is a bounded closed subset of $\mathcal{O}^{k,\infty}$ with $k \geq 2$.

Remark 1 *It should be pointed out that for the case of HVI of type (7) the regularity of mappings in the space* $\mathcal{F}^{k,\infty}$ *is not sufficient to get an existence of optimal domain (cf. [9], [5]).*

3. Hemivariational inequality with nonlinear laws in Ω

In this section we consider a class of HVIs in vector-valued function spaces. We present a result on existence of solutions, we show a priori estimate for solutions and we give a theorem on the dependence of the solution set on the domain.

Let us fix a set Ω in $\mathcal{B}$ where $\mathcal{B}$ is a bounded closed subset of $\mathcal{O}^{k,\infty}$ with $k \geq 2$. Given two positive integers M, N, we denote by $V = V(\Omega)$ the space $H_0^1(\Omega; \mathbb{R}^M)$ which is endowed with the norm $||v|| = ||\nabla v||_{L^2(\Omega;\mathbb{R}^{MN})}$.

By a hemivariational inequality we mean the following problem:

$$\begin{cases} \text{find } u \in V(\Omega) \text{ such that} \\ \displaystyle\int_\Omega [w(\nabla v) - w(\nabla u)]\, dx + \int_\Omega j^0(u; v-u)\, dx \geq \langle f, v-u \rangle_{V' \times V}\,, \quad \forall\, v \in V(\Omega). \end{cases} \tag{8}$$

We admit the following hypotheses:

$\underline{H(w)}$: $w: \mathbb{R}^{MN} \to \mathbb{R}$ is a convex, differentiable function such that $c_1|\xi|^2 - c_2 \leq w(\xi) \leq c_3(1 + |\xi|^2)$ for all $\xi \in \mathbb{R}^{MN}$, where $0 < c_1 \leq c_3 < +\infty$ and $0 \leq c_2 < +\infty$.

$\underline{H(j)}$: $j: \mathbb{R}^M \to \mathbb{R}$ is a locally Lipschitz function satisfying

(i) the growth condition: $|\eta| \leq c_0(1 + |\xi|)$ for every $\eta \in \partial j(\xi)$ and $\xi \in \mathbb{R}^M$ with $c_0 > 0$,

(ii) the generalized sign condition: $j^0(\xi; -\xi) \leq c_4|\xi| + c_5$ for $\xi \in \mathbb{R}^M$ with $c_4, c_5 > 0$.

Above, $j^0(\cdot;\cdot)$ is the generalized directional derivative of $j: \mathbb{R}^M \to \mathbb{R}$ given by

$$j^0(\xi;\eta) = \limsup_{h\to 0,\ t\downarrow 0} \frac{j(\xi + h + t\eta) - j(\xi + h)}{t} \quad \text{for all } \xi, \eta \in \mathbb{R}^M$$

and $\partial j: \mathbb{R}^M \to 2^{\mathbb{R}^M}$ denotes the Clarke's generalized subdifferential of j (see [3]) defined as follows

$$\partial j(\xi) = \{\eta \in \mathbb{R}^M \mid j^0(\xi;\gamma) \geq \eta \cdot \gamma, \ \ \forall\, \gamma \in \mathbb{R}^M\} \quad \text{for all } \xi \in \mathbb{R}^M.$$

Definition 1 *An element $u \in V(\Omega)$ is said to be a solution to (8) if there exists $\chi \in L^2(\Omega;\mathbb{R}^M)$ such that*

$$\int_\Omega [w(\nabla v) - w(\nabla u)]\, dx + \int_\Omega \chi \cdot (v - u)\, dx \geq \langle f, v - u\rangle_{V' \times V}\,, \quad \forall\, v \in V(\Omega) \qquad (9)$$

and

$$\chi(x) \in \partial j(u(x)) \text{ for a.e. } x \in \Omega. \qquad (10)$$

For a justification of this definition we refer to [11] and [10].

Theorem 2 *If hypotheses $H(w)$, $H(j)$ hold and $f \in V'(\Omega)$, then problem (8) admits a solution.*

For the proof of the above result, we note that it is a consequence of Theorem 4.25 of [10] because the operator $Av = -\operatorname{div}(\nabla w(\nabla v))$ from V into V' is maximal monotone and coercive. Its coerciveness on $V(\Omega)$ follows from the estimate

$$(\nabla w(\xi), \xi) \geq c_1|\xi|^2 - (c_2 + c_3) \quad \text{for every} \quad \xi \in \mathbb{R}^{MN}$$

which in turn follows from $H(w)$.

In the following the solution set of problem (8) will be denoted by $S(\Omega)$.

Lemma 2 *Under the hypotheses of Theorem 2, if $u \in S(\Omega)$, then*

$$||u|| \leq c\,(1 + meas(\Omega) + ||f||_{V'}) \quad \text{with} \quad c > 0.$$

Proof: It is enough to observe that taking $v = 0$ in (8) and then using $H(w)$ and the generalized sign condition of j, we easily get

$$c_1||u||^2 \le c_2 + (c_3 + c_5)meas(\Omega) + (c_6 + ||f||_{V'})||u||$$

with $c_6 > 0$. Hence the desired estimate follows. ∎

The following result on the dependence of solutions of HVI (8) with respect to the domain is a crucial point in obtaining the existence of optimal shapes.

Proposition 1 *Let us assume that* (H_0), $H(w)$, $H(j)$ *hold and* $f \in L^2(\mathbb{R}^N; \mathbb{R}^M)$. *Then the map* $\mathcal{B} \ni \Omega \mapsto S(\Omega) \subset V(\Omega)$ *has a closed graph in the following sense: if* $\Omega_0, \Omega_m \in \mathcal{B}$, $\Omega_m \to \Omega_0$ *in* $\mathcal{O}^{k,\infty}$ *with* $k \ge 1$, $u_m \in S(\Omega_m)$, $\widehat{u}_m = u_m \circ T_m$, $\widehat{u}_m \to u^*$ *weakly in* $V(C) = H_0^1(C; \mathbb{R}^M)$, *then* $u^* = u_0 \circ T_0$ *for some* $u_0 \in S(\Omega_0)$, *where* $\Omega_m = T_m(C)$ *and* $\Omega_0 = T_0(C)$.

Proof: Let $\Omega_0, \Omega_m \in \mathcal{B}$ be such that $\Omega_m \to \Omega_0$ in $\mathcal{O}^{k,\infty}$, where $\Omega_m = T_m(C)$, $\Omega_0 = T_0(C)$ and $T_m, T_0 \in \mathcal{F}^{k,\infty}$ ($k \ge 1$). By Lemma 1 (d), we know that $T_m \to T_0$, $T_m^{-1} \to T_0^{-1}$ in $W^{k,\infty}(\mathbb{R}^N; \mathbb{R}^N)$. Without loss of generality, we assume that $\det J_{T_m} > 0$ and $\det J_{T_0} > 0$ on $\mathbb{R}^N$. Let $u_m \in S(\Omega_m)$, i.e. $u_m \in V(\Omega_m)$ and there exists $\chi_m \in L^2(\Omega_m; \mathbb{R}^M)$ such that $\chi_m(x) \in \partial j(u_m(x))$ for a.e. $x \in \Omega_m$ and

$$\int_{\Omega_m} [w(\nabla v) - w(\nabla u_m)]\, dx + \int_{\Omega_m} \chi_m \cdot (v - u_m)\, dx \ge \int_{\Omega_m} f \cdot (v - u_m)\, dx$$

for every $v \in V(\Omega_m)$. By using the transformation $T_m: C \ni X \mapsto x = T_m(X) \in \Omega_m$, we rewrite the above problem as the following equivalent one on the set C:

$$W_{T_m}(v) - W_{T_m}(\widehat{u}_m) + (\widehat{\chi}_m, v - \widehat{u}_m) \ge (\widehat{f}_m, v - \widehat{u}_m), \quad \forall\, v \in V(C) \tag{11}$$

and

$$\widehat{\chi}_m(X) \in \partial j(\widehat{u}_m(X)) \text{ for a.e. } X \in C, \tag{12}$$

where $\widehat{u}_m = u_m \circ T_m$, $\widehat{\chi}_m = \chi_m \circ T_m$, $\widehat{f}_m = f \circ T_m$ and

$$W_{T_m}(v) = \int_C w\left(J_{T_m}^{-t} \nabla v\right) \det J_{T_m}\, dX,$$

$$(\widehat{\chi}_m, v) = \int_C \widehat{\chi}_m v \det J_{T_m}\, dX, \quad (\widehat{f}_m, v) = \int_C \widehat{f}_m v \det J_{T_m}\, dX.$$

From Lemma 1 (a), we deduce that $\widehat{u}_m \in V(C)$, $\widehat{\chi}_m \in L^2(C; \mathbb{R}^M)$ and $\widehat{f}_m \in L^2(\mathbb{R}^N; \mathbb{R}^M)$. The goal is now to pass to the limit, as $m \to +\infty$, in (11) and (12). By hypothesis

$$\widehat{u}_m \to u^* \quad \text{weakly in } V(C) \tag{13}$$

and since the embedding $V(C) \subset L^2(C; \mathbb{R}^M)$ is compact, we have (at least for subsequences)

$$\widehat{u}_m \to u^* \quad \text{in } L^2(C; \mathbb{R}^M) \text{ and a.e. in } C. \tag{14}$$

From (13) and the fact that

$$J_{T_m}^{-t} \to J_{T_0}^{-t} \quad \text{in } L^\infty(\mathbb{R}^N; \mathbb{R}^{2N}) \tag{15}$$

(cf. Lemma 1 (c)), we obtain

$$J_{T_m}^{-t} \nabla \widehat{u}_m \to J_{T_0}^{-t} \nabla u^* \quad \text{weakly in} \quad L^2(C; \mathbb{R}^{MN}). \tag{16}$$

Exploiting $H(j)(i)$ we can show, after passing to a subsequence, if necessary, that

$$\widehat{\chi}_m \to \chi^* \quad \text{weakly in } L^2(C; \mathbb{R}^M) \tag{17}$$

with some $\chi^* \in L^2(C; \mathbb{R}^M)$. Analogously as in Proposition 3.1 in [5], we can obtain the following two convergences:

$$\begin{cases} (\widehat{f}_m, v - \widehat{u}_m) \to (\widehat{f}_0, v - u^*), \\ (\widehat{\chi}_m, v - \widehat{u}_m) \to (\chi^*, v - u^*) \end{cases} \tag{18}$$

with $\widehat{f}_0 = f \circ T_0$. Taking the *liminf* in (11), using (18) and the inequality $\liminf(\alpha_m - \beta_m) \leq \liminf \alpha_m - \liminf \beta_m$, we get

$$(\widehat{f}_0, v - u^*) \leq (\chi^*, v - u^*) + \liminf_m W_{T_m}(v) - \liminf_m W_{T_m}(\widehat{u}_m) \tag{19}$$

for every $v \in V(C)$. Further, we are going to show that

$$W_{T_0}(u^*) \leq \liminf_m W_{T_m}(\widehat{u}_m), \tag{20}$$

$$W_{T_0}(v) = \lim_m W_{T_m}(v), \quad \forall\, v \in V(C). \tag{21}$$

First we show (20). To this end we observe that due to the hypothesis $H(w)$, the functional $L^2(C; \mathbb{R}^{MN}) \ni z \mapsto \int_C w(z(X)) \det J_{T_0}(X)\, dX \in \mathbb{R}$ is l.s.c. in the weak topology of $L^2(C; \mathbb{R}^{MN})$ (see e.g. Example 1.23 of [4]). Hence and from (16), we have

$$W_{T_0}(u^*) \leq \liminf_m \int_C w\left(J_{T_m}^{-t} \nabla \widehat{u}_m\right) \det J_{T_0}\, dX \leq \tag{22}$$

$$\leq \liminf_m W_{T_m}(\widehat{u}_m) + \limsup_m \int_C |w\left(J_{T_m}^{-t} \nabla \widehat{u}_m\right) (\det J_{T_0} - \det J_{T_m})\, |\, dX.$$

Using the following inequality (see $H(w)$)

$$\int_C |w\left(J_{T_m}^{-t} \nabla \widehat{u}_m\right)|\, |\det J_{T_m} - \det J_{T_0}|\, dX \leq$$

$$\leq \overline{c} \int_C \left(1 + |J_{T_m}^{-t} \nabla \widehat{u}_m|^2\right) dX\, ||\det J_{T_m} - \det J_{T_0}||_{L^\infty(\mathbb{R}^N)} \leq$$

$$\leq \overline{c} \left(meas(C) + ||J_{T_m}^{-t}||_{L^\infty}^2\, ||\widehat{u}_m||_{V(C)}^2 \right) ||\det J_{T_m} - \det J_{T_0}||_{L^\infty},$$

(for some $\overline{c} > 0$), together with (13), (15) and the convergence

$$\det J_{T_m} \to \det J_{T_0} \quad \text{in} \quad L^\infty(\mathbb{R}^N) \tag{23}$$

(see Lemma 1 (c)), we readily obtain (20) owing to (22). In order to prove (21), we note that (15) implies $J_{T_m}^{-t} \nabla v \to J_{T_0}^{-t} \nabla v$ in $L^2(C; \mathbb{R}^{MN})$ and subsequently also a.e. in C.

Using the continuity of w and (23), and applying the dominated convergence theorem, we immediately get (21). Combining (19), (20) and (21), we have

$$(\widehat{f}_0, v - u^*) \leq (\chi^*, v - u^*) + W_{T_0}(v) - W_{T_0}(u^*), \quad \forall\, v \in V(C). \tag{24}$$

Then, by employing the transformation $X = T_0^{-1}(x)$, we can write down the above inequality in the equivalent form

$$\int_{\Omega_0} [w(\nabla v) - w(\nabla u_0)]\, dx + \int_{\Omega_0} \chi_0 \cdot (v - u_0)\, dx \geq \int_{\Omega_0} f \cdot (v - u_0)\, dx$$

for every $v \in V(\Omega_0)$, where $u_0 = u^* \circ T_0^{-1}$ and $\chi_0 = \chi^* \circ T_0^{-1}$. Similarly as in [5], using (14), (17), the properties of ∂j and applying the convergence theorem (see Theorem 1 in Chapter 1.4 of [1]), from (12) we deduce that $\chi^*(X) \in \partial j(u^*(X))$ for a.e. $X \in C$. Hence $\chi_0(X) \in \partial j(u_0(X))$ for a.e. $X \in \Omega_0$. Since $u_0 \in V(\Omega_0)$ and $\chi_0 \in L^2(\Omega_0; \mathbb{R}^M)$, we conclude that $u_0 \in S(\Omega_0)$. This completes the proof. ∎

4. A domain optimization problem

The goal of this section is to give the main result of the paper on existence of solutions (optimal domains) to the following optimal shape design problem:

$$\begin{cases} \text{find } (\Omega^*, u^*) \in \bigcup\limits_{\Omega \in \mathcal{B}} (\Omega \times S(\Omega)) \text{ such that} \\ \mathcal{J}(\Omega^*, u^*) = \min\limits_{\Omega \in \mathcal{B}} \min\limits_{v \in S(\Omega)} \mathcal{J}(\Omega, v), \end{cases} \tag{25}$$

where $S(\Omega) \subset V(\Omega)$ denotes the set of solutions to the hemivariational inequality (8), $\mathcal{B} \subset \mathcal{O}^{k,\infty}$ with $k \geq 2$ and the cost functional is of the form

$$\mathcal{J}(\Omega, v) = \int_{\Omega} g(x, v(x))\, dx.$$

We need the following hypothesis:

$\underline{H(g)}$: $g\colon \mathbb{R}^N \times \mathbb{R}^M \to [0, +\infty)$ is a Borel function such that
(i) $g(x, \cdot)$ is l.s.c. on $\mathbb{R}^M$,
(ii) $g(x, \zeta) \leq c(1 + |\zeta|^2)$ for all $x \in \mathbb{R}^N, \zeta \in \mathbb{R}^M$ with $c > 0$.

Theorem 3 *If hypotheses* (H_0), $H(w)$, $H(j)$, $H(g)$ *hold and* $f \in L^2(\mathbb{R}^N; \mathbb{R}^M)$, *then problem (25) admits at least one solution.*

Proof: We prove the existence of minimizers by applying the direct method of the calculus of variations. Let $(\Omega_m, u_m) \in \bigcup_{\Omega \in \mathcal{B}} (\Omega \times S(\Omega))$ be a minimizing sequence for (25). Since $\mathcal{B}$ is compact in $\mathcal{O}^{k-1,\infty}$ (cf. Theorem 1(3)), we find a subsequence of Ω_m (denoted in the same way) and a set $\Omega_0 \in \mathcal{B}$ such that $\Omega_m \to \Omega_0$ in $\mathcal{O}^{k-1,\infty}$. Since $u_m \in S(\Omega_m)$, from Lemma 2, we have

$$||u_m||_{V(\Omega_m)} \leq c\,(1 + meas(\Omega_m) + ||f||)\,, \tag{26}$$

with $c > 0$ independent of m. Since the convergence of $\{\Omega_m\}$ in $\mathcal{O}^{k-1,\infty}$ entails the convergence of characteristic functions $1_{\Omega_m} \to 1_{\Omega_0}$ in $L^2(\mathbb{R}^N)$ (see [7]), we get from the latter that $meas(\Omega_m) \leq c_7$, where $c_7 > 0$ is independent of m. So (26) gives the boundedness of u_m in $V(\Omega_m)$. Taking $\widehat{u}_m = u_m \circ T_m$ and using the inequalities in Remark 4.1 of [9] (see also Section 2 of [5]), we deduce that $\{\widehat{u}_m\}$ lies in a bounded subset of $V(C)$. Therefore, there exists $u^* \in V(C)$ such that

$$\widehat{u}_m \to u^* \text{ weakly in } V(C). \tag{27}$$

From Proposition 1, it follows that $u^* = u_0 \circ T_0 = \widehat{u}_0$ with $u_0 \in S(\Omega_0)$. Hence the pair (Ω_0, u_0) is admissible for (25).

In order to prove that (Ω_0, u_0) is an optimal pair for (25), it is enough to show that

$$\mathcal{J}(\Omega_0, u_0) \leq \liminf_m \mathcal{J}(\Omega_m, u_m). \tag{28}$$

To this end, let $\underline{\widehat{u}_m}$ and $\underline{u^*}$ denote the functions in $L^2(\mathbb{R}^N; \mathbb{R}^M)$ obtained from $\widehat{u}_m$ and u^*, respectively, by extending them by zero outside C. From (27) and the compactness of the embedding $V(C) \subset L^2(C; \mathbb{R}^M)$, it follows that

$$\underline{\widehat{u}_m} \to \underline{\widehat{u}_0} \text{ in } L^2(\mathbb{R}^N; \mathbb{R}^M). \tag{29}$$

Let $\underline{u_m}$ and $\underline{u_0}$ be functions in $L^2(\mathbb{R}^N; \mathbb{R}^M)$, defined by

$$\underline{u_m}(x) = \begin{cases} u_m(x), & \text{if } x \in \Omega_m \\ 0, & \text{if } x \in \mathbb{R}^N \setminus \Omega_m, \end{cases} \qquad \underline{u_0}(x) = \begin{cases} u_0(x), & \text{if } x \in \Omega_0 \\ 0, & \text{if } x \in \mathbb{R}^N \setminus \Omega_0. \end{cases}$$

From the estimate

$$||\underline{u_m} - \underline{u_0}||_{L^2(\mathbb{R}^N,\mathbb{R}^M)} = ||\underline{\widehat{u}_m} \circ T_m^{-1} - \underline{\widehat{u}_0} \circ T_0^{-1}|| \leq ||\underline{\widehat{u}_m} \circ T_m^{-1} - \underline{\widehat{u}_0} \circ T_m^{-1}|| +$$
$$+ ||\underline{\widehat{u}_0} \circ T_m^{-1} - \underline{\widehat{u}_0} \circ T_0^{-1}|| \leq ||\det J_{T_m^{-1}}||^{1/2}_{L^\infty(\mathbb{R}^N)} ||\underline{\widehat{u}_m} - \underline{\widehat{u}_0}|| + ||\underline{\widehat{u}_0} \circ (T_m^{-1} - T_0^{-1})||,$$

Lemma 1(b), (c) and (29), we easily obtain that $\underline{u_m} \to \underline{u_0}$ in $L^2(\mathbb{R}^N, \mathbb{R}^M)$. By passing to subsequences, if necessary, we may assume that

$$\underline{u_m}(x) \to \underline{u_0}(x), \quad 1_{\Omega_m}(x) \to 1_{\Omega_0}(x) \text{ for a.e. } x \in \mathbb{R}^N. \tag{30}$$

Hence and from $H(g)(i)$, we subsequently get

$$g(x, \underline{u_0}(x)) \leq \liminf_m g(x, \underline{u_m}(x)) \text{ for a.e. } x \in \mathbb{R}^N. \tag{31}$$

Exploiting (31), $H(g)(ii)$ and (30) again, we have

$$\liminf_m g(x, \underline{u_m}(x)) 1_{\Omega_m}(x) \geq \liminf_m g(x, \underline{u_m}(x))(1_{\Omega_m}(x) - 1_{\Omega_0}(x)) +$$
$$+ \liminf_m g(x, \underline{u_m}(x)) 1_{\Omega_0}(x) \geq g(x, \underline{u_0}(x)) 1_{\Omega_0}(x)$$

for a.e. $x \in \mathbb{R}^N$. Integrating the last inequality and applying Fatou's lemma, we immediately obtain (28) which completes the proof. ■

Remark 2 *Existence results for the problems of domain optimization for systems governed by hemivariational inequalities (6) and (7) with cost functionals of the form $\mathcal{J}(\Omega, v) = \int_\Omega g(x, v(x), \nabla v(x))\, dx$ have been recently obtained in [5].*

References

[1] Aubin, J.-P.; Cellina, A.: *Differential Inclusions*, Springer-Verlag, Berlin – Heidelberg – New York – Tokyo, 1984.

[2] Buttazzo, G.; Dal Maso, G.: *An existence result for a class of shape optimization problems*, Arch. Rat. Mech. Anal., 122 (1993), 183–195.

[3] Clarke, F. H.: *Optimization and Nonsmooth Analysis*, John Wiley & Sons, New York, 1983.

[4] Dal Maso, G.: *An Introduction to* Γ*-convergence*, Birkhäuser, Boston, 1993.

[5] Denkowski, Z.; Migórski, S.: *Optimal shape design problems for a class of systems described by hemivariational inequality*, J. Global Optim., to appear.

[6] Haslinger, J.; Neittaanmäki, P.; Tiihonen, T.: *Shape optimization in contact problems based on penalization of the state inequality*, Aplikace Mat., 31 (1986), 54–77.

[7] Liu W. B.; Rubio, J. E.: *Optimal Shape Design for Systems Governed by Variational Inequalities, Part 1: Existence Theory for Elliptic Case* , J. Optim. Th. Appl., 69 (1991), 351–371.

[8] Moreau, J. J. *Sur les lois de frottement, de plasticité et de viscosité*, C. R. Acad. Sci. Paris, 271A (1970), 608–611.

[9] Murat, F.; Simon, J.: *Sur le Controle par un Domaine Geometrique*, Preprint no. 76015, University of Paris 6, 1976.

[10] Naniewicz, Z.; Panagiotopoulos, P. D.: *Mathematical Theory of Hemivariational Inequalities*, Dekker, New York, 1995.

[11] Panagiotopoulos, P. D.: *Variational-hemivariational inequalities in nonlinear elasticity. The coercive case*, Aplikace Mat., 33 (1988), 249–268.

[12] Panagiotopoulos, P. D.: *The nonmonotone skin effects in plane elasticity problems obeying to linear elastic and subdifferential material laws*, Z. Angew. Math. Mech., 70 (1990), 13–21.

[13] Panagiotopoulos, P. D.: *Hemivariational Inequalities, Applications in Mechanics and Engineering*, Springer-Verlag, Berlin, 1993.

[14] Pironneau, O.: *Optimal Shape Design for Elliptic Systems*, Springer-Verlag, New York, 1984.

International Series of Numerical Mathematics
Vol. 124, © 1998 Birkhäuser Verlag, Basel

A Discretization for Control Problems with Optimality Test

Ursula Felgenhauer*

Abstract. The paper deals with a RITZ type discretization for constrained optimal control problems. The approach starts from a primal-dual formulation containing the HAMILTON-JACOBI inequality in integrated form. For the discrete problems there are given conditions guaranteeing the optimality of the limit solution. They take the form of a discrete analogy to certain matrix RICCATI differential inequality.

1. Introduction

Recently there have been obtained essential results concerning second order sufficient optimality criteria for control problems, cf. e.g. [16], [18] [4] (and references given there). An important tool suitable for a unified analysis of different classes of problems (including also the case of multiple integrals in the cost function) was proposed by R. KLÖTZLER, [10], (see also [11], [17]). It is closely connected to the investigation of the so-called HAMILTON-JACOBI inequality. In the paper we use a weak form of it to derive discrete subproblems for the numerical treatment. This is a main difference to former discrete techniques [1], [8] or to the direct EULER approach in [14]. The paper is directed on the proof of a discrete version of sufficient optimality criteria. The main result is a local quadratic growth estimation for the discrete and the continuous objective functionals. Its derivation had been inspired by related results of ROBINSON [19] for nonlinear mathematical programs. The obtained results may be seen in one line with the very recent papers [14] and [2] on EULER's method in the optimal control context.

Notations. Let the EUCLIDean norm in the vector space $I\!R^\nu$ be given by $|\cdot|$; further denote by the superscript T the transposition of a vector, and the scalar product of u and v from $I\!R^\nu$ by u^Tv. The symbol I stands for the unit matrix.

Throughout the paper L_p, L_∞ denote spaces of measurable functions with the norms $\|\cdot\|_p = \left[\int_0^1 |\cdot(t)|^p\,dt\right]^{1/p}$, $\|\cdot\|_\infty = \sup_{t\in[0,1]} ess\,|\cdot(t)|$; W_p^l are the related SOBOLEW spaces of functions having l generalized derivatives belonging to L_p.

The duality map for a BANACH (function) space X and its dual X^* will be written as $< \cdot, \cdot >_X$.

*BTU Cottbus, Institut für Mathematik, Universitätsplatz 3-4, D-03013 Cottbus, e-mail: felgenh@math.tu-cottbus.de

2. Characterization of optimal controls via duality

Consider the nonlinear constrained optimal control problem

$$\textbf{(P)} \qquad J(x,u) \; = \; k(\pi x) \; + \int_0^1 r(t,x(t),u(t))\,dt \; \longrightarrow \; \min\,, \tag{1}$$

subject to:

$$\dot{x}(t) \; = \; f(t,x(t),u(t)) \quad \text{for } a.e.\, t \ in \ [0,1]; \tag{2}$$

$$\beta(\pi x) \; = \; 0\,; \quad \gamma(\pi x) \; \leq \; 0; \tag{3}$$

$$g(t,x(t),u(t)) \leq 0 \qquad \text{for } a.e.\, t \ in \ (0,1); \tag{4}$$

where $x \in W^1_\infty([0,1];\mathbb{R}^n)$, $u \in L_\infty([0,1];\mathbb{R}^m)$, and π stands for the trace operator on W^1_∞ : $\pi x = (x(0),x(1))$. The functions r,f,g are defined on $\mathbb{R}_+ \times \mathbb{R}^n \times \mathbb{R}^m$ with values in $\mathbb{R}$, $\mathbb{R}^n$ and $\mathbb{R}^l$ respectively; $\beta : \mathbb{R}^{2n} \to \mathbb{R}^r$, $\gamma : \mathbb{R}^{2n} \to \mathbb{R}^s$ with $r{+}s \leq 2n$, and $k : \mathbb{R}^{2n} \to \mathbb{R}$. We will assume that the problem data functions are twice continuously FRÉCHET differentiable with LIPSCHITZ continuous second derivatives in all arguments.
Feasible sets occuring in problem **(P)** we will denote by

$$W \; = \; \{(t,x,u) : \, t \in [0,1], \; g(t,x,u) \; \leq \; 0\}, \quad W(t) \; = \; \{(x,u) : \; (t,x,u) \in W\},$$

$$\Gamma \; = \; \{\xi : \, \beta(\xi) = 0, \gamma(\xi) \leq 0\}\,.$$

Further we use the HAMILTONian H and the augmented HAMILTONian $\hat{H}$ in the form

$$H(t,x,u,\eta) \; = \; r(t,x,u) + \eta^T f(t,x,u)\,, \quad \hat{H}(t,x,u,\eta,\mu) \; = \; H(t,x,u,\eta) + \mu^T g(t,x,u)\,.$$

Characterizations of optimal solutions for problem **(P)** may be derived from a duality approach proposed by KLÖTZLER, [10]. It has been successfully used for the investigation of several control problems in a series of papers by KLÖTZLER, PICKENHAIN et al., see [11], [12], [16] and [17].
In the particular situation of a problem with boundary and mixed inner point constraints a dual problem may be derived as follows: Given $S : \mathbb{R}_+ \times \mathbb{R}^n$, define first $\theta(S,\xi) = S(0,\xi_0) - S(1,\xi_1) + k(\xi_0,\xi_1)$, and consider

$$\textbf{(D)} \qquad T(S) \; = \; \inf_{\xi\in\Gamma} \theta(S,\xi)\,dt \; \longrightarrow \; \max\,, \tag{5}$$

subject to:

$$H(t,x,u,\nabla_x S(t,x)) \; + \; S_t(t,x) \; \geq \; 0 \qquad a.e. \quad \text{on } W. \tag{6}$$

Here we assume that $S(\cdot,x)$ is LIPSCHITZ continuous on $[0,1]$, $S(t,\cdot)$ and $S_t(t,\cdot$ are supposed to be C^2 functions on $\mathbb{R}^n$. If $z = (x,u)$ is feasible for **(P)** and S for **(D)** resp., then

$$J(x,u) \; \geq \; T(S)\,.$$

Indeed,

$$\begin{aligned} J(x,u) &= k(\pi x) + \int_0^1 H(t,x,u,\nabla_x S(t,x))\,dt - \int_0^1 \nabla_x S(t,x)^T f(t,x,u)\,dt \\ &= \int_0^1 (H + S_t)\,dt - \int_0^1 \frac{d}{dt} S(t,x(t))\,dt + k(\pi x) \geq T(S)\,. \end{aligned}$$

The duality gap becomes zero iff the following conditions hold for x^*, u^* and S^*:

$$\int_0^1 (H^* + S_t^*)\,dt = 0\,, \quad \theta(S^*,\pi x^*) = \inf_{\xi\in\Gamma} \theta(S^*,\xi)\,. \tag{7}$$

These conditions represent at the same time sufficient optimality conditions for the solution (x^*,u^*) of **(P)**.
Similiar characterizations have been found e.g. by MAURER and PICKENHAIN [15] for weak local minimizers (cf. [16] for definitions). The respective neighborhoods are defined by means of the balls

$$B_\epsilon(x,u) = \{(y,v) \in I\!R^n \times I\!R^m : |y-x| \leq \epsilon,\ |v-u| \leq \epsilon\}\,,$$

$$\text{and} \qquad B'_\epsilon(\pi x^*) = \{\xi \in I\!R^{2n} : |\xi - \pi x^*| \leq \epsilon\}\,.$$

The next Theorem was proved in [15] (Theor 3.1). The formulation given here only additionally allows general boundary constraints, and it contains the HAMILTON-JACOBI inequality (cf. (a) below) in a slightly weakened form:

Theorem 1 *A feasible pair (x^*,u^*) is a weak strict local minimizer, if there exists a function $S = S(t,x)$ and positive constants c_1, c_2 such that*

(a) $\int_0^1 [H(t,x(t),u(t),\nabla_x S(t,x(t))) + S_t(t,x(t))]\,dt \geq c_1\,(\|x-x^*\|_2^2 + \|u-u^*\|_2^2)$ *for all (x,u) with $(x(t),u(t)) \in W_\epsilon(t) = W(t) \cap B_\epsilon(x^*(t),u^*(t))$ a.e. on $[0,1]$;*

(a′) $\int_0^1 [H(t,x^*(t),u^*(t),\nabla_x S(t,x^*(t))) + S_t(t,x^*(t))]\,dt = 0$;

(b) $\theta(S,\xi) - \theta(S,\pi x^*) \geq c_2\,|\xi - \pi x^*|^2$ *on $\Gamma_\epsilon = \Gamma \cap B'_\epsilon(\pi x^*)$.*

For every feasible $x = x(t)$, $u = u(t)$ then there will hold the estimate:

$$J(x,u) - J(x^*,u^*) \geq c_3 \left(\|x-x^*\|_2^2 + \|u-u^*\|_2^2\right) \tag{8}$$

with some positive c_3 if only $(x(t),u(t)) \in W_\epsilon(t)$ a.e. on $[0,1]$.

The proof repeats the calculation of the duality relation, for details compare [15].

3. Auxiliary problem and optimality criteria

Theorem 1 suggests investigating the auxiliary problem

$$\textbf{(OP)}\quad \Phi(x,u) = \int_0^1 [H(t,x(t),u(t),\nabla_x S(t,x(t))) + S_t(t,x(t))]\,dt \longrightarrow \min\,, \tag{9}$$

subject to: $\quad G(x,u) \in -K\,,$

where $G(x,u)(\cdot) = g(\cdot, x(\cdot), u(\cdot))$ and K consists of all L_∞ functions with nonnegative essential infimum.
With $L(x,u,\mu) = \Phi(x,u) + <\mu, G(x,u)>_\infty$, the KARUSH-KUHN-TUCKER system may be obtained:

$$\begin{aligned} 0 = \frac{\partial L}{\partial x} &= r_x + \mu^T g_x + \nabla_x S^T f_x + \nabla_x^2 S\, f + \nabla_x S_t\,, \\ 0 = \frac{\partial L}{\partial u} &= r_u + \mu^T g_u + \nabla_x S^T f_u\,, \\ g = \frac{\partial L}{\partial \mu} &\in -K\,,\ \mu \in K\,,\ <\mu, G> = 0\,. \end{aligned} \tag{10}$$

Under the assumption that there exists a solution (x^*, u^*) which is also feasible for problem **(P)**, in the first expression (L_x) we may use the chain rule

$$\nabla_x S_t + \nabla_x^2 S\, f = \nabla_x S_t + \nabla_x^2 S\, \dot{x} = (d/dt)\, \nabla_x S(t, x(t))$$

to simplify the system. Denoting $\nabla_x S(t, x^*(t)) = p(t)$, the equality parts in (10) may be rewritten then as

$$L_x = \dot{p} + \hat{H}_x = 0\,, \quad L_u = \hat{H}_u = 0\,,$$

Particularly we see that $p = p(t)$ coincides with the adjoint variable for **(P)**. The question of the existence and regularity of the adjoint and the multiplier functions p, μ in the last years could be in essence clarified. Several authors formulated assumptions of constraint qualification type suited for control problems, see e.g. [13] or for a survey [4]. The most common form of conditions are the invertibility of nearly-active constraints w.r.t. u, and the so-called controllability assumption on the linearized state equation. These conditions guarantee $p \in W_\infty^1$ and $\mu \in L_\infty$ (cf. [9], [14] or [7]). As a consequence, one may obtain the so-called transversality conditions for $\pi\, p$ (for instance, from requirement (b) in Theor.1):

$$(-1)^a p(a) - \nabla_a k(\pi\, x) + \rho_1^T \nabla_a \beta(\pi\, x) + \rho_2^T \nabla_a \gamma(\pi\, x) = 0\,,\ a = 0, 1\,,$$

which complete the system of necessary first order conditions. Our further interest is directed to second order sufficient optimality conditions. To this end consider the second FRÉCHET derivatives of L for primarily feasible stationary (x,u): As in the case of the first order derivatives, we may use the chain rule (here for $(d/dt)\, \nabla_x^2 S$) to simplify particularly the expression for L_{xx}, and with the abbreviation $\nabla_x^2 S(t, x^*(t)) = Q(t)$ and after some calculation we get

$$\begin{aligned} L_{xx} &= \hat{H}_{xx} + Q\, f_x + f_x^T Q + \dot{Q}\,, \\ L_{xu} &= \hat{H}_{xu} + Q\, f_u\,, \quad L_{uu} = \hat{H}_{uu}\,. \end{aligned}$$

Consequently, second order sufficient optimality criteria using coercivity properties of $\nabla_{(x,u)}^2 L$ may be expressed in terms of a certain matrix RICCATI differential inequality for Q (see [16], [14], [15] for a detailed discussion, for comparison cf. [20]).

Notice that an analogous second order analysis of θ in (b), Theorem 1, allows finishing with boundary restrictions on Q, namely

$$\hat{\pi}\, Q \;-\; \rho_1^T \nabla_\xi^2 \beta(\pi x^*) \;-\; \rho_2^T \nabla_\xi^2 \gamma(\pi x^*) \;\geq\; c\, I\;, \quad c > 0\,, \tag{11}$$

on $T_\Gamma \;=\; \{\zeta \,:\, \nabla_\xi \beta^T \zeta = 0\,,\; (\nabla_\xi \gamma^i)^T \zeta = 0 \text{ for } i \text{ with } \gamma^i = 0, \rho_2^i > 0\}$;
$\hat{\pi}\, Q \;=\; \operatorname{diag}(Q(0), -Q(1))$.
Particular cases for such conditions have been considered in [15] e.g.

4. The discrete problem

Let any regular sequence of discretizations $\omega = \omega(N)$ of $[0,1]$ be given, i.e. $0 = t_0 < t_1 \ldots < t_N = 1$; $\omega_j = [t_{j-1}, t_j]$ with $h = \min_j |\omega_j| \;\to\; 0$ for $N \to \infty$ and $\{\max_j |\omega_j| / \min_j |\omega_j|\}$ – uniformly bounded.
We consider finite-dimensional subspaces $Z_h \subset Z = W_\infty^1 \times L_\infty$ of the structure $Z_h = D_\omega^1 \times D_\omega^0$ according to ω : D_ω^1 consists of all continuous on $[0,1]$ functions which are linear on every interval $\omega_j \in \omega$, and D_ω^0 of all elementwise constant functions respectively. For functions $z_h = (x_h, u_h) \in Z_h$ denote $x_j = x_h(t_j)\,,\; u_j = u_h(\bar{t}_j)$ and $\bar{x}_j = x_h(\bar{t}_j)$ where $\bar{t}_j = 0.5\,(t_{j-1} + t_j)$, $j = (0), 1, \ldots, N$.
Further, the projection operator $P_h :\; Z \to Z_h$ will be used with

$$\begin{aligned} P_h^1 \,&:\; W_\infty^1 \to D_\omega^1, \quad (P_h^1 x)(t_j) \;=\; x(t_j)\,, \\ P_h^0 \,&:\; L_\infty \to D_\omega^0, \quad (P_h^0 u)(\bar{t}_j) \;=\; |\omega_j|^{-1} \int_{\omega_j} u(t)\, dt\,. \end{aligned}$$

This operator is the best approximation operato from $\hat{Z} = W_2^1 \times L_2$ to $Z_h \subset \hat{Z}$ if W_2^1 is supplied with the norm $\|x\| = |x(0)|^2 \,+\, \|\dot{x}\|_2^2$.
Next, we are going to apply the RITZ approach to problem **(OP)**. For this purpose, in addition to the projection operator we have to choose an appropriate summation formula. The structure of Z_h suggests utilizing the rectangular rule, i.e.:

$$\begin{aligned} \Phi_\omega(x,u) \;&=\; \int_0^1 R_h\, \phi(t, P_h^1 x, P_h^0 u)\, dt \\ &=\; \sum_j |\omega_j|\, \phi(\bar{t}_j, \bar{x}_j, u_j) \;=:\; \sum_j |\omega_j|\, \phi^j(\bar{x}_j, u_j) \end{aligned}$$

where ϕ is the abbreviation of $(S_t \,+\, H)$, and $(R_h \,\cdot)|_{\omega_j} \;=\; (\,\cdot\,)\,(\bar{t}_j)$. The constraints will be handled similiarly, i.e.

$$G_\omega(x,u)(\cdot) \;=\; R_h\, g(\cdot, P_h^1 x(\cdot), P_h^0 u(\cdot))\;,$$

so that we arrive at the discrete version of **(OP)**

$$\textbf{(DP)} \qquad \Phi_\omega(x,u) \;\longrightarrow\; \min \qquad \text{s.t.} \quad G_\omega(x,u) \in -K_\omega;\; K_\omega = K \cap Z_h \tag{12}$$

The special choice of the approximation allows considering **(DP)** in a first step in decomposed form ($\bar{z}_j = (\bar{x}_j, u_j)$)

$$\phi^j(\bar{z}_j) \;\to\; \min \qquad \text{s.t.} \quad G^j(\bar{z}_j) = g(\bar{t}_j, \bar{z}_j) \;\leq\; 0\;;\;\; j = 1, \ldots, N.$$

Let the corresponding LAGRANGE functions be given by

$$l^j(\xi,\eta,\mu) = \phi^j(\xi,\eta) + \mu^T G^j(\xi,\eta), \tag{13}$$

the KKT systems read as $(j = 1,\ldots,N)$:

$$\phi^j_\xi + \mu^T G^j_\xi = 0\,, \quad \phi^j_\eta + \mu^T G^j_\eta = 0\,, \tag{14}$$

$$\mu \geq 0\,,\ G^j \leq 0\,,\quad \mu_i G^j_i = 0\,,\ i = 1,\ldots,l. \tag{15}$$

Suppose that there exists $(x_h, u_h) \in Z_h$ and $\mu_h \in D^0_\omega$ satisfying (14), (15), the boundary conditions (3) and the state equation in discrete form,

$$\dot{x}_h(\bar{t}_j) = |\omega_j|^{-1}(x_j - x_{j-1}) = f(\bar{t}_j, \bar{x}_j, u_j)\,;\ j = 1,\ldots,N\,. \tag{16}$$

In analogy to section 3 we will use for $S = S_h(t,x)$ an expansion

$$S_h(t,x) = S^0_h(t) + p_h(t)^T(x - x_h) + 0.5\,(x - x_h)^T Q_h(t)(x - x_h) + \delta_h(t,x)$$

where $|\delta_h| = \mathrm{o}(|x - x_h|^2)$ for all t, and p_h, Q_h are D^1_ω functions. Substituting S_h into (14) we get

$$\dot{p}_h(\bar{t}_j) = -\hat{H}_x(\bar{t}_j)\,;\quad 0 = \hat{H}_u(\bar{t}_j)\,, \tag{17}$$

where the right-hand sides are evaluated at $(\bar{t}_j, \bar{x}_j, u_j, \bar{p}_j, \mu_j)$.
The analysis of the asymptotic behavior of the discretizations requires relatively strong assumptions on the system (15) - (17):

(A1) For every $\omega = \omega(N)$ with sufficiently small h there exists a solution $z_h = (x_h, u_h) \in Z_h$ and $\lambda_h = (p_h, \mu_h) \in Z_h$ satisfying (17), (16),(3) and (15). Further, (z_h, λ_h) are uniformly bounded in the $Z \times Z$ norm and converge in $\hat{Z} \times \hat{Z}$ to $(\bar{z}, \bar{\lambda})$; additionally (3) holds for the limit function $\bar{x}$.

As usual in finite element type discretizations the asymptotic behavior depends on additional regularity properties of the approximated functions. In our situation we introduce an $\{\omega(N)\}$-dependent regularity class by the following

Definition 1 *$y \in L_\infty(0,1;\mathbb{R}^\nu)$ is called ω-regular if there exists a function $\epsilon(h)$ tending to zero with $h \to 0$ such that for every sufficiently small h one can find an index set $J(h)$ with the properties*

(i) $\sum_{j\notin J(h)} |\omega_j| \leq \epsilon(h)$,

(ii) $\max_{j\in J(h)} \max_{t\in\omega_j} |y(t) - y(\bar{t}_j)| \leq \epsilon(h)$.

We require:

(A2) The functions $\bar{z}$, $\dot{\bar{x}}$, $\bar{\lambda}$ and $\dot{\bar{p}}$ are ω-regular.

An important consequence now consists in the approximation characteristic:

Lemma 1 *If $y_h \to \bar{y}$ in $L_2(0,1;\mathbb{R}^\nu)$ and $\bar{y}$ - ω-regular, then for arbitrary $\epsilon > 0$ and h - sufficiently small there exists $J'(h)$ such that*

$$\sum_{j\notin J'(h)} |\omega_j| \leq \epsilon\,,\quad \max_{j\in J'(h)} \max_{t\in\omega_j} |\bar{y}(t) - y_h(t)| \leq \epsilon.$$

The proof is technical and therefore will be omitted here. Cf. [6], Lemma 1 for details. The last regularity assumption we need for $(\bar{z}, \bar{\lambda})$ is a stable formulation of the invertibility condition mentioned in the previous section:

(A3) For given $\sigma > 0$ define $I^\sigma(t) = \{i : g^i(t, \bar{x}, \bar{u}) \geq -\sigma\}$; $G^\sigma(t) = (g^i(t) : i \in I^\sigma(t))$. There exist $\sigma > 0$, $\kappa > 0$ such that

$$|\,\eta^T \nabla_u G^\sigma(t)\,| \geq \kappa\,|\,\eta\,| \qquad \text{for arbitrary } \eta, \ \textit{a.e.} \text{ on } [0,1]\,.$$

Under the assumptions **(A1)** and **(A2)** the approximation **(DP)** is consistent to **(OP)**, and the same is true for the related first order systems. The consistency has been analyzed in [5]. One particular consequence then is a constraint qualification valid for G_ω at the solution z_h of **(DP)**:

Lemma 2 *Denote by I_+^j the index set of $i \in (1, \ldots, l)$ with $G_i^j(z_{h,j}) = 0$ and $\mu_i^j > 0$. Let further $(A^j)^+$ be the pseudoinverse of the matrix $A^j = (\nabla_z G_i^j : i \in I_+^j)$. If* **(A1)** *-* **(A3)** *are fulfilled then for arbitrary $\epsilon > 0$ and sufficiently small h there exist $\tilde{J}(h)$ such that*

$$\|\,(A^j)^+\| \leq c/\kappa^2 \qquad \textit{for all} \quad j \in \tilde{J}(h)$$

and $\sum_{j \notin \tilde{J}(h)} |\omega_j| \leq \epsilon$. *The constant c depends only on $\|\nabla_{(x,u)} g(\bar{z})\|_\infty$.*

Proof: First remark that for any matrix M with the bloc structure $M^T = (M_1^T \; M_2^T)$ and full row rank the relation $\|M_1^+\| \leq \|M^+\|$ holds. Indeed, the definition of the pseudoinverse yields $\|A^+\| = sup_{w \neq 0} \min_{Ay=w} (|y|/|w|)$; thus

$$\|M_1^+\| = \sup_{w_1 \neq 0} \min_{M_1 y = w_1} \frac{|y|}{|w_1|} \leq \sup_{w_1 \neq 0} \min_{M_1 y = w_1,\, M_2 y = 0} \frac{|y|}{|w_1|} \leq \sup_{w \neq 0} \min_{My = w} \frac{|y|}{|w|} = \|M^+\|\,.$$

Consider now the matrices $A = \bar{A}^j = \nabla_z G^\sigma(\bar{t}_j)$. ¿From condition **(A3)** it follows that $\|(AA^T)^{-1}\| \leq \kappa^{-2}$ holds independently of h for every j. But then we have: $\|A^+\| = \|A^T(AA^T)^{-1}\| \leq \kappa^{-2}\|\nabla_z G^\sigma(\bar{t}_j)\| \leq \kappa^{-2} m$ for $m = \max_j \|\nabla_z G_j^\sigma\|$.
Denote by A_1^j the matrix with rows $(\nabla_z g^i(\bar{t}_j) : i \in I_+^j)$.
We will use the consistency lemma now to complete the proof: Let ϵ from $(0, \sigma)$ be given, then there exists h_0 such that for every $j \in J'(h)$ (see Lemma 1) we have

$$\|g(\bar{t}_j, \bar{z}_j) - G^j(z_{h,j})\| < \epsilon; \; \|A_1^j - A^j\| < \epsilon\,.$$

The first estimate guarantees that $I_+^j \subseteq I^\sigma(\bar{t}_j)$. But then we may rearrange the rows of $\bar{A}^j$ so that the first rows build the bloc A_1^j and get $\|(A_1^j)^+\| \leq \|(\bar{A}^j)^+\| \leq \kappa^{-2} m$. Using finally te continuity of the pseudoinverse, for small ϵ the assertion follows.

■

5. Optimality test criteria in discrete form

This section deals with second order optimality conditions for **(DP)** resp. its decomposition. Starting from (13) we consider the HESSE matrices $H^j = \nabla^2_{\xi,\eta} l^j$, $j = 1, \ldots, N$ at the point $(z_h, \mu_h)|_{\bar{t}_j}$. The properties assumed for the discrete solution points are

given in **(A1)**. Particularly, $z_h = (x_h, u_h)$ then satisfies the discrete state equation (16).
With the notation $\nabla_x^2 S_h(t, x_h(t)) = Q_h(t)$ the second order derivatives of l^j take the form:

$$\begin{aligned} C_j &= \frac{\partial^2 l^j}{\partial \xi^2} = |\omega_j|^{-1}(Q_j - Q_{j-1}) + \bar{Q}_j\,(f_x)_j + (f_x^T)_j \bar{Q}_j + (\hat{H}_{xx})_j \\ B_j &= \frac{\partial^2 l^j}{\partial \xi \partial \eta} = \bar{Q}_j\,(f_u)_j + (\hat{H}_{xu})_j\,, \quad D_j = \frac{\partial^2 l^j}{\partial \eta^2} = (\hat{H}_{uu})_j\,;. \end{aligned}$$

(The subscript j in the right-hand side expressions stands for the evaluation at $(\bar{t}_j, \bar{z}_j, \bar{\lambda}_j)$.)
A discrete version of second order optimality conditions for **(OP)** (and also for **(P)** then) should enclose an uniform coercivity requirement to H^j as well as certain boundary restrictions on Q_h (cf. (11)) on appropriate subspaces. Those are the tangent spaces to active constraints

$$T_+^j(\sigma) = \{\zeta : \nabla_z G_i^j(z_{h,j})^T \zeta = 0 \ \forall\, i \in I_+^j(\sigma)\}\,, \qquad I_+^j(\sigma) = \{i \in I_+^j : \mu_i^j \geq \sigma\}$$

with I_+^j as in Lemma 2, and T_Γ from (11) for the boundary constraints respectively.

(DSSC) There exist positive γ_1, γ_2 and h_1 such that for given $\sigma > 0$ and for every $h < h_1$ a matrix function $Q_h = Q_h(t) \in D_\omega^1(0,1; \mathbb{R}^{n\times n})$ exists for which

$$(y,v)^T \begin{pmatrix} C_j & B_j \\ B_j^T & D_j \end{pmatrix} \begin{pmatrix} y \\ v \end{pmatrix} \geq \gamma_1\,(\,|y|^2 + |v|^2) \tag{18}$$

for all $\zeta = (y, v) \in T_+^j(\sigma)$; and

$$d^T \left[\hat{pi}\, Q_h - \rho_1^T \nabla_\xi^2 \beta(\pi x_h) - \rho_2^T \nabla_\xi^2 \gamma(\pi x_h)\right] d \geq \gamma_2\,|\,d\,|^2$$

holds for arbitrary $d \in T_\Gamma$. Further, $\{Q_h\}$ is uniformly bounded in $L_\infty(0,1; \mathbb{R}^{n\times n})$.

In a preliminary step we will show that **(DSSC)** together with **(A3)** yields the coercivity of H_j on perturbed cones

$$T_\alpha^j = \{\,\zeta : \ |\nabla_z G_i^j(z_{h,j})^T \zeta\,| \leq \alpha\,|\zeta| \quad \forall i \in I_+^j(\sigma)\,\}\,,$$

where α is a small positive number.

Lemma 3 *For any given $q \in (0,1)$ there exists $\alpha = \alpha(q, \kappa)$ such that for sufficiently small h*

$$\zeta^T H^j \zeta \geq (q\,\gamma_1)\,|\,\zeta\,|^2 \qquad \textit{for all } \zeta \in T_\alpha^j$$

holds for all $j \in \tilde{J}(h)$ with $\tilde{J}(h)$ from Lemma 2.
In addition, $\sum_{j\notin \tilde{J}(h)} |\omega_j| \to 0$ as $h \to 0$.

Proof: In analogy to Lemma 2 denote $\tilde{A}^j = (\,\nabla_z G_i^j \ : \ i \in I_+^j(\sigma)\,)$; this matrix consists of a choice of rows of A^j so that $\|(\tilde{A}^j)^+\| \leq c/\kappa^2$ holds uniformly on $\tilde{J}(h)$. If $I_+^j(\sigma) = \emptyset$

or $\tilde{A}^j = 0$, the sets $T^j_+(\sigma)$ and T^j_α coincide with the full vector space $\mathbb{R}^{n+m}$ so that the assertion follows from **(DSSC)**. Consider the case $|\, I^j_+(\sigma)\,| = e > 0$, $\tilde{A}^j \neq 0$:
Let ζ belong to T^j_α. Then, $\|\tilde{A}^j\zeta\| \leq e\,\alpha^2\,|\,\zeta\,|^2$. Consequently, for the projection $\Pi^j_+ : \mathbb{R}^{n+m} \to T^j_+(\sigma)$ we have

$$\begin{aligned}(I - \Pi^j_+)\,\zeta &= (\tilde{A}^j)^+\,(\tilde{A}^j\zeta)\,;\\ \|\,(I - \Pi^j_+)\,\zeta\,\| &\leq \|\,(\tilde{A}^j)^+\|\, e^{1/2}\,\alpha\,|\,\zeta\,| \leq \alpha\,\nu'\,|\,\zeta\,|\end{aligned}$$

for some $\nu' > 0$ uniformly for all $j \in \tilde{J}(h)$.
Set $\Pi^j_+\zeta =: w$, $(I - \Pi^j_+)\,\zeta =: v$. Then $\|\zeta\|^2 = \|w\|^2 + \|v\|^2$, and abbreviating $\alpha\,\nu' = \nu$ we obtain

$$\begin{aligned}\zeta^T H^j \zeta &\geq (w+v)^T H^j (w+v)\\ &\geq w^T H^j w - \|\,H^j\|\,\|v\|\,(2\,\|w\| + \|v\|\,)\\ &\geq \gamma_1\|w\|^2 - \|\,H^j\|\,(\nu\,\|\zeta\|)\,(3\,\|\zeta\|\,)\\ &\geq \Big[\gamma_1\,(1-\nu^2) - 3\,\nu\,\|\,H^j\|\Big]\,\|\zeta\|^2\,.\end{aligned}$$

The terms $\|\,H^j\|$ due to the data smoothness and **(A1)** are uniformly bounded for all $j = 1, \dots, N$. We may choose α small enough to guarantee $\nu < 1$ and $\nu < (1-q)/(\,\gamma_1 + 3\,\max_j \|H^j\|)$ so that

$$\zeta^T H^j \zeta \geq \Big[\gamma_1\,(1-\nu) - 3\,\nu\,\|\,H^j\|\Big]\,\|\zeta\|^2 \geq (q\,\gamma_1)\,\|\zeta\|^2\,.$$

■

After obtaining this stable coercivity property we are able now to prove the basic local growth estimation for the sequence $\{\Phi_\omega\}$:

Theorem 2 *Suppose that for the discretization* **(DP)** *of* **(OP)** *conditions* **(A1)-(A3)** *are fulfilled. If the discrete second order sufficient conditions* **(DSSC)** *hold, then for any given $q \in (0,1)$ and arbitrary $\epsilon > 0$ there exist $\rho = \rho(q)$ and $h_2 > 0$ such that for every $\omega = \omega(N)$ with $h < h_2$ the functional Φ_ω near z_h satisfies the estimate*

$$\Phi_\omega(z) - \Phi_\omega(z_h) \geq 0.5\,(q\,\gamma_1) \sum_{j=1}^{N} |\omega_j|\,\Big|(z - z_h)|_{\bar{t}_j}\Big|^2 - c\,\rho\,\epsilon$$

for every z feasible w.r.t. **(DP)** *with $\|\,z - z_h\|_\infty \leq \rho$, where the constant $c > 0$ is independent of q, ϵ and h.*

Proof: Set $q_1 = 0.5(1+q)$. Further, choose h_2 small enough so that Lemma 1 holds with the given ϵ and Lemma 2 and 3 may be applied resp. Let $J(h)$ stand for the index set according to Lemma 1 and 3.

Part 1: Estimate for ϕ^j, $j \in J(h)$.
Denote $\zeta = \zeta^j = z(\bar{t}_j)$, $\zeta_h = \zeta^j_h = z_{h,j}$; $\mu^j = \mu^j_h$, and define $w = w^j = \zeta^j - \zeta^j_h$.
Without loss of generality assume that $|w| \leq \rho_0 = \min\{1, \|z_h\|_\infty\}$. From Lemma 3 we deduce that there exists $\alpha > 0$ corresponding to q_1 so that H^j is coercive on T^j_α uniformly w.r.t. j. Thus, we may distinguish between two cases:

case (i) $w \in T^j_\alpha$

Lemma 3 yields: $w^T H^j w \geq q_1\gamma_1 \,|\, w \,|^2$.

Consider $\phi^j(\zeta)$ using the TAYLOR expansion in ζ_h. The smoothness properties of the data ensure that all l^j together with their first and second derivatives are uniformly LIPSCHITZ continuous w.r.t. ζ on $B_{\rho_0}(\zeta_h)$, thus for some $m' > 0$

$$\begin{aligned}\phi^j(\zeta) - \phi^j(\zeta_h) &\geq l^j(\zeta, \mu_h) - l^j(\zeta_h, \mu_h) \\ &\geq 0.5\,(q_1\gamma_1)\, w^T H^j w - m'|w|^3 \geq 0.5\, q\, \gamma_1\, |w|^2\end{aligned}$$

if only $|w| < \rho_1 = (1-q)\,\gamma_1\,/\,(2\,m')$.

case (ii) $w \notin T^j_\alpha$, i.e $\exists \iota \in I^j_+(\sigma)$: $\left|\nabla_z G^j_\iota(\zeta_h)^T w\right| > \alpha\,|\,w\,|$.

The smoothness assumptions on g, G and the feasibility of ζ and ζ_h allow finding $m'' > 0$ to estimate

$$\nabla_z G^j_\iota(\zeta_h)^T w \leq G^j_\iota(\zeta) - G^j_\iota(\zeta_h) + m''|w|^2 \leq m''|w|^2 \tag{19}$$

for all $i \in I^j_+$ and uniformly w.r.t. j. Consequently, $\nabla_z G^j_\iota(\zeta_h)^T w < -\alpha\,|\,w\,| \leq 0$. Since the KARUSH-KUHN-TUCKER conditions hold in ζ_h from (19) one may derive the relation

$$\begin{aligned}\nabla_z \phi^j \zeta_h)^T w &= -\sum_{i\in I^j_+} (\mu_h)^j_i \nabla_z G^j_i(\zeta_h)^T w \\ &\geq (\mu_h)^j_\iota\, \alpha\,|w| - l\, m''|w|^2 \\ &\geq \sigma\,\alpha\,|\,w\,|\,(1 - lm''|w|) \geq 0.5\, q\,\gamma_1\,|\,w\,|^2\end{aligned}$$

if $|w| \leq \rho_2 = \min\{(2l\,m'')^{-1},\, (\sigma\,\alpha)\,/\,(q\,\gamma_1)\}$.

Part 2: Estimate for Φ_ω.

Taking $\rho = \min\{\rho_0, \rho_1, \rho_2\}$ we require $\|z - z_h\|_\infty \leq \rho$. While $\{\phi^j\}$ are uniformly LIPSCHITZ continuous, there exists $c' > 0$ with $|\,\phi^j(\zeta^j) - \phi^j(\zeta^j_h)\,| \leq c'\rho$ for all $j = 1, \ldots, N$. Together with part 1 this leads to:

$$\begin{aligned}\Phi_\omega(z) - \Phi_\omega(z_h) &= \sum_{j=1}^N |\omega_j| \left(\phi^j(\zeta^j) - \phi^j(\zeta^j_h)\right) \\ &\geq \sum_{j\in J(h)} |\omega_j| \left(\phi^j(\zeta^j) - \phi^j(\zeta^j_h)\right) - c'\rho \sum_{j\notin J(h)} |\omega_j| \\ &\geq 0.5\,(q\,\gamma_1) \sum_{j=1}^N |\omega_j| \left|\zeta^j - \zeta^j_h\right|^2 - (1+c')\,\rho\,\epsilon,\end{aligned}$$

$(\zeta^j - \zeta^j_h) = (\,z - z_h)|_{\bar{t}_j}$. ■

Our final result is related to the continuous problem formulations **(OP)** and **(P)**:

Corollary 1 *Let for* $\bar{z} = \lim_{h\to 0} z_h$ *the assumptions* **(A1)-(A3)** *hold together with* **(DSSC)**. *Then for any* $q \in (0,1)$ *there exists a positive* ρ' *such that*

$$\Phi(z') - \Phi(\bar{z}) \geq 0.5\,(q\,\gamma_1)\,\|\, z' - \bar{z}\,\|_2^2$$

for every z' *feasible for* **(OP)** *with* $\|z' - \bar{z}\|_\infty \leq \rho'$.
If in addition z' *is feasible for* **(P)** *then it follows that*

$$J(z') - J(\bar{z}) \geq c(q,\gamma_1,\gamma_2)\,\|\, z' - \bar{z}\,\|_2^2$$

holds, i.e. the limit of the discretizations $\bar{z}$ *is a weak local minimizer to* **(P)**.

Proof: Notice first that for any bounded sequence $\{y_h\}$ in D^0_ω (and in D^1_ω analogously) with $y_h \to \bar{y}$ in L_2

$$\lim_{\omega=\omega(N),N\to\infty} \sum_{j=1}^{N} |\omega_j|\,|\,y_{h,j}|^2 = \|\,\bar{y}\,\|_2^2\,.$$

Suppose that $\rho' \leq 0.5\,\rho$. Introduce the test function

$$z = P_h'(z') = \begin{cases} P_h\, z' & \text{on } \omega_j,\ j \in J(h)\,, \\ z_h & \text{on } \omega_j,\ j \notin J(h)\,. \end{cases}$$

Obviously, $\lim_{h\to 0} \|P_h' z' - '\|_2 = 0$, so that for sufficiently small h
$\|z - z_h\| \leq \|z' - \bar{z}\| + \|z - z'\| + \|z_h - \bar{z}\| \leq \rho' + 2\epsilon < \rho$.
Due to condition **(A1)** in the statement of Theorem 2 we may tend $h \to 0$ and arrive at

$$\Phi(z') - \Phi(\bar{z}) \geq 0.5\,(q\,\gamma_1)\,\|\, z' - \bar{z}\,\|_2^2 - c\rho\epsilon\,,$$

ϵ – an arbitrary positive number. Tending ϵ to zero then leads to the desired result. The second part follows immediately from Theorem 1 when the second relation in **(DSSC)** is taken into account together with the above quadratic growth estimation. ∎

References

[1] Budak, B. M.; Berkovich, E. M.; Solov'eva, E. N.: *On the convergence of finite-difference approximations for optimal control problems*, U.S.S.R. Comput.Maths.Math.Phys., (Russ. edition: Zhurnal vych. mat. i mat. fiziki, vol. 9, 1969, no. 3, 522–547).

[2] Dontchev A. L.: *An a priori estimate for discrete approximations in nonlinear optimal control*, SIAM J. Contr. Optim., vol. 34, 1996, 1315–1328.

[3] Dontchev, A. L.; Hager W. W.: *Lipschitzian stability in nonlinear .control and optimization* , SIAM J. Contr. Optim., vol. 31, 1993, 569–603.

[4] Dontchev, A. L.; Hager W. W.; Poore A. B.; Yang, B.: *Optimality, Stability, and convergence in Nonlinear Control*, J. Appl. Math. and Optim., vol 31, 1995, 297–326.

[5] Felgenhauer, U.: *Numerical optimality test for control problems*, in: Proc. IV. Conference on "Parametric Optimization and Related Topics", Enschede 1995; eds.: J.Guddat, H.Th.Jongen, G.Still, F.Twilt; Peter Lang publ., 1996, (to appear).

[6] Felgenhauer, U.: *Discretization based optimality test for certain parametric problems*, Preprint, BTU Cottbus, Reihe Mathematik, M-01/1996.

[7] Felgenhauer, U.: *On optimality criteria for control problems. Part I: Theory*, Preprint, BTU Cottbus, Reihe Mathematik, M-04/1996.

[8] Grachev, I. I.; Evtushenko, Yu. G.: *A library of programs for solving optimal control problems*, U.S.S.R. Comput.Maths.Math.Phys., vol.19, 1980, 99–119.

[9] Hager, W. W.: *Lipschitz continuity for constrained processes*, SIAM J.Contr. Optim., vol.17, 1979, 321–338.

[10] Klötzler, R.: *On a general conception of duality in optimal control* , Lect. Notes Math. **703**, Springer Verlag, New York - Heidelberg - Berlin 1979, 189–196.

[11] Klötzler, R.; Pickenhain, S.: *Pontryagin's maximum principle for multidimensional control problems*, Int. Series of Numer. Math, vol. 111, Birkhäuser Basel, 1993, 21–30.

[12] Klötzler, R.; Pickenhain, S.: *Stability and maximum principle for multiple integral control problems*, FB Mathematik/Informatik, Univ. Leipzig, Report No. 505, 1994. analysis of

[13] Malanowski, K.: *Stability and sensitivity analysis of solutions to nonlinear optimal control problems*, J. Appl. Math. and Optim., vol. 32, 1995, 111–141.

[14] Malanowski, K.; Büskens, C.; Maurer, H.: *Convergence of approximations to nonlinear control problems*, in: Mathematical Programming with Data Perturbation, ed.: A.V.Fiacco, Marcel Dekker Inc., New York 1996, (to appear).

[15] Maurer, H.; Pickenhain, S.: *Second Order Sufficient Conditions for Optimal control problems with Mixed Control-State Constraints*, J. Optim. Theor. Appl., 86, 1995, 649–667.

[16] Pickenhain, S.: *Sufficiency Conditions for Weak Local Minima in Multidimensional Optimal Control Problems with Mixed Control-State Restrictions*, Z. Anal. Anw. **11**, 1992, 559–568.

[17] Pickenhain, S.: *A pointwise maximum principle in optimal control with multiple integrals*, Optimization, 38, 1996, 343–355.

[18] Pickenhain, S.; Tammer, K.: *Sufficient Conditions for Local Optimality in Multidimensional Control Problems with State Restrictions*, Z. Anal. Anw. **10**, 1991, 3, 397–405.

[19] Robinson, S. M.: *Strongly regular generalized equations*, Math. Oper. Res. 5, 1980, 43–62.

[20] Zeidan, V.: *The Riccati Equation for Optimal Control Problems with Mixed State-Control Constraints: Necessity and Sufficiency*, SIAM J. Contr. Optim., vol. 32, 1994, 5, 1297–1321.

International Series of Numerical Mathematics
Vol. 124, © 1998 Birkhäuser Verlag, Basel

Smooth and Nonsmooth Optimal Lipschitz Control – a Model Problem

Manfred Goebel*

Abstract. For a semilinear second order ordinary differential equation a control problem with pointwise state constraints is considered. The control of the system is realized via the outer function of the nonlinear term in the state equation. We prove a simple theorem stating the existence of optimal controls within the class of Lipschitz continuous functions and derive two necessary optimality conditions depending on their smoothness properties. For a smooth optimal control function the given optimality condition looks somehow like Pontryagin's maximum principle. In the general case the optimality condition involves Clarke's generalized directional derivative. Furthermore a modified problem is considered, for which by means of Ekeland's variational principle a suboptimality condition can be proved.

1. Introduction

In our contribution we discuss mainly necessary optimality conditions for the following control problem (**P**), in which the behaviour is monitored by a semilinear second order ordinary differential equation.
Find

$$\inf \mathcal{J}(y), \qquad \mathcal{J}(y) = \int_0^1 (y(x) - h(x))^2 \, dx,$$

subject to the state equation

$$y''(x) + \sigma(y(x)) = 0, \quad x \in (0,1), \qquad y(0) = 0, \quad y'(1) = 0, \tag{1}$$

the state constraints

$$0 \le y(x) \le a \quad \forall x \in [0,1] \tag{2}$$

and the set of admissible controls

$$\Sigma_{ad} = \big\{\sigma \in C[0,a] \;: 0 \le \sigma(x) \le b \quad \forall x \in [0,a], \\ |\sigma(x_1) - \sigma(x_2)| \le l|x_1 - x_2| \quad \forall x_1, x_2 \in [0,a]\big\}. \tag{3}$$

The function $h \in C[0,1]$ and the constants $a, b, l \in \mathbb{R}$ satisfying $0 < a$, $0 < b \le 2a$ and $0 < l < 2$ are given before. The difficulties of the problem are not caused by the state

*Martin-Luther-Universität Halle-Wittenberg, FB Mathematik und Informatik, D–06099 Halle (Saale), Germany, goebel@maja.mathematik.uni-halle.de

constraints (2), but by the fact that the outer function σ of the nonlinear term in (1) acts as the control function.
Problems of such a type are studied fairly seldom. We refer to the papers by R. Kluge and his coworkers, see for example R. KLUGE [1], to the doctoral thesis A. RÖSCH [2], and to the very recent papers V. BARBU, K. KUNISCH [3, 4], in which similar problems for partial differential equations were considered (mainly from another point of view). In our paper [5] we have also dealt with an optimal control problem of the above type, however for a nonlinear singular integral equation. Here, essentially the same ideas as in [5] are used to derive necessary optimality conditions in case of smooth optimal controls (see [6]). Another set of necessary optimality conditions is found for nonsmooth optimal controls using Clarke's generalized directional derivative. Moreover, for a modified problem, for which the infimum of the cost functional is the same as for (**P**), a suboptimality condition is given via Ekeland's variational principle. In a forthcoming paper [7] these results will be proved for a more general second order ordinary differential equation.
State equations of the form (1) play an very important role in different applications. The control problem (**P**) to be considered has its origin in [8]; it is a least squares formulation of an inverse problem occuring in a certain one-dimensional electrothermal system, which has been solved via Abel integral equations within the class of piecewise continuous functions. In this class, considering control problems of the type above seems to be very challenging.

2. State equation and existence of optimal controls

Let $G = G(x,\xi)$ be the uniquely determined Green's function to the boundary value problem

$$y''(x) = 0, \quad x \in (0,1), \qquad y(0) = 0, \quad y'(1) = 0,$$

that means

$$G(x,\xi) = -\xi \text{ for } 0 \le \xi \le x \le 1 \quad \text{and} \quad G(x,\xi) = -x \text{ for } 0 \le x \le \xi \le 1.$$

G is continuous and symmetric on $[0,1] \times [0,1]$; it holds

$$0 \le -G(x,\xi) \le 1 \quad \forall\, x,\xi \in [0,1], \qquad 0 \le -2\int_0^1 G(x,\xi)\, d\xi \le 1 \quad \forall\, x \in [0,1].$$

We put (cf. (2))

$$B_+(0,a) = \{y \in C[0,1] \,:\, 0 \le y(x) \le a \quad \forall x \in [0,1]\}$$

and take a fixed $\sigma \in \Sigma_{ad}$. Evidently, $y \in C^2[0,1] \cap B_+(0,a)$ is a classical solution to the nonlinear boundary value problem (1) if and only if $y \in B_+(0,a)$ is a solution to the Hammerstein integral equation

$$y(x) = -\int_0^1 G(x,\xi)\sigma(y(\xi))\, d\xi, \quad x \in [0,1]; \tag{4}$$

in the following we will work with (4) instead of (1). The only things which will be used from the state equation (1) are the properties of the Green's function mentioned above. Also for other (linear or nonlinear) boundary value problems occuring in this paper a solution is always to be understood as a solution of the related integral equation obtained by means of the above Green's function G. Using the Banach fixed point theorem we can prove that for each $\sigma \in \Sigma_{ad}$ the state equation (1) has a unique solution $y = y(\sigma) \in C[0,1]$. Thereby, of course, the assumption $0 < l < 2$ is essential. Without supposing this, due to the Schauder fixed point theorem, we have only the solvability of (1). It is quite interesting that the needed contractivity of the integral operator occuring in (4) requires necessarily Lipschitz continuity of the looked for function σ (see [6]). Concerning the map $y \mapsto \sigma(y)$ we see that the estimate

$$\|y(\sigma_1) - y(\sigma_2)\|_{C[0,1]} \leq \frac{1}{2-l}\|\sigma_1 - \sigma_2\|_{C[0,a]} \tag{5}$$

holds for all $\sigma_1, \sigma_2 \in \Sigma_{ad}$, and hence the cost functional $j(\sigma) = \mathcal{J}(y(\sigma))$, $\sigma \in \Sigma_{ad}$, is Lipschitz continuous, too:

$$|j(\sigma_1) - j(\sigma_2)| \leq \frac{2}{2-l}\left(a + \|h\|_{C[0,1]}\right)\|\sigma_1 - \sigma_2\|_{C[0,a]} \quad \forall\, \sigma_1, \sigma_2 \in \Sigma_{ad}.$$

Since Σ_{ad} is compact in C[0,a], we get immediately the existence of optimal controls.

Theorem 1 (Existence) *The optimal control problem* (**P**) *has an optimal solution* $\sigma_0 \in \Sigma_{ad}$.

Thus, here the existence of optimal solutions is proved via the classical Weierstrass theorem using strong convergence, which does not very often happen.

3. Necessary optimality conditions

3.1 Preliminaries

Let $\sigma_0 \in \Sigma_{ad}$ be an optimal control to (**P**) and $\sigma \in \Sigma_{ad}$ another admissible control. The respective states are denoted by $y_0 = y(\sigma_0)$ and $y = y(\sigma)$. For any $\lambda \in [0,1]$ we put $\sigma_\lambda = \sigma_0 + \lambda(\sigma - \sigma_0)$ and $y_\lambda = y(\sigma_\lambda)$. From (5) we see that

$$\|y_\lambda - y_0\|_{C[0,1]} \leq \frac{\lambda}{2-l}\|\sigma - \sigma_0\|_{C[0,a]} \quad \forall\, \lambda \in [0,1], \tag{6}$$

such that from

$$0 \leq \frac{j(\sigma_\lambda) - j(\sigma_0)}{\lambda} = 2\int_0^1 \frac{y_\lambda - y_0}{\lambda}(y_0 - h)\, dx + \frac{1}{\lambda}\int_0^1 (y_\lambda - y_0)^2\, dx \quad \forall\, \lambda \in (0,1]$$

our first necessary optimality condition follows.

Lemma 1 *If* $(\lambda_n) \subset (0,1]$ *is a zero sequence and* $\widetilde{y} \in C[0,1]$ *with*

$$(*) \qquad \frac{y_{\lambda_n} - y_0}{\lambda_n} \longrightarrow \widetilde{y} \quad \textit{in} \quad C[0,1] \qquad \textit{then} \qquad 0 \leq \int_0^1 \widetilde{y}(y_0 - h)\, dx.$$

In this simple condition the integral essentially represents the directional derivative of $j = j(\sigma)$ at σ_0 in direction $\sigma - \sigma_0$, provided $(*)$ holds for any zero sequence $(\lambda_n) \subset (0,1]$ with a fixed function $\widetilde{y} \in C[0,1]$. However, this property could be verified only under an additional smoothness assumption concerning the optimal control σ_0. In general, because of (6), we have only the uniform boundedness of the quotient $(y_\lambda - y_0)\lambda^{-1}$ for all λ belonging to the interval $(0,1]$. To find an appropriate zero sequence $\lambda_n \subset (0,1]$ with $(*)$ becomes the more difficult the less nonsmooth the optimal control σ_0 is.

3.2 Smooth optimal control

Now, let us suppose the optimal control σ_0 to be an element of $C^1[0,a]$. Then, for an arbitrary right-hand side $f \in C[0,1]$ the linear boundary value problem

$$z''(x) + \sigma_0'(y_0(x))z(x) = f(x), \quad x \in (0,1), \qquad z(0) = 0, \quad z'(1) = 0,$$

has a unique solution $z \in C[0,1]$, which again follows from the Banach fixed point theorem applied to the corresponding (linear) Fredholm integral equation. This allows the limit function in $(*)$ to be characterized as a solution of such a boundary value problem with a particular right-hand side.

Lemma 2 *It holds*

$$\frac{y_\lambda - y_0}{\lambda} \longrightarrow \widetilde{y} \quad \textit{in} \quad C[0,1] \quad \textit{as} \quad \lambda \longrightarrow 0+,$$

where $\widetilde{y} = \widetilde{y}(\sigma_0, \sigma) \in C[0,1]$ *denotes the unique solution to*

$$z''(x) + \sigma_0'(y_0(x))z(x) = -\sigma(y_0(x)) + \sigma_0(y_0(x)), \quad x \in (0,1), \quad z(0) = 0, \quad z'(1) = 0.$$

Thus, Lemma 1 gives the necessary optimality condition

$$0 \le \int_0^1 \widetilde{y}(y_0 - h)\, dx \quad \forall\, \sigma \in \Sigma_{ad},$$

which after introducing the adjoint state $z_0 \in C[0,1]$ as the unique solution to the linear boundary value problem

$$z''(x) + \sigma_0'(y_0(x))z(x) = 2(h(x) - y_0(x)), \quad x \in (0,1), \qquad z(0) = 0, \quad z'(1) = 0,$$

yields eventually the following necessary optimality condition.

Theorem 2 *If* $\sigma_0 \in \Sigma_{ad} \cap C^1[0,a]$ *is an optimal control to* (**P**), $y_0 \in C[0,1]$ *the related optimal state and* $z_0 \in C[0,1]$ *the related adjoint state, then it holds*

$$0 \le \int_0^1 (\sigma(y_0(x)) - \sigma_0(y_0(x)))z_0(x)\, dx \quad \forall\, \sigma \in \Sigma_{ad}.$$

This optimality condition is very clear and perhaps cannot be improved. In particular, it seems to be impossible to derive a pointwise optimality condition. However, since there are no known criteria assuring the needed smoothness of the optimal control, we have also to consider the case $\sigma_0 \notin C^1[0,a]$, which will be done in the next subsection. It is clear that in this case we cannot expect such a nice optimality condition as given in the last theorem.

3.3 Nonsmooth optimal control

Here we briefly deal with the more interesting case that the optimal control σ_0 does not belong to $C^1[0,a]$. Since the map $j : \Sigma_{ad} \to \mathbb{R}$ is Lipschitz continuous (but surely not convex), we have the optimality condition

$$0 \in \partial j(\sigma_0) + N_{\Sigma_{ad}}(\sigma_0), \tag{7}$$

where j is now a Lipschitz continuous extension of $j : \Sigma_{ad} \to \mathbb{R}$ to the whole space $C[0,a]$, $\partial j(\sigma_0)$ denotes Clarke's subdifferential of j at σ_0 and $N_{\Sigma_{ad}}(\sigma_0)$ is the normal cone to Σ_{ad} at σ_0 (see [9, p. 52]). Unfortunately, we failed to interpret (7) in terms of given datas describing (**P**), so that we have to go another way. Clearly, we try to verify (∗) of Lemma 1. To this end we put

$$\Phi(\sigma) = \left\{ \varphi_\lambda \in C[0,1] \, : \, \varphi_\lambda = \frac{y_\lambda - y_0}{\lambda}, \quad \lambda \in (0,1] \right\}$$

and define two functions $\varphi_\mp$ by setting

$$\varphi_-(x) = \liminf_{\lambda \to 0+} \varphi_\lambda(x), \quad \varphi_+(x) = \limsup_{\lambda \to 0+} \varphi_\lambda(x), \quad x \in [0,1].$$

Note, both functions depend also on the element $\sigma \in \Sigma_{ad}$ chosen before. The set $\Phi(\sigma)$ is bounded (see (6)), equi-continuous, and therefore relatively compact in $C[0,1]$. As a first consequence of this, the functions φ_- and φ_+ are continuous over $[0,1]$. By

$$\Phi_0(\sigma) = \Big\{ \varphi \in C[0,1] \, : \, \varphi = \lim_{n \to \infty} \varphi_{\lambda_n} \quad \text{in} \quad C[0,1], \quad \text{where}$$
$$(\varphi_{\lambda_n}) \subset \Phi(\sigma) \quad \text{and} \quad (\lambda_n) \subset (0,1] \quad \text{with} \quad \lambda_n \to 0+ \Big\}$$

we introduce another set in $C[0,1]$, which, due to the relative compactness of $\Phi(\sigma)$, is nonempty and bounded. Using these notations we may formulate the following optimality condition.

Theorem 3 *Let $\sigma_0 \in \Sigma_{ad}$ be an optimal control to* (**P**) *and $y_0 \in C[0,1]$ the related optimal state. Then for all $\sigma \in \Sigma_{ad}$ and all $\varphi \in \Phi_0(\sigma)$ it holds*

(i) $\quad \varphi(x) \in [\varphi_-(x), \varphi_+(x)] \quad \forall x \in [0,1],$

(ii) $\quad \varphi(x) \leq -\int_0^1 G(x,\xi)\left[\sigma_0^0(y_0(\xi);\varphi(\xi)) - (\sigma(y_0(\xi)) - \sigma_0(y_0(\xi)))\right] d\xi \quad \forall x \in [0,1],$

(iii) $\quad 0 \leq \int_0^1 \varphi(x)(y_0(x) - h(x))\, dx.$

The assertions (i) and (iii) are obvious. In (ii) $\sigma_0^0(x;\xi)$ denotes Clarke's directional derivative of $\sigma_0 : \mathbb{R} \to \mathbb{R}$ at x in direction ξ, where now σ_0 is a Lipschitz continuous extension of the originally given σ_0 to all of $\mathbb{R}$. Of course, the proof of (ii) requires some knowledge of subgradient calculus as given, for example, in [9, 10]. Particularly,

instead of the classical mean value theorem, which we have applied in the proof of Lemma 2, now we have to make use of the mean value theorem for subgradients. The unexpected inequality sign in (ii) comes just from this theorem.
For several reasons Theorem 3 seems to be only an interime result. First, we could not find an appropriate adjoint state, by means of which the optimality condition could be somehow simplified. Secondly, it is desirable that an optimality condition for a nonsmooth optimal control (like in Theorem 3) reduces in case of a smooth optimal control to the optimality condition just proved under this stronger assumption (like in Theorem 2). But here we don't have such a relationship. Only going step by step through the proof of Theorem 3 can it be seen that it generalizes Theorem 2.
Note, since we know that there exist optimal controls $\sigma_0 \in \Sigma_{ad}$ to (**P**) Theorem 3 includes also the statement that the sublinear "Fredholm integral inequality" in (ii) is solvable in $C[0,1]$.

4. A modified optimal control problem

In the foregoing section we have given necessary optimality conditions for an existing optimal control to (**P**). However, in the first condition we had to assume smoothness of the optimal control and the second one does not seem to be very convenient. In this section we announce some results for a slightly modified optimal control problem, for which the infimum of the cost functional is the same as for the originally given control problem (see (8) below). Since we cannot verify that for the modified problem there exists an optimal control, we restrict our consideration to suboptimal solutions, for which we describe a necessary condition. The results to be presented rely heavily on Ekeland's variational principle quoted as the following Lemma 3 (see, for example, [11]).

Lemma 3 *Let (Σ, d) be a complete metric space and $j : \Sigma \longrightarrow \mathbb{R}$ a lower semicontinuous functional bounded from below. For any $\alpha > 0$ there is an element $\sigma_\alpha \in \Sigma$ with:*

$$\text{(a)} \quad j(\sigma_\alpha) \le \inf\big\{j(\sigma) \,:\, \sigma \in \Sigma\big\} + \alpha,$$
$$\text{(b)} \quad j_\alpha(\sigma_\alpha) \le j_\alpha(\tau) \quad \forall\, \tau \in \Sigma, \quad \textit{where} \quad j_\alpha(\tau) = j(\tau) + \alpha d(\sigma_\alpha, \tau).$$

We modify the control problem (**P**) simply by taking another set of admissible controls. So, let Σ_{ad} be replaced by Σ^1_{ad} defined by

$$\Sigma^1_{ad} = \Big\{\sigma \in C^1[0,a] \,:\, 0 \le \sigma(x) \le b, \quad |\sigma'(x)| \le l \quad \forall\, x \in [0,a]\Big\} \subset \Sigma_{ad}.$$

The new optimal control problem will be shortly denoted by ($\mathbf{P_1}$). First of all we note that

$$\inf\big\{j(\sigma) \,:\, \sigma \in \Sigma^1_{ad}\big\} = \inf\big\{j(\sigma) \,:\, \sigma \in \Sigma_{ad}\big\} = j(\sigma_0) \;\text{ with }\; \sigma_0 \in \Sigma_{ad}. \tag{8}$$

Here the last equality sign is due to Theorem 1. To verify the first equality sign we introduce the Steklov function σ^h_0 corresponding to σ_0 by

$$\sigma^h_0(x) = \frac{1}{2h}\int_{x-h}^{x+h} \sigma_0(\xi)\,d\xi, \quad x \in [0,a], \quad h > 0,$$

where before we have extended σ_0 to a Lipschitz continuous function on the whole space $\mathbb{R}$ having the same Lipschitz constant as before (namely 1). Then $\sigma_0^h \in \Sigma_{ad}^1$ for all $h > 0$ and $\sigma_0^h \to \sigma_0$ in $C[0,a]$ as h tends to zero. Substituting $\sigma = \sigma_0^h$ into the obvious estimate

$$j(\sigma) \geq \inf\left\{j(\sigma) \,:\, \sigma \in \Sigma_{ad}^1\right\} \geq j(\sigma_0) \quad \forall \sigma \in \Sigma_{ad}^1,$$

and letting $h \to 0+$ we get the first equality sign in (8).
The modified problem has the great advantage that all admissible controls are smooth, so that in principle we may apply the same analytical methods as in section 3.2. Particularly, the classical directional derivative $\delta j(\sigma_\alpha, \sigma - \sigma_\alpha)$ of the cost functional j can be evaluated for any $\sigma_\alpha \in \Sigma_{ad}^1$ as

$$\delta j(\sigma_\alpha, \sigma - \sigma_\alpha) = 2 \int_0^1 \widetilde{y}(y_\alpha - h)\, dx \quad \forall\, \sigma \in \Sigma_{ad}^1,$$

where $y_\alpha = y(\sigma_\alpha)$ is the state related to σ_α and $\widetilde{y} = \widetilde{y}(\sigma_\alpha, \sigma) \in C[0,1]$ denotes the unique solution to the linear boundary value problem

$$z''(x) + \sigma_\alpha'(y_\alpha(x))z(x) = -\sigma(y_\alpha(x)) + \sigma_\alpha(y_\alpha(x)),\ x \in (0,1),\ \ z(0) = 0,\ \ z'(1) = 0.$$

Besides Lemma 3 this is the crucial point in proving Theorem 4 below. Since, furthermore, (Σ_{ad}^1, d) with $d(\sigma,\tau) = \|\sigma - \tau\|_{C^1[0,a]}/(b + l)$, $\sigma, \tau \in \Sigma_{ad}^1$, is a complete metric space with $d(\sigma,\tau) \leq 2$ for all $\sigma, \tau \in \Sigma_{ad}^1$ and since the cost functional $j : \Sigma_{ad}^1 \longrightarrow \mathbb{R}$ is Lipschitz continuous and hence also bounded from below, Lemma 3 can be applied to the control problem $(\mathbf{P_1})$. This gives the condition

$$0 \leq \int_0^1 \widetilde{y}(y_\alpha - h)\, dx + \alpha \quad \forall\, \sigma \in \Sigma_{ad}^1.$$

Here y_α is the state corresponding to the suboptimal solution $\sigma_\alpha \in \Sigma_{ad}^1$ for the modified problem $(\mathbf{P_1})$. After some further steps like in subsection 3.2 we obtain the following necessary suboptimality condition.

Theorem 4 *For any $\alpha > 0$ there exists an element $\sigma_\alpha \in \Sigma_{ad}^1$, such that with the associated state $y_\alpha = y(\sigma_\alpha) \in C[0,1]$ it holds:*

(a) $\quad j(\sigma_\alpha) \leq \inf\left\{j(\sigma) \,:\, \sigma \in \Sigma_{ad}^1\right\} + \alpha,$

(b) $\quad 0 \leq \displaystyle\int_0^1 \big(\sigma(y_\alpha(x)) - \sigma_\alpha(y_\alpha(x))\big) z_\alpha(x)\, dx + \alpha \quad \forall\, \sigma \in \Sigma_{ad}^1,$

where $z_\alpha \in C[0,1]$ denotes the unique solution to

$$z''(x) + \sigma_\alpha'(y_\alpha(x))z(x) = 2(h(x) - y_\alpha(x)), \quad x \in (0,1), \qquad z(0) = 0, \quad z'(1) = 0.$$

Condition (b) is anlogous to the optimality condition given in Theorem 2. At least under some additional assumptions it should be possible to let α tend to zero.

References

[1] Kluge, R.: *Zur Parameterbestimmung in nichtlinearen Problemen*, volume 81 of *Teubner-Texte zur Mathematik*. BSB B. G. Teubner Verlagsgesellschaft, Leipzig, 1985.

[2] Rösch, A.: *Identifikation nichtlinearer Wärmeübergangsgesetze mit Methoden der Optimalen Steuerung*. Diss., Techn. Univ. Chemnitz-Zwickau, 1995.

[3] Barbu, V.; Kunisch, K.: *Identification of Nonlinear Elliptic Equations*. Appl. Math. Optim., 33, 139–167, 1996.

[4] Barbu, V.; Kunisch, K.: *Identification of Nonlinear Parabolic Equations*. Preprint, Techn. Univ. Graz, Inst. Math., 1996.

[5] Goebel, M.; Oestreich, D.: *Optimal Control of a Nonlinear Singular Integral Equation Arising in Electrochemical Machining*. Z. Anal. Anwend., 10(1), 73–82, 1991.

[6] Recknagel, G.: *Zur Lösung und optimalen Steuerung einer gewöhnlichen nichtlinearen Randwertaufgabe zweiter Ordnung*. Diplomarb., Martin-Luther-Univ. Halle–Wittenberg, FB Math./Inf., 1995.

[7] Goebel, M.: *Smooth and Nonsmooth Optimal Lipschitz Control*, manuscript to be published.

[8] Schleiff, M.: *Eindimensionale elektrothermische Prozesse und Randwertaufgaben für eine Differentialgleichung 2. Ordnung*. Z. Angew. Math. Mech., 66(10), 483–488, 1986.

[9] Clarke, F. H.: *Optimization and nonsmooth analysis*. SIAM, Philadelphia, 1990.

[10] Mäkelä, M. M.; Neittaanmäki, P.: *Nonsmooth optimization*. World Scientific Publishing Co., Singapore, New Jersey, London, Hong Kong, 1992.

[11] Ekeland, I.: *Nonconvex Minimization Problems*. Bull. Am. Math. Soc. (N. S.), 1(3), 443–474, 1979.

International Series of Numerical Mathematics
Vol. 124,

Suboptimality Theorems in Optimal Control

Andreas Hamel*

Abstract. Necessary optimality conditions are statements about nothing if an optimal solution does not exist. Therefore it makes sense to embed such conditions in a general framework which excludes the possibility of empty (even false) assertions all the more as the assumptions for existence results are much stronger than those for necessary optimality conditions. We call such a framework *suboptimality theorem* and exemplify it by a simple problem of optimal control. Furthermore, we investigate the set of suboptimal solutions by means of a new theorem about the approximation of measurable by simple functions.

1. Introduction

In variational calculus and optimal control theory, there is a peculiar gap between necessary optimality conditions and existence results for optimal solutions, see [7], [8].

The classical way for proving existence results consists in restricting the class of problems in order to ensure compactness properties. Compare e. g. [7], chapter 9.

Another way consists in enlarging the class of admissible functions in order to ensure compactness properties in a quite weaker toplogy. This way leads to the concept of generalized controls and has been gone by Young [12] and more recently, by Klötzler [8]. Balder makes use of generalized controls as well and a new extremum principle for proving existence results in the absence of convexity, [1].

In both ways there are no changes in the cost functional on principle.

With Ekeland's variational principle known since 1972 we have a new tool at hand for connecting existence results and necessary optimality conditions - without any compactness property. The price of this way is a change of the cost functional. Instead of the original one, it is necessary to consider perturbations of the functional with the consequence that we obtain existence and necessary conditions not for optimal but for suboptimal solutions which approximate the infimum as well as one wishes.

We denote by the expression "*suboptimality theorem*" a mathematical assertion which includes

- existence of suboptimal solutions (minimizing sequence property),
- necessary conditions for suboptimal solutions,
- a continuity property with respect to the optimal solution if one exists.

The first point is neither new nor surprising: The existence of minimizing sequences is trivial for a functional with a finite infimum. But the second point selects special minimizing sequences which can be characterized by means of Ekeland's principle.

*Martin-Luther-Universität Halle-Wittenberg, Fachbereich Mathematik u. Informatik, Theodor-Lieser-Str. 5, 06099 Halle/S., e-mail: hamel@mathematik.uni-halle.d400.de

The idea to do that for a simple problem of optimal control goes back to Ekeland, [2]. The third point seems to be new and requires the application of a special variant of Ekeland's principle which is due to Penot [9] .

In this paper we demonstrate the derivation of a suboptimality theorem using as example an optimal control problem. Similar results but without the third point have been given for different types of control problems by Plotnikov/Sumin [10], Fattorini [3], Fattorini/Frankowska [4] and the author [5], [6]. Furthermore, we investigate the structure of the set of suboptimal solutions by the means of an approximation theorem for measurable by simple functions (in the sense of measure theory) which satisfy an additional admissibility condition.

Let us remark that the procedure indicated in this note can be applied to more complicated problems, e. g. of the type considered in [10], [5], [6].

2. A suboptimality theorem in optimal control

2.1 Problem statement and preliminary results

Let be $T > 0$ a fixed number, $g : R^n \to R$, $f : R \times R^n \times R^m \to R^n$ two functions, $x_0 \in R^n$ a given point and $U \subset R^m$ a set. We consider the following optimal control problem: Minimize the functional

$$g\left(x\left(T\right)\right) \tag{1}$$

under the conditions

$$\dot{x}\left(t\right) = f\left(t, x\left(t\right), u\left(t\right)\right), \qquad x\left(0\right) = x_0 \tag{2}$$

$$u\left(t\right) \in U. \tag{3}$$

We define the set of admissible controls as

$$\mathcal{U} = \left\{u : R \to R^m, u \text{ } Lebesgue - measurable, u\left(t\right) \in U \text{ } a.e.\right\}$$

and a distance function on this set

$$d\left(u, v\right) = meas\left\{t \in [0, T] : \, u\left(t\right) \neq v\left(t\right)\right\}.$$

By $meas\,\{E\}$we denote the Lebesgue-measure of the set E. The assumptions are the following:

(A1) The functions f and $\frac{\partial f}{\partial x_i}, i = 1, \ldots, n$ are well defined and continuous on $[0, T] \times R^n \times U$.

(A2) There exists a $\gamma > 0$ such that $\langle x, f\left(t, x, u\right)\rangle \leq \gamma\left(1 + \|x\|^2\right)$ for all $(t, u) \in [0, T] \times U$.

(A3) The set $U \subset R^m$ is compact.

For a given control function $u \in \mathcal{U}$, we denote an absolute continuous function which satisfies the differential equation in (2) almost everywhere by $x\left(u\right)$ and its value at the point $t \in [0, T]$ by $x\left(u\right)\left(t\right)$. According to the assumptions (A1) and (A2), there exists for every $u \in \mathcal{U}$ a unique solution $x\left(u\right)$ of (2).

In a usual way the function $H : R^n \times U \times R^n \times R \to R$ defined by $H(x, u, p, t) := p^T f(x, u, t)$ is designated as the Hamiltonian of the problem.

Lemma 1 *The pair $(\mathcal{U}, d)$ is a complete metric space.*

Lemma 2 *The functional $\varphi : \mathcal{U} \to R$, defined by $\varphi(u) = g(x(u)(T))$ is continuous with respect to the metric d.*

The proofs of the two lemmas above can be found e. g. in [2].

Remark 1 *It is not possible to give up the boundedness assumption on the set U and to define the set of admissible controls by*

$$\mathcal{V} = \{u : R \to R^m, u \text{ Lebesgue} - \text{measurable}, \ u(t) \in U \text{ a.e.}, \ u \in L_\infty(0,T)\}.$$

The reason is that if U is only closed but not bounded the set $\mathcal{V}$ is not a closed subset of $\mathcal{U}$ with respect to the metric d. To see this consider the case $m = 1$, $T = 1$, $U = [0, \infty)$ and the sequence of functions $u_n \in \mathcal{V}$ defined by

$$u_n(t) = \begin{cases} 0: & t \in \left[0, \frac{1}{n+1}\right] \\ l: & t \in \left(\frac{1}{l+1}, \frac{1}{l}\right], \ l = 1, 2, \ldots, n \end{cases}$$

which converges to

$$u_0(t) = \begin{cases} 0: & t = 0 \\ l: & t \in \left(\frac{1}{l+1}, \frac{1}{l}\right], \ l = 1, 2, \ldots \end{cases}$$

with respect to the metric d. But $u_0 \notin L_\infty(0,1)$. Hence $(\mathcal{V}, d)$ is not a complete metric space.

2.2 The suboptimality theorem

Theorem 1 *Let be $\{\alpha_n\}_{n=1,2,\ldots}$ a sequence of positive numbers with $\lim_{n\to\infty} \alpha_n = 0$ and $\{u_n\}_{n=1,2,\ldots} \subset \mathcal{U}$ a minimizing sequence for the control problem with*

$$\varphi(u_n) \le \inf_{u \in \mathcal{U}} \varphi(u) + \alpha_n. \tag{4}$$

Then there exists a second minimizing sequence $\{v_n\}_{n=1,2,\ldots} \subset \mathcal{U}$ with

$$\varphi(v_n) + \alpha_n d(v_n, u_n) \le \varphi(u_n) \tag{5}$$

$$\varphi(v_n) < \varphi(u) + \alpha_n d(u, v_n) \quad \forall u \in \mathcal{U}, \ u \neq v_n. \tag{6}$$

Proof: We apply Penot's formulation of Ekeland's principle (see the appendix) to the functional φ. ■

Theorem 2 *For the sequence $\{v_n\}_{n=1,2,\ldots} \subset \mathcal{U}$ of theorem 1 it holds*

$$\varphi(v_n) \le \varphi(u_n) \le \inf_{u \in \mathcal{U}} \varphi(u) + \alpha_n \tag{7}$$

and

$$H(x(v_n)(t), v_n(t), p(v_n)(t), t) \le H(x(v_n)(t), w, p(v_n)(t), t) + \alpha_n \quad \forall w \in U \tag{8}$$

almost everywhere in $[0,T]$, *where the functions* $p(v_n)$ *are solutions of the adjoint equation*

$$\dot{p}(t) = -f'_x(t, x(v_n)(t), v_n(t))\, p(t) \tag{9}$$

with the end condition

$$p(T) = g_x(x(v_n)(T)). \tag{10}$$

Proof. The proof is a transcription of the proof for theorem 7.1 in [2]. ■

Theorem 3 *If we suppose additionally that there exists an optimal control* $u_0 \in \mathcal{U}$ *then we have*

$$H(x(u_0)(t), u_0(t), p(u_0)(t), t) \leq H(x(u_0)(t), w, p(u_0)(t), t) \quad \forall w \in U$$

where $p(u_0)$ *is the solution of the system*

$$\dot{p}(t) = -f'_x(t, x(u_0)(t), u_0(t))\, p(t), \quad p(T) = g_x(x(u_0)(T))$$

Proof. Because $\varphi(u_0) = \inf_{u\in\mathcal{U}} \varphi(u)$ we can choose $u_n = u_0\ \forall n = 1,2,\ldots$ as minimizing sequence in (4) and get from (5) $\varphi(v_n) + \alpha_n d(v_n, u_0) \leq \varphi(u_0)$ and hence $\varphi(v_n) = \varphi(u_0)$ as well as $v_n = u_0\, \forall n = 1,2,\ldots$ From theorem 2 it follows that

$$H(x(u_0)(t), u_0(t), p(u_0)(t), t) \leq H(x(u_0)(t), w, p(u_0)(t), t) + \alpha_n \quad \forall w \in U$$

almost everywhere in $[0,T]$ and for all $n = 1,2,\ldots$ This yields the assertion. ■

Remark 2 *Theorem 3 yields the classical necessary optimality conditions from the theorem about suboptimal solutions if we additionally assume the existence of an optimal control.*

Remark 3 *The simple procedure for deriving theorem 3 depends mostly on the variational principle, not on the concrete (control) problem. The essence is to establish Fermat's rule via Ekeland's principle. See the appendix for some details.*

2.3 Further remarks on sets of suboptimal controls

In view of theorem 2 it might be worth describing the set of controls which satisfy the condition (8). We define for $\alpha > 0$

$$\mathcal{U}_\alpha := \left\{ \begin{array}{r} u \in \mathcal{U} : H(x(u)(t), u(t), p(u)(t), t) - H(x(u)(t), w, p(u)(t), t) \leq \alpha \\ \forall w \in U \text{ a.e. in } [0,T] \end{array} \right\}$$

and observe that $\mathcal{U}_{\alpha_1} \subset \mathcal{U}_{\alpha_2}$ for $0 < \alpha_1 \leq \alpha_2$. Theorem 2 states that $\mathcal{U}_\alpha \neq \emptyset$ for every $\alpha > 0$. The question arises naturally, how to compute an element of $\mathcal{U}_\alpha$. Does the set $\mathcal{U}_\alpha$ contain simple structured functions which can be calculated easier than generally measurable functions? The following procedure gives an answer to the last questions. Theorem 4 is a generalization of the well-known approximation theorem for measurable by simple functions and might be of independent interest. For the standard definitions and the background of measure theory compare [11]. Especially, we call a function $u : \Omega \to R^m$ where $(\Omega, \mathcal{F})$ is a measure space measurable iff $u^{-1}(E) = \{t \in \Omega : u(t) \in E\} \in \mathcal{F}$ for every Borel-measurable set $E \subset R^m$.

Theorem 4 *Let $(\Omega, \mathcal{F}, \mu)$ be a measure space, $U \subset R^m$ a compact set, $u : \Omega \to R^m$ a measurable function with $u(t) \in U$ for $\mu-$almost all $t \in \Omega$. Then there exists a sequence $\{v_n\}_{n=1,2,\dots}$ of simple functions $v_n : \Omega \to R^m$ with*
(i) $v_n(t) \in U$ for all $t \in \Omega$,
(ii) $\lim_{n\to\infty} v_n(t) = u(t)$ for all $t \in \Omega$ with $u(t) \in U$.

Proof. We want to construct the function v_n. We divide R^m into hypercubes of egde length $\frac{1}{2^n}$ and consider the sets

$$U_n^{(M_1,\dots,M_m)} = \left\{ w \in R^m : \frac{M_i}{2^n} \le w_i < \frac{M_i+1}{2^n}, M_i \in Z, i = 1,\dots,m \right\} \tag{11}$$

where Z denotes the set of integers. Every point $w \in R^m$ is contained in exactly one set of the form (11). We denote by $U_n^1, \dots, U_n^{l(n)}$ the finite number of sets of the form (11) with the property $U_n^k \cap U \neq \emptyset$, $k = 1,\dots,l(n)$. Then $U \subset \bigcup_{k=1}^{l(n)} U_n^k$. Because of the compactness of U, we can always find a finite number $l(n)$ of such sets. Next we can define the sets

$$Q_n^k := \left\{ t \in \Omega : u(t) \in U_n^k \right\}, \ k = 1,\dots,l(n),$$

$$Q_0 := \{ t \in \Omega : u(t) \notin U \} = \Omega \setminus \bigcup_{k=1}^{l(n)} Q_n^k$$

which are all $\mu-$measurable, and the points z_n^k as the centers of the closed hypercubes $\overline{U_n^k}$. We define the function v_n by

$$v_n(t) := \begin{cases} P_U\left(z_n^k\right) : t \in Q_n^k, \ k = 1,\dots,l(n) \\ u_0 : t \in Q_0 \end{cases}$$

where $u_0 \in U$ is arbitrary, but fixed and $P_U(z)$ denotes one of the solutions of the minimization problem $\inf_{u\in U} \|u - z\|$. There exists at least one such solution because of the continuity of the Euclidean norm $\|\cdot\|$ and the compactness of the set U.

The function v_n is well defined because $Q_n^i \cap Q_n^j = \emptyset$ for $i \neq j$, $i,j = 1,\dots,l(n)$ and $Q_n^k \cap Q_0 = \emptyset$ for $k = 1,\dots,l(n)$. It is measurable because of the measurability of the sets Q_n^k, $k = 1,\dots,l(n)$ and Q_0. Clearly, it holds $v_n(t) \in U$ for all $t \in \Omega$.

It remains to prove that $u(t)$ is the pointwise limit of $v_n(t)$. For this purpose, fix any $t \in \Omega \setminus Q_0$. Then $\|u(t) - v_n(t)\| \le \sqrt{m\left(\frac{1}{2^n}\right)^2} = \frac{1}{2^n}\sqrt{m}$ because $u(t)$ and $v_n(t)$ are contained in the same set U_n^k for a certain $k \in \{1,2,\dots,l(n)\}$. This yields $\lim\limits_{n\to\infty} \|u(t) - v_n(t)\| = 0$ for all $t \in \Omega \setminus Q_0$ and the proof is complete. ■

Theorem 5 *For every $\alpha > 0$, the set $\mathcal{U}_\alpha$ contains a simple function.*

Proof. According to theorem 2, there exists a control function $u_\alpha \in \mathcal{U}_\alpha$ which can be approximated , due to theorem 4, by a sequence of simple functions $\{v_n\}_{n=1,2,\dots} \in \mathcal{U}$. We want to prove that there exists an integer n_0 such that $v_n \in \mathcal{U}_{2\alpha}$ $\forall n \ge n_0$.

With the same argument as in ([2], proof of lemma 7.3), we can show that $x(v_n) \to x(u_\alpha)$ and $p(v_n) \to p(u_\alpha)$ uniformly in $[0,T]$. Because of the continuity of the function $H(x,u,p,t)$ with respect to all arguments, we conclude the pointwise convergences

$$H(x(v_n)(t), v_n(t), p(v_n)(t), t) \to H(x(u_\alpha)(t), u_\alpha(t), p(u_\alpha)(t), t)$$

$$H(x(v_n)(t), w, p(v_n)(t), t) \to H(x(u_\alpha)(t), w, p(u_\alpha)(t), t) \quad \forall\, w \in U$$

almost everywhere in $[0,T]$. Fix any $t \in [0,T]$ for which both convergences take place. Then for all $\epsilon > 0$ there exists a $n_0 \in N$ with

$$\begin{aligned} -\epsilon \;\leq\; & H(x(v_n)(t), v_n(t), p(v_n)(t), t) - H(x(v_n)(t), w, p(v_n)(t), t) \\ & - \left(H(x(u_\alpha)(t), u_\alpha(t), p(u_\alpha)(t), t) - H(x(u_\alpha)(t), w, p(u_\alpha)(t), t)\right) \\ \leq\; & +\epsilon \end{aligned}$$

for all $n \geq n_0$ and for all $w \in U$. This yields

$$H(x(v_n)(t), v_n(t), p(v_n)(t), t) - H(x(v_n)(t), w, p(v_n)(t), t) \leq \alpha + \epsilon \quad \forall\, w \in U$$

if $n \geq n_0$. If we choose $\epsilon = \alpha$, we have proved that $v_n \in \mathcal{U}_{\epsilon\alpha}$ if $n \geq n_0$ for some n_0 which depends on α. Replacing α by $\frac{\alpha}{2}$ and starting the above procedure with a function $u_{\frac{\alpha}{2}} \in \mathcal{U}_{\frac{\alpha}{2}}$ we find a simple function in $\mathcal{U}_\alpha$. ■

3. Appendix

3.1 Ekeland's variational principle in Penot's formulation

Theorem 6 *Let (X,d) be a complete metric space, $\varphi : X \to R \cup \{+\infty\}$ a lower semicontinuous functional which is not improper and bounded from below. Then there exists for all $\alpha > 0$ and for all $x_0 \in X$ an $x_\alpha \in X$ with*

$$\varphi(x_\alpha) + \alpha d(x_\alpha, x_0) \leq \varphi(x_0)$$

$$\varphi(x_\alpha) < \varphi(x) + \alpha d(x, x_\alpha) \quad \forall x \in X, x \neq x_\alpha \tag{12}$$

For the proof and the equivalence with the classical formulation of Ekeland's principle see the paper [9] of Penot.

3.2 Gateaux-differentiable functions on a Banach space

Theorem 7 *Let $(X, \|\cdot\|)$ be a Banach space, $\varphi : X \to R \cup \{+\infty\}$ a lower semicontinuous functional which is not improper and bounded from below. Furthermore, let φ be Gateaux-differentiable at all points of the closed and convex set $M \subset X$. Then there exists for all $\alpha > 0$ and for all $x_0 \in M$ an $x_\alpha \in M$ with*

$$\varphi(x_\alpha) + \alpha \|x_\alpha - x_0\| \leq \varphi(x_0)$$

$$\varphi'(x_\alpha)(x - x_\alpha) + \alpha \|x - x_\alpha\| \geq 0 \quad \forall x \in M \tag{13}$$

Proof: We apply Ekeland's principle to the complete metric space $(M, \|\cdot\|)$ and get immediately the first inequality. For proving the second, we fix an arbitrary $x \in M$. Then $y := (1-\lambda)\, x_\alpha + \lambda x \in M$ for all $\lambda \in (0,1)$ because of the convexity of M. Substituting y instead of x in (12) and rearranging the terms we get

$$\frac{1}{\lambda}\left[\varphi\left(x_\alpha + \lambda\left(x - x_\alpha\right)\right) - \varphi\left(x_\alpha\right)\right] \geq -\alpha \left\|x - x_\alpha\right\|.$$

Since the right-hand side does not depend on λ, we can carry out the limit process $\lambda \longrightarrow +0$ which completes the proof. ∎

Remark 4 *If $x_\alpha \in \operatorname{int} M$ then $x := x_\alpha \pm th \in M$ for every $h \in X$ with $\|h\| = 1$ and sufficiently small $t > 0$. From (13) we obtain*

$$\pm\varphi'\left(x_\alpha\right) h \geq -\alpha,$$

hence $|\varphi'(x_\alpha)\, h| \leq \alpha$ for all $\|h\| = 1$ and $\|\varphi'(x_\alpha)\|_* \leq \alpha$ where $\|\cdot\|_*$ denotes the norm in the topological dual X^* of X. This is a well-known consequence of Ekeland's principle, compare [2], theorem 2.2.

3.3 Minimizing sequence formulation

Theorem 8 *Let the assumptions of theorem 7 be in force. Let $\{\alpha_n\}_{n=1,2,\ldots}$ be a sequence of positive numbers with $\lim_{n\to\infty} \alpha_n = 0$ and $\{x_n\}_{n=1,2,\ldots} \subset X$ a minimizing sequence for φ with $\varphi(x_n) \leq \inf_{x\in M} \varphi(x) + \alpha_n$. Then there exists a second minimizing sequence $\{y_n\}_{n=1,2,\ldots} \subset X$ with*

$$\varphi\left(y_n\right) + \alpha_n \left\|y_n - x_n\right\| \leq \varphi\left(x_n\right)$$
$$\varphi'\left(y_n\right)\left(y - y_n\right) + \alpha_n \left\|y - y_n\right\| \geq 0 \quad \forall y \in M, \quad n = 1, 2, \ldots$$

The proof is clear from theorem 7.

3.4 Fermat's rule

Theorem 9 *Let the assumptions of theorem 7 be in force. Let $x_0 \in M$ be a minimal point of φ on the set M, i.e. $\varphi(x_0) = \inf_{x\in M} \varphi(x)$. Then*

$$\varphi'\left(x_0\right)\left(x - x_0\right) \geq 0 \quad \forall x \in M.$$

Proof: Take $x_n = x_0$ in the theorem 8 as minimizing sequence. We get a sequence $\{y_n\}_{n=1,2,\ldots} \subset M$ with $\varphi(y_n) + \alpha_n \|y_n - x_0\| \leq \varphi(x_0)$ for all n. Because x_0 is a minimizer so is y_n, and it follows that $y_n = x_0$ for all n. Therefore, we have

$$\varphi'\left(x_0\right)\left(x - x_0\right) \geq -\alpha_n \left\|x - x_0\right\| \quad \forall x \in M$$

for all n and hence the result. ∎

Remark 5 *If $x_0 \in \operatorname{int} M$, then the same argument as in remark 4 yields*

$$\varphi'\left(x_0\right) = 0 \quad \text{in } X^*.$$

We understand Ekeland's principle as a generalization of Fermat's rule in the sense that it yields the classical (old or new) necessary optimality conditions in a specific optimization problem if we assume the existence of a minimizer.

References

[1] Balder, E. J.: *New Existence results for Optimal Controls in the Absence of Convexity: the Importance of Extremality*, SIAM J. Control Optimization 32 (1994) No.3, 890–916.

[2] Ekeland, I.: *On the Variational Principle*, J. Math. Anal. Appl. 47 (1974) 324–353.

[3] Fattorini, H. O.: *The Maximum Principle for Nonlinear Nonconvex Systems in Infinite Dimensional Spaces*, in: Lecture Notes in Control and Information Science 75 (Springer-Verlag New York 1985) 162–178.

[4] Fattorini, H. O.; Frankowska, H.: *Necessary Conditions for Infinite-Dimensional Control Problems*, Math. Control Signals Systems 4 (1991) 41–67.

[5] Hamel, A.: *Anwendungen des Variationsprinzips von Ekeland in der Optimalen Steuerung,* Doct. Diss., Halle/Saale 1996.

[6] Hamel, A.: *Suboptimal Solutions of Control Problems for Distributed Parameter Systems*, to appear in Proceedings of the 8th French German Conference on Optimization, LN in Economics and Math. Systems, Springer-Verlag 1997.

[7] Ioffe, A. D.; Tichomirov, V. M.: *Theorie der Extremalaufgaben*, Deutscher Verlag der Wissenschaften Berlin 1979.

[8] Klötzler. R.: *Optimal Transportation Flows*, ZAA 14 (1995) No. 2, 391–401.

[9] Penot, J. P.: *The Drop Theorem, the Petal Theorem and Ekeland's Variational Principle*, Nonlinear Analysis, Theory, Methods & Appl. 10 (1986) No. 9, 813–822.

[10] Plotnikov, V. I.; Sumin, M. I.: *The Construction of Minimizing Sequences in Problems of the Control of Systems with Distributed Parameters*, U.S.S.R. J. Comput. Math. Math. Phy. 22, No. 1 (1982) 49–57.

[11] Taylor, S. J.: *Introduction to Measure and Integration*, Cambrigde University Press 1966.

[12] Young, L. C.: *Calculus of Variations and Optimal Control Theory*, Chelsea Publishing Company New York 1980.

International Series of Numerical Mathematics
Vol. 124,

A Second Order Sufficient Condition for Optimality in Nonlinear Control – the Conjugate Point Approach

Andrzej Nowakowski*

Abstract. Second order optimality conditions in terms of conjugate points are stated. A finite number of discontinuities of optimal control is admitted. The aim of the article is to show that the classical approach to second order optimality conditions may also be successful.

1. Preliminaries and assumptions

Let us consider the simplest control problem. The dynamic of control is described by differential equation

$$\dot{x}(t) = f(t, x(t), u(t)), \quad t \in [0, T], \quad x(0) = 0, \quad x(T) = b. \tag{1}$$

Here $x \in R^n$ is a state vector, $b \in R^n$ fixed, $u \in R^q$ is a control parameter subject to the following constraint $u \in Q \subset R^q$, where Q is a compact set. Denote by V the set of all measurable functions $u : [0, T] \to Q$. Our aim is to find minimizers in V for the functional

$$J(x, u) = \int_0^T L(t, x(t), u(t))dt, \tag{2}$$

where $x : [0, T] \to R^n$ is a trajectory of the system (1) generated by control $u(\cdot) \in V$.

We assume that f, L satisfy the following hypotheses:

H1 *function* $(f, L) : R \times R^n \times Q \to R^n \times R$ *is continuous*

H2 *there exist the following partial derivatives:* f_x, L_x, f_{xx}, L_{xx} *and they are continuous, there exists* $C > 0$ *such that for all* $t \in [0, T]$, $x \in R^n$, $u \in Q$ *the conditions are satisfied:* $\|f(t, x, u)\| \leq C(1 + \|x\|)$, $|L(t, x, u)| \leq C(1 + \|x\|)$

H3 *for each* $t \in [0, T]$, $x \in R^n$ *the sets* $F(t, x) = \{(f(t, x, u), L(t, x, u)) : u \in Q\} \subset R^{n+1}$ *are either strongly convex or they are graphs of strongly convex functions.*

Consequences of these assumptions are (see [2]):

1) For each $(t, x, y) \in R^{2n+1}$ there exists a unique vector $(\bar{f}(t, x, y), \bar{L}(t, x, y)) \in F(t, x)$ such that

$$\begin{aligned} \langle y, \bar{f}(t, x, y)\rangle + \bar{L}(t, x, y) &= \min_{(f(t,x,u), L(t,x,u)) \in F(t,x)} [\langle y, f(t, x, u)\rangle + L(t, x, u)] \\ &= \min_{u \in Q} [\langle y, f(t, x, u)\rangle + L(t, x, u)] = H(t, x, y) \end{aligned} \tag{3}$$

*Faculty of Math., University of Lodz, Banacha 22, 90-238 Lodz, Poland, annowako@imul.uni.lodz.pl

2) The Hamilton function H from (3) has continuous partial derivatives: $H_y(t,x,y) = \bar{f}(t,x,y)$, $H_x(t,x,y) = \bar{f}_x(t,x,y)\,y + \bar{x}(t,x,y)$, $H_{xy}(t,x,y) = \bar{x}(t,x,y)$, and the functions $y \to H_y(t,x,y)$, $x \to H_x(t,x,y)$ are locally Lipschitz continuous.

3) There exist global solutions of the following Hamilton equations:

$$\begin{aligned} \frac{dx}{dt} &= H_y(t,x,y), \quad \frac{dy}{dt} = -H_x(t,x,y), \\ x(0,\varsigma) &= \varsigma, \quad x(T,\zeta) = \zeta, \end{aligned} \tag{4}$$

where ζ, ς belong to some open sets which will be defined later.
We will name them canonical extremals for our problem. We will distinguish one of them, namely that for which $x(0,0) = 0, x(T,0) = b$, denoting it by $\overline{x}\ (t)$ and corresponding to it canonical trajectory $\overline{y}(t)$, and suitable control function $\overline{u}(t)$, (i.e. $\overline{x}\ (t)$, $\overline{u}(t)$ satisfy (1)). The existence of the control $\overline{u}(t)$ corresponding to $\overline{x}\ (t)$ and $\overline{y}(t)$ follows from the following arguments: by (4) and 2) $d\overline{x}\ (t)/dt = \bar{(}t,\overline{x}(t),\overline{y}(t))$ and by 3) $d\overline{x}\ (t)/dt \in f(t,\overline{x}(t),Q)$; the last and the assumptions on f allow us to use [1, Corollary 1 p.91] to assert the existence of $\overline{u}(t)$. Of course we can not assert that the solutions of (4) are unique and that they depend in a smooth way upon any parameters e.g. initial conditions. The latter will cause us to assume a little more about our family of solutions of (4). Because we consider control problems it is unnatural to propose hypotheses which would be general enough and imply the same regularity on the family of solutions as in the classical calculus of variations. However we need one more hypothesis along $\overline{x}\ (t)$:

H4 *the control function $\overline{u}(t)$ is piecewise continuous and the generalized Jacobian (in the sense of Clarke) $\partial_y H_y(t,\overline{x}\ (t),\overline{y}(t))$, $t \in [0,T]$ has a maximal rank.*

By the generalized Jacobian $\partial_y H_y(t,x,\cdot)$ at the point y we mean (see e.g. [5, p. 563]) the convex hull of all matrices B of the form $B = \lim_{i\to\infty}\{D_y H_y(t,x,y_i)\}$, where y_i converges to y and the usual Jacobian $D_y H_y(t,x,y_i)$ exists for each i.

The last hypothesis allows us to state a local one to one and smooth embedding theorem. We formulate it in a form convenient for us.

Theorem 1 *There exists $\delta > 0$ and a neighborhood N of (T,b) such that extremals $x(t)$ of (4) restricted to $(T-\delta,T)$ which meet N cover N simply and may be smoothly parametrized.*

Proof: Setting $\dot{x} = H_y(t,x,y)$ by **H4** there exists a neighborhood K of $(T,\overline{x}\ (T),\overline{y}(T))$, such that K is mapped by $(t,x,y) \to (t,x,\dot{x})$ one-to-one onto a neighborhood M of $(T,\overline{x}\ (T),\dot{\overline{x}}(T))$. Consider, for a small canonical neighborhood for constant $t = T$ of $(\overline{x}\ (T),\dot{\overline{x}}(T))$ the canonical extremals $x(t,w,v)$, $y(t,w,v)$, $t \in (T-\delta,T)$, with the end values $x = w$, $y = v$ when $t = T$, where (w,v) lies in the neighborhood in question. The canonical extremals exist and are unique, and they are suitably smooth. Shrinking K we infer that the equations $t = t$, $x = x(t,w,v)$, $y = y(t,w,v)$ have, for $(t,x,y) \in K$, unique solutions t, w, v. Thus we obtain our second map $(t,w,v) \to (t,x,y)$. The same argument allows us to define our third map $(t,x,v) \to (t,w,v)$ in suitable domains. All three maps are one-to-one. If we shrink the domain for (t,x,v) sufficiently the image of each map can be mapped by the previous map, and the final map will be in M. If we compose the three maps in reverse order then we obtain a map which is given by equations $t = t, \quad x = x, \quad \dot{x} = p(t, x, v)$, and here, for given $(t, x, \dot{x}) \in M$, there is just one v for which the third equation holds. Hence we infer the assertion of the theorem (compare also [4, pp. 57,58]). ■

Following the same way as in [4] we state the existence of a solution of the boundary-value problem in a local angle for (4). We shall term local pencil the family of solutions of (4) defined on $[T-\delta, T]$ whose derivatives $\dot{x}$ at T satisfy $\left|\dot{x} - \dot{\overline{x}}(T)\right| \leq \delta$. We term local angle about $(T, \overline{x}\ (T), \dot{\overline{x}}(T))$ the set of points (t, x) for which $t \in [T-\delta, T]$, and $\left|(x - \overline{x}(T))/(t-T) - \dot{\overline{x}}(T)\right| < \delta$.

Theorem 2 *Let (t_1, x_1) lie in the local $(\delta/3)$- angle about $(T, \overline{x}\ (T), \dot{\overline{x}}(T))$. Then the points $(T, \overline{x}\ (T)), (t_1, x_1)$ can be joined by a trajectory of the local pencil, and by no trajectory not of that pencil.*

Both above theorems shows that assumptions **H3, H4** look a little strong in optimal control theory. They can be weakened but then proofs of the final theorems are much more complicated and longer. This is why we assume that they hold in this paper. As a direct consequence of Theorems 1 and 2 we have classical local result (see [4, p.62].

Theorem 3 *There exist a neighborhood N of (T, b) and a local angle about $(T, \overline{x}\ (T), \dot{\overline{x}}(T))$ such that for any extremal C_0 in N,with one end at (T, b), and with the derivative at some relevant t in the local angle, and for any other admissible trajectory C lying in N with the same ends as C_0 we have $J(C_0) \leq J(C)$ where $J(C)(J(C_0))$ denotes the value of functional (2) restricted to the trajectory C (C_0).*

2. The secondary Hamiltonian

The main tools in studying second order optimality conditions are secondary notions (see e.g. the classical book on the calculus of variations [4]). The papers of V. Zeidan [5, 6] try to omit that classical approach – it is related to the fact that in optimal control problems we can not expect sufficient regularity of dates to defined properly secondary notions and develop approaches using certain types of Riccati equations or inequalities. We try to show that the classical approach to second order optimality conditions may also be successful.

To this effect we define first a generalized secondary Hamiltonian - generalized because we are not able to assert that H_{xx}, H_{yy} exist. However because of 2) there exist generalized Jacobians (see the definition of it below **H4**) $\partial_x H_x, \partial_y H_y$ and they are bounded. Thus let us write **A,B,C** for the generalized quadratic form $X \partial_x H_x X = \{XAX : for\ all\ matrices\ A \in \partial_x H_x\}$, the quadratic form $X H_{xy} Y$, and the generalized quadratic form $Y \partial_y H_y Y = \{YCY : for\ all\ matrices\ C \in \partial_y H_y\}$, where X, Y are vectors in R^n which will be specified later. We term generalized secondary Hamiltonian generally a multifunction $\mathbf{H}(t, X, Y)$ defined by setting $2\mathbf{H} = \mathbf{A} + 2\mathbf{B} + \mathbf{C}$.

Let us assume for the moment that we are given a one parameter smooth family of trajectories satisfying (4):

$$x(t, \alpha), \quad y(t, \alpha), \quad t_1(\alpha) \leq t \leq t_2(\alpha), \tag{5}$$

which reduce to the solution $\overline{x}(t)$ when $\alpha = 0$. Here we suppose that α is a real parameter whose range includes 0 in its interior. We write $X, Y, \dot{X}, \dot{Y}$, for the derivatives in α of $x, y, \dot{x}, \dot{y}$ as a functions of (t, α). Substituting (5) in (4) and differentiating both sides of them with respect to α we find for $\alpha = 0$

$$\frac{\partial}{\partial \alpha} H_{y_i} \stackrel{df}{=} \sum_j (X_j H_{x_j y_i} + \partial_{y_i} H_{y_j} Y_j) = \mathbf{H}_{Y_i},$$

which may be written $\frac{\partial}{\partial \alpha} H_y = \mathbf{H}_Y$. By entirely similar calculations we see that $\frac{\partial}{\partial \alpha} H_x = \mathbf{H}_X$. Of course, $\mathbf{H}_Y$ and $\mathbf{H}_X$ are generally non single-element sets. From these we see that X, Y satisfy inclusions:

$$\dot{X} \in \mathbf{H}_Y, \quad \dot{Y} \in -\mathbf{H}_X. \tag{6}$$

Therefore X, Y satisfy generalized secondary Hamiltonian inclusions. Further we easily check that they are Euler inclusions for secondary problem with the generalized Lagrangian $\mathbf{L} = Y\dot{X} - \mathbf{H}(t, X, Y)$.

This result is worth to be stressed; from a family of canonical extremals $x(t, \alpha)$, $y(t, \alpha)$ (satisfying equations (4)) of the original problem, we derive by partial differentiation in α at $\alpha = 0$, a secondary extramal, whose canonical inclusions are (6).

Inclusions (6), which are linear and homogeneous in X, Y with coefficients depending on t, we name the Jacobi differential inclusions. Conversely, any solution $X(t), Y(t)$ of (6) can be derived by partial differentiation in α at $\alpha = 0$ from the family of extremals $x(t, \alpha)$, $y(t, \alpha)$ determined by the original canonical Euler equation (Pontryagin maximum principle) together with the end conditions $x(t_0) + \alpha X(t_0)$, $y(t_0) + \alpha Y(t_0)$ at $t = t_0$ $(t_1(0) = t_1 < t_0 < t_2 = t_2(0))$. Along a secondary extremal, we have, since $\mathbf{H}$ is a homogeneous generalized quadratic form in $\dot{X}, Y$, $2\mathbf{H} = X\mathbf{H}_X + Y\mathbf{H}_Y \ni -X\dot{Y} + Y\dot{X}$, and therefore, using (6), $2\mathbf{L} \ni \dot{X}Y + \dot{Y}X = \frac{d}{dt}(XY)$. Let a secondary extremal X, Y be given. Then we set $\int_{t_1}^{t_2} 2\mathbf{L}dt = \left\{\int_{t_1}^{t_2} l(t)dt : l \text{ is an integrable selection from } 2\mathbf{L}\right\}$ (compare [1, p.98]). Hence, by integrating along a secondary $X(t)$, $Y(t)$, at whose ends t takes the values t_1, t_2, we find that

$$\int_{t_1}^{t_2} 2\mathbf{L}dt \ni X(t_2)Y(t_2) - X(t_1)Y(t_1). \tag{7}$$

This simple way of carrying out the integration explicitly will greatly assist our study.

3. Geometrical interpretation of exactness and distinguished families

Because of formula (7) we can repeat following L.C. Young [4] p. 76–81 the general considerations concerning exactness and distinguished families. If we look carefully on the right-hand side of (7) then we see that it is nothing more than the Lagrange brackets of some functions, in fact (5). We are interested in making it vanish.

We consider (see [4]) complex vectors $X + iY$ with n complex numbers as components. We shall term $X + iY$ a pure complex vector if X and Y are real multiples of a same real unit e; the letter is unique in that case, unless X and Y both vanish, and we term it the direction of the pure complex vector. A set of pure complex vectors will be said to be independent if their directions are orthogonal. The set of real linear combinations of n independent pure complex vectors will be termed a distinguished hyperplane.

We term enlarged scalar product of two complex vectors $X+iY$ and X^*+iY^*, the complex number

$$XX^* + YY^* + i(YX^* - XY^*), \tag{8}$$

where the expressions XX^* and so on denote ordinary scalar products of real vectors. The expression (8) is in general not commutative. Clearly it is only commutative if the Lagrange bracket vanishes.. In that case we term (8) the exact scalar product. A set of complex vectors, every pair of which have an exact scalar product, will be termed a set of exactness. In [4] we find:

Theorem 4 *In order that a set of complex vectors be a set of exactness, it is necessary and sufficient that it lie in some distinguished hyperplane.*

Now again a pair (X, Y) or $(X+iY)$, will denote a canonical secondary extremal, possibly one which depends on a further parameter σ. The superscript ^{0} then denotes the "end" values at some $t = t_0$.

Because of the linearity of the Jacobi inclusions the canonical secondary extremals constitute a $2m$-dimensional linear family, where members are determined by their end values.

This means also that the correspondence between end values and members preserves linear relations: we express this by saying that our $2m$-dimensional linear family is isomorphic with $2m$-dimensional Euclidean space. Thus any family of canonical secondaries can be pictured by a subset of this $2m$-dimensional space. We shall be mainly concerned with linear subfamilies, which are pictured by hyperplanes and particularly with the case of an m-parameter family, which then consists of members of the form

$$X + iY = \sum_k \sigma_k(X_k + iY_k), \tag{9}$$

where the σ_k are the components, for $k = 1, \ldots, m$ of an m-dimensional parameter σ and where the (X_k, Y_k) are m canonical secondary extremals, whose end values at some $t = t_0$, are linearly independent. We shall write $X(t,\sigma), Y(t,\sigma)$ for the family thus defined by (9).

One consequence of this linearity is that any two canonical extremals (X, Y), (X^*, Y^*) can always be embedded in a family of the form $(\alpha X + \alpha^* X^*, \alpha Y + \alpha^* Y^*)$, in which the Lagrange bracket $[\alpha, \alpha^*]$ has for the secondary problem, the same expression as for the first, namely $XY^* - YX^*$ and we term it the Lagrange bracket of two canonical secondaries. We note that $\frac{d}{dt}(XY^* - YX^*) = 0$. Or we may simply observe that

$$\frac{d}{dt}(XY^*) \in -X\mathbf{H}_X + Y^*\mathbf{H}_Y,$$

where the right-hand side is the generalized bilinear form, which by symmetry must also contain the derivative of X^*Y.

From these it follows that secondary families in which the Lagrange bracket vanishes are isomorphic to sets of exactness of their end values $X^0 + iY^0$. Such family will be termed families of exactness, and if it is an m-parameter then we term it a distinguished family.

If F is a distinguished family of canonical secondaries and t_0 is any value of t for which F is defined, there exist orthogonal unit vectors e_k $(k = 1, \ldots, m)$ and corresponding pairs of real a_k, b_k not both 0, such that the members of F are the linear combinations of (9) of m canonical secondaries $X_k + iY_k$ whose end values are

$$X_k^0 + iY_k^0 = a_k e_k + i b_k e_k. \tag{10}$$

We shall write $c_k = 0$ when $a_k = 0$ and $c_k = b_k/a_k$ otherwise. By F^* we denote linear combinations of a second system of m canonical secondaries $X_k^* + iY_k^*$ whose end values are

$$X_k^{*0} + iY_k^{*0} = e_k + i c_k e_k. \tag{11}$$

Note also (see (10)) that scalar product of XY at t_0 for members of F is $X^0 Y^0 = \sum_k a_k b_k (\sigma_k)^2$, where $X_k^0 = (X^0 e_k)$ depends only on X^0 for each k.

Denote by G, G^* the families of real parts of members $X + iY$ of F, F^*, respectively.

Lemma 1 *There exists $\delta > 0$ such that the strip $|t - t_0| < \delta$ of (t, X)-space is simply covered by G^*.*

Proof: As solutions of Jacobi inclusions the members of F^* have the form $X^* + iY^* = X^*(t,\sigma) + iY^*(t,\sigma) = \sigma(\mathbf{A}(t) + i\mathbf{B}(t))$ where σ is as in (9) and $\mathbf{A}, \mathbf{B}$ are $m \times n$ matrices. The end values are $X^{*0} + iY^{*0} = \sum_k \sigma_k e_k + i \sum_k c_k \sigma_k e_k$. Thus $\mathbf{A}$ contains the $m \times m$ unit matrix for $t = t_0$ and therefore it is of this type in some interval $|t - t_0| < \delta$. In that interval we can solve the equation $X^*(t, \sigma) = v$ and thus we get the assertion of the lemma. ■

In [2] a generalization of the field of extremals is considered. Its the simplest form as well as the nearest to the field of extremals is a spray of flights.

Corollary 1 *G^* is a spray of flights in the strip.*

Proof: In [2] it is assumed that at t_0 the Lagrange brackets vanish. By (10) in our case they vanish too. From the above lemma and our earlier assumptions we obtain that all other hypotheses of [2] are satisfied too. Therefore by [2] G^* is a spray of flights. ■

Theorem 5 *With the same δ as in the lemma the strip $t_0 - \delta < t < t_0$ of (t, X)-space is simply covered by G.*

Proof: Since the map provided by G is again linear for each t, we need verify that no two members can intersect in the strip, or, what comes to the same by substriction, that in the strip the t-axis does not meet any other member of G. We shall suppose the contrary.

There is then an arc γ of a member of G that joins two points, one of the form (t_0, X^0), the other $(t_1, 0)$ where $t_1 - t_0 < \delta$, and moreover γ is not a segment of the t-axis. According to (11) and (7),

$$\mathbf{I}(\gamma) = \frac{1}{2}\sum_k c_k (X^0 e_k)^2.$$

Further, we can define in the strip $t_0 - \delta < t < t_0$, a function $I(t, X)$, vanishing at $(t_0, 0)$, by means of the Hilbert independence integral for the family F^*. Let the Hilbert independence integral be calculated along a straight line from $(t_0, 0)$. We find

$$I(t_0, X) = \frac{1}{2}\sum_k c_k (X e_k)^2.$$

Further, on any arc γ^* of a member of G^* in the strip $t_0 - \delta < t < t_0$, the difference of I at the ends is $\mathbf{I}(\gamma^*)$. In particular, by taking γ^* on the t-axis, we find that $I(t_1, 0) = 0$. We can thus calculate the difference ΔI of I at the two ends of our previous arc γ. We find in this way that $\Delta I = \mathbf{I}(\gamma)$. This contradicts the formula given in the proof of Theorem 4 in [2, p.735], as the right-hand side in this formula must have negative value along γ, which is not an arc of G^*. The proof is completed. ■

Corollary 2 *In the same strip, if $X(t,\sigma) + iY(t,\sigma)$ is the family F, we have $X(t,\sigma) \neq 0$ unless $\sigma = 0$.*

Proof: In fact, if $X(t,\sigma) = 0$ for some t, then by Theorem 4 $X(t,\sigma) = 0$ for all t of the segment of the t-axis in the strip. From Jacobi inclusions and **H4** we infer that then must be $Y(t,\sigma) = 0$, and so $\sigma = 0$. ■

4. Conjugate points

We continue with the same assumptions as in section 1. However we need one more hypothesis to define focal points and then conjugate points. The reason is that the hypotheses **(H1–H4)** do ensure the existence of solutions to the differential equations (1) but do not ensure any global smoothness dependence on parameters which is necessary in conjugate point theory (see section 1). The situation is even worse in control theory as in general it is impossible to impose assumptions on the data of our problem to obtain any reasonable regularity. In order our problem (1),(2) does not specify further we simply assume the following hypothesis:

H5 *there exists a division of the interval $[0,T]$ on subinterval $[t_i, t_{i+1}], i = 0, \ldots, q$, such that in each of this interval a family of solutions to (4) can be parametrized smoothly i.e. for each $i = 0, \ldots, q$, there exists an open set $Q_i \subset R^{m_i}$ of parameters σ_i, containing zero, smooth functions $t_i(\sigma^i), t_{i+1}(\sigma^i)$ $(t_i(0) = t_i, t_{i+1}(0) = t_{i+1})$, canonical extremals*

$$x(t,\sigma^i), \quad y(t,\sigma^i), \quad t \in [t_i(\sigma^i), t_{i+1}(\sigma^i)], \quad \sigma^i \in Q_i \tag{12}$$

which are smooth functions of both variables and $x(t,0) = \overline{x}\ (t)$. *Moreover we assume that at* $\sigma_i = 0$ *the* $2n \times m_i$ *Jacobian matrix* $(x_{\sigma^i}, y_{\sigma^i})$ *has rank* m_i *for some* $t_0 \in (t_i, t_{i+1})$, $i = 0, \dots, q$.

When a family of solutions to (4) satisfies **H5** and in addition the Lagrange bracket vanishes at the points $(t_{i+1}(\sigma^i), x(t_{i+1}(\sigma^i), \sigma^i))$, $i = 0, \dots, q$ then we call a such embedding of $\overline{x}(t)$ canonical.

Let us consider now any subinterval $[t_i, t_{i+1}]$ and canonical family (12) restricted to it. For convenience denote it by

$$x(t,\sigma), \quad y(t,\sigma) \tag{13}$$

To a canonical embedding (13) we associate the family F of canonical secondaries

$$X(t,\sigma) + iY(t,\sigma) = \sum_k \sigma_k (X_k(t) + iY_k(t)),$$

where X_k, Y_k are the partial derivative in σ_k, $k = 1, \dots, m_i$, at $\sigma = 0$ of the functions $x(t,\sigma)$, $y(t,\sigma)$. The Jacobian matrix (x_σ, y_σ) is thus the matrix whose rows are the vectors (X_k, Y_k). Since the latter are thus linearly independent at t_0 and have vanishing Lagrange bracket in the whole time interval (the proof of that is the same as in [2, Lemma 4]), it follows that F is distinguished family. Any distinguished family defined on the interval provides, in its turn, a canonical embedding for the secondary problem of the canonical secondary $X(t) = Y(t) = 0$, $(X(t) = X(t,0),\ Y(t) = Y(t,0))$ which reduces to that interval of the t-axis.

Let further γ denote the extremal $\overline{x}\ (t)$ restricted to $[t_i, t_{i+1}]$. The extremal γ is then embedded in the family of extremals $x(t,\sigma)$.

Definition 1 *We term focal point of our embedding a point of* γ *at which the real Jacobian matrix* x_σ *has rank less than* m_i. *The corresponding value of* t *determines on the* t*-axis a focal point of the secondary canonical embedding, provided by the family* F*; this can be defined as a point* t *for which there exists a linear combination*

$$X(t,\sigma) = \sum_k \sigma_k X_k(t) \quad (\sigma \neq 0)$$

such that $X(t,\sigma)$ *vanishes at this value of* t.

Thus, by Corollary 2 , we have:

Theorem 6 *If* **H4** *holds along* γ, *then the focal points of each canonical embedding of (13) are isolated.*

Theorem 7 *Let* γ *have a canonical embedding without focal points. Let* Gr *denote the set of* (t,x)*-space consisting of points belonging to graphs of trajectories of the real part of our embedding. Then there exists a set* W *of* γ *such that for any admissible trajectory* $x(t)$ *whose graph lies in* $Gr \cap W$ *with the same ends as* γ *we have*

$$\int_{t_i(0)}^{t_{i+1}(0)} L(t, \overline{x}(t), \overline{u}(t))dt \leq \int_{t_i(0)}^{t_{i+1}(0)} L(t, x(t), u(t))dt.$$

Proof: By the assumptions the matrix $x_\sigma(t,0)$ has rank m_i for each $t \in [t_i, t_{i+1}]$, thus we can solve, near $\sigma = 0$, the equation $x(t,\sigma) = x$ uniquely in σ. Hence, if we shrink the σ-domain, if necessary, our family of extremals will cover descriptively (see [4, p.266] a set $W \supset \gamma$. This family, restricted to W, then provides a spray of flights from [2, p.729], since Lagrange brackets vanish. The final inequality thus follows from [2, Corollary 2, see also Theorem 4 there]. ■

It is clear that nonexistence of focal points for embedding (12) means that there are no focal points in each subinterval $[t_i, t_{i+1}]$, $i = 0, \ldots, q$, in the sense of Definition 1. Therefore we can formulate the global version of Theorem 7.

Theorem 8 *Let $\overline{x}(t), t \in [0,T]$, have a canonical embedding without focal points. Let Gr denote the set of (t,x)-space consisting of points belonging to graphs of trajectories of the real part of our embedding. Then there exists a set W of $\overline{x}(t)$ such that for any admissible trajectory $x(t)$, $t \in [0,T]$, $x(0) = 0$, $x(T) = b$, whose graph lies in $Gr \cap W$ we have*

$$\int_0^T L(t, \overline{x}(t), \overline{u}(t))dt \le \int_0^T L(t, x(t), u(t))dt.$$

Proof: In view of Theorem 7 our canonical embedding, if we shrink it if necessary - obtaining in this way a W, consists of finite number of spray of flights. What we need to do now is to join them together. But this procedure is described in [2, p.732]. Thus the assertion of the theorem follows from [2, Theorem 2]. ■

In the sequel the most important case of embedding (12) will be that in which all the extremals pass through the same point (T,b). In that case we speak of a pencil of extremals and the point (T,b) will be termed its vertex. Thus we shall further consider only a such embedding of $\overline{x}(t)$, $t \in [0,T]$, for which a family of canonical extremals in subinterval $[t_q, t_{q+1}] = [t_q, T]$ is of the form $x(t,\sigma^q)$, $y(t,\sigma^q)$, $\sigma^q \in Q_q \subset R^n$ subject at T to the end conditions

$$x(T,\sigma^q) = b, \quad y(T,\sigma^q) = \overline{y}(T) + \sigma^q. \tag{14}$$

We shall limit ourselves here to a sufficiently small open set Q_q. We note that the matrix $(x_{\sigma^q}, y_{\sigma^q})$ has the required rank (Theorems 1–3) n for $\sigma^q = 0$, and the Lagrange brackets vanish at T, as well as hypothesis (H6) from [2] is satisfied.

The vertex of a pencil is clearly a focal point of the embedding. The other focal points, if any, on $\overline{x}(t)$, constitute the conjugate set of the point (T,b).

Theorem 9 (Jacobi) *Assume hypotheses (**H1**)–(**H5**) to be satisfied. Suppose that $\overline{x}(t)$, $t \in [0,T]$ contains no conjugate point of (T,b). Then there exists a set W_0 containing the graph of $\overline{x}(t)$, $t \in [0,T]$ such that, for every other admissible trajectory $x(t)$, $t \in [0,T]$, $x(0) = 0$, $x(T) = b$, whose graph lies in W_0 we have*

$$\int_0^T L(t, \overline{x}(t), \overline{u}(t))dt \le \int_0^T L(t, x(t), u(t))dt.$$

Proof: We wish to apply results of [2] to obtain the assertion of the theorem. To this effect we have to show that embedding (12) subject to the end conditions (14), after shrinking it if necessary, is a canonical embedding without conjugate point thus in terms of the paper [2] it means that it is a chain of flights i.e. a suitable joined finite number of spray of flights.

First note that we can apply Theorems 1 and 3. We then get that the family $x(t,\sigma_q)$, $y(t,\sigma_q)$, $t \in [t_q, t_{q+1}] = [t_q, T]$, $\sigma^q \in Q_q \subset R^n$, if we shrink it if necessary, is a spray of flights. This means in particular that at the points $(t_q(\sigma^q), x(t_q(\sigma^q),\sigma^q))$, $\sigma^q \in Q_q$, the Lagrange brackets vanish. By Theorem 2, if we shrink it if necessary, the family

$$x(t,\sigma^{q-1}), \quad y(t,\sigma^{q-1}), \quad t \in [t_{q-1}(\sigma^{q-1}), t_q(\sigma^q)], \quad \sigma^q \in Q_q \tag{15}$$

then each its member can be join with a member of the family $x(t,\sigma_q)$, $y(t,\sigma_q)$, $t \in [t_q, t_{q+1}] = [t_q, T]$, $\sigma^q \in Q_q$, therefore at the ends of $x(t,\sigma_q)$ the Lagrange brackets will vanish. Thus the family (15) is also a spray of flights (by Theorem 7). We can continue this procedure throughout all subintervals $[t_i, t_{i+1}]$, $i = 0, \ldots, q-2$.

Our assumption (14) implies directly hypothesis (H6) from [2]. Thus all hypotheses of [2, Theorem 2] are satisfied. The set W_0 we obtain as a set covered by graphs of all extremals being members of our modified canonical embedding. Therefore the assertion of the theorem we get as a consequence of [2, Theorem 2, Theorem 4]. ■

References

[1] Aubin, J.-P.; Cellina, A.: *Differential inclusions.* Springer, Berlin, 1984.

[2] Nowakowski, A.: *Field theories in the modern calculus of variations.* Trans. Amer. Math. Soc. **309**, (1988), 725–752.

[3] Subbotina, N.N.: *Unified optimality conditions in control problems.* (Russian) Trudy Inst. Mat. Mekh. (Ekaterinburg) **1** (1992), 147–159.

[4] Young, L.C.: *Lecture on the calculus of variations and optimal control theory.* Saunders, Philadelphia, Pa., 1969.

[5] Zeidan, V.: *Sufficient conditions for the generalized problem of Bolza.* Trans. Amer. Math. Soc . **275,** 1983, 561–586.

[6] Zeidan, V.: *First and second order sufficient conditions for optimality control and the calculus of variations.* Appl. Math. Optim. **11** (1984), 209–226.

International Series of Numerical Mathematics
Vol. 124,

Extremal Problems for Elliptic Systems

Uldis Raitums *

Abstract. The specific properties of optimal control problems for elliptic systems, if compared with the case of a single equation, are described. Within them are: strong closures of sets of feasible states; the relaxability via convexification; the type of necessary optimality conditions.

1. Introduction

Despite the great progress during last three decades in the theory of optimal control problems with distributed parameters, the case of elliptic systems is still unclear. Still unknown, for instance, is a possible type of necessary optimality conditions if the main part of differential operators depends on controls from nonconvex sets.

In this lecture, we consider the case of state equations in the form

$$\operatorname{div} A(x)\nabla\overline{u} = \operatorname{div}\overline{f}(x,\sigma),\ x \in \Omega \subset \mathcal{R}^n,$$

$$\overline{u} = (u_1, \dots, u_m) \in \left[H_0^1(\Omega)\right]^m,$$

where matrices A and functional parameters σ from nonconvex sets play the role of controls.

If $m = 1$, i.e. the case of a single equation, then for typical optimal control problems without additional constraints it is known, see for instance Raitums [1], that

(i) the convexification preserves the price of the problem even for non weakly continuous functionals;

(ii) the convexification gives the strong closure of the set of feasible states;

(iii) an analogue of the Pontryagin's maximum principle (or the Lagrange principle) is valid as a necessary optimality condition.

The precise formulations of these statements will be given in the section 2.

We shall show (in section 3) that these important features do not remain true for the case of elliptic systems with $m \geq n$.

After that (in sections 4 and 5) we shall give a description of the strong closure of the set of feasible states for the case with $m = n$ where the set of admissible matrices has only two elements, and shall discuss a possible type of necessary optimality conditions for the case $m = n$ with a linear (with respect to $\nabla\overline{u}$) cost functional.

*University of Latvia, LV-1459 Riga, 29 Rainis boulevard, Latvia, e-mail: raitums@acad.latnet.lv

2. Preliminaries

Let $r \geq 1, n \geq 2$ and $m \geq 1$ be integers, let $\Omega \subset \mathcal{R}^n$ be a bounded Lipschitz domain, let $S_0 \subset \mathcal{R}^r$ be a bounded set and let $M_0 \subset \mathcal{R}^{(n\times n)(m\times m)}$ be a bounded set of symmetric positive definite matrices.

For a fixed subset $D \subset \Omega$ we introduce the following sets S and M of admissible controls

$$S := \{\sigma \in [L_2(\Omega)]^r \mid \sigma(x) \in S_0 \; if \; x \in D, \;\; \sigma(x) = \sigma_*(x) \text{ otherwise}\}$$

$$M := \{A \in [L_2(\Omega)]^{(n\times n)(m\times m)} \mid A(x) \in M_0 \; if \; x \in D, \;\; A(x) = A_*(x) \text{ otherwise.}\}$$

Here the function σ_* and the symmetric matrix function A_* are given. We assume that there exist constants $0 < \nu < \mu$ such that for all $x \in \Omega$ and all $\sigma \in S, \; A \in M$

$$| \sigma(x) | \leq \mu, \; | A(x) | \leq \mu$$

$$\langle A(x)\overline{\xi}, \overline{\xi} \rangle \geq \nu | \overline{\xi} |^2 \;\; \forall \overline{\xi} \in \mathcal{R}^{n\times m}.$$

Let us denote

$$L := [L_2(\Omega)]^{n\times m},$$

$$H := [H_0{}^1(\Omega)]^m,$$

and let $F = F(x, \sigma, \overline{u}, \overline{\xi})$ and $\overline{f} = \overline{f}(x, \sigma)$ be Caratheodory functions such that

1. $F := \Omega \times \mathcal{R}^r \times \mathcal{R}^m \times \mathcal{R}^{n\times m} \to \mathcal{R}$,

$$\overline{f} : \Omega \times \mathcal{R}^n \to \mathcal{R}^{n\times m};$$

2. The function F is twice differentiable with respect to $(\overline{u}, \overline{\xi})$ and these derivatives are continuous with respect to $(\sigma, \overline{u}, \overline{\xi})$;

3. There exist a constant μ_1 and a function $h \in L_2(\Omega)$ such that for all arguments

$$| F(x, \sigma, \overline{u}, \overline{\xi}) | \leq \mu_1(h(x) + | \overline{u} |^2 + | \overline{\xi} |^2),$$

$$| \overline{f}(x, \sigma) | \leq h(x)$$

$$| \nabla_{\overline{u}} F(x, \sigma, \overline{u}, \overline{\xi}) | + | \nabla_{\overline{\xi}} F(x, \sigma, \overline{u}, \overline{\xi}) \leq \mu_1(h(x) + | \overline{u} | + | \overline{\xi} |)$$

4. The second derivatives of the function F with respect to $(\overline{u}, \overline{\xi})$ are bounded.

5. The functions F and $\overline{f}$ are affine with respect to σ.

Remark 1 *The last assumption is only for the sake of simplicity. If the functions F and $\overline{f}$ are affine then the convexification can be described in terms of closed convex hulls of the sets S and M. In general, one has to pass to the convex hull of all admissible pairs of operators and functionals.*

Consider the optimal control problem

$$I(\sigma,\overline{u}) := \int_{\Omega} F(x,\sigma,\overline{u},\nabla\overline{u})dx \to \min, \tag{1}$$

$$(\sigma,A) \in S \times M,\ \overline{u} \in H, \tag{2}$$

$$\operatorname{div} A(x)\nabla\overline{u} = \operatorname{div}\overline{f}(x,\sigma),\ x \in \Omega. \tag{3}$$

We introduce the following notations.
The Lagrangian associated with the problem (1)–(3) is

$$\begin{aligned} &\mathcal{L} = \mathcal{L}(\sigma,A,\overline{u},\overline{\psi}) := \int_{\Omega} F(x,\sigma,\overline{u},\nabla\overline{u})dx - \int_{\Omega}\langle A(x)\nabla\overline{u} - \overline{f}(x,\sigma),\nabla\overline{\psi}\rangle dx,\\ &(\sigma,A,\overline{u},\overline{\psi}) \in S \times M \times H \times H, \end{aligned} \tag{4}$$

where by $\langle .,. \rangle$ we denote the scalar product in Euclidean spaces.

The set of feasible states $Z(S \times M,\overline{g})$ is defined for arbitrary fixed element $\overline{g} \in L$ and

$$Z(S\times M,\overline{g}) := \{\overline{u} \in H \mid \operatorname{div} A\nabla\overline{u} - \operatorname{div}\overline{f}(.,\sigma) - \operatorname{div}\overline{g} = 0 \text{ in } \Omega, (\sigma,A) \in S\times M\}. \tag{5}$$

We will say that a triple $(\sigma_0,\overline{u}_0,\overline{\psi}_0)$ satisfies the Lagrange principle if

(i) $\overline{u}_0$ is the solution of the equation (3) with $(\sigma,A) = (\sigma_0,A_0)$ and $(\sigma_0,A_0) \in S\times M$;

(ii) $\overline{\psi}_0$ is the solution of the adjoint equation

$$\begin{aligned} &\operatorname{div} A_0\nabla\overline{\psi}_0 = \operatorname{div} F'_{\overline{\xi}}(x,\sigma_0,\overline{u}_0,\nabla\overline{u}_0) - F'_{\overline{u}}(x,\sigma_0,\overline{u}_0,\nabla\overline{u}_0) \text{ in } \Omega,\\ &\overline{\psi}_0 \in H; \end{aligned} \tag{6}$$

(iii) $\mathcal{L}(\sigma_0,A_0,\overline{u}_0,\overline{\psi}_0) \le \mathcal{L}(\sigma,A,\overline{u}_0,\overline{\psi}_0)\ \forall(\sigma,A) \in S \times M$.

By $\overline{u}(\sigma,A)$ we will denote the solution of the equation (3) corresponding to chosen controls $(\sigma,A) \in S \times M$.

3. Difference between cases $m \le n-1$ and $m \ge n$

For the case $m \le n-1$ the problem (1)-(3) has the following properties, see for instance Raitums [1], [2] for the case $m = 1$.

Proposition 1 *Let $m \le n-1$ and let the hypotheses 1–5 hold. Then for every fixed $\overline{g} \in L$ the closure of the set $Z(S \times M,\overline{g})$ in the strong topology of H coincides with the set $Z(\overline{co}(S \times M),\overline{g})$.*

Here the set $Z(\overline{co}(S \times M),\overline{g})$ is defined in the same way as $Z(S \times M,\overline{g})$, only with the set $\overline{co}(S \times M)$ instead of the set $S \times M$.

Proposition 2 *Let $m \le n-1$ and let the hypotheses 1–5 hold. Then the infimum of the cost functional in the original problem (1)–(3) is the same as in the convexificated problem where the sets S and M are replaced by $\overline{co}S$ and $\overline{co}M$ respectively.*

Proposition 3 *Let $m \leq n-1$ and let the hypotheses 1–5 hold. If the triple $(\sigma_0, A_0, \overline{u}_0)$ is a solution of the problem (1)–(3) then the Lagrange principle holds, i.e.*

$$\mathcal{L}(\sigma_0, A_0, \overline{u}_0, \overline{\psi}_0) \leq \mathcal{L}(\sigma, A, \overline{u}_0, \overline{\psi}_0) \;\; \forall (\sigma, A) \in S \times M,$$

where $\overline{\psi}_0$ is the solution of the adjoint equation (6).

The proof of these properties for $1 < m \leq n-1$ is practically the same as for $m = 1$.

On the other hand, counterexamples, see for instance Raitums [2], show that none of these properties remains true for the cases with $m \geq n$.

4. Closure of the set of feasible states for the case $m = n$

Consider a special case of the set M. Let $A_1, A_2 \in \mathcal{R}^{(n\times n)(n\times n)}$ be given symmetric positive definite matrices and let

$$S_1 := \{\theta \in L_2(\Omega) \mid \theta(x) = 0 \text{ or } 1, \; x \in \Omega\},$$

$$M_1 := \{A \mid A(x) = \theta(x)A_1 + (1-\theta(x))A_2, \; x \in \Omega\}.$$

With the set M_1 we associate the corresponding family of equations

$$\begin{aligned} &\operatorname{div}(A_2 + \theta(x)(A_1 - A_2))\nabla\overline{u} = \operatorname{div}\overline{g}, \; x \in \Omega, \\ &\overline{u} \in H. \end{aligned} \tag{7}$$

It can be shown that for every fixed $\overline{g} \in L$ the strong closure in H of the set $Z(M_1, \overline{g})$ of all solutions of (7) with $A \in M_1$ is equal to the set of all solutions $\overline{u} \in H$ of the following problem:

Find a pair $(\theta, \overline{u}) \in \overline{co}S_1 \times H$ such that

(i) $\theta(x) = 0$ or 1 at $x \in \Omega$ if $\operatorname{rank}(A_1 - A_2)\nabla\overline{u}(x) = n$;

(ii) $0 \leq \theta(x) \leq 1$ at $x \in \Omega$ if $\operatorname{rank}(A_1 - A_2)\nabla\overline{u}(x) \leq n-1$;

(iii) $\operatorname{div}(A_2 + \theta(x)(A_1 - A_2))\nabla\overline{u} = \operatorname{div}\overline{g}$ in Ω.

If matrices A_1 and A_2 have the following specific property that

$$\operatorname{rank}(A_1 - A_2)\overline{\xi} \leq n-1$$

implies

$$(A_1 - A_2)\overline{\xi} = 0$$

then the set $Z(M_1, g)$ of solutions of (7) with $A \in M_1$ is closed.

As an example, for which this property holds, we can mention the case where the matrices A_1 and A_2 constitute the Hook's law for two elastic materials with one and the same shear moduli μ.

These results show that for the case of elliptic systems with $m = n$ (or $m \geq n$) the structure of the set of feasible states $Z(M, \overline{g})$ (or as in our example $Z(M_1, \overline{g})$) can be very similar to the structure of the set of characteristic functions (the set S_1). There exist examples which show that, in general, the strong closure of the set $Z(M, \overline{g})$ does not contain approximative families of the kind

$$\overline{u}_\varepsilon = \overline{u}_0 + \varepsilon \overline{v} + o(\varepsilon)$$

with a nonzero $\overline{v} \in H$. This means that for the case of elliptic systems one can not expect, in general, to obtain sensitivity analysis by using cone-like approximations in the space of states.

5. On necessary optimality conditions

In this section we assume that $m = n$ and consider the optimal control problem

$$I(\sigma, \overline{u}) := I_0(\sigma) + \int_\Omega \langle \overline{g}(x, \sigma) \nabla \overline{u} \rangle dx \to \min, \tag{8}$$

$$\sigma \in S,\ A \in M,\ \overline{u} \in H, \tag{9}$$

$$\operatorname{div} A(x) \nabla \overline{u} = \overline{f}(x, \sigma) \text{ in } \Omega, \tag{10}$$

where the sets S and M are defined in the section 2 and both functions $\overline{g}$ and $\overline{f}$ satisfy the assumption 3 from the section 2.
Let us introduce the following subspaces of H:

$$G := \{\overline{\varphi} \in L \mid \overline{\varphi} = \nabla \overline{u} \in H\},$$

$$N := L \ominus G.$$

Since the functional I is affine with respect to $\nabla \overline{u}$ then analogously as in Raitums [3] it can be shown that

$$\begin{aligned} I(\sigma, \overline{u}(\sigma, A)) = \min_{\overline{v} \in G} \min_{\overline{\eta} \in N} \{ & I_0(\sigma) \\ & + \int_\Omega [\langle A(x)\overline{v}, \overline{v} \rangle - \langle \overline{f}(x, \sigma) - \overline{g}(x, \sigma), \overline{v} \rangle] dx \\ & + \tfrac{1}{4} \int_\Omega \langle A^{-1}(x)(\overline{\eta} + \overline{f}(x, \sigma) + \overline{g}(x, \sigma), \overline{\eta} + \overline{f}(x, \sigma) + \overline{g}(x, \sigma) \rangle dx \}, \end{aligned} \tag{11}$$

where by $\overline{u}(\sigma, A)$ we denote the solution of the equation (10) corresponding to a chosen pair $(\sigma, A) \in S \times M$.

Let us fix a pair $(\sigma_0, A_0) \in S \times M$. Let $u_0 = \overline{u}(\sigma_0, A_0)$ and let $\overline{\psi}_0 \in H$ is the solution of the adjoint equation

$$\operatorname{div} A_0(x) \nabla \overline{\psi}_0 = \operatorname{div} \overline{g}(x, \sigma_0) \text{ in } \Omega.$$

If the triple $(\sigma_0, A_0, \overline{u}_0)$ is a solution of the problem (8)–(10) then from (11) it follows that for the functional

$$\begin{aligned}
&J(\sigma, A) := \tfrac{1}{4}\int_\Omega \langle A(x)(\nabla\overline{u}_0 - \nabla\overline{\psi}_0), \nabla\overline{u}_0 - \nabla\overline{\psi}_0\rangle dx \\
&-\tfrac{1}{2}\int_\Omega [\langle \nabla\overline{u}_0 - \nabla\overline{\psi}_0, \overline{f}(x,\sigma) - \overline{g}(x,\sigma)\rangle dx \\
&+\tfrac{1}{4}\int_\Omega \langle A^{-1}(x)[A_0(x)(\nabla\overline{u}_0 + \nabla\overline{\psi}_0) + \overline{g}(x,\sigma) + \overline{f}(x,\sigma) - \overline{g}(x,\sigma_0) - \overline{f}(x,\sigma_0)], \\
&A_0(x)(\nabla\overline{u}_0 + \nabla\psi_0) + \overline{g}(x,\sigma) + \overline{f}(x,\sigma) - \overline{g}(x,\sigma_0) - \overline{f}(x,\sigma_0)\rangle dx
\end{aligned} \tag{12}$$

there is

$$J(\sigma, A) \le J(\sigma_0, A_0)\ \forall(\sigma, A) \in S \times M. \tag{13}$$

The relationship (13) plays the role of a necessary optimality condition. The connection of (13) with the classic Lagrange principle is more easy to see if there is only one control-matrices $A \in M$ or parameters $\sigma \in S$.

If the set M consists of only one matrix A_0 then the relationship (13) gives

$$\begin{aligned}
&\int_\Omega [\langle \nabla\overline{\psi}_0, \overline{f}(x,\sigma) - \overline{f}(x,\sigma_0)\rangle + \langle \nabla\overline{u}_0, \overline{g}(x,\sigma) - \overline{g}(x,\sigma_0)\rangle] dx \\
&+\tfrac{1}{4}\int_\Omega \langle A_0^{-1}(x)(\overline{g}(x,\sigma) - \overline{g}(x,\sigma_0) + \overline{f}(x,\sigma) - \overline{f}(x,\sigma_0)), \\
&\overline{g}(x,\sigma) - \overline{g}(x,\sigma_0) + \overline{f}(x,\sigma) - \overline{f}(x,\sigma_0)\rangle dx \ge 0\ \forall\sigma \in S.
\end{aligned} \tag{14}$$

If, in turn, the set S consists of only one element σ_0, then (13) gives

$$\begin{aligned}
&-\int_\Omega \langle (A(x) - A_0(x))\nabla\overline{u}_0, \nabla\overline{\psi}_0\rangle dx \\
&+\tfrac{1}{4}\int_\Omega \langle [A_0(x)(A^{-1}(x) - A_0^{-1}(x)) + A(x) - A_0(x)](\nabla\overline{u}_0 + \nabla\overline{\psi}_0), \\
&\nabla\overline{u}_0 + \nabla\overline{\psi}_0\rangle dx \ge 0\ \forall A \in M.
\end{aligned} \tag{15}$$

It is easy to see that the first integrals in (14) and (15) coincides with analogical terms in the Lagrange principle. On the other hand, the second integrals in (14) and (15) are new and are "quadratic" with respect to increments of controls $\sigma - \sigma_0$ and $A - A_0$. We suggest that they are principal for the cases with $m \ge n$. As a simple analysis of this hypothesis consider the following problem:

$$\begin{gathered}
I(\sigma, \overline{u}) := \int_\Omega \sigma \sum_{i=1}^n b_i u_{ix_i} dx \to \min, \\
\sigma \in S_2 := \{\sigma \in L_2(\Omega) \mid \sigma(x) = \alpha \text{ or } \beta,\ x \in \Omega\}, \\
\overline{u} = (u_1, \dots, u_n) \in H, \\
\gamma_i \Delta u_i = \tfrac{\partial}{\partial x_i}\sigma a_i \text{ in } \Omega,\ i = 1, \dots, n,
\end{gathered} \tag{16}$$

where $\alpha, \beta, \gamma_1, \dots, \gamma_n$ are fixed positive constants and $\overline{a} = (a_1, \dots, a_n)$ and $\overline{b} = (b_1, \dots, b_n)$ are fixed functions from $[L_2(\Omega)]^n$.

For standard variations of controls

$$\sigma_\varepsilon(x) = \begin{cases} \sigma_1(x), & x \in E_\varepsilon, \\ \sigma_0(x), & x \in \Omega \setminus E_\varepsilon, \end{cases}$$

where $\sigma_0, \sigma_1 \in S_2$ and E_ε are selfsimilar ellipsoids with volume ε the increment of the cost functional I is

$$\begin{aligned} I(\sigma_\varepsilon, \overline{u}(\sigma_\varepsilon)) - I(\sigma_0, \overline{u}) &= \int_{E_\varepsilon} (\sigma_\varepsilon - \sigma_0)[\sum_{i=1}^{n} (b_i u_{ix_i} + a_i \psi_{ix_i})] dx \\ &+ \int_{E_\varepsilon} (\sigma_\varepsilon - \sigma_0)^2 \sum_{i=1}^{n} \frac{1}{\gamma_i} n_i a_i b_i dx + o(\varepsilon), \end{aligned} \tag{17}$$

where $\overline{u}$ and $\overline{\psi}$ are the solutions of the state equation and the adjoint equation with $\sigma = \sigma_0$ respectively. The constants n_i depend only on the shape of ellipsoids E_ε and

$$n_1 + \cdots + n_n = 1, \; n_i > 0, i = 1, \ldots, n.$$

In turn, the relationship (14) gives

$$\begin{aligned} &\int_{E_\varepsilon} (\sigma_\varepsilon - \sigma_0) \sum_{i=1}^{n} (b_i u_{x_i} + a_i \psi_{ix_i}) dx \\ &+ \frac{1}{4} \int_{E_\varepsilon} (\sigma_\varepsilon - \sigma_0)^2 \sum_{i=1}^{n} \frac{1}{\gamma_i} (a_i + b_i)^2 dx \geq 0. \end{aligned} \tag{18}$$

The first integrals in (17) and (18) coincide and the second integrals converge to the same limit if, for instance,

$$a_1 = b_1 \neq 0, \; a_2 = b_2 = \cdots = a_n = b_n = 0,$$

and ellipsoids E_ε collapse in the direction of the axis Ox_1 (then $n_1 \to 1$, $n_i \to 0$, $i = 2, \ldots, n$).

This example shows that the "quadratic" terms in (14) or (15) are principal for sensitivity analysis or for the type of necessary optimality conditions for optimal control problems governed by elliptic systems with $m \geq n$.

Remark 2 *If the functions*

$$a_i(x) \neq 0 \; a.e. \; x \in \Omega, \; i = 1, \ldots, n,$$

then the set of feasible states in the problem (16) is closed in the strong topology of H.

References

[1] Raitums, U.: *Optimal control problems for elliptic equations*, Zinatne, Riga (1989) (in Russian).

[2] Raitums, U.: *The maximum principle and the convexification of optimal control problems*, Control and Cybernetics, vol. 23, No 4, p. 745–760 (1994).

[3] Raitums, U.: *On the minimization of quadratic functionals on the set of solutions of a family of linear equations*, Optimization, Bd. 17, No 3, p. 349–354 (1986).

International Series of Numerical Mathematics
Vol. 124, © 1998 Birkhäuser Verlag, Basel

Existence Results for Some Nonconvex Optimization Problems Governed by Nonlinear Processes

Tomáš Roubíček*

Abstract. Optimal control problems with nonlinear equations usually do not possess optimal solutions. Nevertheless, if the cost functional is uniformly concave with respect to the state, the solution may exist. Using the Balder's technique based on a Young-measure relaxation, Bauer's extremal principle and investigation of extreme Young measures, the existence is demonstrated here for the case of nonlinear ordinary and partial differential equations.

1. Introduction

The optimal control problems usually fail to have solutions unless the cost functional to be minimized is convex and the controlled system is linear with respect to the controls. This failure is typically due to oscillations effects: minimizing sequences tend to oscillate faster and faster which eventually prevent them to be convergent in a norm topology so that the limit passage through the involved Nemytskiĭ mappings is impossible. A typical example of this sort is the following optimal control problem:

$$\left.\begin{array}{lll} \text{Minimize} & \displaystyle\int_0^T (u(t)^2-1)^2+y(t)^2\,\mathrm{d}t & \text{(cost functional)} \\ \text{subject to} & \mathrm{d}y/\mathrm{d}t=u(t) \ \text{ for a.a. } t\in(0,T),\ y(0)=0, & \text{(state equation)} \\ & -1\le u(t)\le 1 \quad \text{for a.a. } t\in(0,T). & \text{(control constraints)} \end{array}\right\} \quad (1.1)$$

Minimizing sequences of controls inevitably oscillate faster and faster around -1 and 1, converging weakly* (but not strongly) to $u=0$. Yet, $u=0$ is not an optimal control.

Sometimes, even problems with nonconvex cost functionals or nonlinear state equations may have a solution. E.g., if we change the sign in the term y^2 in (1.1) so that we deal with the cost functional

$$\int_0^T (u(t)^2-1)^2-y(t)^2\mathrm{d}t,$$

the problem (1.1) does have a solution! Using constructive methods, this phenomenon for cost functionals and systems linear with respect to the state y has already been observed by Gabasov and Kirillova ([8]; Section 5.3), Macki and Strauss ([10]; Section 4.59, Neustadt [12] or Olech [13]. For more complicated problems (typically convex cost functionals and linear systems or, in a multidimensional case, the variational

*Mathematical Institute, Charles University, Institute of Information Theory and Automation, Academy of Sciences, email: roubicek@karlin.mff.cuni.cz

constraint $\nabla y = u$) see also Cellina and Colombo [5], Cesari [6] and ([7]; Chapter 16), Mariconda [11] or Raymond [14]–[16].

Recently, Balder [1] proposed another scheme based on the following three steps: first to relax the original problem to ensure an existence of so-called optimal relaxed controls, then to use a Bauer extremal principle [2] to show that at least one optimal relaxed control is an extreme point of the set of all admissible relaxed controls, and then to show that such extreme points are essentially the original controls, which eventually yields an optimal control for the original problem. However, Bauer's principle requires the relaxed problem to be concave. Therefore, the existence investigations for the original problem are reduced basically to two questions:

- Which data qualification guarantees the concave structure of the relaxed problem?
- When is every extreme point of the set of admissible relaxed controls the original control?

In Section 2 of this contribution, we want to illustrate the basic situation on the simplest optimal control problem for the ordinary differential equations with additively separated cost functional as well as the state equation as in (1.1) with $S(t)$ uniformly bounded. Considering problems having a uniformly concave cost functional but a "slightly" nonlinear state equation, both above questions can be affimatively answered; cf. respectively Lemmas 1-2 and 3. The methods used for the first question are intimately related with a sufficiency of the Pontryagin maximum principle, which requires basically a convexity of the relaxed problem (at least "at the optimal point"); see Gabasov and Kirillova ([8]; Section VII.2 or, for the case of general integral processes, also Schmidt [19]. By this technique one can show the concave structure of the relaxed problem even if the controlled system is "slightly" nonlinear with respect to the state on the assumption that the cost functional is "enough" uniformly concave with respect to the state. As to the second question, we will use (and modify) the results by Berliocchi and Lasry [3] and Castaing and Valadier [4].

Then, in Sections 3 and 4, we want to expose briefly the generalizations to problems with unbounded controls and governed by partial differential equations.

2. Main results

Avoiding problems with state-space constraints as well as problems with non-additively coupled states and controls (which would cause considerable complications), we consider the optimal control problem for a (system of) ordinary differential equations in the form:

$$
(\mathsf{P})\quad \left\{\begin{array}{lll}
\text{Minimize} & \displaystyle\int_0^T g(t,y(t)) + h(t,u(t))\,\mathrm{d}t & \text{(cost functional)}\\
\text{subject to} & \displaystyle\frac{\mathrm{d}y}{\mathrm{d}t} = G(t,y(t)) + H(t,u(t)) \quad \text{for a.a. } t\in(0,T) & \text{(state equation)}\\
& y(0) = 0, & \text{(initial condition)}\\
& u(t) \in S(t) \qquad \text{for a.a. } t\in(0,T), & \text{(control constraints)}\\
& y \in W^{1,q}(0,T;\mathbb{R}^n), \quad u \in L^\infty(0,T;\mathbb{R}^m)\ ,
\end{array}\right.
$$

where $g : (0,T) \times \mathbb{R}^n \to \mathbb{R}$, $G : (0,T) \times \mathbb{R}^n \to \mathbb{R}^n$, $h : (0,T) \times \mathbb{R}^m \to \mathbb{R}$, $H : (0,T) \times \mathbb{R}^m \to \mathbb{R}^n$ are Carathéodory functions satisfying the growth conditions

$$\exists a \in L^q(0,T)\ \exists b \in \mathbb{R}: \quad |G(t,r)| \le a(t) + b|r|, \qquad |H(t,s)| \le a(t), \tag{2.1a}$$
$$\exists a \in L^1(0,T): \quad |g(t,r)| \le a(t), \quad |h(t,s)| \le a(t) \tag{2.1b}$$

with some $q \in (1,+\infty)$. Moreover, $G(t,\cdot)$ is Lipschitz continuous in the sense

$$\exists a \in L^1(0,T): \quad |G(t,r_1) - G(t,r_2)| \le a(t)|r_1 - r_2|. \tag{2.1c}$$

As to the multivalued mapping $S : (0,T) \rightrightarrows \mathbb{R}^m$, we suppose it in this section as bounded, measurable, and in the form $S(t) = M(t, S_0)$ for some $S_0 \subset \mathbb{R}^m$ compact and $M : (0,T) \times \mathbb{R}^m \to \mathbb{R}^m$ a Carathéodory mapping such that both $M(t,\cdot)$ and $M(t,\cdot)^{-1}$ are Lipschitz continuous uniformly with respect to $t \in (0,T)$. Note that (P) covers also (1.1) for S qualified appropriately.

Following ideas by Young [21], we extend the set of admissible controls $U_{\rm ad} := \{u \in L^\infty(0,T;\mathbb{R}^m);\ u(t) \in S(t) \text{ for a.a. } t \in (0,T)\}$ to the set of admissible relaxed controls $\bar U_{\rm ad} := \{\nu \in \mathcal{Y}(0,T;\mathbb{R}^m);\ \mathrm{supp}(\nu_t) \subset S(t) \text{ for a.a. } t \in (0,T)\}$, where $\mathcal{Y}(0,T;\mathbb{R}^m) := \{\nu : t \mapsto \nu_t : (0,T) \to \mathrm{rca}(\mathbb{R}^m) \text{ weakly measurable; } \nu_t \text{ is a probability measure for a.a. } t \in (0,T)\}$ denotes the set of the so-called Young measures and $\mathrm{rca}(\mathbb{R}^m) \cong C_0(\mathbb{R}^m)^*$ stands for Radon measures on $\mathbb{R}^m$. It is known that $U_{\rm ad}$ is weakly* dense in $\bar U_{\rm ad}$ if embedded via the mapping $i : u \mapsto \nu$ with $\nu_t = \delta_{u(t)}$ where $\delta_s \in \mathrm{rca}(\mathbb{R}^m)$ denotes the Dirac measure supported at $s \in \mathbb{R}^m$; cf. [17] or, for S constant, also Cesari [7] or Warga [20]. The relaxed problem is then created by the continuous extension of the original problem (P) from $U_{\rm ad}$ to $\bar U_{\rm ad}$, which gives:

$$(\mathsf{RP}) \quad \begin{cases} \text{Minimize} & \displaystyle\int_0^T \Big(g(t,y(t)) + \int_{\mathbb{R}^m} h(t,s)\nu_t(\mathrm{d}s) \Big)\, dt \\ \text{subject to} & \displaystyle\frac{\mathrm{d}y}{\mathrm{d}t} = G(t,y(t)) + \int_{\mathbb{R}^m} H(t,s)\nu_t(\mathrm{d}s) \quad \text{for a.a. } t \in (0,T), \\ & y(0) = 0, \\ & \mathrm{supp}(\nu_t) \subset S(t) \qquad \text{for a.a. } t \in (0,T), \\ & y \in W^{1,q}(0,T;\mathbb{R}^n), \quad \nu \in \mathcal{Y}(0,T;\mathbb{R}^m)\,. \end{cases}$$

It is known (see e.g. Warga [20] or also [17]) that (RP) is actually a correct relaxation of (P) in the sense that (RP) always posseses a solution and $\min(\mathsf{RP}) = \inf(\mathsf{P})$. Moreover, if ν solves (RP) and $\nu = i(u)$ for some $u \in U_{\rm ad}$, then u solves (P).

Obviously, (RP) just represents minimization over $\bar U_{\rm ad}$ of the extended cost functional Φ defined by $\Phi(\nu) := \int_0^T (g(t,[y(\nu)](t)) + \int_{\mathbb{R}^m} h(t,s)\nu_t(\mathrm{d}s))\, dt$ with $y = y(\nu) \in W^{1,q}(0,T;\mathbb{R}^n)$ being the unique solution to the initial-value problem in (RP). To investigate the geometrical properties of Φ, we will calculate its Gâteaux differential with respect to the geometry coming from $L^1(0,T;C_0(\mathbb{R}^m))^* \supset \bar U_{\rm ad}$. This is, in fact, a standard task undertaken within derivation of the Pontryagin maximum principle for the relaxed controls. This principle usually needs Fréchet differentiability with respect to y which can be guaranteed by the following assumptions on the partial derivative of

g and G with respect to the variable r, denoted respectively by $g'(t,r)$ and $G'(t,r)$:

$$\exists a \in L^1(0,T) \quad \exists b : \mathbb{R} \to \mathbb{R} \text{ continuous}: \quad |g'(t,r)| \le a(t) + b(|r|), \tag{2.2a}$$
$$|g'(t,r_1) - g'(t,r_2)| \le (a(t) + b(|r_1|) + b(|r_2|))|r_1 - r_2|,$$
$$\exists a \in L^{1+\varepsilon}(0,T) \quad \exists b : \mathbb{R} \to \mathbb{R} \text{ continuous}: \quad |G'(t,r)| \le a(t) + b(|r|), \tag{2.2b}$$
$$|G'(t,r_1) - G'(t,r_2)| \le (a(t) + b(|r_1|) + b(|r_2|))|r_1 - r_2|,$$

The maximum principle involves the so-called adjoint equation

$$\frac{\mathrm{d}\lambda}{\mathrm{d}t} = -\lambda(t)G'(t,y(t)) - g'(t,y(t)), \quad \lambda(T) = 0. \tag{2.3}$$

The assumption (2.2) ensures that the terminal-value problem (2.3) possesses precisely one solution $\lambda \in W^{1,1}(0,T;\mathbb{R}^n)$. Likewise in a procedure by Gabasov and Kirillova ([8]; Section VII.2) or (for the general integral processes) by Schmidt [19] developed to prove sufficiency of the maximum principle for optimal control problems, we can establish the following incrementation formula.

Lemma 1 *Let (2.1) and (2.2) be satisfied, let $\nu, \tilde{\nu} \in \bar{U}_{\rm ad}$, $y, \tilde{y} \in W^{1,q}(0,T;\mathbb{R}^n)$ solve the initial-value problem in* (RP) *with ν and $\tilde{\nu}$ respectively, and let $\lambda \in W^{1,1}(0,T;\mathbb{R}^n)$ solve (2.3). Then*

$$\Phi(\tilde{\nu}) - \Phi(\nu) = \int_0^T \int_{\mathbb{R}^m} [\lambda(t)H(t,s) + h(t,s)]\,[\tilde{\nu}_t - \nu_t](\mathrm{d}s)\mathrm{d}t + \int_0^T \Delta_g(t) + \lambda(t)\Delta_G(t)\mathrm{d}t \tag{2.4}$$

with the second-order correcting terms $\Delta_g(t)$ and $\Delta_G(t)$ defined by

$$\Delta_g(t) := g(t,\tilde{y}(t)) - g(t,y(t)) - g'(t,y(t))(\tilde{y}(t) - y(t)), \tag{2.5a}$$
$$\Delta_G(t) := G(t,\tilde{y}(t)) - G(t,y(t)) - G'(t,y(t))(\tilde{y}(t) - y(t)). \tag{2.5b}$$

Proof: Using the extended state equation both for y and $\tilde{y}$, the per-partes integration and the adjoint equation (2.3), we can calculate:

$$\begin{aligned}
&\Phi(\tilde{\nu}) - \Phi(\nu) - \int_0^T \int_{\mathbb{R}^m} [\lambda(t)H(t,s) + h(t,s)]\,[\tilde{\nu}_t - \nu_t](\mathrm{d}s)\mathrm{d}t \\
&= \int_0^T \Big(g(t,\tilde{y}(t)) - g(t,y(t)) - \int_{\mathbb{R}^m} \lambda(t)H(t,s)[\tilde{\nu}_t - \nu_t](\mathrm{d}s) \Big)\,\mathrm{d}t \\
&= \int_0^T g(t,\tilde{y}(t)) - g(t,y(t)) + \lambda(t)\left(G(t,\tilde{y}(t)) - G(t,y(t)) - \frac{\mathrm{d}(\tilde{y}(t) - y(t))}{\mathrm{d}t} \right)\mathrm{d}t \\
&= \int_0^T g(t,\tilde{y}(t)) - g(t,y(t)) + \lambda(t)(G(t,\tilde{y}(t)) - G(t,y(t))) + \frac{\mathrm{d}\lambda}{\mathrm{d}t}(\tilde{y}(t) - y(t))\mathrm{d}t \\
&= \int_0^T \Big(g(t,\tilde{y}(t)) - g(t,y(t)) - g'(t,y(t))(\tilde{y}(t) - y(t)) \\
&\qquad\qquad + \lambda(t)(G(t,\tilde{y}(t)) - G(t,y(t)) - G'(t,y(t))(\tilde{y}(t) - y(t))) \Big)\,\mathrm{d}t \\
&=: \int_0^T \Delta_g(t) + \lambda(t)\Delta_G(t)\mathrm{d}t.
\end{aligned}$$

■

The formula (2.4) enables us to investigate concavity of the extended cost functional Φ. Let us take a sufficiently large radius R so that $|[y(u)](t)| \leq R$ for any $u \in U_{\rm ad}$ and any $t \in (0,T)$, where $y(u) \in W^{1,q}(0,T;\mathbb{R}^n)$ denotes the unique solution to the initial-value problem in (P). Concretely, let us take

$$R = \sup_{u\in U_{\rm ad}} \sup_{t\in(0,T)} \Big|[y(u)](t)\Big| \,. \tag{2.6}$$

Furthermore, let $a(t) := \sup_{|r|\leq R} |g'(t,r)|$ and $A(t) := \sup_{|r|\leq R} |G'(t,r)|$. As (2.2) ensures $a, A \in L^1(0,T)$, we can define

$$b(t) := \int_t^T a(\tau)\mathrm{d}\tau \,, \quad B(t) := \int_t^T A(\tau)\mathrm{d}\tau \,. \tag{2.7}$$

Lemma 2 *Let us assume (2.1), (2.2) and $G(t,\cdot)$ twice continuously differentiable, and let $g(t,\cdot)$ be uniformly concave on the ball of the radius R from (2.6) in the sense*

$$\begin{aligned} \forall r,\tilde r \in \mathbb{R}^n: \quad & \max(|r|,|\tilde r|) \leq R \implies \\ & g(t,\tilde r) - g(t,r) - g'(t,r)(\tilde r - r) \leq -\alpha(t)|\tilde r - r|^2 \end{aligned} \tag{2.8}$$

with the modulus $\alpha \geq 0$ satisfying, for b and B defined by (2.7), the condition

$$\alpha(t) \geq \frac{b(t)}{2}\mathrm{e}^{B(t)} \sup_{|r|\leq R} |G''(t,r)| \,. \tag{2.9}$$

Then Φ is concave on $\bar U_{\rm ad}$.

Proof: From the adjoint equation (2.3) we can estimate $\mathrm{d}|\lambda|/\mathrm{d}t \leq A(t)|\lambda(t)| + a(t)$ so that by the Gronwall inequality one gets

$$|\lambda(t)| \;\leq\; \left(\int_t^T a(\tau)\mathrm{e}^{-\int_t^T A(\theta)\mathrm{d}\theta}\mathrm{d}\tau\right)\mathrm{e}^{\int_t^T A(\tau)\mathrm{d}\tau} \,. \tag{2.10}$$

To simplify the notation, we can also (a bit more pessimistically) estimate

$$|\lambda(t)| \;\leq\; b(t)\mathrm{e}^{B(t)} \,. \tag{2.11}$$

By the Taylor expansion, we can estimate $|G(t,\tilde y(t)) - G(t,y(t)) - G'(t,y(t))(\tilde y(t) - y(t))| \leq \sup_{|r|\leq R} \frac{1}{2}|G''(t,r)|\,|\tilde y(t) - y(t)|^2$. Then (2.8) with (2.9) and (2.11) ensure the following inequality

$$\begin{aligned} \Delta_g(t) + \lambda(t)\Delta_G(t) &\leq -\alpha(t)|\tilde y(t) - y(t)|^2 \\ &\qquad + \frac{1}{2}|\lambda(t)|\;|G''(t,y(t))|\;|\tilde y(t) - y(t)|^2 \\ &\leq \left(-\alpha(t) + \frac{b(t)}{2}\mathrm{e}^{B(t)} \sup_{|r|\leq R} \frac{1}{2}|G''(t,r)|\right)|\tilde y(t) - y(t)|^2 \leq 0. \end{aligned} \tag{2.12}$$

so that the second right-hand term in (2.4) is non-positive. Since the first right-hand term in (2.4) represents just the Gâteaux differential of Φ, i.e. $[\nabla\Phi(\nu)](\tilde\nu - \nu) = \int_0^T \int_{\mathbb{R}^m} [\lambda(t)H(t,s) + h(t,s)]\,[\tilde\nu_t - \nu_t](\mathrm{d}s)\mathrm{d}t$, we obtained

$$\forall \nu, \tilde\nu \in \bar U_{\rm ad}: \qquad \Phi(\tilde\nu) - \Phi(\nu) - [\nabla\Phi(\nu)](\tilde\nu - \nu) \leq 0 \,, \tag{2.13}$$

which just says that Φ is concave on $\bar U_{\rm ad}$. ∎

Lemma 3 *If $\nu \in \bar{U}_{\rm ad}$ is an extreme point of $\bar{U}_{\rm ad}$ (i.e. $\nu = \frac{1}{2}\nu^1 + \frac{1}{2}\nu^2$ for some $\nu^1, \nu^2 \in \bar{U}_{\rm ad}$ implies $\nu_1 = \nu_2$), then $\nu = i(u)$ for some $u \in U_{\rm ad}$.*

Sketch of the proof. If ν_t were not a Dirac mass for a.a. $t \in (0,T)$, then ν_t would not be an extreme point in the set of probability measures on $S(t)$, and thus the multivalued mapping $C : (0,T) \rightrightarrows {\rm rca}(\mathbb{R}^m)^2$ defined by

$$C(t) := \Big\{ \ (\mu^1, \mu^2) \in {\rm rca}(\mathbb{R}^m)^2; \\ \mu^1, \mu^2 \geq 0, \ \mu^1(S(t)) = 1 = \mu^2(S(t)), \ \frac{1}{2}\mu^1 + \frac{1}{2}\mu^2 = \nu_t \ \Big\} \tag{2.14}$$

is not a singleton for a.a. $t \in (0,T)$. Since S is measurable, ν is weakly measurable, and $C_0(\mathbb{R}^m)$ is separable, the multivalued mapping C is measurable. Then this mapping admits a measurable selection $t \to (\nu_t^1, \nu_t^2)$ which is not equal to (ν_t, ν_t) for a.a. $t \in (0,T)$. This shows that $\nu^1 \neq \nu^2$ but $\nu = \frac{1}{2}\nu^1 + \frac{1}{2}\nu^2$ so that ν is not an extreme point in $\bar{U}_{\rm ad}$, a contradiction. ∎

Proposition 1 *Let (2.1), (2.2), and (2.8)–(2.9) be satisfied. Then* (P) *has an optimal solution.*

Proof: The relaxed problem (RP) consists in minimization of the weakly* continuous functional Φ on the convex weakly* compact set $\bar{U}_{\rm ad}$ so that it certainly has a solution by the standard compactness arguments. By Lemma 2, Φ is concave, so that by Bauer's extremal principle [2] at least one solution to (RP) is an extreme point of $\bar{U}_{\rm ad}$. By Lemma 3, this extreme optimal control for (RP) can be represented by the ordinary control from $U_{\rm ad}$ which is obviously optimal also for (P). ∎

Remark. If the controlled system is linear with respect to the state, i.e. $G(t,\cdot)$ is affine, then obviously $G'' \equiv 0$ and one can take $\alpha \equiv 0$ in (2.8) which then just requires $g(t,\cdot)$ to be concave.

3. A generalization: unbounded controls

If the multivalued mapping $S : (0,T) \rightrightarrows \mathbb{R}^m$ acting as control constraints in (P) is not bounded, several sophisticated approaches must be still incorporated. First, we must suppose certain (for simplicity polynomial) coercivity of the problem, say:

$$c_0|s|^p \leq g(t,r) + h(t,s) \leq a_0(t) + C_1|s|^p \ , \tag{3.1}$$

$$|G(t,r) + H(t,s)| \leq a_1(t) + b_1|r| + c_1|s|^{p-\varepsilon} \tag{3.2}$$

for some $a_0 \in L^1(0,T)$, $a_1 \in L^q(0,T)$, $c_0, C_0, b_1, c_1 \in \mathbb{R}$, $p \in [1,+\infty)$, $q \in (1,+\infty)$, and $\varepsilon > 0$. Then the modified formulation of the original problem (P) involves $L^p(0,T;\mathbb{R}^m)$ in place of $L^\infty(0,T;\mathbb{R}^m)$. As to the measurable multivalued mapping S, it is now natural to suppose it again in the form $S(t) := M(t, S_0)$ with some $S_0 \in \mathbb{R}^m$ closed and the Carathéodory mapping M satisfying $\max(|M(t,s)|, |M(t,s)^{-1}|) \leq a(t) + b|s|$ for some $a \in L^p(0,T)$ and $b \in \mathbb{R}$. Note that this growth condition makes the Nemytskiĭ mapping $\mathcal{N}_M$ generated by M a homeomorphism on $L^p(\Omega;\mathbb{R}^m)$ whose inverse transforms $U_{\rm ad}$ onto the set $\mathcal{N}_M^{-1}(U_{\rm ad}) = \{u \in L^p(0,T;\mathbb{R}^m);\ u(t) \in S_0$ for a.a. $t \in (0,T)\}$ which uses a fixed constraint S_0.

Then the correct relaxed problem looks like (RP) but with $\mathcal{Y}(0,T;\mathbb{R}^m)$ replaced by $\mathcal{Y}^p(0,T;\mathbb{R}^m)$ defined as

$$\mathcal{Y}^p(\Omega;\mathbb{R}^m) = \left\{\nu \in \mathcal{Y}(0,T;\mathbb{R}^m);\ \textstyle\int_0^T\int_{\mathbb{R}^m}|s|^p\nu_t(\mathrm{d}s)\mathrm{d}t < +\infty\right\}.$$

This set contains just those Young measures, called L^p-Young measures, that can be attained by sequences bounded in $L^p(0,T;\mathbb{R}^m)$; cf. [17]. The quite nontrivial fact that such (RP) has a solution and $\inf(\mathsf{P}) = \min(\mathsf{RP})$ relies on a nonconcentration of energy of any minimizing sequence $\{u_k\}_{k\in\mathbb{N}}$ for (P), i.e. the relative weak compactness of the set $\{|u_k|^p;\ k \in \mathbb{N}\}$ in $L^1(0,T)$; cf. [17].

By the assumed coercivity (3.1), all solutions to (RP) must belong to the set

$$\mathcal{Y}^p_{\varrho_0}(0,T;\mathbb{R}^m) = \left\{\nu \in \mathcal{Y}(0,T;\mathbb{R}^m);\ \left(\textstyle\int_0^T\int_{\mathbb{R}^m}|s|^p\nu_t(\mathrm{d}s)\mathrm{d}t\right)^{1/p} \le \varrho_0\right\}$$

for some $\varrho_0 \in \mathbb{R}$ sufficiently large.

Proposition 2 *Let (2.1c), (2.2), (3.1), (3.2), and (2.8)–(2.9) be satisfied for the radius*

$$R = \sup_{u\in U_{\rm ad},\ \|u\|_{L^p(0,T;\mathbb{R}^m)}\le 2^{1/p}\varrho_0} \|y(u)\|_{C(0,T;\mathbb{R}^n)}\,. \tag{3.3}$$

Then (P) *has an optimal solution.*

Proof: Let us consider a problem $(\mathsf{RP})_\varrho$ with $\mathcal{Y}^p_\varrho(0,T;\mathbb{R}^m)$ in place of $\mathcal{Y}^p(0,T;\mathbb{R}^m)$. By the coercivity (3.1), $(\mathsf{RP})_\varrho$ has the same set of solutions as (RP) provided $\varrho \ge \varrho_0$. By (2.8) valid for R satisfying (3.3), the problem $(\mathsf{RP})_{2^{1/p}\varrho_0}$ has the concave cost functional Φ so that, by Bauer's extremal principle, it must admit at least one optimal control ν which is an extreme point in the convex weakly* compact set $\mathcal{Y}^p_{2^{1/p}\varrho_0}(0,T;\mathbb{R}^m)$. By the coercivity, we also know that even $\nu \in \mathcal{Y}^p_{\varrho_0}(0,T;\mathbb{R}^m)$.

However, then $\nu = i(u)$ for some $U_{\rm ad}$. Indeed, in the opposite case the Young measures $\nu^1, \nu^2 \in \mathcal{Y}(0,T;\mathbb{R}^m)$ obtained in the proof of Lemma 3 would also satisfy the estimate

$$\int_0^T\int_{\mathbb{R}^m}|s|^p\nu^1_t(\mathrm{d}s)\mathrm{d}t = \int_0^T\int_{\mathbb{R}^m}|s|^p(2\nu_t - \nu_t^2)(\mathrm{d}s)\mathrm{d}t \le 2\int_0^T\int_{\mathbb{R}^m}|s|^p\nu_t(\mathrm{d}s)\mathrm{d}t \le 2\varrho_0^p$$

because $\frac{1}{2}\nu_t^1 + \frac{1}{2}\nu_t^2 = \nu_t$. In other words, $\nu^1 \in \mathcal{Y}^p_{2^{1/p}\varrho_0}(0,T;\mathbb{R}^m)$. Replacing the role of ν^1 and ν^2, we get $\nu^2 \in \mathcal{Y}^p_{2^{1/p}\varrho_0}(0,T;\mathbb{R}^m)$, as well. This would show that ν is not an extreme point in $\mathcal{Y}^p_{2^{1/p}\varrho_0}(0,T;\mathbb{R}^m)$, a contradiction.

Then obviously u is the sought optimal control for (P). ■

4. A generalization: distributed-parameter problems

The presented method readily extends for distributed-parameter controlled systems which we want to illustrate here briefly for systems governed by elliptic partial differential equations; for Fredholm integral equations see [18]. In fact, the only peculiarity is that both the state y and the adjoint state λ are required to be bounded in L^∞-norm, which may require certain additional regularity.

We will consider a bounded convex domain $\Omega \subset \mathbb{R}^n$, $n \leq 3$, with a Lipschitz boundary $\partial\Omega$ and the following optimal control problem for one elliptic equation with homogeneous Dirichlet boundary conditions:

$$
(\mathsf{P}') \quad \begin{cases}
\text{Minimize} & \int_\Omega g(x,y(x)) + h(x,u(x))\,\mathrm{d}x & \text{(cost functional)}\\
\text{subject to} & \operatorname{div}(\nabla y(x)) = G(x,y(x)) + H(x,u(x)) & \text{(state equation)}\\
& y|_{\partial\Omega} = 0, & \text{(boundary condition)}\\
& u(x) \in S(x) \quad \text{for } x \in \Omega, & \text{(control constraints)}\\
& y \in W^{1,2}(\Omega), \quad u \in L^p(\Omega;\mathbb{R}^m)\ ,
\end{cases}
$$

where $g, G : \Omega \times \mathbb{R} \to \mathbb{R}$, $h, H : \Omega \times \mathbb{R}^m \to \mathbb{R}$ are Carathéodory functions satisfying (3.1)–(3.2) and additionally

$$
G(x,\cdot) \text{ nondecreasing for a.a. } x \in \Omega \tag{4.1}
$$

$$
|G(x,r)| \leq a(x) + b|r|^{c/q}\ , \quad |H(x,s)| \leq a(x) + b|s|^{p/q} \tag{4.2}
$$

for some $a \in L^q(\Omega)$, $b \in \mathbb{R}$, and $q \geq 1$ and $c < +\infty$ such that $q > n/2$ and, if $n = 3$, also $c \leq 2n/(n-2)$. Note that the involved boundary-value problem has, for any control $u \in L^p(\Omega;\mathbb{R}^m)$, a unique weak solution $y = y(u) \in W_0^{1,2}(\Omega)$.

Besides, the multivalued mapping $S : \Omega \rightrightarrows \mathbb{R}^m$ is subjected to the same assumptions as in Section 2 with Ω in place of $(0,T)$.

Let us note that, for u ranging a bounded set in $L^p(\Omega;\mathbb{R}^m)$, $y(u)$ ranges a bounded set in $W^{1,2}(\Omega) \subset L^c(\Omega)$. By (4.2), $G(x,y(u)) + H(x,u)$ then ranges a bounded set in $L^q(\Omega) \subset W^{-1+\varepsilon,2}(\Omega)$ for $\varepsilon \in [0,1]$ such that $q > 2n/(n+2-2\varepsilon)$; the last embedding is just adjoint to $L^q(\Omega)^* \equiv L^{q/(q-1)}(\Omega) \supset W_0^{1-\varepsilon,2}(\Omega) \equiv W^{-1+\varepsilon,2}(\Omega)^*$. Moreover, the standard elliptic regularity (see Grisward [9]) shows that $y(u)$ is bounded $W^{2,2}(\Omega)$ provided $G(x,y(u)) + H(x,u)$ ranges a bounded set in $L^2(\Omega)$. By interpolation for the linear operator $f \mapsto y$ with y solving the problem $\operatorname{div}(\nabla u) = f$ and $y|_{\partial\Omega} = 0$, the solution $y(u)$ ranges a bounded set in $W^{1+\varepsilon,2}(\Omega)$ so that, using a well-known embedding theorem, $W^{1+\varepsilon,2}(\Omega) \subset C^0(\Omega)$ provided $\varepsilon > (n-2)/2$, and thus the solution $y(u)$ lives in $C^0(\Omega)$, which is essential for our theory. Note that $1 \geq \varepsilon > (n-2)/2$ and $q > 2n/(n+2-2\varepsilon)$ yield respectively the mentioned restrictions $n \leq 3$ and $q > n/2$.

Again, we make relaxation by a continuous extension of (P') from U_{ad} on $\bar{U}_{\mathrm{ad}}$ (defined as previously but with Ω in place of $(0,T)$):

$$
(\mathsf{RP}') \quad \begin{cases}
\text{Minimize} & \int_\Omega \Big(g(x,y(x)) + \int_{\mathbb{R}^m} h(x,s)\nu_x(\mathrm{d}s)\Big)\,\mathrm{d}x\\
\text{subject to} & \operatorname{div}(\nabla y(x)) = G(x,y(x)) + \int_{\mathbb{R}^m} H(x,s)\nu_x(\mathrm{d}s)\ ,\\
& \operatorname{supp}(\nu_x) \subset S(x) \qquad \text{for a.a. } x \in \Omega,\\
& y \in W_0^{1,2}(\Omega), \quad \nu \in \mathcal{Y}^p(\Omega;\mathbb{R}^m)\ .
\end{cases}
$$

Using Green's formula instead of per-partes integration, as in Lemma 1 one can derive the incrementation formula for $\Phi(\nu) := \int_\Omega (g(x,[y(\nu)](x)) + \int_{\mathbb{R}^m} h(x,s)\nu_x(\mathrm{d}s))\,\mathrm{d}x$ with

$y = y(\nu) \in W_0^{1,2}(\Omega)$ satisfying $\operatorname{div}(\nabla y(x)) = G(x, y(x)) + \int_{\mathbb{R}^m} H(x,s)\nu_x(\mathrm{d}s)$ in the weak sense, which now looks like

$$\Phi(\tilde{\nu}) - \Phi(\nu) = \int_\Omega \int_{\mathbb{R}^m} [\lambda(x)H(x,s) + h(x,s)]\,[\tilde{\nu}_x - \nu_x](\mathrm{d}s)\mathrm{d}x + \int_\Omega \Delta_g(x) + \lambda(x)\Delta_G(x)\mathrm{d}x$$

with the adjoint state $\lambda \in W_0^{1,2}(\Omega)$ satisfying (in the weak sense)

$$\operatorname{div}(\nabla\lambda(x)) = G'(x, y(x))\lambda + g'(x, y(x)). \tag{4.3}$$

We will assume, beside (2.2), the following growth conditions on G' and g':

$$|G'(x,r)| \le b(|r|) \qquad |g'(x,r)| \le a(x) + b(|r|) \tag{4.4}$$

with some $a \in L^q(\Omega)$ and $b : \mathbb{R} \to \mathbb{R}$ continuous. Note that, by similar regularity arguments as used for y, the adjoint state λ lives in $W^{1+\varepsilon,2}(\Omega) \subset C^0(\Omega)$, which allows us to establish the criterion (4.5) below.

Proposition 3 *Let (2.2), (2.8), (3.1), (3.2), (4.1)–(4.4) be fulfilled. Then* (P′) *has an optimal solution provided the uniform-concavity coefficient α of g satisfies*

$$\alpha(x) \ge \frac{1}{2} \sup_{\substack{u\in U_{\mathrm{ad}} \\ \|u\|_{L^p(\Omega;\mathbb{R}^m)} \le 2^{1/p}\varrho_0}} \|\lambda(u)\|_{C^0(\Omega)} \sup_{|r|\le R} |G''(x,r)| \tag{4.5}$$

with $R = \sup_{u\in U_{\mathrm{ad}},\, \|u\|_{L^p(\Omega;\mathbb{R}^m)}\le 2^{1/p}\varrho_0} \|y(u)\|_{C^0(\Omega)}$, where ϱ_0 is a sufficiently large radius of the ball in $L^p(\Omega;\mathbb{R}^m)$ where every minimizing sequence for (P′) *eventually lives.*

Acknowledgement

This research was partly covered by the grant No. 201/96/0228 of the Grant Agency of the Czech Republic. The author also appreciates his visit at Ernst-Moritz-Arndt-Universität in Greifswald and useful discussions with Professor Werner H. Schmidt.

References

[1] Balder, E. J.: *New existence results for optimal controls in the absence of convexity: the importance of extremality.* SIAM J. Control Optim. **32** (1994), 890–916.

[2] Bauer, H.: *Minimalstellen von Funktionen und Extremalpunkte.* Archiv d. Math. **9** (1958), 389–393, **11** (1960), 200–205.

[3] Berliocchi, H.; Lasry, J.-M.: *Intégrandes normales et mesures paramétrées en calcul des variations.* Bull. Soc. Math. France **101** (1973), 129–184.

[4] Castaing, C.; Valadier, M.: *Convex Analysis and Measurable Multifunctions.* Lecture Notes in Math. **580**, Springer, Berlin, 1977.

[5] Cellina, A.; Colombo, G.: *On a classical problem of the calculus of variations without convexity assumptions.* Annales Inst. H. Poincaré, Anal. Nonlin. **7** (1990), 97–106.

[6] Cesari, L.:*An existence theorem without convexity conditions.* SIAM J. Control **12** (1974), 319–331.

[7] Cesari, L.: *Optimization: Theory and Applications.* Springer, New York, 1983.

[8] Gabasov, R.; Kirillova, F.: *Qualitative Theory of Optimal Processes.* Nauka, Moscow, 1971.

[9] Grisvard, P.: *Elliptic problems in nonsmooth domains.* Pitman, Boston, 1985.

[10] Macki, J.; Strauss, A.: *Introduction to Optimal Control Theory.* Springer, New York, 1982.

[11] Mariconda, C.: *On a parametric problem of the calculus of variations without convexity assumptions.* J. Math. Anal. Appl. **170** (1992), 291–297.

[12] Neustadt, L. W.: *The existence of optimal controls in the absence of convexity conditions.* J. Math. Anal. Appl. **7** (1963), 110–117.

[13] Olech, C.: *Integrals of set-valued functions and linear optimal control problems.* Clloque sur la Théorie Math. du Contrôle Optimal, C.B.R.M., Vander Louvain, 1970, pp.109–125.

[14] Raymond, J.-P.: *Existence theorems in optimal control theory without convexity assumptions.* J. Optim. Theory Appl. **67** (1990), 109–132.

[15] Raymond, J.-P.: *Existence theorems without convexity assumptions for optimal control problems governed by parabolic and elliptic systems.* Appl. Math. Optim. **26** (1992), 39–62.

[16] Raymond, J.-P.: *Existence and uniqueness results for minimization problems with non-convex functionals.* J. Optim. Theory Appl. **82** (1994), 571–592.

[17] Roubíček, T.: *Relaxation in Optimization Theory and Variational Calculus.* W. de Gruyter, Berlin, 1997.

[18] Roubíček, T.; Schmidt, W. H.: *Existence of solutions to nonconvex optimal control problems governed by nonlinear Fredholm integral equations.* (submitted)

[19] Schmidt, W. H.: *Maximum principles for processes governed by integral equations in Banach spaces as sufficient optimality conditions.* Beiträge zur Analysis **17** (1981), 85–93.

[20] Warga, J.: *Optimal Control of Differential and Functional Equations.* Academic Press, New York, 1972.

[21] Young, L. C.: *Generalized curves and the existence of an attained absolute minimum in the calculus of variations.* Comptes Rendus de la Société des Sciences et des Lettres de Varsovie, Classe III **30** (1937), 212–234.

International Series of Numerical Mathematics
Vol. 124,

Multiobjective Optimal Control Problems

Christiane Tammer *

Abstract. From an Ekeland-type variational principle for vector optimization problems we derive an ϵ-minimum principle in the sense of Pontrjagin for suboptimal controls using classical results for differential equations.

1. Introduction

In our present paper we study suboptimal controls of a class of multiobjective optimal control problems.
In control theory often one has the problem to minimize more than one objective function, for instance a cost functional as well as the distance between the final state and a given point.
To realize this task usually one takes as objective function a weighted sum of the different objectives. However, the more natural way would be to study the set of efficient points of a vector optimization problem with the given objective functions. It is well known that the weighted sum is only a special surrogate problem to find efficient points, which has the disadvantage that in the nonconvex case one can not find all efficient elements in this way.
Necessary conditions for solutions of multiobjective dynamic programming or control problems were derived by several authors, see Klötzler [14], Benker and Kossert [2], Breckner [3], Gorochowik and Kirillowa [9], Gorochowik [8], [10] and Salukvadze [19]. It is difficult to show the existence of an optimal control (see Klötzler [15]), whereas suboptimal controls exist under very weak assumptions. So it is important to derive some assertions for suboptimal controls.
Ekeland's variational principle [5], [6], [1] is a very deep result in optimization theory and says that there exists an exact solution of a slightly perturbed optimization problem in a neighbourhood of an approximate solution of the original problem. This principle can be used in order to derive necessary conditions for solutions and to obtain approximate solutions of optimization problems.
Many authors have published extensions and applications of Ekeland's variational principle as well as equivalent statements.
Extensions of Ekeland's variational principle to vector optimization have been given by several authors. Loridan (1984) [16] has presented a vector-valued variational principle for the finite-dimensional case where he could apply Ekeland's original result directly by taking a suitable scalarization. Vector-valued variational principles for an objective function, which takes its values in general spaces have been derived by Nemeth (1986)

*Martin-Luther-University Halle-Wittenberg, Department of Mathematics and Informatics, D-06099 Halle, Germany, e-mail: tammer@mathematik.uni-halle.d400.de

[17], Khanh (1986) [13], Tammer (1992) [20], [23], Dentscheva and Helbig (1996) [4] and Isac (1996) [12]. In our papers [22] and [11] we have used the vector-valued variational principle of [20] in order to show ϵ-variational inequalities (or ϵ-Kolmogorov conditions) for approximate solutions of a general approximation problem.
The aim of this paper is to derive an ϵ-minimum principle for suboptimal controls of multicriteria control problems from the vector-valued variational principle.

2. Solution concepts in multicriteria optimization

We formulate a multicriteria control problem with an objective function which takes its values in the m-dimensional Euclidean space R^m.
In the following we denote the topological interior of a set $C \subset R^m$ by $int\ C$, the topological boundary of C by $bd\ C$ and the topological closure of C by $cl\ C$.
Let us assume:

(A1): (V, d) is a complete metric space,
$K \subset R^m$ is a convex cone with $k^0 \in int\ K$,
$B \subset R^m$ is a pointed convex cone with $int\ B \neq \emptyset$ such that
$cl\ B + (K \setminus \{0\}) \subset int\ B$.

(A2): $F : V \longrightarrow R^m$ is
lower semicontinuous with respect to k^0 and B in the sense that
$M_r = \{ v \in V \mid F(v) \in rk^0 - cl\ B \}$ is closed for each $r \in R$
and bounded from below, i. e., $F[V] \subset y + B$ for a certain $y \in R^m$.

Now we consider the following vector optimization problem to determine the efficient point set of $F[V]$ with respect to K :

$$(P): \qquad \text{Compute the set} \qquad Eff(F[V], K),$$

where

$$Eff(F[V], K) = \{ F(\bar{v}) \mid \bar{v} \in V \quad \text{and} \quad F[V] \cap (F(\bar{v}) - (K \setminus \{0\})) = \emptyset \}.$$

Furthermore, we will introduce approximately efficient elements of vector optimization problems.
The reason for introducing approximately efficient solutions is that numerical algorithms usually generate only approximate solutions anyhow; moreover, the efficient point set may be empty in the general noncompact case, whereas approximately efficient points always exist under very weak assumptions (see Tammer [21], where existence results for approximate solutions of a vector optimization problem have been shown, especially under the assumption that the objective function is bounded from below).

Definition 1 *An element $F(v_\epsilon) \in F[V]$ is called an approximately efficient point of $F[V]$ with respect to K, $k^0 \in int\ K$ and $\epsilon > 0$, if*

$$F[V] \cap (F(v_\epsilon) - \epsilon k^0 - (K \setminus \{0\})) = \emptyset.$$

The approximately efficient point set of $F[V]$ with respect to K, k^0 and ϵ is denoted by $Eff(F[V], K_{\epsilon k^0})$, where $K_{\epsilon k^0} := \epsilon k^0 + K$.
Moreover, we will study approximately efficient elements with respect to the cone B from assumption $(A1)$ instead of K.

3. Variational principles in partially ordered spaces

Beginning with the paper of Loridan [16] various authors proved variational principles of Ekeland's type for ϵ-efficient solutions of multicriteria optimization problems by scalarization. In [20] we have shown a vector-valued variational principle by using a separation theorem for nonconvex sets and Ekeland's original result. Here we will use the following variational principle which was shown in [23] and which is a sharper result than in [20].

The following assertion can be considered as an extension of Ekeland's variational principle ([5], [6]) to vector optimization.

Theorem 1 *([23]) Assume (A1) and (A2).*
Then for any $\epsilon > 0$, $\lambda > 0$ and any $F(v_0) \in Eff(F[V], B_{\epsilon k^0})$ there exists an element $v_\epsilon \in V$ such that

1. *$F(v_\epsilon) \in f(v_0) - \lambda d(v_0, v_\epsilon)k^0 \; - \; cl\ B$ and $F(v_\epsilon) \in Eff(F[V], D_{\epsilon k^0})$, where D is an open subset of R^m with $K \setminus \{0\} \subset D$, $0 \in bd\ D$ and $cl\ D + (K \setminus \{0\}) \subset D$,*
2. *$d(v_0, v_\epsilon) \leq \epsilon/\lambda$,*
3. *$F_{\lambda k^0}(v_\epsilon) \in Eff(F_{\lambda k^0}[V], K)$*

are fulfilled, where $F_{\lambda k^0}(v) := F(v) + \lambda d(v, v_\epsilon)\, k^0$.

The following theorem follows immediately from Theorem 1 regarding the fact that under the given assumptions there always exists an approximately efficient element $v_0 \in V$ with

$$F(v_0) \in Eff(F[V], B_{\epsilon k^0})$$

(cf. Tammer [21]).

Theorem 2 *Assume* $(A1)$, $(A2)$.
Then for every $\epsilon > 0$ there exists some point $v_\epsilon \in V$ such that

1. *$F[V] \cap (F(v_\epsilon) - \epsilon k^0 - (K \setminus \{0\})) = \emptyset$,*
2. *$F_{\epsilon k^0}[V] \cap (F_{\epsilon k^0}(v_\epsilon) - (K \setminus \{0\})) = \emptyset$,*
   where
   $F_{\epsilon k^0}(v) := F(v) + d(v, v_\epsilon)\epsilon k^0$.

Remark 1: Theorem 1 is slightly stronger than Theorem 2. The main difference concerns condition (ii) in Theorem 1, which gives the whereabouts of point x_ϵ in V.

Remark 2: The main result of the last theorems (statement 3 Theorem 1 and statement 2 in Theorem 2) says that v_ϵ is an efficient solution of a slightly perturbed vector optimization problem. This statement can be used in order to derive necessary conditions for approximately efficient elements. In our papers [22] and [11] we have shown

ϵ-Kolmogorov-conditions for approximately efficient solutions of abstract approximation problems applying the third condition in Theorem 1. In the next chapter we will use Theorem 2 in order to derive an ϵ-minimum-principle in the sense of Pontrjagin for suboptimal solutions of multicriteria control problems.

4. An ϵ-minimum principle for multiobjective optimal control problems

In this chapter we will give an application of the multicriteria variational principle in Theorem 2 to control problems.

Consider the system of differential equations

$$\left.\begin{aligned} \frac{dx}{dt}(t) &= \varphi(t, x(t), u(t)), \\ x(0) &= x_0 \in R^n \end{aligned}\right\} \tag{1}$$

and the control restriction

$$u(t) \in U,$$

which must hold almost everywhere on $[0, T]$ with $T > 0$.
We assume that

(C1) $\varphi : [0, T] \times R^n \times U \longrightarrow R^n$ is continuous and U is compact.

The vector $x(t)$ describes the state of the system, $u(t)$ is the control at time t and belongs to the set U.

Furthermore, suppose that

(C2) $\frac{\partial \varphi}{\partial x_i}$, i=1,...,n, are continuous on $[0, T] \times R^n \times U$,
(C3) $(x, \varphi(t, x, u)) \leq c(1 + ||x||^2)$ for some $c > 0$.

Remark 3: Let $u : [0, T] \longrightarrow U$ be a measurable control. Condition (C2) and the continuity of φ ensure that there exists a unique solution x of the differential equation (1) on $[0, \tau]$ for a sufficiently small $\tau > 0$. By using Gronwall's inequality condition (C3) implies

$$||x(t)||^2 \leq (||x_0||^2 + 2cT)e^{2cT},$$

and, hence, ensures the existence of the solution on the whole time interval $[0, T]$. Moreover, the last inequality yields

$$||\frac{dx(t)}{dt}|| \leq max\{\varphi(t, x, u) \mid (t, x, u) \in [0, T] \times B^0 \times U\},$$

where B^0 denotes the ball of radius $(||x_0||^2 + 2cT)^{\frac{1}{2}} e^{cT}$. Applying Ascoli's theorem, we see that the family of all trajectories x of the control system (1) is equicontinuous and bounded, and hence relatively compact in the uniform topology (compare Ekeland [5]).

In order to formulate the multicriteria control problem we introduce the objective function $f : R^n \longrightarrow R^m$ and suppose that

(C4) f is a differentiable vector-valued function,

(C5) $K \subset R^m$ is a convex cone with nonempty interior, $B \subset R^m$ is a pointed convex cone with $int\ B \neq \emptyset$ such that $cl\ B + (K \setminus \{0\}) \subset int\ B$.

Now we formulate the **multiobjective optimal control problem** under the assumptions $(C1)$–$(C5)$

(P): Find some measurable control $\bar{u}$ such that the corresponding trajectory $\bar{x}$ satisfies

$$f(x(T)) \notin f(\bar{x}(T)) - (K \setminus \{0\})$$

for all solutions x of (1).

It is well known that it is difficult to show the existence of optimal (or efficient) controls of (P), whereas suboptimal controls exist under very weak conditions. So it is important to derive some assertions for suboptimal controls.
An application of a variational principle for vector optimization problems yields an ϵ-minimum principle for (P), which is closely related to Pontryagin's minimum principle (for $\epsilon = 0$).
Now we apply Theorem 2 in order to derive an ϵ-minimum principle for the multiobjective optimal control problem (P).
We introduce the space V of controls, defined as the set of all measurable functions $u : [0,T] \longrightarrow U$ with the metric

$$d(u_1, u_2) = meas\{t \in [0,T] : u_1(t) \neq u_2(t)\}.$$

In order to prove our main result we need the following lemmata:

Lemma 1 *(Ekeland [5])*
(V, d) is a complete metric space.

Lemma 2 *(Ekeland [5])*
The function $F : u \longrightarrow f(x(T))$ is continuous on V, where $x(.)$ is the solution of (1) depending on $u \in V$.

Theorem 3 *Consider the multicriteria control problem under the assumptions $(C1)$–$(C5)$.*
For every $\epsilon > 0$, there exists a measurable control u_ϵ with the corresponding admissible trajectory x_ϵ such that

1. $f(x(T)) \notin f(x_\epsilon(T)) - \epsilon k^0 - (K \setminus \{0\})$
   for all solutions x of (1),
2.

$$\begin{pmatrix} (\varphi(t, x_\epsilon(t), u(t)) \ ,\ p_\epsilon^1(t)) \\ \cdots \\ (\varphi(t, x_\epsilon(t), u(t)) \ ,\ p_\epsilon^m(t)) \end{pmatrix} \notin \begin{pmatrix} (\varphi(t, x_\epsilon(t), u_\epsilon(t)) \ ,\ p_\epsilon^1(t)) \\ \cdots \\ (\varphi(t, x_\epsilon(t), u_\epsilon(t)) \ ,\ p_\epsilon^m(t)) \end{pmatrix} - \epsilon k^0 - int\ K$$

for any $u \in U$ and almost all $t \in [0,T]$, where $p_\epsilon(.) = (p_\epsilon^1(.), \ldots, p_\epsilon^m(.))$ is the solution of the linear differential system:

$$\left.\begin{array}{rcl} \frac{dp_{\epsilon i}^s}{dt}(t) & = & -\sum_{j=1}^n \frac{\partial \varphi_j}{\partial x_i}(t, x_\epsilon(t), u_\epsilon(t)) p_{\epsilon j}^s, \\ i & = & 1, \ldots, n; \\ p_\epsilon^s(T) & = & f_s'(x_\epsilon(T)), \\ s & = & 1, \ldots, m. \end{array}\right\} \qquad (2)$$

Proof:
We consider the vector-valued function $F \; : \; u \longrightarrow f(x(T))$ and the space V of measurable controls $u \; : \; [0,T] \longrightarrow U$.
From Lemma 1 we get that (V,d) is a complete metric space. Lemma 2 and Remark 3 yield that $F \; : \; u \longrightarrow f(x(T))$ is lower semicontinuous with respect to k^0 and B and bounded from below on V. So we can conclude that the assumptions of Theorem 2 are fulfilled and we can apply Theorem 2 to the vector-valued function F. This yields a measurable control $u_\epsilon \in V$ such that

(i) $F(u_\epsilon) \in Eff(F[V], K_{\epsilon k^0})$,
(ii) $F(u_\epsilon) \in Eff(F_{\epsilon k^0}[V], K)$,

where

$$F_{\epsilon k^0}(u) := F(u) + \epsilon k^0 d(u, u_\epsilon).$$

The corresponding admissible trajectory x_ϵ to u_ϵ satisfies

$$\frac{dx_\epsilon(t)}{dt} = \varphi(t, x_\epsilon(t), u_\epsilon(t)) \qquad (3)$$

for almost all $t \in [0,T]$ and

$$x_\epsilon(0) = x_0.$$

So we derive from (i)

$$f(x(T)) \notin f(x_\epsilon(T)) - \epsilon k^0 - (K \setminus \{0\})$$

for all solutions x of (1), i.e., statement 1 is fulfilled.
In order to prove statement 2 we take $t_0 \in (0,T)$, where equality (3) holds, $u_0 \in U$, and define $v_\tau \in V$ for $\tau \geq 0$ by: :

$$v_\tau(t) := \begin{cases} u_0 & : \; t \in [0,T] \cap (t_0 - \tau, t_0) \\ u_\epsilon(t) & : \; t \notin [0,T] \cap (t_0 - \tau, t_0) \end{cases}$$

For a sufficiently small τ it holds:

$$d(u_\epsilon, v_\tau) = meas\{t \; : \; v_\tau(t) \neq u_\epsilon(t)\} \; \leq \; \tau.$$

Furthermore, for the corresponding admissible trajectory x_τ it holds: (compare Pallu de la Barriere [18])

$$\frac{d}{d\tau} f_s(x_\tau(T))_{|\tau=0} = (\varphi(t_0, x_\epsilon(t_0), u_0) - \varphi(t_0, x_\epsilon(t_0), u_\epsilon(t_0)) \ , \ p_\epsilon^s(t_0)), \tag{4}$$

$$s = 1, \ldots, m,$$

where $p_\epsilon = (p_\epsilon^1, \ldots, p_\epsilon^m)$ satisfies (2).

Now, we can conclude from (ii) for $u = v_\tau$

$$F(u_\tau) + \epsilon k^0 d(u_\tau, u_\epsilon) \notin F(u_\epsilon) - (K \setminus \{0\})$$

and

$$f(x_\tau(T)) \notin f(x_\epsilon(T)) - \epsilon k^0 d(u_\tau, u_\epsilon) - (K \setminus \{0\}).$$

For suffiently small τ we get

$$f(x_\tau(T)) \notin f(x_\epsilon(T)) - \epsilon\tau k^0 - (K \setminus \{0\}), \qquad \tau > 0.$$

This implies

$$\frac{f(x_\tau(T)) - f(x_\epsilon(T))}{\tau} \notin -\epsilon k^0 - (K \setminus \{0\}) \qquad \tau > 0$$

and

$$\lim_{\tau \to +0} \frac{f(x_\tau(T)) - f(x_\epsilon(T))}{\tau} \notin -\epsilon k^0 - (K \setminus \{0\}) \qquad \tau > 0,$$

i.e.,

$$\frac{d}{d\tau} f(x_\tau(T))_{|\tau=0} \notin -\epsilon k^0 - int\ K.$$

Finally, (4) yields

$$\begin{pmatrix} (\varphi(t_0, x_\epsilon(t_0), u_0) - (\varphi(t_0, x_\epsilon(t_0), u_\epsilon(t_0)) \ , \ p_\epsilon^1(t_0)) \\ \cdots \\ (\varphi(t_0, x_\epsilon(t_0), u_0) - (\varphi(t_0, x_\epsilon(t_0), u_\epsilon(t_0)) \ , \ p_\epsilon^m(t_0)) \end{pmatrix} \notin -\epsilon k^0 - int\ K$$

for an arbitrary $u_0 \in U$ and almost all $t_0 \in [0, T]$.
Hence,

$$\begin{pmatrix} (\varphi(t, x_\epsilon(t), u(t)) \ , \ p_\epsilon^1(t)) \\ \cdots \\ (\varphi(t, x_\epsilon(t), u(t)) \ , \ p_\epsilon^m(t)) \end{pmatrix} \notin \begin{pmatrix} (\varphi(t, x_\epsilon(t), u_\epsilon(t)) \ , \ p_\epsilon^1(t)) \\ \cdots \\ (\varphi(t, x_\epsilon(t), u_\epsilon(t)) \ , \ p_\epsilon^m(t)) \end{pmatrix} - \epsilon k^0 - int\ K$$

for an arbitrary $u \in U$ and almost all $t \in [0, T]$. ∎

Remark 4: If we put $\epsilon = 0$ then Theorem 3 coincides with the following assertion: Whenever there is a measurable control u_ϵ and the corresponding admissible trajectory x_ϵ with

1. $f(x(T)) \notin f(x_\epsilon(T)) - (K \setminus \{0\})$
   for all solutions x of (1),
   then
2. $$\begin{pmatrix} (\varphi(t, x_\epsilon(t), u(t)) \, , \, p_\epsilon^1(t)) \\ \cdots \\ (\varphi(t, x_\epsilon(t), u(t)) \, , \, p_\epsilon^m(t)) \end{pmatrix} \notin \begin{pmatrix} (\varphi(t, x_\epsilon(t), u_\epsilon(t)) \, , \, p_\epsilon^1(t)) \\ \cdots \\ (\varphi(t, x_\epsilon(t), u_\epsilon(t)) \, , \, p_\epsilon^m(t)) \end{pmatrix} - int\ K$$
   for any $u \in U$ and almost all $t \in [0, T]$, where $p_\epsilon(.)$ is the solution of the linear differential system (2).

This means that u_ϵ fulfills a minimum principle for multicriteria control problems in the sense of Pontrjagin (compare Gorochowik and Kirillowa [9] and Gorochowik [8], [10]).

Remark 5: Now let us study the special case $Y = R^1$ and put $\epsilon = 0$. Then Theorem 3 coincides with the following assertion:
Whenever

(i) $f(x_\epsilon(T)) \leq inf\ f(x(T))$ holds,

then

(ii) $(\varphi(t, x_\epsilon(t), u_\epsilon(t)) \, , \, p_\epsilon(t)) \leq \min_{u \in U}(\varphi(t, x_\epsilon(t), u(t)) \, , \, p_\epsilon(t))$

almost everywhere on [0,T], where $p_\epsilon(.)$ is the solution of (2).

This is the statement of Pontrjagin's minimum principle.
However, Theorem 3 holds even if optimal solutions do not exist.

In Theorem 3 we have shown an ϵ-minimum principle without any scalarization of the multiobjective optimal control problem. Finally, we derive an ϵ-minimum principle using a suitable scalarization with a functional $z : R^m \longrightarrow R^1$ defined for $y \in R^m$ by:

$$z(y) := inf\{t \in R \mid y \in - \, cl\ K + tk^0\}. \tag{5}$$

Definition 2 *Let D be a nonempty subset of R^m and $y^1, y^2 \in R^m$. A functional $z : R^m \longrightarrow R$ is called D-monotone, if $y^1 \in y^2 + D$ implies $z(y^1) \geq z(y^2)$. A D-monotone functional z is called strictly D-monotone, if $y^1 \in y^2 + (D \setminus \{0\})$ implies $z(y^1) > z(y^2)$.*

In [7], Corollary 2.1, under the assumption that K is a convex cone with nonempty interior, we have shown that the functional z in (5) is continuous, sublinear and strictly ($int\ K$)-monotone.

Theorem 4 *Consider the multicriteria control problem under the assumptions $(C1)$–$(C5)$.*
For every $\epsilon > 0$, there exists a measurable control u_ϵ with the corresponding admissible trajectory x_ϵ such that

1. $f(x(T)) \notin f(x_\epsilon(T)) - \epsilon k^0 - (K \setminus \{0\})$
for all solutions x of (1),

2.

$$z\left(\begin{pmatrix} (\varphi(t,x_\epsilon(t),u(t)) - \varphi(t,x_\epsilon(t),u_\epsilon(t))\ ,\ p_\epsilon^1(t)) \\ \cdots \\ (\varphi(t,x_\epsilon(t),u(t)) - \varphi(t,x_\epsilon(t),u_\epsilon(t))\ ,\ p_\epsilon^m(t)) \end{pmatrix}\right) \geq -\epsilon,$$

for any $u \in U$ and almost all $t \in [0,T]$, where $z : R^m \longrightarrow R^1$ is the continuous, sublinear, strictly (int K)-monotone functional defined by (5) *and $p_\epsilon = (p_\epsilon^1, \ldots, p_\epsilon^m)$ is the solution of the linear differential system (2).*

Proof:
Under the given assumptions we can conclude from the statement 2 of Theorem 3

$$\begin{pmatrix} (\varphi(t,x_\epsilon(t),u(t)) - \varphi(t,x_\epsilon(t),u_\epsilon(t))\ ,\ p_\epsilon^1(t)) \\ \cdots \\ (\varphi(t,x_\epsilon(t),u(t)) - \varphi(t,x_\epsilon(t),u_\epsilon(t))\ ,\ p_\epsilon^m(t)) \end{pmatrix} \notin -\epsilon k^0 - int\ K \qquad (6)$$

for any $u \in U$ and almost all $t \in [0,T]$, where $p_\epsilon = (p_\epsilon^1, \ldots, p_\epsilon^m)$ is the solution of (2). The functional z in (5) has for $y \in R^m$ and $t \in R$ the property (cf. [7], Theorem 2.1)

$$z(y) \geq t \iff y \notin -\ int\ K + tk^0,$$

such that we can conclude from (6)

$$z\left(\begin{pmatrix} (\varphi(t,x_\epsilon(t),u(t)) - \varphi(t,x_\epsilon(t),u_\epsilon(t))\ ,\ p_\epsilon^1(t)) \\ \cdots \\ (\varphi(t,x_\epsilon(t),u(t)) - \varphi(t,x_\epsilon(t),u_\epsilon(t))\ ,\ p_\epsilon^m(t)) \end{pmatrix}\right) \geq -\epsilon,$$

for any $u \in U$ and almost all $t \in [0,T]$, where $p_\epsilon(.)$ is the solution of (2). ■

References

[1] Aubin, J. P.; Ekeland, I.: *Applied Nonlinear Analysis.* Wiley, New York, 1984.

[2] Benker, H.; Kossert, S.: *Remarks on quadratic optimal control problems in Hilbert spaces.* Zeitschr. Analysis und ihre Anwendungen 1 (3), 1982, 13–21.

[3] Breckner, W. W.: *Derived sets for weak multiobjective optimization problems with state and control variables* (to appear), 1996.

[4] Dentscheva, D.; Helbig, S.: *On variational principles in vector optimization*, Lecture on the 4th Workshop Multicriteria Decision, Holzhau, Germany, 1994.

[5] Ekeland, I.: *On the variational principle.* J. Math. Anal. Appl. 47, 1974, 324–353.

[6] Ekeland, I.: *Nonconvex minimization problems.* Bull. Amer. Mathem. Society 1 (3), 1979, 443–474

[7] Gerth (Tammer), C.; Weidner, P.: *Nonconvex separation theorems and some applications in vector optimization.* J. Optim. Theory Appl. 67 (2), 1990, 297–320.

[8] Gorochowik, B. B.: *About multiobjective optimization problems* (in Russian). AN Soviet Union (6), 1972.

[9] Gorochowik, B. B.; Kirillowa, F. M.: *About scalarization of vector optimization problems* (in Russian). Docl. AN Soviet Union (19), No. 7, 1975, 588–591.

[10] Gorochowik, B. B.: *Conditions for weakly efficient elements in multiobjective optimization* (in Russian). Minsk, IN-T Matem. AN Soviet Union, Preprint, 1976.

[11] Henkel, E.-C.; Tammer, C.: *ϵ-variational inequalities in partially ordered spaces.* Optimization 36, 1996, 105–118.

[12] Isac, G.: *A variant of Ekeland's principle for Pareto ϵ-efficiency.* In prep, 1994.

[13] Khanh, P. Q.: *On Caristi-Kirk's theorem and Ekeland's variational principle for Pareto extrema.* Inst. of Mathem., Polish Acad. Sciences, Preprint 357, 1986.

[14] Klötzler, R.: *Multiobjective dynamic programming.* Math. Operationsforschung, Statist., Ser. Opt. 9 (3), 1978, 423–426.

[15] Klötzler, R.: *On a general conception of duality in optimal control.* Lecture Notes in Math. 703, 1979, 183–196.

[16] Loridan, P.: *ϵ-solutions in vector minimization problems.* Journ. Optimization Theory Applications 43 (2), 1984, 265–276.

[17] Nemeth, A. B.: *A Nonconvex Vector Minimization Proble.* Nonlin. Anal. Theory, Methods and Applications 10 (7), 1986, 669–678.

[18] Pallu de la Barriere: *Cours d'automatique theorique.* Dunod, 1965.

[19] Salukvadze, M. E.: *Vector optimization problems in control theory.* Tiblisi, 1975.

[20] Tammer, C.: *A generalization of Ekeland's variational principle.* Optimization 25, 1992, 129–141.

[21] Tammer, C.: *Existence results and necessary conditions for ϵ-efficient elements.* In: Brosowski, Ester, Helbig, Nehse (Eds): Multicriteria Decision. Lang Verlag Frankfurt/Main Bern, 1993, 97–110.

[22] Tammer, C.: *Necessary conditions for approximately efficient solutions of vector approximation.* Proc. 2nd Conf.Approximation and Optimization, Havana, 1993.

[23] Tammer, C.: *A variational principle and applications for vectorial control approximation problems.* To appear: Math. Journ. Univ. Bacau (Romania), 1994.

International Series of Numerical Mathematics
Vol. 124,

Existence Principles and the Theory of Extremal Problems

Vladimir Tikhomirov*

1. Introduction

1.1 Formalization of extremal problems

Extremal problems arising in mathematics, in the natural sciences, or in practical activities are usually stated initially without formulas, using the terminology of the field in which they arise.

In order to be able to utilize the analytical tools it is necessary to translate the statesment of the problem from each specific language to the language of mathematics. Such a translation is called *a formalization.*

To formalize an extremal problem means to describe a functional f_0 ($f_0: X \to \overline{\mathbb{R}}$, $\overline{\mathbb{R}} = \mathbb{R} \cup \pm\infty$) to be minimized or maximized (together with its domain X) and a constraint $C \subset X$. The constraint is usually given by equalities and inequalities.

We shall use the following abbreviated notation for the formalized problem:

$$f_0(x) \to \min\ (\max), \quad x \in C. \tag{P}$$

The points $x \in C$ are called *admissible points.* An admissible point $\widehat{x}$ is said to be an *absolute minimum* (*maximum*) of the problem (P) if $f(x) \geq f(\widehat{x})$ ($f(x) \leq f(\widehat{x})$) for every $x \in C$. We can change a problem of maximization into a minimization problem by changing the sign of the functional.

An absolute minimum of an extremum problem is said to be *a solution of the problem.* Our goal is to find a solution of a given problem, but at the beginning (as a rule) we find local extrema of these problems. Let X be a topological space. An admissible point $\widehat{x}$ is called a *local minimum in* (P) if there exists a neighborhood U of x such that $\widehat{x}$ is a solution of the problem $f(x) \to \min$, $x \in C \cap U$.

1.2 Main chapters of the theory

Mathematical programming with the equality constraints studies the problems (P) with $X = \mathbb{R}^n$ and $C = \{x \mid F(x) = 0,\ F: \mathbb{R}^n \to \mathbb{R}^m\}$.

*Faculty of Mathematics, Moscow State University, 117333 Moscow, Russia,
e-mail: TIKHOM@nw.math.msu.su

In finite-dimentional *convex programming* $X = \mathbb{R}^n$, $C = \{x \mid f_i(x) \le 0,\ 1 \le i \le m,\ x \in A\}$, where all the functons f_i, $1 \le i \le m$, are convex and A is a convex set in $\mathbb{R}^n$.

In the *calculus of variations* the domains of functionals are *functional spaces* (C^r, W_p^r, H^r and so on) and the constraints are defined by *differential equations.* Here is a typical example of a problem of the calculus of variations (so-called *Lagrange problem*):

$$f_0(x(\cdot)) = \int_\Delta L(t, x(t), \dot{x}(t))\, dt \to \min\ (\max);\ \ M(t, x, \dot{x}) = 0,\ \ x_{|\partial\Delta} = \xi,$$

where Δ is a segment $[t_0, t_1]$ and, in multidimensional case, Δ is a domain in $\mathbb{R}^n$.

In *optimal control* the unknown variables are divided in two parts, the state (or phase) coordinates (x) and the control coordinates (u). A typical problem of optimal control is the following:

$$f_0(x(\cdot), u(\cdot)) = \int_{t_0}^{t_1} f(t, x(t), u(t))\, dt + l(x(t_0), x(t_1)) \to \min;$$

$$\dot{x} = \varphi(t, x(t), u(t)),\ \ u(t) \in U.$$

The crucial distinction of this problem from the problems in the calculus of variation consists of the constraint $u(t) \in U$ or in *inequality constraints in control coordinates.*

The following questions could be asked about each extremal problem:

1. What are the conditions of extremum in the problem (necessary, sufficient, necessary and sufficient)?
2. How to describe the evolution of solutions if the problem is perturbed?
3. Does the solution of the problem exist or not?

The goal of the article is to discuss some important general approaches to these questions.

2. Existence principles

Among the most important results in the general theory of extremal problems are the existence theorems of Lagrange multipliers, of fields of extremals, of solutions and so on. Besides this the investigation of the problems of calculus of variations and optimal control requires the use of existence theorems from the theory of differential equations (existence of a solution of the Cauchy problem, existence of the global solution for linear systems, dependence of solutions on the parameters, and so on). Many of these classical existence theorems can be deduced from a few general principles of existence, which are based on considerations of compactness, contractivity, monotonicity and on different topological ideas.

Most of the theorems of the theory of extremal problems are corollaries of the theorems we speak about in this subsection.

Topological principles

Theorem 1 (The Weierstrass–Lebesgue compactness principle) *Let X be a compact topological space and let f be a lower semicontinuous function on X. Then there exists a point $\widehat{x}$ in X at which f attains its absolute minimum.*

Theorem 2 (Brouwer's principle) *Let $F \in C(B^n, \mathbb{R}^n)$, $B^n = \{x \in \mathbb{R}^n \mid |x| \leq 1\}$ where $|x - F(x)| < 1 - \delta$, $0 < \delta < 1$. Then for all $y \in \mathbb{R}^n$, $|y| < \delta$ there exists an $\widehat{x} \in B^n$ such that $y = F(\widehat{x})$.*

Iterative principle

Theorem 3 (The generalized iterative principle) *Let (X, d_X) be a complete metric space, (Y, d_Y) be a metric space, F be a continuous (or a closed multivalued) mapping from X to Y, $y \in Y$, $x_0 \in X$, $y \in Y$, $r > 0$, $0 < \theta < 1$ and let there exist a mapping $M: X \to X$ such that*

$$a)\ d_Y(y, F(x_0)) \leq (1-\theta)r; \quad b)\ d_X(x, M(x)) \leq d_Y(y, F(x)),$$

and either

$$c)\ d_Y(y, F(M(x))) \leq \theta d_X(x, M(x)) \quad \forall x,$$

or

$$c')\ d_Y(y, F(M(x))) \leq \theta d_Y(y, F(X)) \quad \forall x.$$

Then there exists a point $\widehat{x}$ such that $y = F(\widehat{x})$ ($y \in F(\widehat{x})$). Moreover, the iterative sequence $x_k = M(x_{k-1})$ tends to $\widehat{x}$.

Here were combined the contraction mapping principle of Picard–Caccioppoli–Banach and the Newton's method for solving equations. For the proof of this theorem see [1].

We want to illustrate the generalized inverse principle with two results very important in the theory of extremal problems.

Corollary 1 (Generalized Banach right inverse theorem) *Let X, Y be Banach spaces, U a closed convex set in Y with an interior point, $0 \in C$, $A: X \to Y$ a linear continuous surjective operator. Then there exists an operator $R: C \to X$ (right inverse operator) such that a) $ARy = y \quad \forall y \in C$ and b) $\|Ry\|_X \leq K\|y\|_Y \quad \forall y \in C$.*

The inf of constants in b) is called the *Banach constant of A with respect to C.*

Corollary 2 (Generalized theorem of inverse function of Lyusternik, Graves) *Suppose X, Y be Banach spaces, $\mathcal{U}$ is a topological space, U is a neighborhood of origin in X, $F: U \times \mathcal{U} \to Y$ a mapping such that $F(0, \widehat{u}) = 0$ and there exists an operator A satisfying conditions of Theorem 1 and such that*

$$\|F(x,u) - F(x',u) - A(x-x')\| \leq \varepsilon\|x - x'\|$$

for all x, x' such that $\|x\| < \delta$, $\|x'\| < \delta$ and u from some neighborhood of $\widehat{u}$ in $\mathcal{U}$, and C is a closed convex set with an interior point, $0 \in C$. Then there exists an operator $\varphi: C \cap c\varepsilon B_Y \times \mathcal{U} \to \delta B_X$ (inverse operator) such that $F(\varphi(y,u),u) = y$, where β is the Banach constant of A and $c < \beta^{-1} - \varepsilon$.

This theorem generalizes many well-known results devoted to the implicit function (for a remarkable survey of all these results, see [2]. I want to express my deep gratitude to A. Dontchev for a very fruitful discussion of these questions).

Proof of Corollary 1. Denote $Y_k = AkB_X$, $kB_X := \{x \mid \|x\| \leq k\}$, $k \in \mathbb{N}$. From Baire's principle it follows that there exist $y_0 \in Y$, $\varepsilon > 0$ and $m \in \mathbb{N}$ such that Y_m is dense in $(\varepsilon B_Y + y_0) \cap C$. Shifting by $-y_0$ we see that Y_s for some $s \in \mathbb{N}$ is dense in $\varepsilon B_Y \cap C$. Changing norms in X, $(\|x\| \to \alpha\|x\|)$ and in Y, $(\|y\| \to \beta\|y\|)$ we see that AB_X is dense in $B_Y \cap C$ (and consequently γAB_X is dense in γB_Y). By definition of density it is possible to find ξ_k such that

$$\|\xi_k\| \leq \|y - Ax_k\|, \quad \|y - A(x_k + \xi_k)\| \leq \frac{1}{2}\|y - Ax_k\|.$$

Applying the generalized iterative principle, $(M_k(x_k) = x_k + \xi_k)$ we construct the right inverse operator for $y \in C \cap B_Y$. The result of Corollary 1 follows from homotety reasons.

Proof of Corollary 2. We must apply the generalized iterative principle with $M(x) = x + R(y - F(x))$, where R is the operator constructed in Corollary 2.4.

Ekeland's principle and generalizations

The term 'monotonicity' has several meanings, which are the same in one-dimensional cases. One meaning is connected with *order*, the other with the derivatives of a convex function. Below we present two corresponding monotonicity principles.

Theorem 4 (Variational principle of Ekeland) *Let (X, d) be a complete metric space, f a lower semicontinuous function bounded from below, and $x_0 \in X$ a point such that $f(x_0) \leq \inf f + \varepsilon$. Then for all $\lambda > 0$ there exists an element $\widehat{x} = \widehat{x}(\lambda)$ such that:*

$$(a)\ f(\widehat{x}) \leq f(x_0);\ (b)\ d(x_0, \widehat{x}) \leq \lambda;\ (c)\ f(x) + (\varepsilon/\lambda)d(x, \widehat{x}) > f(\widehat{x})\ \forall x \neq \widehat{x}$$

(i. e. the perturbed function $f(\cdot) + (\varepsilon/\lambda)d(\cdot, \widehat{x})$ attains its strong absolute minimum at the point $\widehat{x}$).

And now we formulate another version of this principle, that generalises variational principles of Ekeland, Borwein-Preiss and DeVille (see [3]–[5]).

A continuous function $\varphi_{x,\alpha}(\cdot)$ $(\alpha > 0)$ on a metric space (X, d) we call a *bump-function* with the support on $B(x, \alpha) := \{u \in X \mid d(x, u) \leq \alpha\}$, if it is bounded, nonnegative on $B(x, \alpha)$ and nonpositive out of this ball.

Theorem 5 *(Constuctive variational principle). Let (X, d) be a complete metric space, let Φ be a complete family of bump-functions on X, let f be a lower semicontinuous and nonnegative function on X and suppose that there exist $w \in X$, $\varepsilon > 0$ such that $f(w) \leq \varepsilon$.*

Then for any $\lambda > 0$ there exists a number $0 < \alpha_* < 1$ such that for any sequences $\{\alpha_n\}_{n \in \mathbf{Z}_+}, \{\beta_n\}_{n \in \mathbf{N}}$ of positive numbers that satisfy conditions

$$\alpha_0 \leq \lambda\alpha_*, \ \sum_{n=0}^{\infty} \alpha_n \leq \lambda, \ \sum_{n=1}^{\infty} \beta_n \leq 1,$$

there exists a sequence $\{x_n\}_{n \in \mathbf{Z}_+}$, $x_n \in X$, which tends to $x_* \in X$ such that

(a) $d(x_n, w) \leq \lambda$;
(b) $f(x_n) \leq f(w)$;
(c) *there exists a sequence* $\{\gamma_n\}_{n\in\mathbb{N}_+}$ *such that* $0 \leq \gamma_n \leq \beta_n$ $\forall n \in \mathbb{N}$ *and the function* $g(\cdot) = f(\cdot) - \varepsilon \sum_{n\in\mathbf{Z}_+} \gamma_n\varphi_n(\cdot)$ *attains its absolute minimum at* x_* *(where we put* $\varphi_n(\cdot) = \varphi_{x_n,\alpha_n}(\cdot)$*).*

For proof of this theorem, see [1].

3. Some principles of theory of extremal problems

3.1 Lagrange principle in necessary conditions

> *On peut les réduire à ce principe général. Lorsqu'une fonction de plusieurs variables doit avoir un maximum ou un minimum, et qu'il y a entre ces variables une ou plusieurs équations, il suffira d'ajouter à la fonction proposée les fonctions qui doivent être nulles, multiplier chacune par une quantité indéterminée, et chercher ensuite le maximum ou minimum comme si les variables étaient indépendantes; les équations qu'on a trouvées serviront à determiner toutes les inconnues.*
>
> *J. L. Lagrange*

The honor of creating the general strategy for the investigation of extremal problems with constraints is due Lagrange. For all the problems that he met, Lagrange used a unified approach expressed in the words quoted as the epigraph to this section.

Lagrange's pivotal idea can be applied to an extremely broad class of extremal problems of a diverse nature. We shall try to explain this phenomenon. The first sketch of this explanation is the following.

In the majority of problems where the idea of Lagrange can be realized the variables are divided into two parts. Functionals and mappings are smooth in the variables of the first group and convex in the second group. Such problems will be called *smooth-convex problems.* Existence principles allow to prove the existence of Lagrange multipliers which fulfil the generalised idea of Lagrange. One of the first realizations of this idea can be seen in [6].

3.2 Lagrange principle for smooth-convex problems

Let X and Y be Banach spaces, U a topological space. Consider the problem:

$$f_0(x) \to \inf, \quad F(x,u) = 0, \ u \in U, \tag{P}$$

where $f_0: W \to \mathbb{R}$, $F: W \times U \to Y$, W is an open subset of X. We say that $(\widehat{x}, \widehat{u})$ is *a strong local minimum of problem* (P) if there exists a $\delta > 0$ and an open set $\mathcal{V}$ in U such that for any pair (x,u) for which $F(x,u) = 0$, $u \in \mathcal{V}$ and $\|x - \widehat{x}\| < \delta$ the following inequality holds true: $f(x) \geq f(\widehat{x})$.

The function

$$\mathcal{L}((x,u), \lambda_0, \lambda) := \lambda_0 f_0(x) + \langle \lambda, F(x,u) \rangle$$

is called the *Lagrange function of problem* (P). The number λ_0 and the element $\lambda \in Y^*$ are called *Lagrange multipliers.*

We call the mapping F in (P) *smooth-convex* at the point $(\widehat{x}, \widehat{u})$, if the mapping $x \to F(x,u)$ belongs to $SD(\widehat{x})$ $\forall u \in U$ and $F(x,U)$ is a convex set $\forall x \in W$. We call F *uniformly smooth* at $(\widehat{x}, \widehat{u})$ if $F(\cdot, \widehat{u}) \in SD(\widehat{x}, Y)$ and if moreover for any $\varepsilon > 0$ there exist $\delta > 0$ and a neighborhood $\mathcal{V}$ of $\widehat{u}$ such that $\|x_i - \widehat{x}\| < \delta$, $i = 1,2$, $u \in \mathcal{V}$, then $\|F(x_2,u) - F(x_1,u) - F_x(\widehat{x},\widehat{u})(x_2 - x_1)\| < \varepsilon\|x_2 - x_1\|$.

We call F *weakly approximative convex* if for each u_1, u_2 from U and $0 < \alpha < 1$ there exists a sequence $\{u_n(\alpha)\}_{n \in \mathbb{N}}$, $u_n \in U$ such that for any $\varepsilon > 0$ and any neighborhood $\mathcal{V}$ of εu there exist $n \in \mathbb{N}$ and $\bar{\alpha} > 0$ for which

$$\|F(x, u_n(\alpha)) - (1-\alpha)F(x,u_1) - \alpha F(x,u_2)\| < \varepsilon, \quad u_n(\alpha) \in \mathcal{V}, \ \forall \alpha \in [0, \alpha].$$

The mapping F is called *regular at a point* $(\widehat{x}, \widehat{u})$ if the space $F_x(\widehat{x},\widehat{u})X$ is closed in X and has finite codimension in Y and *totally regular* if $\operatorname{codim} F_x(\widehat{x},\widehat{u})X = 0$ or in other words if $F_x(\widehat{x},\widehat{u})X = Y$.

Theorem 6 (Lagrange principle for smooth-convex problems)

(a) Suppose that in problem (P) f_0 *is a Frechet differentiable function and let the mapping* F *be smooth-convex and regular. Then if* $(\widehat{x}, \widehat{u})$ *attains a local minimum for the problem* (P)*, then the Lagrange principle at* $(\widehat{x}, \widehat{u})$ *holds true; if* F *is a totally regular, then* $\lambda_0 \neq 0$.

(b) The same assertion is true if F *is a totally regular, uniformly smooth and weakly approximative convex mapping.*

The Lagrange principle for the problem (P) means that there exists Lagrange multipliers $\lambda \in Y^*$, $\lambda_0 \in \mathbb{R}$ (not equal to zero simultaneously) such that in the smooth problem (where $\mathcal{L}((x,u),\lambda,\lambda_0) = \lambda_0 f_0(x) + \langle \lambda, F(x,u)\rangle$)

$$\mathcal{L}((x,u),\lambda,\lambda_0) \to \min$$

when "les variables $[x]$ étaient indépendantes" Fermat's theorem holds:

$$\mathcal{L}_x((\widehat{x},\widehat{u}),\lambda,\lambda_0) = 0$$

(*stationary condition*)

$$(\Longleftrightarrow \lambda_0 f_0'(\widehat{x}) + (F_x(\widehat{x},\widehat{u}))^*\lambda = 0\,).$$

And in the convex problem

$$\mathcal{L}((\widehat{x},\widehat{u}),\lambda,\lambda_0) \to \min_{u \in U}$$

the *minimum principle* is satisfied

$$\min_{u \in U} \mathcal{L}((\widehat{x},\widehat{u}),\lambda,\lambda_0) = \mathcal{L}(\widehat{x},\widehat{u},\lambda,\lambda_0) \quad (\Longleftrightarrow \langle \lambda, F(\widehat{x},u)\rangle \geq 0 \quad \forall f \in U\,).$$

Proof 1. a) Denote

$$\Lambda := F_x(\widehat{x},\widehat{u}), \quad L_0 := I_m\Lambda = F_x(\widehat{x},\widehat{u})X, \quad A := F(x,U), \quad C := L_0 + A,$$

and let $\pi \colon Y \to Y/L_0$ be the canonical projection.

Without loss of generality we can put $f(\widehat{x}) = 0$. There are two possibilities: (I) $0 \notin \operatorname{int} \pi C$ (degenerate case) and (II) $0 \in \operatorname{int} \pi C$ (nondegenerate case).

The quotient-space $Z := Y/L_0$ by the condition of regularity is finite-dimensional. From the finite-dimensional separation theorem, it follows that there exists a nonzero functional $z^{(*)} \in \mathbf{Z}$ such that

$$\langle z, z^* \rangle \geq 0 \qquad \forall z \in \pi C. \tag{i}$$

Denote by π^* the conjugate operator $\pi^*\colon Z^* \to Y^*$ and put $\lambda := \pi^* z^* \neq 0$. Then it is evident that $\lambda \neq 0$ (because π is a surjective operator) and then for all $x \in X$ and $u \in U$

$$\langle \lambda, \Lambda x + F(\widehat{x}, u) \rangle \overset{\text{def}}{=} \langle \pi^* z^*, \Lambda x + F(\widehat{x}, u) \rangle \overset{\text{Id}}{=} \langle z^*, \pi(\Lambda x + F(\widehat{x}, u) \rangle \geq 0.$$

Putting $u = \widehat{u}$ here and then setting $x = 0$, we obtain that $\Lambda^* \lambda = 0$ and $\langle \lambda, F(\widehat{x}, u) \rangle \geq 0$ $\forall u \in U$, i.e., the stationary condition with $\lambda_0 = 0$ and minimum principle hold.

Consider the nondegenerate case. If $0 \in \operatorname{int} \pi C$, then there exists an $m \in \mathbb{N}$, there exist $\{z_i\}_{i=1}^m$, $\bar{\alpha}_i > 0$, such that $\sum_{i=1}^m \bar{\alpha}_i z_i = 0$, $z_i = \pi F(\widehat{x}, v_i)$ and a $\operatorname{cone}\{z_i\}_{i=1}^M = Z$. Then by definition $\sum_{i=1}^m \bar{\alpha}_i F(\widehat{x}, v_i) \in L_0$, hence there exists an element $\bar{\xi} \in X$ such that $\Lambda \bar{\xi} + \sum_{i=1}^m \bar{\alpha}_i F(\widehat{x}, v_i) = 0$ (i'). For each $v_0 \in U$ we define the map $\varphi_{v_0}\colon X \times \mathbb{R}^{m+1} \to Y$ by setting:

$$\varphi_{v_0}(x, \alpha) = (1 - \alpha_i) F(\widehat{x} + x, \widehat{u}) + \sum_{i=1}^m \alpha_i F(\widehat{x} + x, v_i) \quad x \in X, \ \alpha \in \mathbb{R}^{m+1}.$$

From the superposition theorem it follows that φ_{v_0} is strictly differentiable and $\varphi'_{v_0}(0,0)[x, \alpha] = \Lambda x + \sum_{i=0}^m \alpha_i F(\widehat{x}, v_i)$ (ii). Let $|\Lambda \bar{x} + F(\widehat{x}, \bar{v}) = 0$ (iii) and $\varepsilon > 0$. Then from (i) we have $\varphi'_{\bar{v}}(0,0)[\bar{x} + \varepsilon, 1, \varepsilon \bar{\alpha}_1, \ldots, \varepsilon \bar{\alpha}_m] = \Lambda(\bar{x} + \varepsilon \bar{\xi}) + F(\widehat{x}, \bar{v}) + \varepsilon \sum_{i=1}^m \bar{\alpha}_i F(\widehat{x}, v_i) = 0$.

From the tangent space theorem (which is a trivial corollary of the implicit function theorem) it follows (because φ_{v_0} is a totally regular operator) that there exist $r(\cdot)\colon [-1,1] \to X$, $\rho_i\colon [-1,1] \to \mathbb{R}$, $0 \leq i \leq m$, $r(t) = o(t)$, $\rho_i(t) = o(t)$ such that

$$\varphi_{\bar{v}}(\widehat{x} + \varepsilon \bar{\xi}, t + \rho_0(t), \varepsilon t \bar{\alpha}_1 + \rho_1(t) + \ldots, \varepsilon t \bar{\alpha}_m + \rho_m(t)) = 0 \quad \forall t \in [-1,1].$$

From the condition of convexity (and using the fact that for small $t > 0$, $t + \rho(t) > 0$, $\varepsilon t \bar{\alpha}_I + \rho_i > 0$) we see that there exists a family $\{u(t)\}_{t \in [-1,1]}$ such that $F(\widehat{x} + t\bar{x} + \varepsilon t \bar{\xi} + r(t), u(t)) = 0$ $\forall t \in [0, \delta]$. This means that the pair $(\widehat{x} + t\bar{x}\varepsilon\bar{\xi}, u(t))$, $t \in [0, \delta]$ is admissible and (because $(\widehat{x}, \widehat{u})$ is a local minimum) there exists $\delta_1 > 0$ such that $f_0(\widehat{x} + t\bar{x} + \varepsilon t \bar{\xi} + r(t)) \geq f_0(\widehat{x})$ $\forall t \in [0, \delta_1]$, hence $\langle f_0'(\widehat{x}), \bar{x} \rangle \geq 0$ (iv).

We use (iii) with $\bar{v} = \widehat{u}$ ($\Leftrightarrow \bar{x} \in \operatorname{Ker} \Lambda$), then from ($iv$) it follows that $f_0'(\widehat{x}) \in (\operatorname{Ker} \Lambda)^\perp$. From the annihilator lemma it follows that there exists $\lambda_1 \in L_0^*$ such that $f_0'(\widehat{x}) + \Lambda^* \lambda_1 = 0$ (v). Let now $F(\widehat{x}, u) \in L_0$. Then we can find $x(u) \in X$ such that $\Lambda x(u) + F(\widehat{x}, u) = 0$ (vi). Consequently

$$0 \overset{(iv)}{\leq} \langle f_0'(\widehat{x}), x(u) \rangle \overset{(v)}{=} -\langle \Lambda^* \lambda_1, x(u) \rangle \overset{\text{Id}}{=} \langle \lambda_1, \Lambda x(u) \rangle \overset{(vi)}{=} \langle \lambda_1, F(\widehat{x}, u) \rangle.$$

And finally from the separation theorem (we take $\bar{C}+\text{co}(C\cap\bar{B})$, where $\bar{B}$ is a ball with center $\eta \in L_0$, $\langle\lambda_1,\eta\rangle > 0$ that does not intersect $\{y \mid \langle\lambda_1, y\rangle = 0\}$, then $\text{int}\,\bar{C} \neq 0$ and it is possible to use the separation theorem) there exists $\lambda \in Y^*$ such that $\lambda_{|L_0} = \lambda_1$ and $\langle\lambda, F(\widehat{x}, u)\rangle \geq 0\rangle$ $\forall u \Leftarrow$ minimum principle. From the first equality we obtain $\langle\lambda, \Lambda x\rangle = \langle\lambda_1, \Lambda x\rangle \overset{\text{Id}}{=} \langle\Lambda^*\lambda_1, x\rangle \overset{(v)}{=} -\langle f_0'(\widehat{x}), x\rangle$ $\forall x$. This is the stationary condition. Thus the Lagrange principle in case a) is proven.

b) Denote $\varphi_v(x,\alpha) = (1-\alpha)F(\widehat{x}+x,\widehat{u}) + \alpha F(\widehat{x}, v)$ and apply the ordinary theorem on implicit functions and obtaining that if $\varphi_{\bar{v}}(0,0)[\bar{x},1] = 0$, then there exist $r(\cdot){:}\,[-1,1] \to X$, $\rho{:}\,[-1,1] \to \mathbb{R}$ so that $(1-t-\rho(t))F(\widehat{x}+t\bar{x}+r(t),\widehat{u}) + (t+\rho(t))F(\widehat{x}+t\bar{x}+r(t),\bar{v}) = 0$. Choose $\delta > 0$ such that for $t \in [0,\delta] \Leftarrow t+\rho(t) > 0$, $\widehat{x}+t\bar{x}+r(t) \in B_X(\widehat{x}, r(t))$ and for such t choose n for which $u_n(t+\rho(t)) \in V$. Applying the general theorem of implicit functions, we obtain

$$F(\xi(\widehat{x}+t\bar{x}+r(t)), u_n(t+\rho(t)) = 0, \quad \|\xi(\widehat{x}+t\bar{x}+r(t) - (\widehat{x}+t\bar{x}+r(t)\| \\ \leq K\|F(\widehat{x}+t\bar{x}+r(t), u_n(t+\rho(t))\| \leq r(t) \Longleftarrow \langle f_0'(\widehat{x}), \bar{x}\rangle \geq 0$$

and as in case a) we obtain the Lagrange principle.

3.3 Perturbations

In many cases it is convenient to include the primary extremal problem into some family of problems and consider (instead of the problem $f(x) \to \min$) the family of problems $F(x,y) \to min(overx), whereF(x,y_0) = f(x)$, or instead of the problem $f_0(x) \to \min, F(x) = 0$, family of problems $f_0(x) \to \min, F(x) = y$ and so on.

Let us consider the last situation (which goes back to Lagrange) in a finite dimensional case: we suppose that f_0 and F are continuously differentiable in a neighborhood W of the origin):

$$f_0(x) \to \min, F(x) = 0 \quad (\Leftrightarrow f_i(x) = 0),\ 1 \leq i \leq m,\ x \in \mathbb{R}^n \qquad (P_0)$$

This is called *the primal problem* and

$$f_0(x) \to \min,\ F(x) = y \qquad (P_y)$$

is its *perturbation.* According to the Lagrange principle, if $F'(0)\mathbb{R}^n = \mathbb{R}^m$ and $0 \in \text{locmin}(P_0)$, then there exists a $\lambda \in \mathbb{R}^*$ such that

$$0 = \mathcal{L}(0,\lambda,1) \Leftrightarrow f_0'(0) + \langle\lambda, F'(0) = 0.$$

If f_0, F belong to C^2, then the following condition is necessary for a minimum at a point): the operator $\mathcal{L}_{xx}(0,\lambda,1)$ restricted to the linear space $\text{Ker}F'(0)$ must be nonnegative. If it is positive, then $0 \in \mathbb{R}^n$ is a local minimum. Moreover, in this case it is possible to construct a family of extremals depending on y. More precisely, there is a smooth mapping $y \to (x(y),\lambda(y))$ such that

$$\mathcal{L}_x(x(y),\lambda(y),1) = 0,\ \mathcal{L}_{xx}(x(y),\lambda(y),1)_{|}\text{Ker}F'(x(y)) > 0$$

(in some neighborhood of $0_{\mathbb{R}^m}$).

This fact is a direct consequence of implicit function theorem.

This result can be extended to a wide class of smooth and smooth-convex problems (see, for example, [7]).

Applied to the problems of the calculus of variations it yields generalizations of the classical theorems about fields of extremals and the Jacobi-Hamilton equations.

A very interesting approach to the investigation of these equations is given in [8].

3.4 Existence, extension, relaxation

Ich bin überzeugt, dass es möglich sein wird, diese Existenzbeweise durch einen allgemeinen Grundgedanken zu führen, auf den das Dirichletsche Prinzip hinweist, und der uns dann vielleicht in den Stand setzen wird, der Frage näherzutreten, ob nicht jedes reguläre Variationsproblem eine Lösung besitzt, sobald hinsichtlich der gegebenen Grenzbedingungen gewisse Annahmen,..., erfüllt sind und nötigenfalls der Begriff der Lösung eine sinngemäße Erweiterung erfährt.

D.Hilbert.

The subject we consider in this section is connected with Hilbert's 20th problem. Let us ask the following question: does an arbitrary extremal problem have a solution? Of course it is easy to construct counter examples. But let us try to understand the main idea of Hilbert, which he expressed in the words of our epigraph.

Usually (and this exactly corresponds to Hilbert's text) the 20th problem is treated in connection with the problem of boundary conditions for elliptic equations. But I want to extend the meaning of the Hilbert's word "regular": may be (in his mind) Hilbert treats this word as "natural", i.e. given from some vital scientific problem.

At the beginning we discuss the following question: what are the main reasons for absence of solutions?

Let us consider some simple examples of nonexistence in the calculus of variations.

$$\mathcal{J}(x(\cdot)) = \int_I L(t, x(t), \dot{x}(t))dt \to \inf, \tag{1}$$

$$I = [0,1], \quad x(0) = x_0, \quad x(1) = x_1.$$

We have given here the functional and the constraints but it is also necessary to define the domain of the functional. In textbooks written at the beginning of the century, usually the space $C^1(I)$ of continuously differentiable functions (with two topologies C^1 and C, which lead to the notions of weak and strong extremum) were considered. But later it was understood that it is more fruitful to search for extrema in wider spaces.

This is the reason to consider our problem in the "widest" space W_1^1 of absolutely continuous functions such that their derivatives belong to L_1 :

$$W_1^1(I) = \left\{x(\cdot) \in AC(I) \| \|x(\cdot)\|_{W_1^1(I)} = \|x(\cdot)\|_{L_1(I)} + \|\dot{x}(\cdot)\|_{L_1(I)}\right\}.$$

What are the reasons which prevent the existence of a solution in $W_1^1(I)$?

Example 1 (Bolza: nonconvexity of the functions $\dot{x} \to L(t, x, \dot{x})$ (i.e. nonquasiregularity of the integrand):

$$J_1(x(\cdot)) = \int_I ((\dot{x}^2 - 1)^2 + x^2) dt \to \inf, \quad x(0) = x(1) = 0.$$

It is clear that

$$J_1(x(\cdot)) > 0 \quad \forall \quad x(\cdot) \in W_1^1(I), \quad x(t) \not\equiv 0.$$

On the other hand,

$$\text{if} \quad \bar{x}(t) \equiv 0, \quad \text{then} \quad J_1(\bar{x}(\cdot)) = 1.$$

But if we take the sequence

$$x_n(t) = \int_0^t u_n(\tau) d(\tau), \quad u_n(t) = \operatorname{sgn} \sin 2\pi n t, \quad n \in \mathbb{N}$$

it is evident that

$$x_n(\cdot) \underset{n\to\infty}{\to} 0$$

uniformly and at the same time

$$|\dot{x}_n(t)| = 1 \quad a.e.$$

and consequently

$$J_1(x_n(\cdot)) \underset{n\to\infty}{\to} 0.$$

This means that the value of the problem is equal to zero, but solutions are absent. The reason is the nonconvexity of the function $\dot{x} \to (\dot{x}^2 - 1)^2$.

Example 2 (Weierstrass: insufficient increase of the integrand):

$$J_2(x(\cdot)) = \int_I (t^2 \dot{x}^2 dt) \to \inf, x(0) = 0, x(1) = 1.$$

This is a famous example of Weierstrass. By means of this example, he explained the insufficiency of Riemann's arguments connected with the Dirichlet principle.

We see that

$$J_2(x(\cdot)) > 0 \quad \text{for all} \quad x(\cdot) \in W_1^1(I), \ x(0) = 0, \ x(1) = 1.$$

But if we take

$$x_n(t) = \begin{cases} nt, & 0 \le t \le 1/n \\ 1, t \ge 1/n, \end{cases}, \qquad n \in \mathbf{N}$$

we obtain

$$J_2(x_n(\cdot)) \underset{n\to\infty}{\to} 0.$$

And again the value of the problem is zero, but solutions are absent. The reason is the absence of increase of the integrand at one point.

Example 3 (harmonic oscillator: unboundness of the functional from below):

$$J_3(x(\cdot)) = \int_0^T (\dot{x}^2 - x^2) dt \to \inf T > \pi, x(0) = x(T) = 0.$$

Here if we consider the sequence

$$x_n(t) = n\sin(\pi t/T), \qquad n \in \mathbb{N},$$

it is easy to understand that

$$J_3(x_n(\cdot)) \underset{n\to\infty}{\rightarrow} -\infty,$$

hence there are no solutions.

Let us return to the quotation of Hilbert.

"The general principle" of proof of the existence theorems in the calculus of variations is without any doubt the compactness principle. It consicts of two components: the semicontinuity of functionals and the compactness of constraints. The absence of semicontinuity can sometimes be eliminated by means of relaxation. And then the functional becames lower-semicontinuous.

Lebesgue remarked that in the simplest problem of Calculus of Variations the functional

$$J(x(\cdot)) = \int_{t_0}^{t_1} L(t, x(t), \dot{x}(t))dt)$$

is lower semicontinuous if

$$\text{a function} \quad z \to L(t,x,z) \quad \text{is convex for all} \quad (t,x) \quad (L : \mathbb{R}^3 \to \mathbb{R}).$$

Later it was understood that lower semicontinuity of similar functionals is equal to convexity of functions $z \to L(t,x,z)$.

Functionals J with integrands convex as functions of z are called *quasiregular integrands.*

Now we'll say a few words about relaxations and extensions.

In 1930 N. Bogolyubov proved the following result:

In the problem (1) let $n = m = 1$ and integrand L be a smooth function satisfying Tonelli's condition

$$(L(t,x,\xi) \geq C_1|\xi|^{1+\delta} \quad (\delta > 0).$$

Then for any function $x(\cdot)$ there exists a sequence $\{x_m(\cdot)\}$ of continuously differentiable functions (it is possible also to suppose that $x(t_i) = x_m(t_i)$, $i = 0, 1$, $m \in N$) converging to $x(\cdot)$ uniformly on $T = [t_0, t_1]$ such that

$$\lim_{m\to\infty} \inf J(x_m(\cdot)) \leq \tilde{J}(x(\cdot)),$$

where $\tilde{J}(x(\cdot)) = \int_{t_0}^{t_1} \tilde{L}(t, x(t), \dot{x}(t))$, and $\tilde{L}$ is the "convexification" of L as a function of the last argument (i.e., $\tilde{L} = L_z^{**}$ is the Legendre transformation of $z \to L(t,x,z)$.)

In other words the functional $\tilde{J}$ is the lower semicontinuous extension of the functional J.

Consider the general situation. Let us have a pair (J, X), where J is a functional defined on a linear space X and let Y be a space embedded into X; then lower semicontinuous extension of the functional J is the functional $\tilde{J}$ defined on X as follows:

$$\tilde{J}(x) = \lim_{m\to\infty} \inf J(y_m), y_m \in Y, y_m \to x.$$

We saw that, from the theoretical point of view, functionals in the calculus of variations and optimal control can be regarded as lower semicontinuous, because we make the initial problem quasiregular by means of relaxation. Such results have been obtained by many authors (Young, McShane, Varga, Gamkrelidze, Ioffe-Tikhomirov, Klötzler, Matov, Gusseinov and others – see [8]–[9]).

References

[1] Ioffe, A. D.; Tikhomirov, V. M.: *Some remarks on variational principles.* Matematicheskie Zametki, 1997, N 1.

[2] Dontchev, A. L.: *The Graves Theorem Revisted.* Journ. of Conv. Anal., 1996, N1.

[3] Borwein, J. M.; Preiss, D.: *A smooth variational principle.* Trans. Amer. Math. Soc., 303, 1987.

[4] DeVille, R.: *Nouveaux principles variationnelles.* Sem. Inst. d'Analyse, 1990/91, N 21.

[5] Ekeland, I.: *Nonconvex Variational Problems.* Bull. AMS, 1979.

[6] Ioffe, A. D.; Tikhomirov, V. M.: *Theory of Extremal Problems.* North-Holland, 1979.

[7] Burzev, S. V.: *Existence theorems of implicit function in the conditions of extremum.* Mat. sbornik, 185, 1994.

[8] Ioffe, A. D.; Tikhomirov, V. M.: *Extension of Variational Problems.* Trudy Moscow Math. Obsh. 18, 1968.

[9] Gusseinov, F.: *Extension of Multidimensional Variational Problems.* Doctoral Dissertation. Baku, 1988.

Analysis and Synthesis of Control Systems and Dynamic Programming

International Series of Numerical Mathematics
Vol. 124,

Hamilton-Jacobi-Bellman Equations and Optimal Control

Italo Capuzzo Dolcetta*

1. Introduction

The aim of this paper is to offer a quick overview of some applications of the theory of *viscosity solutions* of *Hamilton-Jacobi-Bellman* equations connected to nonlinear optimal control problems.
The central role played by *value functions* and Hamilton-Jacobi equations in the Calculus of Variations was recognized as early as in C. Caratheodory's work, see [1] for a survey. Similar ideas reappeared under the name of Dynamic Programming in the work of R. Bellman and his school and became a standard tool for the synthesis of feedback controls for discrete time systems.
However, the lack of smoothness of value functions, even in simple problems, was recognized as a severe restriction to the range of applicability of Hamilton-Jacobi-Bellman theory (in short, HJB from now on) to continuous time processes.
The main reasons for this limitation are twofold:

(i) the very basic difficulty to give an appropriate global meaning to the HJB equation (a fully non linear partial differential equation) satisfied at all points of differentiability by the value function,

(ii) to identify value function as the unique solution of that equation; a related important issue is that of stability of value functions, specially in connection with approximation procedures required for computational purposes.

Several non-classical notions of solutions have been therefore proposed to overcome these difficulties. Let us mention, in this respect, the Kruzkov theory which applies, in the case of sufficiently smooth Hamiltonians, to semiconvex functions satisfying the HJB equation almost everywhere (see [2, 3] and also [4, 5] for recent results on semiconcavity of value functions).

Only in the 80's, however, a decisive impulse to the setting of a satisfactory mathematical framework to Dynamic Programming came from the introduction by Crandall-Lions [6] of the notion of viscosity solutions of Hamilton-Jacobi equations.

The presentation here, which is mainly based on material contained in the forthcoming book [7], to which we refer for detailed proofs, will be focused on optimization problems for controlled ordinary differential equations and discrete time systems.

*Dipartimento di Matematica, Università di Roma "La Sapienza", Piazzale A.Moro 2 - I 00185 Roma, Italy; e-mail: capuzzo@mat.uniroma1.it

The content of the paper is as follows:

2. Some examples of HJB equations in optimal control

Let us consider the *control system*, whose solution will be denoted by y_x^α,

$$\dot{y}(t) = f(y(t), \alpha(t)) \quad (t > 0), \quad y(0) = x. \tag{2.1}$$

In (2.1), $f : \mathbb{R}^N \times A \longrightarrow \mathbb{R}^N$, A is a topological space and

$$\alpha \in M(A) = \{\alpha : [0, +\infty) \longrightarrow A, \, \alpha \text{ measurable}\}$$

We assume:

(A_0) A is compact, f is continuous on $\mathbb{R}^N \times A$.

(A_1) $\exists L_f \geq 0 : |f(x,a) - f(y,a)| \leq L_f |x - y|, \forall a \in A.$

Here we list a few classical examples of optimal control problems and associated HJB equations (complemented in some cases by initial or boundary conditions) for system (2.1).

2.1 The infinite horizon discounted regulator

For this problem the *value function* is

$$v(x) = Inf_{\alpha \in M(A)} \int_0^{+\infty} l(y_x^\alpha(t), \alpha(t)) \, e^{-\lambda t} dt, \tag{2.2}$$

where the running cost l is a real valued function on $\mathbb{R}^N \times A$ and the discount factor λ is a positive number.
The corresponding HJB equation is

$$\lambda v(x) + Sup_{a \in A}[-f(x,a) \cdot Dv(x) - l(x,a)] = 0 \quad \text{in } \mathbb{R}^N. \tag{2.3}$$

2.2 The Mayer problem

In this case the value function is

$$v(x,t) = Inf_{\alpha \in M(A)} \quad g(y_x^\alpha(t)),$$

where the terminal cost is a given real valued function on $\mathbb{R}^N$.
The HJB equation takes now the form of a Cauchy problem, namely:

$$\frac{\partial v}{\partial t} + Sup_{a \in A}[-f(x,a) \cdot Dv(x)] = 0 \quad \text{in} \quad \mathbb{R}^N \times (0, +\infty)$$

$$v(x,0) = g(x).$$

2.3 Exit time problems

In these problems a subset $\mathcal{T}$ of $\mathbb{R}^N$ (the *target*) is given. If t_x^α denotes the first time when the trajectory y_x^α of system (2.1) hits the target, we consider the value function

$$v(x) = Inf_{\alpha \in M(A)} J(x,\alpha).$$

Here we set:

$$J(x,\alpha) = \int_0^{t_x^\alpha} l(y_x^\alpha(t),\alpha(t))\, e^{-\lambda t} dt + e^{-\lambda t_x^\alpha} g(y_x^\alpha({t_x}^\alpha)), \quad \text{if } t_x^\alpha < +\infty,$$

and

$$J(x,\alpha) = \int_0^{+\infty} l(y_x^\alpha(t),\alpha(t))\, e^{-\lambda t} dt, \quad \text{otherwise.}$$

The corresponding HJB equation is now the Dirichlet problem:

$$\lambda v(x) + Sup_{a\in A}[-f(x,a)\cdot Dv(x) - l(x,a)] = 0 \text{ in } \mathbb{R}^N \setminus \mathcal{T},$$
$$v(x) = g(x) \quad \text{if} \quad x \in \partial\mathcal{T}.$$

2.4 State-constrained problems

For this kind of problem the value function is

$$v(x) = Inf_{\alpha\in M_x(A)} \int_0^{+\infty} l(y_x^\alpha(t),\alpha(t))\, e^{-\lambda t} dt.$$

Here, the feasible controls are taken in the set

$$M_x(A) = \{\alpha \in M(A) : y_x^\alpha(t) \in \overline{\Omega}, \quad \forall t \geq 0\}.$$

The set $\overline{\Omega}$ plays here the role of a constraint on the states of system (2.1).
For the present problem the HJB equation (2.3) is complemented with a quite unusual boundary condition, namely

$$\lambda v(x) + Sup_{a\in A}[-f(x,a)\cdot Dv(x) - l(x,a)] \geq 0 \quad \text{in} \quad \partial\Omega.$$

2.5 The monotone control problem

In this example the control set A is the interval $[0,1]$ and the value function $v : \mathbb{R}^N \times [0,1] \longrightarrow \mathbb{R}$ is

$$v(x,a) = Inf_{\alpha\in M_a^m(A)} \int_0^{+\infty} l(y_x^\alpha(t),\alpha(t))\, e^{-\lambda t} dt.$$

Here we set:

$$M_a^m(A) = \{\alpha \in M(A), \quad \alpha \text{ nondecreasing}, \alpha(0) \geq a\}.$$

The HJB equation for this problem takes the form of the evolutionary variational inequality:

$$Max[\lambda v(x,a) - f(x,a)\cdot D_x v(x,a) - l(x,a); -\frac{\partial v}{\partial a}] = 0 \quad \text{in} \quad \mathbb{R}^N \times [0,1),$$

$$v(x,1) = \int_0^{+\infty} l(y_x^1(t),1)\, e^{-\lambda t} dt \quad \text{in } \mathbb{R}^N.$$

3. Some basic facts about viscosity solutions

It is well known that the value functions of the various optimal control problems described above satisfy the corresponding HJB equations at all points of differentiability. This fact can be proved by means of the *Dynamic Programming Principle*, a functional equation relating the value of v at the initial point x to its value at some point reached later by a trajectory of system (2.1).

Let us just indicate that, in the simplest case of the infinite horizon discounted regulator problem, the Dynamic Programming Principle is expressed by the identity

$$v(x) = Inf_{\alpha \in M(A)} \int_0^T l(y_x^\alpha(t), \alpha(t))\, e^{-\lambda t} dt + e^{-\lambda T} v((y_x^\alpha(T)), \tag{3.4}$$

which holds for all $x \in \mathbb{R}^N$ and $T > 0$.

It is well known as well that everywhere differentiability of v cannot hold in general. Indeed, simple examples show that two different optimal trajectories for some initial point x may exist implying non differentiability of v at x.

A concept of solution which allows to understand HJB equations globally, in a weak sense, is provided by the notion of viscosity solution.

The definition is based on the notion of first order *semidifferentials* of a function $u : \Omega \longrightarrow \mathbb{R}$, Ω being an open subset of $\mathbb{R}^N$.

These are the convex sets

$$D^+u(x) = \left\{ p \in \mathbb{R}^N : \limsup_{\Omega \ni y \to x} \frac{u(y) - u(x) - p \cdot (y - x)}{|x - y|} \leq 0 \right\}$$

$$D^-u(x) = \left\{ p \in \mathbb{R}^N : \liminf_{\Omega \ni y \to x} \frac{u(y) - u(x) - p \cdot (y - x)}{|x - y|} \geq 0 \right\}.$$

Consider the partial differential equation

$$F(x, u(x), Du(x)) = 0 \quad x \in \Omega \tag{3.5}$$

where $F : \Omega \times \mathbb{R} \times \mathbb{R}^N \longrightarrow \mathbb{R}$ is a continuous function.

Definition 1 *Let $u : \Omega \longrightarrow \mathbb{R}$. Then,*

(i) u is a viscosity subsolution of (3.5) if u is upper semicontinuous on Ω and

$$F(x, u(x), p) \leq 0 \quad \forall p \in D^+u(x), \forall x \in \Omega;$$

(ii) u is a viscosity supersolution of (3.5) if u is lower semicontinuous on Ω and

$$F(x, u(x), p) \geq 0, \quad \forall p \in D^-u(x), \forall x \in \Omega;$$

(iii) u is a viscosity solution of (3.5) if u satisfies both (i) and (ii).

Here we list some facts and remarks about this definition:

- if u is continuous on Ω then the sets $A^{\pm} = \{x \in \Omega : D^{\pm}u(x) \neq \emptyset\}$ are dense in Ω;
- if u is a viscosity solution of (3.5) then $F(x, u(x), Du(x)) = 0$ at any x where u is differentiable;
- if u is Lipschitz continuous and satisfies (3.5) in the viscosity sense then, $F(x, u(x), Du(x)) = 0$ almost everywhere;
- if $u \in C^1(\Omega)$ satisfies (3.5) at all points then u is a viscosity solution of (3.5); on the other hand, if $u \in C^1(\Omega)$ is a viscosity solution of (3.5) then u is a classical solution of (3.5);
- the equations (3.5) and $-F(x, u(x), Du(x)) = 0$ are not equivalent in the viscosity sense.
- a Lipschitz continuous function is a solution of equation (3.5) in the *extended sense* if

$$Sup_{p \in \partial u(x)} F(x, u(x), p) = 0$$

where ∂u is the Clarke's gradient of u; if u is a solution in the extended sense then u is both a viscosity subsolution of (3.5) and a supersolution of

$$-F(x, u(x), Du(x)) = 0;$$

on the other hand, a Lipschitz continuous viscosity solution u of (3.5) is a solution in the extended sense as well.

Although the notion of viscosity solution is a weak one, good *comparison*, *uniqueness* and *stability* properties hold. Sample results in these directions are the following ones where, for simplicity, we take

$$F(x, r, p) = r + H(x, p).$$

Theorem 1 *Assume that H is continuous and satisfies*

$$|H(x, p) - H(y, p)| \leq \omega(|x - y|(1 + |p|))$$

and

$$|H(x, p) - H(x, q)| \leq \omega(|p - q|)$$

for all $x, y \in \Omega$, and $p, q \in \mathbb{R}^N$, where $\omega : \mathbb{R}^+ \longrightarrow \mathbb{R}^+$ is continuous, nondecreasing, $\omega(0) = 0$.

If u,w are bounded continuous viscosity sub- and super-solutions of

$$u(x) + H(x, Du(x)) = 0 \quad in \quad \mathbb{R}^N,$$

then

$$u \equiv w \quad in \quad \mathbb{R}^N.$$

Let us just sketch the proof of the inequality $u \leq w$, the argument to prove the reverse being completely similar. Assume, by contradiction, the existence of some x_0 such that:

$$u(x_0) - w(x_0) := \delta > 0.$$

Define then

$$\Phi(x,y) = u(x) - w(y) - \frac{|x-y|^2}{2\varepsilon} - \beta((g(x)+g(y)),$$

where $\varepsilon > 0$ and

$$g(x) = \frac{1}{2} log(1+|x|^2).$$

The parameter β is chosen as to satisfy

$$\beta \leq \frac{\delta}{4g(x_0)}, \quad \omega(2\beta) \leq \frac{\delta}{6}.$$

The above choices yield:

$$Sup_{\mathbb{R}^N \times \mathbb{R}^N} \Phi \geq \Phi(x_0, x_0) \geq \frac{\delta}{2}. \tag{3.6}$$

By the assumptions made on u, w it is not hard to prove the existence of $(x_\varepsilon, y_\varepsilon)$ such that

$$\Phi(x_\varepsilon, y_\varepsilon) = Sup_{\mathbb{R}^N \times \mathbb{R}^N} \Phi$$

and that $(x_\varepsilon, y_\varepsilon)$ remain uniformly bounded with respect to ε. The key point is to observe that

$$p_\varepsilon := \frac{x_\varepsilon - y_\varepsilon}{\varepsilon} + \beta Dg(x_\varepsilon) \in D^+ u(x_\varepsilon),$$

and

$$q_\varepsilon := \frac{x_\varepsilon - y_\varepsilon}{\varepsilon} - \beta Dg(y_\varepsilon) \in D^- w(y_\varepsilon).$$

By definition of viscosity sub and super solution, then

$$u(x_\varepsilon) + H(x_\varepsilon, p_\varepsilon) \leq 0 \leq w(y_\varepsilon) + H(y_\varepsilon, q_\varepsilon).$$

By the assumptions on H this implies:

$$u(x_\varepsilon) - w(y_\varepsilon) \leq \omega(|p_\varepsilon - q_\varepsilon|) + \omega(|x_\varepsilon - y_\varepsilon|(1+|p_\varepsilon|)).$$

Therefore, by the choice of β and the fact that $|Dg| \leq 1$,

$$\Phi(x_\varepsilon, y_\varepsilon) \leq \frac{\delta}{6} + \omega(\varepsilon^{-1}|x_\varepsilon - y_\varepsilon|^2 + \frac{\delta}{4g(x_0)}|x_\varepsilon - y_\varepsilon|). \tag{3.7}$$

Observe now that the inequality

$$\Phi(x_\varepsilon, x_\varepsilon) + \Phi(y_\varepsilon, y_\varepsilon) \leq 2\Phi(x_\varepsilon, y_\varepsilon) \tag{3.8}$$

yields, since u and w are bounded, the following estimate

$$|x_\varepsilon - y_\varepsilon| \leq (\varepsilon C)^{\frac{1}{2}} \quad \text{for some } C > 0\ . \tag{3.9}$$

Thus (3.6) yields easily

$$\beta(g(x_\varepsilon) + g(y_\varepsilon)) \leq sup\, u + sup\, (-w)\ .$$

Since u,w are bounded while g is unbounded, the preceding implies the existence of some $R > 0$ such that $|x_\varepsilon| \le R$, $|y_\varepsilon| \le R$ for all $\varepsilon > 0$. Observe also that the inequality (3.8) is equivalent to

$$\frac{|x_\varepsilon - y_\varepsilon|^2}{\varepsilon} \le u(x_\varepsilon) - u(y_\varepsilon) + w(x_\varepsilon) - w(y_\varepsilon) \ .$$

Therefore, using the uniform continuity of u and v in the closed ball $B(0, R)$ and (3.9), it follows that

$$\frac{|x_\varepsilon - y_\varepsilon|^2}{\varepsilon} \to 0 \quad \text{as } \varepsilon \to 0.$$

At this point it is easy to realize that inequality (3.7) is contradictory with (3.6). This concludes the proof that $u \le w$.

As for stability we have:

Theorem 2 *Assume that H_n is continuous on $\Omega \times \mathbb{R}^N$ for each $n = 1, 2, \dots$ and that*

$$u_n(x) + H_n(x, Du_n(x)) = 0 \quad \textit{in } \Omega, \quad \textit{in the viscosity sense.}$$

Assume also that, as $n \to +\infty$,

$$u_n \to u, \quad H_n \to H \quad \textit{locally uniformly in} \quad \Omega \times \mathbb{R}^N.$$

Then,

$$u(x) + H(x, Du(x)) = 0 \quad \textit{in } \Omega, \quad \textit{in the viscosity sense.}$$

The uniform convergence of u_n to u guarantees that for any $x \in \Omega$ and $p \in D^+u(x)$ there exist $x_n \in \Omega$ and $p_n \in D^+u_n(x_n)$ such that

$$x_n \to x, \quad p_n \to p.$$

From this fact it follows easily that u is a subsolution of the limit equation.
A completely similar argument, with D^+ replaced by D^-, shows that u is a supersolution as well.

The theory outlined up to now does not depend on the *convexity* of the map

$$p \longrightarrow H(x, p).$$

When this property holds true (this fact is typical of Hamiltonians occurring in optimal control problems), then some special results are valid. Let us just mention the non obvious fact that in this case a function u is a viscosity solution of equation (3.5) if and only if

$$F(x, u(x), p) = 0 \quad \forall p \in D^-u(x)$$

(see [14, 15]).

4. Necessary and sufficient conditions

The value functions in the examples 2.1, 2.2, 2.5 of Section 2 are continuous under the assumptions (A_0), (A_1) plus some uniform continuity conditions on the costs l, g. Continuity of v in problems 2.3 and 2.4 is guaranteed under an additional restriction involving the behaviour of the dynamics f on the boundary of Ω or of $\mathcal{T}$. For problem 2.4 this condition is

$$Inf_{a \in A}\, f(x,a) \cdot n(x) < 0 \quad \forall x \in \partial\Omega$$

where $n(x)$ denotes the outward normal to $\partial\Omega$ at x, see ([16, 17]).

The link between the optimal control problem and the HJB equation is provided by the Dynamic Programming Principle. In all the examples presented the value function turns out to be the viscosity solution of the corresponding HJB equation.
Let us be more specific on this point with reference to Example 2.1 ; similar results hold, however, for the various examples shown in Section 2.
For the infinite horizon discounted regulator problem the Dynamic Programming Principle is expressed by the identity

$$v(x) = Inf_{\alpha \in M(A)} \int_0^T l(y_x^\alpha(t), \alpha(t))\, e^{-\lambda t} dt + e^{-\lambda T} v((y_x^\alpha(T)), \tag{4.10}$$

which holds for all $x \in \mathbb{R}^N$ and $T > 0$.

We have:

Theorem 3 *Assume (A_0),(A_1); assume also l continuous and bounded on $\mathbb{R}^N \times A$ and*

$$(A_2) \quad \exists L_l \geq 0 : |l(x,a) - l(y,a)| \leq L_l |x-y|, \quad \forall a \in A.$$

Then, the value function v of the infinite horizon discounted regulator problem is a bounded, continuous viscosity solution of (2.3).
Moreover, v is the unique viscosity solution of (2.3) in the class of bounded, continuous functions on $\mathbb{R}^N$.

For the proof, note that the second statement follows from the first and the uniqueness Theorem 1.
Let just indicate how to prove that v is a viscosity solution of (2.3). Observe that (4.10) with $\alpha(t) \equiv a \in A$ yields

$$-\frac{v(y_x^a(T) - v(x)}{T} \leq \frac{1}{T} \int_0^T l(y_x^a(t), a)\, e^{-\lambda t} dt + \frac{e^{-\lambda T} - 1}{T} v(y_x^a(T)) \tag{4.11}$$

for all $T > 0$.
On the other hand, $p \in D^+ v(x)$ implies

$$\frac{v(y_x^a(T)) - v(x)}{T} \leq \frac{1}{T} p \cdot (y_x^a(T) - x) + \frac{o(T)}{T}, \quad \text{as } T \to 0^+.$$

Hence, taking (2.1) into account,

$$\frac{v(y_x^a(T)) - v(x)}{T} \leq p \cdot \frac{1}{T} \int_0^T f(y_x^a(t), a)\, dt + \frac{o(T)}{T}, \quad \text{as } T \to 0^+.$$

for all $p \in D^+ v(x)$.

This and (4.11) imply, by continuity of f, l and v that

$$p \cdot f(x,a) \geq -l(x,a) + \lambda v(x)$$

for all $p \in D^+v(x)$ and $a \in A$.
This shows that v is a viscosity subsolution of (2.3); a similar, but slightly more difficult, argument shows that v is a supersolution as well.

From the above result, which characterizes the value function of the infinite horizon problem in terms of the Hamilton-Jacobi-Bellman equation (2.3), one can deduce necessary and sufficient conditions of optimality. Under the assumptions of Theorem 3 we get the following weak formulation of the Pontryagin Maximum Principle.

Theorem 4 *Assume $\alpha^\star = \alpha^\star(t)$ is an optimal control, corresponding to the initial position x, for the infinite horizon discounted regulator, i.e.*

$$v(x) = \int_0^{+\infty} l(y_x^{\alpha^\star}(t), \alpha^\star(t))\, e^{-\lambda t} dt.$$

Then the following hold for all $p \in D^+v(y_x^{\alpha^\star}(t)) \bigcup D^-v(y_x^{\alpha^\star}(t))$:

$$p \cdot \dot{y}_x^{\alpha^\star}(t) + l(y_x^{\alpha^\star}(t), \alpha^\star(t)) = \lambda v(y_x^{\alpha^\star}(t)) \quad a.e.\, t > 0$$

$$p \cdot f(y_x^{\alpha^\star}(t), \alpha^\star(t)) + l(y_x^{\alpha^\star}(t), \alpha^\star(t)) = Min_{a \in A}[p \cdot f(y_x^{\alpha^\star}(t), a) + l(y_x^{\alpha^\star}(t), a)]$$

As for sufficient conditions we have:

Theorem 5 *Under the assumptions of Theorem 3 and*

$$D^+v(x) \bigcup D^-v(x) \neq \emptyset,$$

if $\alpha^\star$ and x satisfy

$$p \cdot f(y_x^{\alpha^\star}(t), \alpha^\star(t)) + l(y_x^{\alpha^\star}(t), \alpha^\star(t)) = \lambda v(y_x^{\alpha^\star}(t)) \quad a.e.\, t > 0$$

for all $p \in D^+v(y_x^{\alpha^\star}(t)) \bigcup D^-v(y_x^{\alpha^\star}(t))$, then $\alpha^\star$ is optimal for the initial position x.

Note that condition $D^+v(x) \bigcup D^-v(x) \neq \emptyset$ is rather restrictive; it is fulfilled, for example, when v is *semiconcave*, i.e.

$$v(x+z) - 2v(x) + v(x-z) \leq C|z|^2$$

for some C and all x,z in $\mathbb{R}^N$.

5. Approximate synthesis

Consider again the infinite horizon problem 2.1, the associated HJB equation

$$\lambda v(x) + Sup_{a\in A}[-f(x,a)\cdot Dv(x) - l(x,a)] = 0 \quad \text{in } \mathbb{R}^N$$

and set for $h>0$:

$$u_h(x) + Sup_{a\in A}[-(1-\lambda h)u_h(x+hf(x,a)) - hl(x,a)] = 0 \quad \text{in} \quad \mathbb{R}^N. \tag{5.12}$$

Under the assumptions of Theorem 3, the Contraction Mapping Principle applies to show that for each $h \in (0, \frac{1}{\lambda})$ there exists of a unique bounded, continuous solution u_h to the above functional equation.

The functions u_h can be interpreted as value functions of a *discrete time* version of the infinite horizon problem. Let us define at this purpose

$$M_h(A) = \{\alpha \in M(A) : \alpha(t) \equiv \text{constant} \quad \forall t \in [kh,(k+1)h)\}$$

and, for $\alpha \in M_h(A)$, the discrete time control system

$$y_h(x,k+1) = y_h(x,k) + hf(y_h(x,k), a(kh)) \quad y_h(x,0) = x, \tag{5.13}$$

where $k = 0, 1, \ldots$.
Define then, a feedback law $a_h^\star : \mathbb{R}^N \longrightarrow A$ by selecting

$$a_h^\star(x) \in argmax_{a\in A}[-(1-\lambda h)u_h(x+hf(x,a)) - hl(x,a)]$$

where u_h is the solution of equation (5.12). Consider now the control $\alpha_h^\star \in M_h(A)$ given by

$$\alpha_h^\star(t) = a_h^\star(y_h^\star(x,[t/h]))$$

where $y_h^\star$ is obtained from (5.13).
It is not hard to check then that the solution u_h of (5.12) is given by

$$u_h(x) = h\sum_{k=0}^{+\infty}(1-\lambda h)^k l(y_h^\star(x,k), \alpha^\star(kh))$$

and also that

$$u_h(x) = Inf_{\alpha\in M_h(A)}\, h\sum_{k=0}^{+\infty}(1-\lambda h)^k l(y_h(x,k), \alpha(kh)).$$

The next result states that u_h converges to the value function v of the infinite horizon problem as the time step $h \to 0^+$.

Theorem 6 *Under the assumptions of Theorem 3 we have:*

$$Sup_K|u_h(x) - v(x)| \longrightarrow 0 \quad as \quad h \to 0^+$$

for all $K \subset\subset \mathbb{R}^N$. Under the further conditions $\lambda > 2L_f$, f smooth and l semiconcave, the estimate

$$Sup_{\mathbb{R}^N}|u_h(x) - v(x)| \le Ch \quad as \quad h \to 0^+$$

holds for some constant C.

As a consequence of this convergence result it follows that any optimal pair $(\alpha_h^\star, y_h^\star)$ for the above described discrete time problem converges weakly to an optimal relaxed pair $(\mu^\star, y^\star)$ for the original problem (2.2), see [11, 12]. Theorem 6 is also the starting point for a numerical approach to the computation of value functions and optimal feedbacks. We refer for example to [18, 19, 20, 21, 22].

6. Final remarks

In this paper we restricted our attention to the role played by viscosity solutions in optimal control problems for systems governed by ordinary differential equations. Only a few examples have been shortly described but many more can be approached in a similar way, see [7] for impulse and switching control problems,the minimum time problem and H_∞ control.
Some important topics we did not touch as well are discontinuous viscosity solutions and their applications to control and game problems with discontinuous value functions (e.g. the classical Zermelo navigation problem). Discontinuous viscosity solutions and the closely related *weak limits technique* are relevant also in the analysis of some asymptotic problems occurring, for example, in connections with ergodic systems (see [23]), large deviations (see [24]) or control of singularly perturbed systems (see [25]).

Let us mention, finally, that the viscosity solutions approach is flexible enough to be applicable to control problems for stochastic and distributed parameters systems as well as to differential games (we refer at this purpose to [8, 9, 10, 13]).

References

[1] Pesch H. J.; Bulirsch, R.: *The Maximum Principle, Bellman's equation, and Caratheodory's work.* J. Optim. Theory Appl. 80, 1994.

[2] Kruzkov, S. N.: *The Cauchy problem in the large for certain nonlinear first order differential equations.* Sov. Math. Dokl. 1, 1960.

[3] Kruzkov, S. N.: *Generalized solutions of the Hamilton-Jacobi equations of eikonal type I.* Math. USSR Sbornik 27, 1975.

[4] Cannarsa P.; Sinestrari, C.: *Convexity properties of the minimum time function.* Calc. Var. 3, 1995.

[5] Sinestrari, C.: *Semiconcavity of solutions of stationary Hamilton-Jacobi equations.* Nonlinear Analysis 24, 1995.

[6] Crandall, M. G.; Lions, P. L.: *Viscosity solutions of Hamilton-Jacobi equations.* Trans. Amer. Math. Soc. 277, 1983.

[7] Bardi, M.; Capuzzo Dolcetta, I.: *Optimal control and viscosity solutions of Hamilton-Jacobi-Bellman equations.* To appear, Birkhäuser 1997.

[8] Lions, P. L.: *Optimal control of diffusion processes and Hamilton-Jacobi equations.* Comm. Partial Differential Equations 8, 1983.

[9] Fleming, W. H.; Soner, M. H.: *Controlled Markov processes and viscosity solutions.* Springer Verlag 1993.

[10] Li X.; Yong, J.: *Optimal control theory for infinite dimensional systems.* Birkhäuser 1995.

[11] Capuzzo Dolcetta, I.: *On a discrete approximation of the Hamilton-Jacobi equation of dynamic programming.* Appl. Math. Optim. 10, 1983.

[12] Capuzzo Dolcetta, I.; Ishii, H.: *Approximate solutions of the Bellman equation of deterministic control theory.* Appl. Math. Optim. 11, 1984.

[13] Bardi, M.: *Some applications of viscosity solutions to optimal control and differential games.* In *Viscosity Solutions and Applications*, I. Capuzzo Dolcetta, P. L. Lions (eds.). To appear in Lecture Notes in Mathematics, Springer 1997.

[14] Barron, E. N.; Jensen, R.: *Semicontinuous viscosity solutions of Hamilton-Jacobi equations with convex hamiltonians.* Comm. Partial Differential Equations 15, 1990.

[15] Barles, G.: *Discontinuous viscosity solution of first order Hamilton-Jacobi equations: a guided visit.* Nonlinear Analysis 20, 1993.

[16] Soner, M. H.: *Optimal control problems with state-space constraints I-II.* SIAM J. Control Optim. 24, 1986.

[17] Capuzzo Dolcetta, I.; Lions, P. L.: *Hamilton-Jacobi equations with state constraints.* Trans. Amer. Math. Soc. 318, 1990.

[18] Gonzale, R.; Rofman, E.: *On deterministic control problems: an approximation procedure for the optimal cost.* SIAM J. Control Optim. 23, 1985.

[19] Falcone, M.: *A numerical approach to the infinite horizon problem of deterministic control theory.* Appl. Math. Optim. 15, 1987.

[20] Rouy, A.: *Numerical approximation of viscosity solutions of Hamilton-Jacobi equations with Neumann type boundary conditions.* Math. Models Methods Appl. Sci. 2, 1992.

[21] Falcone, M.; Ferretti, R.: *Discrete high-order schemes for viscosity solutions of Hamilton-Jacobi equations.* Numer. Math. 67, 1994.

[22] Bardi, M.; Bottacin, S.; Falcone, M.: *Convergence of discrete schemes for discontinuous value functions of pursuit-evasion games.* In G. J. Olsder, editor,*New Trends in Dynamic Games and Applications.* Birkhäuser 1995.

[23] Arisawa, M.: *Ergodic problem for the Hamilton-Jacobi-Bellman equation I and II.* Cahiers du CEREMADE. 1995.

[24] Barles, G.: *Solutions de Viscosite des Equations de Hamilton-Jacobi* Vol. 17. Mathematiques et Applications. Springer 1994.

[25] Bardi, M.; Bagagiolo, F.; Capuzzo Dolcetta, I.: *A viscosity solutions approach to some asymptotic problems in optimal control.* In J. P. Zolesio, editor, *Proceedings of the Conference "PDE's Methods in Control, Shape Optimization and Stochastic Modelling".* M. Dekker 1996.

International Series of Numerical Mathematics
Vol. 124, © 1998 Birkhäuser Verlag, Basel

Output Target Control and Uncertain Infinite-Dimensional Systems

Zbigniew Emirsajlow*

Abstract. The paper considers two output target control problems for an uncertain linear infinite-dimensional system with bounded input and output operators. Uncertainty in the system description is modelled by an unknown bounded perturbation of the system operator. We present an approach to computing estimates for the deviation of the terminal output of the perturbed system from the terminal output of the unperturbed system. This approach involves differential Liapunov equations and a concept of the so-called *composite semigroup*.

1. Introduction

The purpose of this paper is to sketch a mathematical framework for an analysis of terminal output target control problems for uncertain linear infinite-dimensional systems with bounded input and output operators. The uncertainty in the model is deterministic and is described by an unknown additive bounded perturbation to the system operator (a semigroup generator). Although we assume that the input and output operators as well as the perturbation are bounded, our approach can be extended to a wide range of classes of systems with unbounded input and output operators as well as unbounded perturbations, e.g., as in Emirsajlow, Pritchard, Townley [5] and Weiss [10]. The present paper extends the results of Emirsajlow [4]. We need to introduce the following notation and basic assumptions.

- H is a real Hilbert space identified with its dual, i.e. a *pivot space* [1], and plays the role of the *state space*. A is a linear operator on H generating a strongly continuous semigroup $\mathbf{T}(t) \in \mathcal{L}(H)$, $t \geq 0$, which describes the free dynamics of the system. The domain of A, denoted by $D(A)$, is a Hilbert space endowed with the scalar product $\langle \cdot, \cdot \rangle_{D(A)} = \langle A\cdot, A\cdot \rangle_H + \langle \cdot, \cdot \rangle_H$. $D(A^*)^*$ is a Hilbert space defined as the dual to the domain $D(A^*)$ of A^*. ($D(A^*)$ is a Hilbert space defined analogously as $D(A)$).
- U, the *control space*, is a real Hilbert space identified with its dual. $B \in \mathcal{L}(U, H)$ is the *input operator*. Y, the *output space*, is a real Hilbert space identified with its dual. $C \in \mathcal{L}(H, Y)$ is the *output operator*.
- $\Delta \in \mathcal{L}(H)$ is an unknown additive *perturbation* of the system operator A.

With the triple $\{A, B, C\}$ we associate a *nominal control system* Σ described by

$$\dot{x}(t) = Ax(t) + Bu(t), \quad x(0) = x_0, \quad t \in [0, \infty), \tag{1}$$
$$y(t) = Cx(t), \quad t \in [0, \infty). \tag{2}$$

*Technical University of Szczecin, 70-313 Szczecin, Poland, e-mail: emirsaj@we.tuniv.szczecin.pl where $x(\cdot)$ is the *state function* and $u(\cdot)$ is the *control*. A solution of the differential equation (1) is understood in the *mild sense*, which means that for all $x_0 \in H$ and $u(\cdot) \in L^2_{loc}(0, \infty; U)$ it is given by the integral formula

$$x(t) = \mathbf{T}(t)x_0 + \int_0^t \mathbf{T}(t-r)Bu(r)dr, \quad t \in [0, \infty), \tag{3}$$

and hence $x(\cdot) \in C([0, \infty); H)$. The output is described by

$$y(t) = Cx(t) = C\mathbf{T}(t)x_0 + C\int_0^t \mathbf{T}(t-r)Bu(r)dr, \quad t \in [0, \infty), \tag{4}$$

and hence $y(\cdot) \in C([0, \infty); Y)$. In order to make the situation more realistic we assume that in fact the system dynamics are *uncertain* and can be modelled as the following *perturbed control system* Σ_Δ

$$\dot{x}_\Delta(t) = A_\Delta x_\Delta(t) + Bu(t), \quad x_\Delta(0) = x_0, \quad t \in [0, \infty), \tag{5}$$
$$y_\Delta(t) = Cx_\Delta(t), \quad t \in [0, \infty), \tag{6}$$

where $A_\Delta = A + \Delta$.

It is a well-known result, e.g. Kato [6], Pazy [8], that for every $\Delta \in \mathcal{L}(H)$ the operator $A_\Delta = A + \Delta$ generates a strongly continuous semigroup $\mathbf{T}_\Delta(t) \in \mathcal{L}(H)$, $t \geq 0$, and $D(A_\Delta) = D(A)$. For every $u(\cdot) \in L^2_{loc}(0, \infty; U)$ and $x_0 \in H$ there exists a mild solution $x_\Delta(\cdot) \in C([0, \infty); H)$ of (5) given by the integral formula

$$x_\Delta(t) = \mathbf{T}_\Delta(t)x_0 + \int_0^t \mathbf{T}_\Delta(t-r)Bu(r)dr, \quad t \in [0, \infty), \tag{7}$$

and then the output is given by

$$y_\Delta(t) = Cx_\Delta(t) = C\mathbf{T}_\Delta(t)x_0 + C\int_0^t \mathbf{T}_\Delta(t-r)Bu(r)dr. \quad t \in [0, \infty), \tag{8}$$

In order to state the two *terminal output target control problems* we are going to consider, we assume that we are given a fixed time interval $[0, \tau]$, where $\tau \in (0, \infty)$, a final output $y_1 \in Y$ and a number $\alpha \in (0, \infty)$. This allows us to define two *output target sets* in the output space Y: (1) – a single point y_1 and (2) – a ball with radius $\sqrt{\alpha}$ centered at y_1. These two output targets determine the following two *sets of feasible controls*

$$\mathcal{U}_0 = \{u(\cdot) \in L^2(0, \tau; U) : \, y(\tau) = y_1\} \tag{9}$$

and

$$\mathcal{U}_\alpha = \{u(\cdot) \in L^2(0, \tau; U) : \, \|y(\tau) - y_1\|_Y^2 \leq \alpha\}. \tag{10}$$

Our terminal output target control problems (the name 'minimum energy control problems' would be also justified), take the following forms.

(E) : Find a control $u_E(\cdot) \in \mathcal{U}_0$ which minimizes the energy

$$E(u) = \int_0^\tau \|u(t)\|_U^2 dt. \tag{11}$$

(A) : Find a control $u_A(\cdot) \in \mathcal{U}_\alpha$ which minimizes the energy (11).

It is clear that for Problems (E) and (A) to make sense it is necessary that the sets of feasible controls $\mathcal{U}_0$ and $\mathcal{U}_\alpha$ are non-empty. Since, in general, non-emptiness of these sets depends on x_0, y_1 and α it is reasonable to impose on the system Σ conditions which will guarantee non-emptiness of sets of feasible controls for all data. For this reason we recall the following notions.

Definition 1 *The set $\mathcal{R} \subset Y$, defined by*

$$\mathcal{R} = \{y \in Y : y = C\int_0^\tau \mathbf{T}(\tau - t)Bu(t)dt,\ u \in L^2(0,\tau;U)\}, \tag{12}$$

is called the output reachability set *of Σ on $[0,\tau]$.*

Definition 2 *A system Σ is said to be* output exactly controllable *on $[0,\tau]$ if $\mathcal{R} = Y$ and* output approximately controllable *on $[0,\tau]$ if $\overline{\mathcal{R}} = Y$.*

Now, the following corollary is obvious.

Corollary 1 *If a system Σ is output exactly controllable on $[0,\tau]$, then for all $x_0 \in H$ and $y_1 \in Y$ Problem (E) posesses a unique solution $u_E(\cdot) \in \mathcal{U}_0$ and if Σ is output approximately controllable on $[0,\tau]$, then for all $x_0 \in H$, $y_1 \in Y$ and $\alpha \in (0,\infty)$ Problem (A) possesses a unique solution $u_A(\cdot) \in \mathcal{U}_\alpha$.*

Since the perturbation $\Delta \in \mathcal{L}(H)$ is unknown it is possible to compute the controls $u_E(\cdot)$ and $u_A(\cdot)$ only for the nominal system Σ. If we now apply these controls to the perturbed system Σ_Δ, then in general $y_\Delta(\tau) \neq y(\tau)$. The main **purpose** of this paper is to develop techniques for estimating the distance

$$\|y_\Delta(\tau) - y(\tau)\|_Y = ? \tag{13}$$

in terms of the norm $\|\Delta\|_{\mathcal{L}(H)}$ and nominal system parameters A, B, C (in fact, norms of some related operators). Then we can easily estimate the distance $\|y_\Delta(\tau) - y_1\|_Y = ?$

A basic estimate for (13) will be derived in Section 7 by combining auxiliary estimates derived in Sections 4 and 6. Before that, in Section 2, we provide explicit formulas for the controls $u_E(\cdot)$ and $u_A(\cdot)$. Then, in Section 3, we make use of these formulas to derive expressions for the difference $y_\Delta(\tau) - y(\tau)$ in both problems, i.e., Problem (E) and (A). Section 5 is devoted to differential Lyapunov equations where the notion of a composite semigroup is used. We complete the paper with Section 8 presenting a simple example and Section 9 containing some concluding remarks.

2. Explicit formulas for $u_E(\cdot)$ and $u_A(\cdot)$

The results presented in this section can be easily obtained by direct extension of the results contained in Emirsajlow [2] and [3]. So, we omit proofs.

For a start let us introduce two operators $\mathbf{X}(t) \in \mathcal{L}(H)$ and $\mathbf{Y}(t) \in \mathcal{L}(Y)$, $t \in [0,\tau]$, defined by

$$\mathbf{X}(t) = \int_t^\tau \mathbf{T}(r - t)BB^*\mathbf{T}^*(r - t)dr, \tag{14}$$

$$\mathbf{Y}(t) = C\mathbf{X}(t)C^*. \tag{15}$$

Theorem 1 *If Σ is output exactly controllable on $[0, \tau]$, then there exists a unique solution of Problem (E) given by*

$$u_E(t) = B^*\mathbf{T}^*(\tau - t)C^*q_0, \tag{16}$$

where $q_0 \in Y$ is a unique solution to the equation

$$\mathbf{Y}(0)q_0 = y_1 - C\mathbf{T}(\tau)x_0. \tag{17}$$

Theorem 2 *If Σ is output approximately controllable on $[0, \tau]$, then there exists a unique solution of Problem (A) given by:*
(a) For $\alpha \in (0, \|y_1 - C\mathbf{T}(\tau)x_0\|_Y^2)$

$$u_A(t) = B^*\mathbf{T}^*(\tau - t)C^*q_\varepsilon, \tag{18}$$

where $q_\varepsilon \in Y$ is a unique solution to the equation

$$(\mathbf{Y}(0) + \varepsilon I)q_\varepsilon = y_1 - C\mathbf{T}(\tau)x_0, \tag{19}$$

where $\varepsilon \in (0, \infty)$ uniquely satisfies the condition

$$\varepsilon^2\|q_\varepsilon\|_Y^2 = \alpha. \tag{20}$$

(b) For $\alpha \in [\|y_1 - C\mathbf{T}(\tau)x_0\|_Y^2, \infty)$

$$u_A(t) \equiv 0. \tag{21}$$

If α satisfies the condition $\alpha \in (0, \|y_1 - C\mathbf{T}(\tau)x_0\|_Y^2)$, then there always exists a unique $\varepsilon \in (0, \infty)$ satisfying the condition (20).

3. Expression for the difference $y_\Delta(\tau) - y(\tau)$

Applying $u_E(\cdot)$, given by the formula (16), to Σ_Δ leads to the following expression for the final output

$$y_\Delta(\tau) = C\int_0^\tau \mathbf{T}_\Delta(\tau - r)BB^*\mathbf{T}^*(\tau - r)C^*q_0 dr + C\mathbf{T}_\Delta(\tau)x_0. \tag{22}$$

Hence, it follows that if we define an operator $\mathbf{X}_\Delta(t) \in \mathcal{L}(H)$, $t \in [0, \tau]$, by

$$\mathbf{X}_\Delta(t) = \int_t^\tau \mathbf{T}_\Delta(r - t)BB^*\mathbf{T}^*(r - t)dr, \tag{23}$$

then we easily obtain the following expression for the difference $y_\Delta(\tau) - y(\tau)$ in Problem (E)

$$y_\Delta(\tau) - y(\tau) = C(\mathbf{X}_\Delta(0) - \mathbf{X}(0))C^*q_0 + C(\mathbf{T}_\Delta(\tau) - \mathbf{T}(\tau))x_0. \tag{24}$$

In turn, in Problem (A) we obtain

$$y_\Delta(\tau) - y(\tau) = C(\mathbf{X}_\Delta(0) - \mathbf{X}(0))C^*q_\varepsilon + C(\mathbf{T}_\Delta(\tau) - \mathbf{T}(\tau))x_0. \tag{25}$$

In the expresssions (17) and (19) $x_0 \in H$, $y_1 \in Y$ are allowed to be arbitrary. This implies that also $q_0, q_\varepsilon \in Y$ in (24) and (25) can be arbitrary. Thus, in order to estimate the norm

$$\|y_\Delta(\tau) - y(\tau)\|_Y \tag{26}$$

in both cases we have to find a method of estimating the operator norms

$$\|\mathbf{X}_\Delta(0) - \mathbf{X}(0)\|_{\mathcal{L}(H)}, \quad \|\mathbf{T}_\Delta(\tau) - \mathbf{T}(\tau)\|_{\mathcal{L}(H)}. \tag{27}$$

4. An estimate for the norm $\|\mathbf{T}_\Delta(\tau) - \mathbf{T}(\tau)\|_{\mathcal{L}(H)}$

In this section we need to introduce three operators

$$\mathbf{L}_\tau \in \mathcal{L}(C([0,\tau];H)) \ : \ (\mathbf{L}_\tau h)(t) = \int_0^t \mathbf{T}(t-r)h(r)dr,\ t \in [0,\tau], \tag{28}$$

$$\mathbf{N}_\tau \in \mathcal{L}(C([0,\tau];H),H) \ : \ \mathbf{N}_\tau h = (\mathbf{L}_\tau h)(\tau) = \int_0^\tau \mathbf{T}(\tau-r)h(r)dr, \tag{29}$$

$$\mathbf{T}_\tau \in \mathcal{L}(H, C([0,\tau];H)) \ : \ (\mathbf{T}_\tau h)(t) = \mathbf{T}(t)h, \quad h \in H,\ t \in [0,\tau], \tag{30}$$

where $h(\cdot) \in C([0,\tau];H)$. In the sequel, norms of these operators will be understood with respect to the above spaces.
The main result of this section reads as follows.

Theorem 3 *For every $\Delta \in \mathcal{L}(H)$ the following holds*

$$\mathbf{T}_\Delta(\tau) - \mathbf{T}(\tau) = \mathbf{N}_\tau \Delta (I - \mathbf{L}_\tau \Delta)^{-1} \mathbf{T}_\tau. \tag{31}$$

If $\Delta \in \mathcal{L}(H)$ is such that $\|\Delta\|_{\mathcal{L}(H)} < \|\mathbf{L}_\tau\|^{-1}$, then

$$\|\mathbf{T}_\Delta(\tau) - \mathbf{T}(\tau)\|_{\mathcal{L}(H)} \leq \frac{\|\mathbf{N}_\tau\|\|\mathbf{T}_\tau\|\|\Delta\|_{\mathcal{L}(H)}}{1 - \|\mathbf{L}_\tau\|\|\Delta\|_{\mathcal{L}(H)}}. \tag{32}$$

Proof: Let us notice that

$$z(t) = \mathbf{T}(t)x_0, \quad \dot{z}(t) = Ax(t),\ t \in [0,\infty],\ z(0) = x_0$$

and

$$z_\Delta(t) = \mathbf{T}_\Delta(t)x_0, \quad \dot{z}_\Delta(t) = (A+\Delta)z_\Delta(t),\ t \in [0,\infty],\ z_\Delta(0) = x_0.$$

Hence

$$z_\Delta(t) = \mathbf{T}(t)x_0 + \int_0^t \mathbf{T}(t-r)\Delta z_\Delta(r)dr, \quad t \in [0,\infty), \tag{33}$$

and making use of (28) and (30), we obtain

$$z_\Delta(t) = (\mathbf{T}_\tau x_0)(t) + (\mathbf{L}_\tau \Delta z_\Delta)(t), \quad t \in [0,\tau].$$

Since the operator $(I - \mathbf{L}_\tau\Delta) \in \mathcal{L}(C([0,\tau];H))$ is boundedly invertible for all $\Delta \in \mathcal{L}(H)$ we obtain

$$z_\Delta(t) = ((I - \mathbf{L}_\tau\Delta)^{-1}\mathbf{T}_\tau x_0)(t), \quad t \in [0,\tau]. \tag{34}$$

It follows from (33) that

$$z_\Delta(\tau) - z(\tau) = \mathbf{N}_\tau \Delta z_\Delta$$

and putting (34) into this expression gives

$$\mathbf{T}_\Delta(\tau)x_0 - \mathbf{T}(\tau)x_0 = \mathbf{N}_\tau\Delta(I - \mathbf{L}_\tau\Delta)^{-1}\mathbf{T}_\tau x_0.$$

Since $x_0 \in H$ can be arbitrary we obtain (31). This is an exact formula for the difference $\mathbf{T}_\Delta(\tau) - \mathbf{T}(\tau)$, valid for all $\Delta \in \mathcal{L}(H)$. Making now use of a standard estimate, see e.g. Trenogin [9], we obtain (32). ∎

5. Differential Lyapunov equations and a composite semigroup

Differentiating (14) and (23) with respect to t leads to the following differential Lyapunov equations

$$\begin{aligned}\dot{\mathbf{X}}(t)h &= -A\mathbf{X}(t)h - \mathbf{X}(t)A^*h - BB^*h, \quad \mathbf{X}(\tau) = 0, &(35)\\ \dot{\mathbf{X}}_\Delta(t)h &= -A\mathbf{X}_\Delta(t)h - \mathbf{X}_\Delta(t)A^*h - BB^*h - \Delta\mathbf{X}_\Delta(t)h, \quad \mathbf{X}_\Delta(\tau) = 0, &(36)\end{aligned}$$

where $\mathbf{X}(t), \mathbf{X}_\Delta(t) \in \mathcal{L}(H)$ for $t \in [0,\tau]$, $h \in D(A^*)$ and (35) and (36) hold in $D(A^*)^*$.

In the sequel we show that (35) and (36) have unique solutions $\mathbf{X}(\cdot), \mathbf{X}_\Delta(\cdot) \in \mathcal{L}(H, C([0,\tau]; H))$, i.e. $\mathbf{X}(t), \mathbf{X}_\Delta(t) \in \mathcal{L}(H)$ for $t \in [0,\tau]$ and are continuous in the strong operator topology. For this purpose we introduce a special type of a one-parameter semigroup which we call a *composite semigroup* (it is also called an *implemented semigroup* in Nagel [7]).

A *composite semigroup* $\mathbb{T}(t) : \mathcal{L}(H) \to \mathcal{L}(H)$, $t \in [0,\infty)$, is defined by

$$\mathbb{T}(t)X = \mathbf{T}(t)X\mathbf{T}^*(t), \quad X \in \mathcal{L}(H),\ t \in [0,\infty), \tag{37}$$

where $\mathbf{T}(t) \in \mathcal{L}(H)$, $t \in [0,\infty)$, is generated by A. The *infinitesimal generator* $\mathbb{A} : \mathcal{L}(H) \supset D(\mathbb{A}) \to \mathcal{L}(H)$ of $\mathbb{T}(t)$ is defined as follows: its *domain*

$$D(\mathbb{A}) = \{X \in \mathcal{L}(H) : \lim_{t\to 0+} \frac{(\mathbb{T}(t)X)h - Xh}{t} \text{ exists }\}, \tag{38}$$

where the limit has to exist in H (in the strong sense) for every $h \in H$, and then

$$(\mathbb{A}X)h = \lim_{t\to 0+} \frac{(\mathbb{T}(t)X)h - Xh}{t}, \quad X \in D(\mathbb{A}) \quad h \in H. \tag{39}$$

It can be shown that the operator $\mathbb{A}$ satisfies

$$(\mathbb{A}X)h = AXh + XA^*h, \quad X \in D(\mathbb{A}), \quad h \in D(A^*). \tag{40}$$

Let us notice that although $\mathbf{T}(\cdot)$ is strongly continuous, the composite semigroup $\mathbb{T}(\cdot)$ is not strongly continuous in the usual sense, i.e. in general we have $\lim_{t\to 0} \mathbb{T}(t)X = X$ only in the strong topology of $\mathcal{L}(H)$ and *not* in the uniform one. For the latter type of continuity of $\mathbb{T}(\cdot)$ we need uniform continuity of the original semigroup $\mathbf{T}(\cdot)$. It is clear that $D(\mathbb{A}) \subset \mathcal{L}(H)$ is a Banach space when equipped with the graph norm $\|X\|_{D(\mathbb{A})} = \|\mathbb{A}X\|_{\mathcal{L}(H)} + \|X\|_{\mathcal{L}(H)}$.

Using the composite semigroup we can rewrite (35) and (36) as follows

$$\begin{aligned}\dot{\mathbf{X}}(t) &= -\mathbb{A}\mathbf{X}(t) - BB^*, \quad \mathbf{X}(\tau) = 0, \quad t \in [0,\tau], &(41)\\ \dot{\mathbf{X}}_\Delta(t) &= -\mathbb{A}\mathbf{X}_\Delta(t) - \Delta\mathbf{X}_\Delta(t) - BB^*, \quad \mathbf{X}_\Delta(\tau) = 0, \quad t \in [0,\tau]. &(42)\end{aligned}$$

Both these differential equations are now understood in $\mathcal{L}(H)$. Adapting standard results on linear differential equations on Banach spaces, e.g. Pazy [8], we can prove the existence and uniqueness of a solution $\mathbf{X}(t) \in D(\mathbb{A})$ and $\mathbf{X}(\cdot) \in \mathcal{L}(H, C^1([0,\tau]; H))$ of (41). In turn, adapting the standard perturbation results for semigroups on Banach spaces, e.g. Kato [6] and Pazy [8], we can prove the existence and uniqueness of a solution $\mathbf{X}_\Delta(t) \in D(\mathbb{A})$ and $\mathbf{X}_\Delta(\cdot) \in \mathcal{L}(H, C^1([0,\tau]; H))$ of (42). In other words, if Δ is regarded as an operator in $\mathcal{L}(\mathcal{L}(H))$, then the operator $\mathbb{A}_\Delta = \mathbb{A} + \Delta$ generates a semigroup $\mathbb{T}_\Delta(t) : \mathcal{L}(H) \to \mathcal{L}(H)$, $t \in [0,\infty)$, with the same continuity properties with respect to t as $\mathbb{T}(t)$ and $D(\mathbb{A}_\Delta) = D(\mathbb{A})$.

6. An estimate for the norm $\|\mathbf{X}_\Delta(0) - \mathbf{X}(0)\|_{\mathcal{L}(H)}$

The results of Section 5 enable us to express $\mathbf{X}(\cdot)$ and $\mathbf{X}_\Delta(\cdot)$, defined by (14) and (23), in the following form

$$\mathbf{X}(t) = \int_t^\tau \mathbb{T}(\tau - r)(BB^*)dr, \quad \mathbf{X}_\Delta(t) = \int_t^\tau \mathbb{T}(\tau - r)(\Delta \mathbf{X}_\Delta(r))dr + \int_t^\tau \mathbb{T}(\tau - r)(BB^*)dr, \tag{43}$$

where $t \in [0, \tau]$. Next, we need to introduce two operators

$$\mathbb{L}_\tau \in \mathcal{L}(\mathcal{L}(H, C([0,\tau]; H))) \ : \ (\mathbb{L}_\tau X)(t) = \int_t^\tau \mathbb{T}(\tau - r)X(r)dr, \quad t \in [0,\tau], \tag{44}$$

$$\mathbb{N}_\tau \in \mathcal{L}(\mathcal{L}(H, C([0,\tau]; H)), \mathcal{L}(H)) \ : \ \mathbb{N}_\tau X = (\mathbb{L}_\tau X)(0) = \int_0^\tau \mathbb{T}(\tau - r)X(r)dr, \tag{45}$$

where $X(\cdot) \in \mathcal{L}(H, C([0,\tau]; H))$. In the sequel, norms of these operators will be understood with respect to the above spaces. The main result of this section reads as follows.

Theorem 4 *For every $\Delta \in \mathcal{L}(H)$ the following holds*

$$\mathbf{X}_\Delta(0) - \mathbf{X}(0) = \mathbb{N}_\tau \Delta (I - \mathbb{L}_\tau \Delta)^{-1} \mathbb{L}_\tau (BB^*). \tag{46}$$

If $\Delta \in \mathcal{L}(H)$ is such that $\|\Delta\|_{\mathcal{L}(H)} < \|\mathbb{L}_\tau\|^{-1}$, then

$$\|\mathbf{X}_\Delta(0) - \mathbf{X}(0)\|_{\mathcal{L}(H)} \le \frac{\|\mathbb{N}_\tau\| \|\mathbb{L}_\tau(BB^*)\| \|\Delta\|_{\mathcal{L}(H)}}{1 - \|\mathbb{L}_\tau\| \|\Delta\|_{\mathcal{L}(H)}}. \tag{47}$$

Proof: Using (43) and (44) we obtain

$$\mathbf{X}(t) = (\mathbb{L}_\tau(BB^*))(t), \quad t \in [0,\tau] \tag{48}$$

and

$$\mathbf{X}_\Delta(t) = ((I - \mathbb{L}_\tau \Delta)^{-1} \mathbb{L}_\tau(BB^*))(t), \quad t \in [0,\tau], \tag{49}$$

where the bounded invertibility of $(I - \mathbb{L}_\tau \Delta) \in \mathcal{L}(C([0,\tau]; \mathcal{L}(H)))$ can be easily justified. Hence,

$$\mathbf{X}_\Delta(t) - \mathbf{X}(t) = (\mathbb{L}_\tau \Delta (I - \mathbb{L}_\tau \Delta)^{-1} \mathbb{L}_\tau(BB^*))(t), \quad t \in [0,\tau], \tag{50}$$

and using (45), we finally get

$$\mathbf{X}_\Delta(0) - \mathbf{X}(0) = \mathbf{N}_\tau \Delta (I - \mathbb{L}_\tau \Delta)^{-1} \mathbb{L}_\tau (BB^*). \tag{51}$$

This is an exact formula for the difference $\mathbf{X}_\Delta(0) - \mathbf{X}(0)$, valid for all $\Delta \in \mathcal{L}(H)$. Now, using the standard estimate on the norm of the operator $(I - \mathbb{L}_\tau \Delta)^{-1}$, see e.g. Trenogin [9], we obtain (47). ■

7. Estimates for the norm $\|y_\Delta(\tau) - y(\tau)\|_Y$

Using (24), (25) and Theorems 4.1 and 6.1 we easily obtain the required estimates for the norm $\|y_\Delta(\tau) - y(\tau)\|_Y$ in both control problems under consideration.

Theorem 5 *Suppose $\Delta \in \mathcal{L}(H)$ is such that $\|\Delta\|_{\mathcal{L}(H)} < \min\{\|\mathbf{L}_\tau\|^{-1}; \|\mathbb{L}_\tau\|^{-1}\}$, and the assumptions of Theorems 2.2 and 2.3 hold for Problems (E) and (A), respectively. Then, the distance $\|y_\Delta(\tau) - y(\tau)\|_Y$ in Problem (E) can be estimated as follows*

$$\|y_\Delta(\tau) - y(\tau)\|_Y \leq \frac{\|C\mathbb{N}_\tau C^*\|\|\mathbb{L}_\tau(BB^*)\|\|\Delta\|_{\mathcal{L}(H)}}{1 - \|\mathbb{L}_\tau\|\|\Delta\|_{\mathcal{L}(H)}}\|q_0\|_Y + \frac{\|C\mathbf{N}_\tau\|\|\mathbf{T}_\tau\|\|\Delta\|_{\mathcal{L}(H)}}{1 - \|\mathbf{L}_\tau\|\|\Delta\|_{\mathcal{L}(H)}}\|x_0\|_H \tag{52}$$

and in Problem (A), as follows

$$\|y_\Delta(\tau) - y(\tau)\|_Y \leq \frac{\|C\mathbb{N}_\tau C^*\|\|\mathbb{L}_\tau(BB^*)\|\|\Delta\|_{\mathcal{L}(H)}}{1 - \|\mathbb{L}_\tau\|\|\Delta\|_{\mathcal{L}(H)}}\|q_\varepsilon\|_H + \frac{\|C\mathbf{N}_\tau\|\|\mathbf{T}_\tau\|\|\Delta\|_{\mathcal{L}(H)}}{1 - \|\mathbf{L}_\tau\|\|\Delta\|_{\mathcal{L}(H)}}\|x_0\|_H, \tag{53}$$

where we assume $q_\varepsilon = 0$ in the case of part (b) of Theorem 2.3.

Actual use of the estimates (52) and (53) requires computation of the operator norms $\|C\mathbb{N}_\tau C^*\|$, $\|\mathbb{L}_\tau(BB^*)\|$, $\|\mathbb{L}_\tau\|$ and $\|C\mathbf{N}_\tau\|$, $\|\mathbf{L}_\tau\|$, $\|\mathbf{T}_\tau\|$ which is not necessarily easy. In general, there always exist constants $M \geq 1$ and $\omega \in (-\infty, \infty)$ such that $\|\mathbf{T}(t)\|_{\mathcal{L}(H)} \leq Me^{\omega t}, \quad t \in [0, \infty)$. Using this fact we can obtain crude estimates for the above operator norms. In particular,

$$\|C\mathbf{N}_\tau\| \leq M\|C\|\frac{e^{\omega\tau} - 1}{\omega}, \qquad \|\mathbf{L}_\tau\| \leq M\frac{e^{\omega\tau} - 1}{\omega}, \qquad \|\mathbf{T}_\tau\| \leq M_\tau \tag{54}$$

and

$$\|C\mathbb{N}_\tau C^*\| \leq M^2\|C\|^2\frac{e^{2\omega\tau} - 1}{2\omega}, \; \|\mathbb{L}_\tau(BB^*)\| \leq M^2\|B\|^2\frac{e^{2\omega\tau} - 1}{2\omega}, \; \|\mathbb{L}_\tau\| \leq M^2\frac{e^{2\omega\tau} - 1}{2\omega}, \tag{55}$$

where M_τ is defined as follows

$$M_\tau = \begin{cases} M & \text{if} \quad \omega \leq 0 \\ Me^{\omega\tau} & \text{if} \quad \omega > 0 \end{cases}. \tag{56}$$

8. A simple numerical example

The purpose of this section is to test on a simple scalar example how conservative are the estimates (52) and (53). In this example we make use of estimates on the operator norms given in in the previous section.

Let the nominal system Σ be described by the following scalar differential equation

$$\dot{x}(t) = ax(t) + bu(t), \quad x(0) = x_0, \; t \in [0, \infty), \tag{57}$$

$$y(t) = cx(t), \tag{58}$$

where $a, b, c, x_0 \in \mathbb{R}$, and the perturbed system Σ_δ - by

$$\begin{aligned} \dot{x_\delta}(t) &= (a+\delta)x_\delta(t) + bu(t), \quad x_\delta(0) = x_0,\ t \in [0,\infty), & (59)\\ y(t) &= cx(t), & (60) \end{aligned}$$

where $\delta \in \mathbb{R}$. Moreover, we assume we are given $\tau \in (0,\infty)$, $y_1 \in \mathbb{R}$ and $\alpha \in (0, |y_1 - ce^{a\tau}x_0|)$. If $a \neq 0$, $b \neq 0$ and $c \neq 0$, then Σ is exactly output controllable on every interval $[0,\tau]$. In this simple case exact output controllability and approximate output controllability coincide. Using Theorems 1 and 2 we obtain the following formulas for the controls $u_E(\cdot)$ and $u_A(\cdot)$

$$u_E(t) = be^{a(\tau-t)}cq_0, \quad q_0 = \frac{2a}{b^2c^2}\frac{y_1 - ce^{a\tau}x_0}{e^{2a\tau}-1}, \tag{61}$$

and

$$u_A(t) = be^{a(\tau-t)}cq_\varepsilon, \quad q_\varepsilon = q_0 - \frac{2a}{b^2c^2}\frac{\sqrt{\alpha}\,\mathrm{sign}(y_1 - ce^{a\tau}x_0)}{e^{2a\tau}-1}. \tag{62}$$

Applying u_E and u_A to Σ and Σ_δ allows us to calculate expressions for the final outputs $y_\delta(\tau)$ and $y(\tau)$. For Problem (E) we obtain

$$y(\tau) = y_1, \quad y_\delta(\tau) = ce^{(a+\delta)\tau}x_0 + \frac{cb^2}{2a+\delta}(e^{(2a+\delta)\tau} - 1)q_0 \tag{63}$$

and for Problem (A) we obtain

$$y(\tau) = y_1 - \sqrt{\alpha}\,\mathrm{sign}(y_1 - ce^{a\tau}x_0), \quad y_\delta(\tau) = ce^{(a+\delta)\tau}x_0 + \frac{cb^2}{2a+\delta}(e^{(2a+\delta)\tau} - 1)q_\varepsilon. \tag{64}$$

Thus we can compute an exact value for the distance $|y_\delta(\tau) - y(\tau)|$ in both problems. In this example, the constants M and ω from Section 7 satisfy $M = 1$, $\omega = a$, and we can use all the estimates (54) and (55). Thus, formulas (52) and (53) allow us to estimate the distance $|y_\delta(\tau) - y(\tau)|$ in Problems (E) and (A) whenever $\delta \in \mathbb{R}$ satisfies

$$|\delta| < \min\left\{\frac{a}{e^{a\tau}-1}; \frac{2a}{e^{2a\tau}-1}\right\}. \tag{65}$$

Below we present results of computations performed for $a = -1$, $b = 1$, $c = 1$, $\tau = 1$, $x_0 = 0$, $y_1 = 1$, $\alpha = 0.01$ and two values of δ satisfying the same bound $|\delta| \leq 0.05$.

- $\delta_a = 0.05$,

$$\begin{aligned} &\text{(E):}\ |y_\delta(\tau) - y(\tau)| = 0.0174, \quad \mathcal{E} = 0.0221,\\ &\text{(A):}\ |y_\delta(\tau) - y(\tau)| = 0.0157, \quad \mathcal{E} = 0.0199. \end{aligned}$$

- $\delta_a = -0.05$,

$$\begin{aligned} &\text{(E):}\ |y_\delta(\tau) - y(\tau)| = 0.0169, \quad \mathcal{E} = 0.0221,\\ &\text{(A):}\ |y_\delta(\tau) - y(\tau)| = 0.0152, \quad \mathcal{E} = 0.0199. \end{aligned}$$

In the above expressions, $\mathcal{E}$-s denote estimates on the distance $|y_\delta(\tau) - y(\tau)|$ provided by Theorem 5, i.e., for the worst case situation. It follows from these computations that the estimates (52) and (53) are reasonably tight and may be of practical use.

9. Concluding remarks

It is clear that better estimates for the norms $\|C\mathbb{N}_\tau C^*\|$, $\|\mathbb{L}_\tau(BB^*)\|$, $\|\mathbb{L}_\tau\|$ and $\|C\mathbf{N}_\tau\|$, $\|\mathbf{L}_\tau\|$, $\|\mathbf{T}_\tau\|$ imply more tight estimates (52), (53). In this respect it is possible to replace the space $C([0,\tau];H)$ by $L^2(0,\tau;H)$ in definitions (28)–(30) and then use the following useful relations

$$\|\mathbf{T}_\tau\|^2 = \|\int_0^\tau \mathbf{T}^*(t)\mathbf{T}(t)dt\|_{\mathcal{L}(H)}, \quad \|C\mathbf{N}_\tau\|^2 = \|C\int_0^\tau \mathbf{T}(t)\mathbf{T}^*(t)dtC^*\|_{\mathcal{L}(Y)}, \tag{66}$$

and if the semigroup $\mathbf{T}(\cdot)$ is exponentially stable, then also

$$\|\mathbf{L}_\tau\| = \sup_{k=0,\pm1,\pm2,\pm3,\ldots} \|(\jmath\frac{k\pi}{\tau} - A)^{-1}\|_{\mathcal{L}(H)}. \tag{67}$$

Another possibility of obtaining tight estimates for the above norms is to use parametrized quadratic optimization problems and related parametrized differential Riccati equations. Exploration of this possiblity as well as the development of similar characterizations for the norms $\|C\mathbb{N}_\tau C^*\|$, $\|\mathbb{L}_\tau(BB^*)\|$ and $\|\mathbb{L}_\tau\|$ require additional amount of research and are far beyond the scope of this paper.

References

[1] Aubin, J.P.: *Applied Functional Analysis.* John Wiley & Sons, New York, 1979.

[2] Emirsajlow, Z.: *A feedback for an infinite-dimensional linear-quadratic control problem with a fixed terminal state.* IMA J. Mathematical Control and Information, Vol. 6, 1989.

[3] Emirsajlow, Z.: *Feedback control in LQCP with a terminal inequality constraint.* J. Optimization Theory and Applications, Vol. 62, No. 3, 1989.

[4] Emirsajlow, Z.: *Terminal target control and uncertain systems.* Proceedings of the 3rd European Control Conference ECC 95, September 5–8, 1995, Roma, Italy, Vol. 3, Part one.

[5] Emirsajlow, Z.; Pritchard, A.J.; Townley, S.: *On structured perturbations for two classes of linear infinite-dimensional systems.* Dynamics and Control, Vol. 6, 1996.

[6] Kato, T.: *Perturbation Theory of Linear Operators.* Springer-Verlag, Berlin 1966.

[7] Nagel, R. (Ed.): *One-Parameter Semigroup of Positive Operators.* Springer-Verlag, Berlin, 1986.

[8] Pazy, A.: *Semigroups of Linear Operators and Applications to Partial Differential Equations.* Springer-Verlag, New York, 1983.

[9] Trenogin, V.A.: *Functional Analysis.* Nauka, Moscow 1980.

[10] Weiss, G.: *Regular linear systems with feedback.* Mathematics of Control, Signals and Systems, Vol 7, 1994.

International Series of Numerical Mathematics
Vol. 124, © 1998 Birkhäuser Verlag, Basel

Sensitivity Analysis of Stiff and Non-Stiff Initial-Value Problems

Martin Kiehl *

Abstract. The solution $y(t, t_0, y_0)$ of an initial-value problem (IVP) $\dot{y}(t) = f(t, y, p)$ with initial value $y(t_0) = y_0$ at a point t is a differentiable function of the initial value y_0 and the parameter vector p, provided f_y and f_p are continuous. The computation of the derivatives of $y(t, t_0, y_0)$ plays an important role in the efficient numerical solution of optimal-control problems and in parameter identification.
Different implementations of numerical algorithms for the computation of $\partial y(t)/\partial y_0$ already exist, but the techniques are often adapted to the special implementations and cannot easily be transferred to other integration methods. Here we chose a new approach that allows understanding of most of the existing implementations and may serve as theoretical basis of many more.
The basic idea is to regard the numerical approximation of $\partial y(t)/\partial y_0$ as the solution of the variational differential equation of a linearised linear IVP with approximation of the linear right-hand side by the difference quotient of the original non-linear f.
This approach shows how integration codes can in general be extended so that the simultaneous and accurate computation of the sensitivity matrix is possible.

1. Introduction

Let us consider an initial-value problem (IVP)

$$\dot{y}(t) = f(t, y, p), \quad y(t_0) = y_0 \in \mathbb{R}^n, \quad p \in \mathbb{R}^m \tag{1}$$

with parameter vector p. The exact solution of (1) is denoted by $y(t, t_0, y_0, p)$. In many applications the unknown initial values y_0 or the parameters p must be identified so that the solution $y(t, t_0, y_0, p)$ fits some other requirements, e.g.:
Boundary value problem: $y(t)$ solves (1) and

$$r(y(t_a, t_0, y_0, p), y(t_b, t_0, y_0, p)) = 0 \in \mathbb{R}^n \tag{2}$$

Parameter identification problem: $y(t)$ solves (1) and

$$\min_{y_0, p} \sum_{j=1,k} \frac{\|M_j(y(t_j, t_0, y_0, p) - m_j\|}{\sigma_j} \tag{3}$$

with t_j being not necessarily different time points; M_j functions modelling a measurement process; m_j actual measurements at time $t = t_j$; σ_j scaling factors regarding accuracy of the measurements.

*Lehrstuhl für Höhere Mathematik und Numerische Mathematik, Technische Universität München, D-80290 München, email: kiehl@mathematik.tu-muenchen.de

If these problems are solved numerically, most of the computing time is spent solving (1) for different y_0 or different p, respectively, to compute the derivatives

$$G(t,t_0,y_0,p) := \frac{\partial y(t,t_0,y_0,p)}{\partial y_0} \tag{4}$$

and

$$\tilde{G}(t,t_0,y_0,p) := \frac{\partial y(t,t_0,y_0,p)}{\partial p} . \tag{5}$$

G is called the sensitivity matrix or transition matrix of the IVP (1) or monodromy matrix if $t = T$ is the period of a periodic solution.

Although the solution of (2) or (3) may be requested to be very accurate, low accuracy of G is often sufficient, because G is only used to correct approximate solutions of (2) or (3).

Problems with parameters can be reduced to those without parameters by defining

$$z := \begin{pmatrix} y \\ p \end{pmatrix} \Rightarrow \dot{z}(t) = F(t,z) := \begin{pmatrix} f(t,y,p) \\ 0 \end{pmatrix}, \quad z(t_0) = \begin{pmatrix} y_0 \\ p \end{pmatrix} \in \mathbb{R}^{n+m} . \tag{6}$$

Therefore we omit the argument p in the next sections.

The sensitivity matrix $G(t,t_0,y_0)$ in case of no parameters is known to be the solution of the linear variational differential equation

$$\dot{G}(t) = f_y(t, y(t,t_0,y_0))G(t) , \quad G(t_0) = I_n , \tag{7}$$

with $I_n \in \mathbb{R}^{n\times n}$ identity matrix. In case of (6), (7) reads

$$\dot{G}(t) = F_z(t, z(t,t_0,z_0))G(t) , \quad G(t_0) = I_{n+m} , \tag{8}$$

with

$$F_z(t, z(t,t_0,z_0)) = \begin{bmatrix} f_y(t,y(t,t_0,y_0,p),p) & f_p(t,y(t,t_0,y_0,p),p) \\ 0 & 0 \end{bmatrix} . \tag{9}$$

Once the solution $y_1 := y(t_1,t_0,y_0)$ at time t_1 is computed, $y(t)$ for $t > t_1$ only depends on y_1, i.e.,

$$y(t,t_0,y_0) = y(t,t_1,y(t_1,t_0,y_0)) \tag{10}$$

and therefore G can be computed by

$$G(t,t_0,y_0) = G(t,t_1,y_1)G(t_1,t_0,y_0) \tag{11}$$

or with $y_k := y(t_k,t_0,y_0)$

$$G(t,t_0,y_0) = G(t,t_k,y_k)G(t_k,t_{k-1},y_{k-1})\cdots G(t_2,t_1,y_1)G(t_1,t_0,y_0) . \tag{12}$$

Another alternative is to solve the initial-value problem

$$\dot{G}(t) = f_y(t,y(t,t_1,y_1))G(t) \quad G(t_1) = G(t_1,t_0,y_0) . \tag{13}$$

If only the sensitivity $G(t,t_0,y_0)v$ of (1) in direction v is needed, then instead of (7) we solve

$$\dot{u}(t) = f_y(t, y(t,t_0,y_0))u(t) \quad u(t_0) = v \ , \tag{14}$$

and (11) reads

$$u(t) = G(t,t_0,y_0)v = G(t,t_1,y_1)u(t_1) \ . \tag{15}$$

The columns $g_j(t,t_0,y_0) := G(t,t_0,y_0)e_j$ are obtained by choosing $v := e_j$ as the j-th unit vector.

The various numerical methods for the computation of G take advantage of all these different representations and differ more or less only in the derivation and the efficiency. (12) leads to highly parallel algorithms Integrating (7), one can take advantage of the linearity. (4) leads to methods like those called differentiated numerical integrators.

In this paper we'd like to introduce a unifying theory for all of the different methods. This will allow us to mix them, to find even more efficient implementations and give simple rules how to provide any integration routine for the solution of (4) with the additional feature computing the sensitivity matrix.

Throughout this paper $\eta(t,t_0,y_0,[h])$ denotes the numerical solution of $y(t,t_0,y_0)$, where $[h] = (h_i)_{i=1,k}$ with $\sum_{i=1}^{k} h_i = t - t_0$ represents the stepsize sequence used by the integration routine to approximate $y(t,t_0,y_0)$. If the stepsizes are constant we omit the brackets and write $\eta(t,t_0,y_0,h)$. If the stepsizes are controlled by a prescribed local discretization error δ we also use $\eta(t,t_0,y_0,\delta)$ instead.

2. Difference approximation

In the multiple shooting code BNDSCO ([8]) a difference approximation was used to compute the j-th column $g_j := G(t,t_0,y_0)e_j$ of G by

$$g_j(t,t_0,y_0) = \frac{\partial y(t,t_0,y_0)}{\partial y_{0,j}} \doteq \frac{y(t,t_0,y_0+\varepsilon\cdot e_j) - y(t,t_0,y_0)}{\varepsilon} \quad \text{for} \quad j=1,\ldots,n \ , \tag{16}$$

where ε is appropriately small and e_j is the j-th unit vector. With the additional approximation

$$\frac{y(t,t_0,y_0+\varepsilon\cdot e_j) - y(t,t_0,y_0)}{\varepsilon} \doteq \frac{\eta(t,t_0,y_0+\varepsilon\cdot e_j,\delta) - \eta(t,t_0,y_0,\delta)}{\varepsilon} \ . \tag{17}$$

The result is an error of order $O(\frac{\delta}{\varepsilon})$ due to the approximation (17) and an error of order $O(\varepsilon)$ due to the approximation (16). As a rule of thumb we then have to choose ε and $\frac{\delta}{\varepsilon}$ close to the desired accuracy δ_G of G. As a result δ must be chosen very small. On the other hand the method can immediately be used together with any integration routine. This is the reason why this method is still widely used although it is extremely inefficient. The cost is about n solutions of (1) with accuracy $\delta \approx \delta_G^2$ $(\varepsilon \approx \delta_G)$.

A variation of this method uses symmetric difference approximation

$$G(t,t_0,y_0)e_j \doteq \frac{\eta(t,t_0,y_0+\varepsilon\cdot e_j,\delta) - \eta(t,t_0,y_0-\varepsilon\cdot e_j,\delta)}{2\varepsilon} \ . \tag{18}$$

Then we choose ε^2 and $\frac{\delta}{2\varepsilon}$ close to δ_G. The cost is then about $2n$ solutions of (1) with accuracy $\delta = \delta_G^{3/2}$.

However, the accuracy of G is not controlled in both methods.

3. Explicit methods

It is known (cf. [4]) that the discretization error of a given integration scheme is a differentiable function of the initial data, if no stepsize control is applied. Moreover, Bock ([1])has already mentioned that the derivative of an integration scheme is closely related to the derivative of the solution.

Let $\eta(t, t_0, y_0, h)$ denote the numerical approximation of $y(t, t_0, y_0)$ of order p obtained by using the stepsize h, and let all the partial derivatives of f up to order $N+2$ be continuous and bounded on a set $S = \{(t,y)|a \le t \le b, y \in \mathbb{R}^n\}$, then the asymptotic expansions satisfy

$$\begin{aligned}\eta(t,t_0,y_0,h_1) &= y(t,t_0,y_0) + \sum_{i=p}^{N} c_i(t,y_0)h_1^i + \mathcal{O}(h_1^{N+1})\\ \eta(t,t_0,y_0,h_2) &= y(t,t_0,y_0) + \sum_{i=p}^{N} c_i(t,y_0)h_2^i + \mathcal{O}(h_2^{N+1})\end{aligned}$$

where the $c_i(t, y_0)$ are differentiable functions. Therefore we find

$$\begin{aligned}&\frac{\eta(t,t_0,y_0+\varepsilon e_j,h_2)-\eta(t,t_0,y_0,h_1)}{\varepsilon} =\\ &= \frac{y(t,t_0,y_0+\varepsilon e_j)-y(t,t_0,y_0)}{\varepsilon} + \sum_{i=p}^{N} \frac{c_i(t,y_0+\varepsilon e_j)h_2^i - c_i(t,y_0+\varepsilon e_j)h_1^i}{\varepsilon} +\\ &+ \sum_{i=p}^{N} \frac{c_i(t,y_0+\varepsilon e_j)h_1^i - c_i(t,y_0)h_1^i}{\varepsilon} + \mathcal{O}(h_1^{N+1})/\varepsilon + \mathcal{O}(h_2^{N+1})/\varepsilon =\\ &= \frac{\partial y(t,t_0,y_0)}{\partial y_{0,j}} + \mathcal{O}(\varepsilon) + \sum_{i=p}^{N}\left[\frac{\partial c_i(t,y_0)}{\partial y_{0,j}} + \mathcal{O}(\varepsilon)\right] h_1^i +\\ &+ \sum_{i=p}^{N} c_i(t,y_0+\varepsilon e_j)\frac{h_2^i-h_1^i}{\varepsilon} + \mathcal{O}(h_1^{N+1})/\varepsilon + \mathcal{O}(h_2^{N+1})/\varepsilon.\end{aligned}$$

In the case $h_1 = h_2 = h$ we get

$$\frac{\eta(t,t_0,y_0+\varepsilon e_j,h)-\eta(t,t_0,y_0,h)}{\varepsilon} = \frac{\partial y(t,t_0,y_0)}{\partial y_{0,j}} + \mathcal{O}(\varepsilon) + \mathcal{O}(h^p) + \mathcal{O}(\frac{h^{N+1}}{\varepsilon}). \quad (19)$$

Thus the difference approximation of the derivative $\partial\eta(t,t_0,y_0)/\partial y_{0,j}$ is a good approximation of $\partial y(t,t_0,y_0)/\partial y_{0,j}$, whereas for $h_1 \ne h_2$ the terms $c_i(t,y_0+\varepsilon e_j)\frac{h_2^i-h_1^i}{\varepsilon}$ of order $\mathcal{O}((h_1^p - h_2^p)/\varepsilon)$ can be dominant, especially for small ε. This leads to the simple modification of the difference approximation described in section 2 which can be called "frozen stepsize method" (FSM) and means that the same stepsize sequence $[h]$ is used to compute $\eta(t,t_0,y_0,[h])$ and $\eta(t,t_0,y_0+\varepsilon e_j,[h])$. This can be done by computing $\eta(t,t_0,y_0,[h])$, storing $[h]$ and transmitting it as additional parameter to the integration routines for the computation of $\eta(t,t_0,y_0+\varepsilon e_j,[h])$, $j=1,\dots,n$. One way to do this is to transmit the stepsize h after each performed step during the computation of $\eta(t,t_0,y_0,\delta)$, and perform a step for $\eta(t,t_0,y_0+\varepsilon e_j,h)$, $j=1,\dots,n$

with the same step size. This was done in the code DOPR8_LI ([2]) a modification of DOPRI8 ([3]) and in the code ODEX_LI ([2]) a modification of ODEX ([6]).

Another approach is to differentiate the expansion formula

$$\eta(t, t_0, y_0, h) = y(t, t_0, y_0) + \sum_{i=p}^{N} c_i(t, y_0)h^i + \mathcal{O}(h^{N+1}) , \tag{20}$$

where the coefficients $c_i(t, y_0)$ are differentiable functions, which yields

$$\frac{\partial \eta(t, t_0, y_0, h)}{\partial y_{0,j}} = \frac{\partial y(t, t_0, y_0)}{\partial y_{0,j}} + \sum_{i=p}^{N} \frac{\partial c_i(t, y_0)}{\partial y_{0,j}} h^i + \mathcal{O}(h^{N+1}) . \tag{21}$$

(21) leads to methods that compute G by differentiating a given integration scheme. E.g., the Euler method $y_{i+1} = y_i + hf(t_i, y_i)$ yields

$$\frac{\partial y_{i+1}}{\partial y_i} = I_n + hf_y(t_i, y_i) . \tag{22}$$

If this method is applied to compute $\bar{g}_j(t, \varepsilon) := \varepsilon g_j(t, t_0, y_0)$ the $(i+1)$-th Euler-step reads

$$\bar{g}_j(t_{i+1}, \varepsilon) = [I_n + hf_y(t_i, \eta(t_i, t_0, y_0))]\bar{g}_j(t_i, \varepsilon) , \tag{23}$$

or

$$\bar{g}_j(t_{i+1}, \varepsilon) \doteq \bar{g}_j(t_i, \varepsilon) + h[f(t_i, \eta(t_i, t_0, y_0) + \bar{g}_j(t_i, \varepsilon)) - f(t_i, \eta(t_i, t_0, y_0))] . \tag{24}$$

A third approach is to solve the variational equation (7) i.e., to solve the system

$$\dot{y} = f(t, y) \quad ; \quad \dot{G} = f_y(t, y)G \quad ; \quad y(t_0) = y_0 \quad ; \quad G(t_0) = I_n , \tag{25}$$

where the approximation

$$f_y(t, y)g_j(t) \doteq \frac{[f(t, y + \tilde{\varepsilon}g_j(t)) - f(t, y)]}{\tilde{\varepsilon}} \tag{26}$$

can be used to avoid the evaluation of the Jacobian f_y.

If the difference approximations (24) and (26) are used with $\tilde{\varepsilon} = \varepsilon$ then differentiated integration methods (DIM) and the integration of the variational equation method (IVM) are equivalent to the frozen step size method (FSM) for all explicite Runge-Kutta methods (including extrapolation methods) if rounding errors are neglected.

If other approximations than (24) and (26) are used, the methods still lead to similar results. The methods differ in efficiency and implementability.

If the evaluation of f_y and the matrix-vector product $f_y g_j$ is cheap, then (24) and (26) are not used, and DIM as well as IVM are more efficient than FSM. IVM then also allows taking advantage of the linearity of (7). Integration methods that evaluate the right-hand side more than once at the same time point can then also use one evaluation of f_y more than once. This can save up to 50 % of the evaluations of f_y. Examples are RKF4_JI or ODEX_D_JI ([2]) (modifications of RKF4 and the extrapolated midpoint rule implementation ODEX_D_JI[5]). However, FSM is much faster to implement.

The main advantage of all three methods is that cancellation in (19) is avoided and therefore it is possible to choose a large δ, or large stepsizes h.

4. Implicit methods

The classical difference approximation as well as FSM, DIM and IVM can of course be used in connection with implicit integration routines. In this case the dominant computational work often is to set up and to decompose a linear equation system. I.e., to compute $A := f_y(t_i, \eta(t_i, t_0, y_0, \delta))$ and to decompose matrices of the form

$$I_n - h\gamma f_y(t_i, \eta(t_i, t_0, y_0, \delta)) = QU \tag{27}$$

with γ coefficient of the integration method. Using such a method for sensitivity analysis can have different consequences if FSM, DIM or IVM are used.

- FSM in this case requires computing $A_j := f_y(t_i, \eta(t_i, t_0, y_0 + \varepsilon e_j, \delta))$ and decomposing $\frac{I_n}{h\gamma} - A_j$, $j = 1, \ldots, n$.
- Differentiating (27) in DIM requires computing $f_{yy}(t_i, \eta(t_i, t_0, y_0, \delta))$.
- Integrating the variational equation (7) in IVM requires computing $[\dot{G}(t)]_z = [F_z(t, z(t, t_0, z_0))G(t)]_z$. which includes $f_{yy}(t_i, \eta(t_i, t_0, y_0, \delta))$.

Some implicit integration methods allow these computations to be avoided, especially those methods that do not need exact Jacobians but only good approximations. In this case f_y is not used to increase the order (like in Rosenbrock-Wanner methods) but to guarantee stability.

For example the semi-implicit Euler method

$$y_{i+1} = y_i + [I/h_{i+1} - f_y(t_i, y_i)]^{-1} f(t_{i+1}, y_i) \tag{28}$$

is usually implemented as

$$y_{i+1} = y_i + [I/h_{i+1} - A_j]^{-1} f(t_{i+1}, y_i) \tag{29}$$

with $A_j = f_y(t_j, y_j)$ with $j \leq i$ such that $A_j \approx f_y(t_i, y_i)$. If for some reason we have a good approximation $\bar{y}_i \approx y_i$ we could also implement the method in the form

$$y_{i+1} = y_i + [I/h_{i+1} - f_y(t_i, \bar{y}_i)]^{-1} f(t_{i+1}, y_i) \tag{30}$$

with fixed $\bar{y}_i$ independent of y_0. (30) then leads to $\bar{y}_i \not\approx y_i$ if $y(t_0) \not\approx y_0$ and therefore possibly $f_y(t_j, \bar{y}_j) \not\approx f_y(t_j, y_j)$. If, however, only initial values $y(t_0) = y_0$ or $y(t_0) = y_0 + \varepsilon e_j \approx y_0$ are used, then (30) uses a sufficient approximation of f_y. This allows use of the matrices $\bar{A}_i := f_y(t_i, \bar{y}_i)$ and the decompositions of $[I/h_{i+1} - \bar{A}_i]$ also for the sensitivity analysis, i.e., only one computation of f_y and one decomposition is necessary in each integration step.

Remark 1 *Note that these methods cannot be obtained by simply differentiating a given integration method.*

Remark 2 *The restriction of these considerations to methods of special type can also be avoided by formulating a new initial-value problem.*

Remark 3 *A remaining problem is how to control the accuracy of G and how to guarantee $\bar{y}_i \approx y_i$.*

5. Sensitivity of the linearized IVP

Let us consider the linearization of the initial-value problem (1).

$$\dot{w}(t) = \hat{f}(t,w) := f(t,y) + f_y(t,y)(w-y), \quad w(t_0) = y_0 \in \mathbb{R}^n \,. \tag{31}$$

The solution of (31) with this special initial value is of course the same as the solution of (1). The sensitivity matrix $\hat{G}$ of (31) according to (7) is given by

$$\frac{\mathrm{d}}{\mathrm{d}t}\hat{G}(t) = \hat{f}_w(t, w(t,t_0,w_0))\hat{G}(t) = f_y(t, y(t,t_0,y_0))\hat{G}(t) \quad \hat{G}(t_0) = I_n \,. \tag{32}$$

Therefore the sensitivity matrices of (31) and (1) are identical i.e.,

$$\hat{G}(t,t_0,y_0) = G(t,t_0,y_0) \tag{33}$$

and we can apply all methods to the linear problem (31).

If the right-hand side f is linear in y, then usual integration formulas are also linear in y. In this case FSM, DIM and IVM are equivalent if applied to (31).

The right-hand side of (32), however, requires the jacobian $f_y(t, y(t,t_0,y_0))$, but we can use an approximation for each column of G

$$\frac{\mathrm{d}}{\mathrm{d}t}g_j(t) = f_y(t, y(t,t_0,y_0))\frac{\bar{g}_j(t,\varepsilon)}{\varepsilon} \approx \frac{f(t, y(t,t_0,y_0) + \bar{g}_j(t,\varepsilon)) - f(t, y(t,t_0,y_0))}{\varepsilon}, \tag{34}$$

The approximation error of (34) is then of order $O(\|\bar{g}_j(t,\varepsilon)\|^2/\varepsilon)$ and the relative rounding error is then of order $O(\frac{\hat{\varepsilon}\|y(t,t_0,y_0)\|}{\|\bar{g}_j(t,\varepsilon)\|})$ with $\hat{\varepsilon}$ the machine precision.

6. Implementation of sensitivity analysis in a given integration method

In this section we will see how rather arbitrary integration routines can be used for efficient sensitivity analysis.

We assume that the user prescribes an accuracy δ_y as local discretization error of η. As the sensitivity matrix G is often required with a different accuracy than the solution of the IVP (1), we assume that the sensitivity of (1) is required in a direction $v(t_0) = v_0$, with local discretization error δ_G, respectively $\delta_1, \ldots, \delta_k$ if the sensitivity is requested in k directions $v_1, \ldots, v_k$.

The user supplies $y_0 \in \mathbb{R}^n$, δ_y, $v \in \mathbb{R}^n$ and δ_G.

Then two different types of implementations are possible to compute the sensitivity

$$v(t) := G(t,t_0,y_0)v_0$$

1) This method is used e.g. if the user supplies efficient implementations of f_y and a sparse-matrix-vector product $f_y v$.

Then we apply an integration routine with step-size sequence $[h_G]$ to the system

$$\frac{\mathrm{d}}{\mathrm{d}t}\begin{pmatrix} y \\ v \end{pmatrix} = \begin{pmatrix} f(t,y) \\ f_y(t,y)v \end{pmatrix}, \quad v(t_0) = v_0, \quad y(t_0) = y_0 \,. \tag{35}$$

The stepsize is controlled due to the accuracy δ_G of v. The accuracy of y is only monitored.

After each integration step we compute $\eta(t_i, t_0, y_0, \delta_y)$ with a (possibly different) integration routine with step-size sequence $[h_y]$ applied to (1). Then we start the next integration step for the system (35) with $y(t_i) = \eta(t_i, t_0, y_0, \delta_y)$ instead of $y(t_i) = \eta(t_i, t_0, y_0, [h_G])$. This prevents computing far away from the exact solution $y(t, t_0, y_0)$ even if the stepsizes $[h_G]$ are large.

2) This method is based on the linearized IVP (31).

At the beginning we scale vector v_0 by $\bar{v}_0 := \bar{\varepsilon} v_0$ e.g.

$$\bar{\varepsilon} := \left(\frac{\|y_0^T\|}{\|v_0\|} + 1 \right) \sqrt{\hat{\varepsilon}} \ , \tag{36}$$

with $\hat{\varepsilon}$ machine precision.

Then we define $u(t) := y(t) + \bar{v}(t)$ which is a solution of

$$\frac{\mathrm{d}}{\mathrm{d}t} u(t) = f(t, y) + f_y(t, y)(u(t) - y(t)) \ ; \quad u(t_0) := y(t_0) + \bar{v}(t_0) \ . \tag{37}$$

We then perform one integration step with a given integration routine for the system

$$\frac{\mathrm{d}}{\mathrm{d}t} \begin{pmatrix} z \\ u \end{pmatrix} = \begin{pmatrix} f(t, z) \\ f(t, z) + f_y(t, y)(u(t) - z(t)) \end{pmatrix} = F(t, z, u) \doteq \begin{pmatrix} f(t, z) \\ f(t, u) \end{pmatrix} , \tag{38}$$

$$u(t_0) = y_0 + \bar{\varepsilon} v_0 \ , \quad z(t_0) = y_0 \ , \tag{39}$$

where y is the exact solution of (1). If solved exactly, also $z(t) = y(t)$ is the exact solution of (1).

The Jacobian of the system (38) has the form

$$F_{z,u} = \begin{pmatrix} f_y(t, z) & 0 \\ f_y(t, z) - f_y(t, y) & f_y(t, y) \end{pmatrix} \doteq \begin{pmatrix} f_y(t, z) & 0 \\ 0 & f_y(t, z) \end{pmatrix} . \tag{40}$$

Because we do not know the exact solution $y(t)$ in (38) and (40) we have to use the approximations.

- The approximation in (38) has an error of order $O(\|y - z\| \|u - z\| + \|u - z\|^2)$. The error in the evaluation of F therefore is not critical.
- The approximation in (40) has an error of order $O(\|y - z\|)$. This results in (27) in an error of the linear system of order $O(h\gamma \|y - z\|)$ which for small h and small $\|y - z\|$ also is acceptable.
  The approximated Jacobian then is a block-diagonal system with diagonal blocks $f_y(t, z)$. If the sensitivity in direction $v_1, \ldots, v_k$ has to be computed, the system consists of $k + 1$ identical blocks. Therefore, the decomposition of $I_n - h\gamma f_y(t, z)$ is sufficient.

Thus the non exact performance of the discretisation scheme of a given integration method for (38) leads to sufficient accuracy and large savings in computing time.

I.e., for the computation of the sensitivity no additional Jacobian and no additional decomposition is necessary.

In the first step we have $\|y(t_0) - z(t_0)\| = 0$ so that the approximations have a small error. After each step we compute

$$v(t_i) := v(t_i, t_0, y_0, \delta_G) := \frac{u(t_i, t_0, y_0, [h_G]) - z(t_i, t_0, y_0, [h_G])}{\bar{\varepsilon}} .$$

The accuracy δ_G of $v(t_i)$ is used to control $[h_G]$.

After each integration step we solve the IVP (1) with high accuracy δ_y and possibly a different method, i.e.

$$y_i := y(t_i, t_0, y_0, \delta_y) \approx y(t_i, t_0, y_0).$$

According to (11) and (15) the remaining problem is to compute $G(t, t_i, y_i)v(t_i)$.

For the next step we use a new scaling

$$\bar{\varepsilon}_i := \left(\frac{\|y(t_i)\|}{\|v(t_i)\|} + 1 \right) \sqrt{\hat{\varepsilon}} . \tag{41}$$

and define

$$u(t_i) := y(t_i, t_0, y_0, \delta_y) + \bar{\varepsilon}_i v(t_i) .$$

This guarantees that $\|y - z\|$ remains small. It means that we always have to additionally compute the solution of the IVP (1) with high accuracy δ_y together with the sensitivity.

The main advantages are:

- Computation of G is possible with low tolerance δ_G.
- Less additional decompositions necessary for the sensitivity than for the computation of $\eta(t, t_0, y_0, \delta_y)$.
- Local discretization error of G is controlled.
- Easy implementation of different methods.
- In case of sparse matrix multiplication and analytic f_y and f_p, further savings when solving the variational equation.

Implementations for non-stiff problems can be found in [2]. For an adaption of DASSL for stiff and differential-algebraic problems see [7]. Efficient use of sparse matrix techniques was made for the simulation of highly oscillating circuits. Further adaptions (e.g. of SEULEX and GRK4A) are in progress.

References

[1] Bock, H.: *Numerical treatment of inverse problems in chemical reaction kinetics.* In K. Ebert, P. Deuflhard, and W. Jäger, editors, *Modelling of Chemical Reaction Systems*, volume 18 of *Springer Series in Chemical Physics.* Springer, Heidelberg, 1981.

[2] Buchauer, O.; Hiltmann, P.; Kiehl, M.: *Sensitivity analysis of initial-value problems with application to shooting techniques.* J Num. Math., 67:151–159, 1994.

[3] Dormand, J.; Prince, P.: *A family of embedded Runge-Kutta formulae.* J. Comp. Appl. Math., 6:19–26, 1980.

[4] Hairer, E.; Nørsett, S.; Wanner, G.: *Solving Ordinary Differential Equations I.* Springer, Berlin Heidelberg New York, 1987.

[5] Hairer, E.; Ostermann, A.: *Dense output for extrapolation methods.* Num. Math., 58:419–439, 1990.

[6] Hairer, E.; Wanner, G.: *Solving Ordinary Differential Equations II.* Springer, Berlin Heidelberg New York, 1991.

[7] Heim, A.: *Parameteridentifikation in differentialalgebraischen Systemen.* diploma thesis, Mathematisches Institut der TU München, 1992.

[8] Oberle, H.; Grimm, G.: *BNDSCO - A Program for the Numerical Solution of Optimal Control Problems. User Guide.* Technical Report DLR IB 515-89/22, DLR, Germany, Oberpfaffenhofen, 1989.

International Series of Numerical Mathematics
Vol. 124, © 1998 Birkhäuser Verlag, Basel

Algorithm of Real-Time Minimization of Control Norm for Incompletely Determined Linear Control Systems

Olga Ivanovna Kostyukova*

Abstract. As a rule, a real control system acts in presence of some indeterminacy (it may be an unknown disturbance or another kind of indeterminacy). The classical control method for such system is using the control of feedback type. In the first papers on optimal control synthesis, it was supposed that the current system state is known exactly at every current moment. In this case, the optimal feedback with respect to the system state was constructed. Later on, the more complicated practical situations were investigated. Now it is assumed, that the available information about control system behaviour consists of incomplete and inexact measurements of system states. In this connection, the problem of constructing the optimal feedback with respect to such incomplete and inexact measurements arises. This problem is complex. It includes the observation problem and control problem connected with each other.
In this paper such a complex problem of linear dynamic system optimization is investigated. The Mathematical model of this problem is considered to be known exactly but the initial state and errors of measuring device (sensor) are supposed to be unknown.
The finite algorithm of constructing the program control for some incompletely determined linear system was suggested in [1]. The aim of this paper is to develop the results [1,2] to feedback control.

1. Problem statement

Let $u(\cdot) = (u(t), t \in T = [0, t^*])$ be a piecewise continuous function (control) satisfying the constraints

$$|u(t)| \le 1, t \in T. \tag{1}$$

Using such a function, we have to transfer the trajectory $x(\cdot) = (x(t), t \in T)$ of linear system

$$\dot{x} = Ax + bu, (x \in R^n, u \in R) \tag{2}$$

from some fixed (but unknown) initial state

$$x(0) = x_0 \in X_0 = \{x \in R^n : d_* \le x \le d^*\} \tag{3}$$

to given terminal set X^*

$$x(t^*) \in X^* = \left\{x \in R^n : h_i' x \ge g_i, i = \overline{1,m}\right\}.$$

*Institute of Mathematics, Academy of Sciences, ul. Surganov 11, Minsk 220072, Belarus, email: imanb@imanb.belpak.minsk.by

The quality of control $u(\cdot)$ is evaluated by its norm functional

$$J_p(u) = \left(\int_T |u(t)|^p dt\right)^{\frac{1}{p}},$$

where $p > 0$ is a given parameter which takes one of the values $p = 1$, $p = 2$, $p = \infty$. Thus the classical norms are obtained:

$$J_1(u) = \int_T |u(t)|dt, \; J_2(u) = \left(\int_T u^2(t)dt\right)^{\frac{1}{2}}, \; J_\infty(u) = \max_{t\in T} |u(t)|.$$

The calculations described below may be implemented for any indicated value of p. For the purpose of definiteness in this paper we consider the functional

$$J_1(u) = \int_0^{t^*} |u(t)|dt \to \min. \tag{4}$$

It should be noted that the initial state $x(0)$ is not an additional parameter in the problem (1)–(4) which is used together with the control $u(\cdot)$ for the system optimization, but it is a vector whose true value is fixed but unknown before-hand. All initial information about the feasible state $x(0)$ is given by its a priori distribution X_0. In this connection the problem (1)–(4) is further interpreted as a guaranteed result problem. It is assumed that there is no possibility to measure the system state $x(t)$ during the control process. At every current moment τ, all available information about the system behaviour is contained in the sensor signal

$$y_\tau(\cdot) = (y(t), t \in [0, \tau]), \tag{5}$$

$$y(t) = k'x(t) + \xi(t), t \in T_\tau = [0, \tau].$$

At any current moment t sensor (5) measures the projection $k'x(t)$ of the state $x(t)$ on the defined vector k with the error $\xi(t)$. The error function $\xi(t)$, $t \in T$, is assumed to be any smooth function satisfying the restrictions

$$\xi_* \leq \xi(t) \leq \xi^*, t \in T. \tag{6}$$

The general (unformal) statement of the problem under investigation consists in the following: to organize (in the sense of problem (1)–(6)) the optimal control in real-time mode, using the sensor signal $y(t)$, that enters continuously at every current moment.

2. Optimal feedback

To organize the control process of system (1)–(6) we will use optimal feedback constructed with respect to the sensor signal.
Represent the solution $x(t)$, $t \in T$, of system (2) in the form

$$x(t) = \chi(t) + z(t), t \in T,$$

where $\chi(t)$, $z(t)$, $t \in T$, are the solutions of the systems

$$\dot{\chi} = A\chi + bu, \chi(0) = 0; \tag{7}$$

$$\dot{z} = Az, z(0) = x_0 \in X_0. \tag{8}$$

Denote by $\chi(t|u(\cdot))$, $t \in T$, the trajectory of the system (7) corresponding to control $u(\cdot)$ and by $z(t|x_0)$, $t \in T$, the trajectory of the system (8) corresponding to the initial state x_0.
Let τ, $\tau \in T$, be a current moment. Consider the left subinterval $T_\tau = [0, \tau]$ of the control interval T. Let

$$u_\tau(\cdot) = (u(t), t \in T_\tau)$$

be a control which has been used till the moment τ in the control process;

$$\chi = \chi(\tau|u_\tau(\cdot))$$

be the state of the system (7) at the moment τ.
In the sensor signal $y_\tau(\cdot)$ (5) we distinguish a component $\eta_\tau(\cdot) = (\eta(t), t \in T_\tau)$ generated by the realized initial state x_0 and the error $\xi_\tau(\cdot) = (\xi(t), t \in T_\tau)$:

$$\eta(t) = y(t) - k'\chi(t|u_\tau(\cdot)) = k'z(t|x_0) + \xi(t), t \in T_\tau. \tag{9}$$

Call the totality of parameters $s = \{\tau, \eta_\tau(\cdot), \chi\}$ *a position* of the control observation system (1)–(6).
At the moment τ using the sensor signal $\eta_\tau(\cdot)$ we can reduce the indeterminacy of initial state x_0.
Denote by $X(\tau, \eta_\tau(\cdot))$ the set of those and only those elements x_0 from X_0 which are able to generate the signal $\eta_\tau(\cdot)$ together with some admissible error function $\xi_\tau(\cdot)$:

$$X(\tau, \eta_\tau(\cdot)) = \{x_0 \in X_0 \ : \exists \xi_\tau(\cdot), \ k'z(t|x_0) + \xi(t) = \eta(t), \ t \in T_\tau\}.$$

The subset $X(\tau, \eta_\tau(\cdot)) \subset X_0$ is an *a posteriori distribution* of the initial state x_0. If our system is in the position s at the moment τ, then we know that the realized initial state x_0^* belongs to $X(\tau, \eta_\tau(\cdot))$.
Now we can introduce an admissible set of system states corresponding to position s by the rule:

$$X_0(s) = \{x \in R^n \ : \ x = \chi + z(\tau|x_0), \ x_0 \in X(\tau, \eta_\tau(\cdot))\}.$$

That is, $X_0(s)$ is the set of those and only those system (2) states at the moment τ, which are generated by control $u_\tau(\cdot)$ and initial states $x_0 \in X(\tau, \eta_\tau(\cdot))$.

Now let us consider a right subinterval $T^\tau = [\tau, t^*]$ of control interval T.
The piecewise continuous function $u(\cdot|s) = (u(t|s), t \in T^\tau)$ is called a *(guaranteed) admissible program control for position* s if the following relations hold

$$|u(t|s)| \leq 1, t \in T^\tau,$$

$$x(t^*|\tau, z, u(\cdot|s)) \in X^*, \forall z \in X_0(s).$$

Here $x(t|\tau, z, u(\cdot))$, $t \in T^\tau$, is a trajectory of the system

$$\dot{x} = Ax + bu, x(\tau) = z.$$

In other words the function $u(\cdot|s)$ is a guaranteed admissible control for position s if it removes any system (2) state $x(\tau) = z$, $z \in X_0(s)$, to the given terminal set X^*.
Denote by S the set of all positions s for which an admissible program control exists.
Estimate the quality of the admissible program control $u(\cdot|s)$ (corresponding to position s) by the number

$$J_1(u(\cdot|s)) = \int_\tau^{t^*} |u(t|s)|dt.$$

In position s *an optimal program control* $u^0(\cdot|s)$ is determined by the relation

$$J_1(u^0(\cdot|s)) = \min_{u(\cdot|s)} J_1(u(\cdot|s)).$$

The function

$$u^0(\tau, \eta_\tau(\cdot), \chi) = u^0(s) = u^0(\tau|s), s \in S, \tag{10}$$

is said to be *an optimal control of feedback* type for the control observation problem (1)–(6).
The problem of constructing the optimal feedback control is a very complicated one. To construct the function (10), it is necessary to solve the nonlinear Bellman's equations or to find the optimal program control $u^0(\cdot|s)$ for all $s \in S$.
However, we can look at this problem from another side. Let us use the approach suggested in [2]. The main idea of this approach consists in the following.
Let us suppose that the optimal feedback (10) has been constructed. Using (10), we close the system (7)–(9). As a result we obtain the closed system

$$\dot{\chi} = A\chi + bu^0(t, \eta_\tau(\cdot), \chi), \chi(0) = 0,$$

$$\dot{z} = Az, z(0) = x_0, \tag{11}$$

$$\eta(t) = k'z(t) + \xi(t), t \in T,$$

with input signals x_0, $\xi(t)$, $t \in T$.
Consider some concrete process in which the initial state x_0^* and error function $\xi^*(t)$, $t \in T$, have been realized.
Denote by $\chi^*(t)$, $\eta^*(t)$, $t \in T$ the solution of the system (11) corresponding to x_0^* and $\xi^*(t)$, $t \in T$.

It is clear that in the process under consideration we do not use all the values of function (10) but only values

$$u^*(t) = u^0(t, \eta_t^*(\cdot), \chi^*(t)), t \in T. \tag{12}$$

In addition, we don't need to know the control value $u^*(\tau)$ at the beginning of control process. The control value $u^*(\tau)$ will be required only at the real time moment τ when the control observation system (7)–(9) will find itself in current position $\{\tau, \eta_\tau^*(\cdot), \chi^*(\tau)\}$.

The function $u^*(t)$, $t \in T$, (12) is said to be *optimal feedback (10) realization* corresponding to x_0^*, $\xi^*(t)$, $t \in T$.

Taking into account the circumstances mentioned above, we are able to work out an algorithm for constructing optimal feedback realization (12) in real-time mode for any concrete control process. The aim of this paper is to describe such an algorithm.

Remark. It was supposed above that the closed system (11) has a classical solution. However it is not true in the general case. Sliding regimes may appear in the system (11). Such a situation is not considered in this paper. The sliding regimes may be investigated on the base of approach suggested in [3].

3. Algorithm of construction of the optimal feedback realization

Consider some concrete control-observation process of system (1)–(6). Suppose that the initial state x_0^* has been realized in the process under consideration.

Let us consider the current moment τ. Denote by $u_\tau^*(\cdot) = (u^*(t), t \in [0, \tau[)$ the control, which has been used till the moment τ; $\xi_\tau^*(\cdot) = (\xi^*(t),\ t \in [0, \tau])$ the realized error function; $s^* = (\tau, \eta_\tau^*(\cdot), \chi^*(\tau))$ the current system position at the moment τ under action of the control $u_\tau^*(\cdot)$ and error function $\xi_\tau^*(\cdot)$.

According to the definition (10)–(12) control value $u^*(\tau)$ at the moment τ is constructed by the rule

$$u^*(\tau) = u^0(s^*) = u^0(\tau|s^*),$$

where

$$u^0(\cdot|s^*) = (u^0(t|s^*), t \in [\tau, t^*])$$

is a guaranteed optimal program control corresponding to position s^*.

It can be shown that the function $u^0(\cdot|s^*)$ is a solution of the following determinated optimal control problem

$$\int_\tau^{t^*} |u(t)|dt \to \min,\ \dot{\chi} = A\chi + bu,\ \chi(\tau) = 0, \tag{13}$$

$$|u(t)| \leq 1,\ t \in T_\tau;\ h_i'\chi(t^*) \geq g_i - \alpha_i^*(\tau),\ i = \overline{1,m}.$$

Here A, b, h_i, g_i, $i = \overline{1,m}$, are given parameters of the initial problem (1)–(6), but numbers $\alpha_i^*(\tau)$, $i = \overline{1,m}$, are solutions of the following observation problems

$$\alpha_i^*(\tau) = \min \quad h_i'z(t^*);\ \dot{z} = Az,\ z(\tau) = x,\ x \in X_0(s^*), \tag{14}$$

$$i = \overline{1,m}.$$

Consequently, to construct the function $u^*(\tau)$, $\tau \in T$, in real-time mode it is necessary to solve m optimal observation problems (14) and one optimal control problem (13) in the same mode along the current system positions $\{\tau, \eta^*_\tau(\cdot), \chi^*(\tau)\}$, $\tau \in T$.
Devices which are able to evaluate solutions of the problems (14) ($i = \overline{1,m}$) and (13) in real-time mode for every concrete control process of system (1)–(6) are said to be optimal *estimators* and optimal *controller* respectively.
Thus the algorithm of constructing the function $u^*(t)$, $t \in T$, (12) consists of algorithms of m estimators and one controller operation.

4. Algorithm of estimator operations

The problems (14), $i = \overline{1,m}$, are not connected with each other. Therefore we can investigate every i-th problem (14) separately.
The aim of i-th estimator operations is to construct the solutions of the i-th problem (14) in real-time mode for $\tau \in [0, t^*]$.
Consider the i-th problem (14). It can be rewritten in the form

$$\alpha_i^*(\tau) = \min_x (\bar{h}_i' \chi^*(\tau) + p^{(i)\prime} x), \qquad \bar{h}_i' = h_i' \Phi(t^*) \Phi^{-1}(\tau),$$
$$\beta_*(t) \le a'(t) x \le \beta^*(t),\ t \in T_\tau;\ d_* \le x \le d^*, \tag{15}$$

where

$$p^{(i)} = (p_j^{(i)}, j \in J) = h_i' \Phi(t^*),\ J = \{1, 2, \ldots, n\},$$
$$a'(t) = (a_j(t), j \in J) = k' \Phi(t);\ \dot{\Phi}(t) = A\Phi(t),\ \Phi(0) = E,$$
$$\beta_*(t) = \eta^*(t) - \xi^*;\ \beta^*(t) = \eta^*(t) - \xi_*,\ t \in T_\tau.$$

For every fixed τ problem (15) is a semi–infinite linear programming problem.
Let us consider current moment τ and denote by $x_\tau^{(i)} = (x_{\tau j}^{(i)}, j \in J)$ an optimal solution of problem (15).
For the sake of simplicity we will omit the index (i), e.g. we will write x_τ, $x_{\tau j}, p_j$ instead of $x_\tau^{(i)}$, $x_{\tau j}^{(i)}, p_j^{(i)}$.
Construct the sets:

$$T^+(\tau) = \{t \in T_\tau \ : \ a'(t) x_\tau = \beta^*(t)\};\ J^+(\tau) = \{j \in J \ : \ x_{\tau j} = d_j^*\},$$
$$T^-(\tau) = \{t \in T_\tau \ : \ a'(t) x_\tau = \beta_*(t)\};\ J^-(\tau) = \{j \in J \ : \ x_{\tau j} = d_{*j}\}.$$

According to the optimality criterion there exist numbers

$$\eta(t,\tau) \ge 0,\ t \in T^+(\tau);\ \eta(t,\tau) \le 0,\ t \in T^-(\tau),$$

such that the relations

$$\Delta_j(\tau) \ge 0,\ j \in J^-(\tau);\ \Delta_j(\tau) \le 0,\ j \in J^+(\tau);$$
$$\Delta_j(\tau) = 0,\ j \in J \backslash (J^-(\tau) \cup J^+(\tau))$$

hold. Here

$$\Delta_j(t) = \sum_{t \in T(\tau)} a_j(t) \eta(t,\tau) + p_j,\ j \in J;\ T(\tau) = T^+(\tau) \cup T^-(\tau).$$

A vector $\eta_\tau = (\eta(t,\tau), t \in T(\tau))$ is called *an optimal dual solution* of problem (15). A couple $\{x_\tau, \eta_\tau\}$ is said to be *an optimal solution* of problem (15). The optimal solution $\{x_\tau, \eta_\tau\}$ is considered to be *nondegenerate* if the following relations take place

$$\begin{gathered} \ddot{a}'(t)x_\tau \neq \ddot{\beta}(t),\ t \in \tilde{T}(\tau);\ \dot{a}'(t)x_\tau \neq \dot{\beta}(t),\ t \in T(\tau) \cap \{0,\tau\}, \\ d_{*j} < x_{\tau j} < d_j^*,\ j \in J_0(\tau);\ \eta(t,\tau) \neq 0,\ t \in T(\tau); \\ rank\ B(\tau) = |T(\tau)|,\ rank\ (B(\tau), C(\tau)) = |J_0(\tau)|. \end{gathered} \tag{16}$$

Here

$$\begin{gathered} \tilde{T}(\tau) = T(\tau) \setminus \{0,\tau\},\ J_0(\tau) = \{j \in J\ :\ \Delta_j(\tau) = 0\}, \\ \beta(t) = \beta_*(t) \text{ if } t \in T^-(\tau); \beta(t) = \beta^*(t) \text{ if } t \in T^+(\tau); \\ B(\tau) = \begin{pmatrix} a_j(t), t \in T(\tau) \\ j \in J_0(\tau) \end{pmatrix}, C(\tau) = \begin{pmatrix} \dot{a}_j(t), t \in \tilde{T}(\tau) \\ j \in J_0(\tau) \end{pmatrix}. \end{gathered}$$

Denote

$$\begin{gathered} \{t_s(\tau), s \in I(\tau)\} = T(\tau),\ I^+(\tau) = \{s \in I(\tau)\ :\ t_s(\tau) \in T^+(\tau)\}; \\ I^-(\tau) = I(\tau) \setminus I^+(\tau); \\ I_0(\tau) = \oslash \text{ if } 0 \notin T(\tau), I_0(\tau) = i_0 \text{ if } t_{i_0}(\tau) = 0 \in T(\tau); \\ I_*(\tau) = \oslash \text{ if } \tau \notin T(\tau),\ I_*(\tau) = i_* \text{ if } t_{i_*}(\tau) = \tau \in T(\tau). \end{gathered}$$

The totality of parameters

$$P(\tau) = \{J_0(\tau), J^+(\tau), J^-(\tau), I^+(\tau), I^-(\tau), I_0(\tau), I_*(\tau)\}$$

is said to be a structure of solution $\{x_\tau, \eta_\tau\}$.

Suppose that for some moment $\tau = \tau_0 \in [0, t^*[$ the solution $\{x_{\tau_0}, \eta_{\tau_0}\}$ of problem (15) is known and this solution is nondegenerate. Let $P(\tau_0)$ be the structure corresponding to $\{x_\tau, \eta_\tau\}$. Denote by $\tau_1, \tau_1 > \tau_0$, the nearest to τ_0 moment at which the relations (16) are violated.

Consider two cases :*I*) $\tau_0 \in T(\tau_0)$, *II*) $\tau_0 \notin T(\tau_0)$. In the case *I*) it can be shown that for $\tau \in [\tau_0, \tau_1]$ parameters

$$x_{\tau j}, j \in J;\ \eta_s(\tau) = \eta(t_s(\tau), \tau),\ t_s(\tau),\ s \in I(\tau) \tag{17}$$

are uniquely defined by equations :

$$\begin{gathered} x_{\tau j} = d_j^*, j \in J^+;\ x_{\tau j} = d_{*j}, j \in J^-; \\ a'(t_s(\tau))x_\tau = \beta(t_s(\tau)), s \in I; \\ \dot{a}'(t_s(\tau))x_\tau = \dot{\beta}(t_s(\tau)), s \in I \setminus (I_0 \cup I_*), \\ \sum_{s \in I} a_j(t_s(\tau))\eta_s(\tau) = -p_j, j \in J_0; t_s(\tau) = 0, s \in I_0;\ t_s(\tau) = \tau, s \in I_*, \end{gathered} \tag{18}$$

where $J^+ = J^+(\tau_0),\ J^- = J^-(\tau_0),\ I = I(\tau_0),\ I_0 = I_0(\tau_0),\ I_* = I_*(\tau_0)$, and initial conditions

$$x_{\tau_0+0} = x_{\tau_0};\ \eta_s(\tau_0 + 0) = \eta(t_s(\tau_0), \tau_0), t_s(\tau_0 + 0) = t_s(\tau_0), s \in I(\tau_0). \tag{19}$$

It is evident that the form of equations (18) is determined by structure $P(\tau_0)$ of problem (15) solution at the moment τ_0.
Thus in the case I) for $\tau \in [\tau_0, \tau_1]$ the construction of problem (15) solution in real-time mode is reduced to the construction of system (18) solution (17) in the same mode. The algorithm for solving such equations in real-time mode is described in [3].
In the case II) for $\tau \in [\tau_0, \tau_1]$ parameters (17) are constructed by rule :

$$x_{\tau j} = x_{\tau_0 j}, j \in J;\ \eta_s(\tau) = \eta_s(\tau_0), t_s(\tau) = t_s(\tau_0), s \in I(\tau_0). \tag{20}$$

Consider the moment τ_1 where relations (16) are violated. At the moment τ_1 structure $P(\tau)$ changes: $P(\tau_1 + 0) \neq P(\tau_0)$ and solution $\{x_\tau, \eta_\tau\}$ may have break-point: $\{x_{\tau_1+0}, \eta_{\tau_1+0}\} \neq \{x_{\tau_1}, \eta_{\tau_1}\}$. To continue its operations for $\tau > \tau_1$ the estimator needs to know the new structure $P(\tau_1 + 0)$ and new solution $\{x_{\tau_1+0}, \eta_{\tau_1+0}\}$. These parameters are constructed by the rules described in [4].
When the new parameters $P(\tau_1 + 0)$, x_{τ_1+0}, η_{τ_1+0} have been found, the estimator constructs the elements (17) by rules (18), (19) or (20) where τ_0 is replaced by $\tau_1 + 0$. Having at every current moment $\tau \in T$ parameters (17), i-th estimator calculates the estimate $\alpha_i^*(\tau) = h_i'\chi^*(\tau) + p^{(i)\prime} x_\tau^{(i)}$ and sends this estimate to the controller.

5. Algorithm of the controller operations

The aim of controller operations is to construct the solution of problem (13) in real-time mode.
We suppose that at every current moment the controller knows the estimate values $\alpha_i^*(\tau)$, $i = \overline{1,m}$, $\tau \in T$, which are calculated by the estimators (in real-time mode).
Let us consider the current moment τ. Denote by $u^\tau(\cdot) = u^0(\cdot|s^*) = (u^\tau(t), t \in T^\tau)$, $x^\tau(\cdot) = (x^\tau(t), t \in T^\tau)$ an optimal control and optimal trajectory of the problem (13) and by $\psi(\lambda, t)$, $t \in T^\tau$, $\lambda \in R^m$, the solution of conjugate system

$$\dot{\psi} = -A'\psi, \psi(t^*) = -\sum_{i=1}^{m} \lambda_i h_i.$$

According to the maximum principle, there exists such a vector $\lambda(\tau) = (\lambda_i(\tau), i = \overline{1,m})$ that the following relations take place:

$$\lambda_i(\tau) \geq 0, \lambda_i(\tau)(h_i' x^\tau(t^*) - g_i + \alpha_i^*(\tau)) = 0, i = \overline{1,m};$$

$$u^\tau(t) = 1\ if\ \Delta^\tau(t) > 1, u^\tau(t) = -1\ if\ \Delta^\tau(t) < -1,$$
$$u^\tau(t) = 0\ if\ |\Delta^\tau(t)| \leq 1, t \in T^\tau,$$

where $\Delta^\tau(t) = \psi'(\lambda(\tau), t)b$, $t \in T^\tau$.
Pair $\{u^\tau(\cdot), \lambda(\tau)\}$ is called the solution of problem (13).
Denote

$$\left\{t_j(\tau), j = \overline{1, p(\tau)}\right\} = \{t \in T^\tau : |\Delta^\tau(t)| = 1\},$$

$$t_j(\tau) < t_{j+1}(\tau), j = \overline{1, p(\tau) - 1}; t_0(\tau) \equiv \tau, t_{p(\tau)+1} \equiv t^*;$$

$$k_j(\tau) = 1\ if\ \Delta^\tau(t) > 1, t \in]t_j(\tau), t_{j+1}(\tau)[;$$

$$k_j(\tau) = -1 \; if \; \Delta^\tau(t) < -1, t \in]t_j(\tau), t_{j+1}(\tau)[;$$
$$k_j(\tau) = 0, \; if \; |\Delta^\tau(t)| < 1, t \in]t_j(\tau), t_{j+1}(\tau)[, j = \overline{0, p(\tau)}.$$
$$I(\tau) = \{i \in \{1, 2, \ldots, m\} : h_i' x^\tau(t^*) = g_i - \alpha_i^*(\tau)\}.$$

The totalities of parameters

$$S(\tau) = \left\{p(\tau), k_j(\tau), j = \overline{0, p(\tau)}; I(\tau)\right\}, \tag{21}$$
$$Q(\tau) = \left\{t_j(\tau), j = \overline{1, p(\tau)}; \lambda(\tau)\right\}$$

are said to be *a structure* and *defining elements* constructed by $\lambda(\tau)$.
It is clear that if we know the finite totality of parameters (21) we can uniquely restore the problem (13) solution $\{u^\tau(\cdot), \lambda(\tau)\}$.
Suppose that the structure $S(\tau_0)$ and defining elements $Q(\tau_0)$ are known at some moment $\tau_0 \in [0, t^*]$. Assume that for $\tau = \tau_0$ the relations (non-degeneracy conditions)

$$0 < t_1(\tau), t_{p(\tau)}(\tau) < t^*, \text{rank} \begin{pmatrix} h_i(t_j(\tau)), j = \overline{1, p(\tau)} \\ i \in I(\tau) \end{pmatrix} = |I(\tau)|, \tag{22}$$
$$\frac{\partial \Delta^\tau(t_j(\tau))}{\partial t} \neq 0, j = \overline{1, p(\tau)},$$

hold. Here $h_i(\tau) = h_i' F(t) b$; $\dot{F}(t) = -F(t)A$, $F(t^*) = E$, $t \in T^\tau$.
Denote by τ_1, $\tau_1 > \tau_0$, the nearest to τ_0 point at which the relations (22) are violated.
It is not difficult to show that for $\tau \in [\tau_0, \tau_1]$ the parameters (21) are uniquely defined by equations

$$S(\tau) = S = \left\{p, k_j, j = \overline{0, p}, I\right\},$$
$$f_i(\tau, t_j(\tau), j = \overline{1, p}) = 0, i \in I; \lambda_i(\tau) = 0, i \in \{1, \ldots, m\} \backslash I, \tag{23}$$
$$q\left(t_j(\tau), \lambda(\tau)\right) = k_j + k_{j-1}, j = \overline{1, p};$$

where $f_i(\tau, t_j, j = \overline{1, p}) = \sum\limits_{j=0}^{p} k_j \int\limits_{t_j}^{t_{j+1}} h_i(t) dt - g_i + \alpha_i^*(\tau)$, $i = \overline{1, m}$; $t_0 = \tau$, $t_{p+1} = t^*$; $S = S(\tau_0)$, $q(t, \lambda) = -\sum\limits_{i=1}^{m} h_i(t) \lambda_i$.
Taking into account reasonings given above we make a conclusion that for $\tau \in [\tau_0, \tau_1]$ the construction of the problem (13) solution $\{u^\tau(\cdot), \lambda(\tau)\}$ in real-time mode is reduced to the construction of the system (23) solution $Q(\tau)$ in the same mode. The algorithm for solving equations of (23) type is described in [3].
Let us consider the moment τ_1 where the relations (22) are violated. At the moment τ_1 the structure $S(\tau)$ changes: $S(\tau_1) \neq S(\tau_1 + 0)$ and m-vector-function $\lambda(\tau)$, $\tau \in T$ may have a breakpoint: $\lambda(\tau_1) \neq \lambda(\tau_1 + 0)$. Therefore to continue its operations for $\tau > \tau_1$ the controller needs to know the new structure $S(\tau_1 + 0)$ and the new defining elements $Q(\tau_1 + 0)$. Using the information of $S(\tau_1)$ and $Q(\tau_1)$, the special rules of constructing $S(\tau_1 + 0)$, $Q(\tau_1 + 0)$ have been worked out [5].
When the new $S(\tau_1 + 0)$, $Q(\tau_1 + 0)$ have been found, the controller constructs the definding elements by solving system of equations (23) where $S = S(\tau_1 + 0)$. Vector $Q(\tau_1 + 0)$ is used as the initial condition.

Having the structure $S(\tau)$ and defining elements $Q(\tau)$ controller regenerates the optimal control $u^\tau(\cdot)$ of problem (13) by the rules

$$u^\tau(t) = k_j(\tau), j \in [t_j(\tau), t_{j+1}(\tau)[, j = \overline{0, p(\tau)}.$$

Consiquently at the moment τ the value $u^*(\tau)$ of optimal feedback realization (12) is defined by relation: $u^*(\tau) = k_0(\tau)$.
Thus, the description of the algorithm for constructing optimal feedback realization (12) is finished.

References

[1] Gabasov, R.; Kirillova, F. M.: *Finite Algorithm of Constructing Program Control for Incompletely Determined Linear Optimal Control Problem.* Automation and Remote Control, #7, 1991.

[2] Gabasov, R.; Kirillova, F. M.; Kostyukova, O. I.: *Construction of Optimal Controls of Feedback Type in a Linear Problem.* Soviet Math. Dokl. vol. 44, #2, 1992.

[3] Gabasov, R.; Kirillova, F. M.; Kostyukova, O. I.: *Optimization Controls of a Linear Control System under Real-Time Conditions.* J. Comput. Syst. Sci. 31, #4, 1993.

[4] Gabasov, R.; Kirillova, F. M.; Kostyukova, O. I.: *Optimal Positional Observation of Linear System.* Doklady RAN, vol. 339, #4, 1994.

[5] Kostyukova, O. I.: *Researching of the Set of Optimal Control Problems Depended on Parameters.* Submitted to J. Differential Equations.

International Series of Numerical Mathematics
Vol. 124, © 1998 Birkhäuser Verlag, Basel

Set-valued Calculus and Dynamic Programming in Problems of Feedback Control

Alexander B. Kurzhanski*

Abstract. The present paper is a concise overview of some problems of feedback control under uncertainty and state constraints. It emphasizes the application of set-valued techniques for these problems (see [1], [9], [10]) and indicates the connections with Dynamic Programming (DP) approaches, particularly with the "nonsmooth" versions of the latter, (see [5], [6], [18]). A constructive technique based on ellipsoidal calculus as collected in monograph [13] is then described for linear systems with convex-valued hard bounds on the controls and state space variables. Namely, the respective convex compact set-valued constructions (see [8], [11], [12]) are described in terms of *ellipsoidal-valued representations.* This "ellipsoidal" move leads to rather effective algorithms with possibility of further computer animation. For the problem of control synthesis it particularly allows to present the solutions in terms of analytical designs rather than algorithms as required in the "exact" theory. [1] The last approach also appears to show connections with control techniques for uncertain systems based on applying Liapunov functions, [14].

1. Target control synthesis

Consider the system

$$\dot{x} = A(t)x + u + v(t), \tag{1.1}$$

Here function $A(t)$ is continuous, $u \in \mathcal{U}(t,x) \subseteq \mathcal{P}(t), \mathcal{U} \in U_{\mathcal{P}}$ where class $U_{\mathcal{P}}$ consists of multivalued maps $\mathcal{U}(t,x)$ such that existence and extendability of solutions to equation (1.1) with $u = \mathcal{U}(t,x)$ (it now turns into a differential inclusion - DI), holds for any Lebesgue-measurable function $v(t)$; $\mathcal{P}(t)$ is a set-valued function with convex compact values, continuous in time.

Definition 1 *The problem of* **target control synthesis** *with given target set* $\mathcal{M}$ *consists in specifying* **a solvability set** $W(\tau, t_1, \mathcal{M})$ *and a set-valued* **feedback control strategy** $u = \mathcal{U}(t,x)$, *such that all the solutions to the differential inclusion*

$$\dot{x} \in A(t)x + \mathcal{U}(t,x) + v(t), \ \tau \leq t \leq t_1, \tag{1.2}$$

that start from any given position $\{\tau, x_\tau\}$, $x_\tau \in W(\tau, t_1, \mathcal{M})$, $\tau \in [t_0, t_1)$, *would reach the terminal set* $\mathcal{M}$ *at time* $t_1 : \ x(t_1) \in \mathcal{M}$.

Definition 1.1 is nonredundant if $W(\tau, t_1, \mathcal{M}) \neq \emptyset$. Taking the set-valued function $W[t] = W(t, t_1, \mathcal{M})$, $\tau \leq t \leq t_1$, we come to *the solvability tube* $W[\cdot]$. (Note that

*Faculty of Comput. Mathematics and Cybernetics, Moscow State University, 119899 Moscow, Russia, e-mail: kurzhan@cs.msu.su

[1]The ellipsoidal calculus mentioned here is aimed at *exact representation* of the respective convex sets by parametrized families of ellipsoids rather than at approximating them by only one or several ellipsoids, as done in other publications [4], [13].

$W(\tau, t_1, \mathcal{M})$ is the largest set of states, relative to inclusion, from which the solution does exist at all). The problem is properly posed if $\mathcal{U} \in U_{\mathcal{P}}$. Set $W[t]$ is similar to the *attainability (reachability) set* of system (1.1), but taken in backward time.
We note that in the formulated problem *there is no optimality criterion.* How does then optimization come in if we deal with attainability and control synthesis ?
Let us first calculate the attainability tube $X[t] = X(t, t_0, X^0)$, (which is the tube of all solutions to the DI

$$\dot{x} \in f(t, x, \mathcal{P}(t)), \quad t_0 \leq t \leq t_1,$$

that start at instant t_0 from set X^0) and also the solvability tube $W[t]$. (Here and in the sequel $f(t, x, \mathcal{P}(t)) = A(t)x + \mathcal{P}(t) + v(t)$, the set$\mathcal{M}$ is taken to be convex and compact.) We shall do this through the DP techniques, by solving two optimization problems: *Problem* P-1 and *Problem* P-2.
P-1 : given position $\{\tau, x\}$, $t_0 \leq t \leq \tau$, find

$$V(\tau, x) = \min_u d^2(x(t_0), X^0)$$

under restriction $x(\tau) = x$.
Here

$$d^2(x, \mathcal{M}) = \min\{(x - z, x - z) | z \in \mathcal{M}\} = h_+^2(x, \mathcal{M}),$$

is the square of the Euclidean distance from x to set $\mathcal{M}$,

$$h_+(\mathcal{Q}, \mathcal{M}) = \max_x \min_z \{(x - z, x - z)^{1/2} | x \in \mathcal{Q}, z \in \mathcal{M}\}$$

is the Hausdorff semidistance between sets $\mathcal{Q}, \mathcal{M}$.
The *Value function* $V(t, x)$ is the viscosity ([5]) or minmax ([18]) solution to the following *forward* Hamilton-Jacobi-Bellman (H-J-B) equation, formally written as:

$$\frac{\partial V}{\partial t} + \max_{u \in \mathcal{P}(t)} \left\{ \left(\frac{\partial V}{\partial x}, f(t, x, u) \right) \right\} = 0,$$

with boundary condition

$$V(t_0, x) = d^2(x, X^0).$$

Then

$$X[\tau] = \{x : V(\tau, x) \leq 0\}$$

is the *level set* for $V(t, x)$. [2].
On the other hand, to find $W[t]$ we have to solve
P-2: given position $\{\tau, x\}$, find *value function*

$$V^*(\tau, x) = \min_u d^2(x(t_1), \mathcal{M})$$

under restriction $x(\tau) = x$.

[2]The forward H-J-B equation for studying attainability domains in the smooth case was introduced by G. Leitmann

Function $V^*(t,x)$ is the viscosity (or minmax) solution to the *backward H-J-B* equation formally written as

$$\frac{\partial V^*}{\partial t} + \min_{u \in \mathcal{P}(t)} \left\{ \left(\frac{\partial V^*}{\partial x}, f(t,x,u) \right) \right\} = 0,$$

with boundary condition

$$V^*(t_1, x) = d^2(x, \mathcal{M}).$$

Then

$$W[\tau] = \{x : V^*(\tau, x) \leq 0\}$$

is *the level set* for $V^*(\tau, x)$.

Returning to the problem of control synthesis we shall first seek the tube of all trajectories $x[t]$ that solve the problem of Definition 1 (call it Problem 1.1).

Denoting $f(t,x,\mathcal{P}(t)) = F(t,x)$, consider the DI

$$\dot{x} \in F(t,x), \tag{1.3}$$

and a set-valued function $Y(t)$ with convex-compact values, Lipschitz-continuous in time.

A solution $x[\cdot]$ to (1.3) is said to be *viable relative to constraint* $Y(t)$ if $x[t] \in Y(t)$ for all $t \in [t_0, t_1]$.

Lemma 1 *Each of the trajectories $x^{(s)}[t] = x^{(s)}(t, t_0, x^0)$, synthesized from position $\{t_0, x^0\}$ due to Problem 1.1, is a solution of (1.3) viable relative to constraint $W[t]$, $(t \in [t_0, t_1])$.*

Assumption 1.1: There exists a solution $x[\cdot]$ to (1.3) $x(t_0) \in X^0$, such that $x[t] \in$ *int*$W[t], \forall t$.

Theorem 1.1 *Under Assumption 1.1 the tube $X^{(s)}[\cdot]$ of all trajectories $x^{(s)}[t] = x^{(s)}(t, t_0, x^0)$, synthesized due to the problem of Definition 1.1, is given by relation*

$$X^{(s)}[t] = X[t] \cap W[t], \quad t \in [t_0, t_1]. \tag{1.4}$$

The last relation means that *the synthesized trajectory must lie both in the attainability tube and in the solvability tube.* Relation (1.4) is true for the particular constraint $x[t] \in W[t]$. To find the tube $X^{(s)}[\cdot]$ we may consider the following *Problem* P-3:

P-3: due to system (1.1) find

$$\min_{u(\cdot)} \left\{ d^2(x(t_0), X^0) + \int_{t_0}^{t} V^{*2}(s, x(s))ds \right\} = V^0(t,x)$$

under condition

$$x(t) = x$$

Then $V^0(t,x)$ is a viscosity (or minmax) solution to the (forward) H-J-B equation for the collided tube (actually a problem with state constraints, [15], [3]), which is formally written as follows:

$$\frac{\partial V^0}{\partial t} + \max_{u \in \mathcal{P}(t)} \left\{ \left(\frac{\partial V^0}{\partial x}, f(t,x,u) \right) - V^{*2}(t,x) \right\} = 0,$$

with boundary condition

$$V^0(t_0, x) = d^2(x, X^0)$$

and

$$X^{(s)}[t] = \{x : V^0(t, x) \leq 0\}$$

is the level set for $V^0(t, x)$.

Here $V^*(t, x)$ is the value function for Problem P-2 being the solution to the backward H-J-B equation for the solvability tube. Note that to find $X^{(s)}[\cdot]$ we *do not need to know the synthesizing control strategy* $\mathcal{U}$. Due to (1.4), we shall refer to tubes $X[t], W[t]$ as *colliding tubes.*

Finally, let us indicate the control strategy that solves Problem 1.1.

If a continuous gradient $\partial V^0/\partial x$ would exist, one could set

$$U_s = \mathcal{U}_s(t, x) = \operatorname*{argmin}_{u \in \mathcal{P}(t)} \left(\frac{\partial V^0}{\partial x}, f(t, x, u) \right)$$

or, if it does not exist

$$U_s = \mathcal{U}_a(t, x) = \{u : \frac{d(d^2(x, W[t]))}{dt}|_u \leq 0\} \;\; u \in \mathcal{P}(t),$$

or

$$U_s = \mathcal{U}_b(t, x) = \{u : \frac{d(d^2(x, X^{(s)}[t]))}{dt}|_u \leq 0\} \;\; u \in \mathcal{P}(t),$$

If substituted in (1.2) with $X[t_0] = X^{(s)}[t_0]$, each of the strategies $\mathcal{U}_a, \mathcal{U}_b$ or $\mathcal{U}_s$ (if it exists) produces the same solution tube $X_a[t] = X_b[t] = X^{(s)}[t]$.

Remarks.

1. The value functions introduced in this section may be calculated directly in terms of dual problems of convex analysis, [17], [10], [13].
2. The results of this section also allow a propagation to nonlinear systems under additional convexity assumptions on the respective Hamiltonians.

2. Guaranteed control synthesis for uncertain systems

We further deal with system (1.1) with *magnitude* ("hard") bounds on the controls u and *the unknown but bounded uncertain items* $v(t), x_0$:

$$u(t) \in \mathcal{P}(t), \;\; v(t) \in \mathcal{Q}(t), \;\; x^0 \in \mathcal{X}_0, \tag{2.1}$$

presuming the constraints on u, v, x^0 to be ellipsoidal-valued. Namely, by introducing the notation $\mathcal{E}(a, K) = \{x : (x - a, K^{-1}(x - a) \leq 1\}, x \in \mathbb{R}^n, \;\; K > 0$, in terms of inclusions we shall then have:

$$u \in \mathcal{E}(p(t), P(t)) \,, v \in \mathcal{E}(q(t), Q(t)) \,, x^0 \in \mathcal{E}(x^*, X_0), \tag{2.2}$$

where the continuous functions $p(t), q(t)$ and the vector x^* are given together with continuous matrix functions $P(t) > 0, Q(t) > 0$ and matrix $X_0 > 0$. We also use the notation

$$\rho(l|\mathcal{P}) = \max\{(l, x)|x \in \mathcal{P}\}$$

for the *support function* of a compact set $\mathcal{P}$.

We shall now present some basic relations for the problem of control synthesis under uncertainty in the form appropriate for further ellipsoidal approximations. Consider system (1.1), (2.1) and a "terminal set" $\mathcal{M} \in comp\mathbb{R}^n$ – the variety of convex compact sets in $\mathbb{R}^n$.

Definition 2 *The problem of* **target control synthesis under uncertainty** *consists in specifying* **a solvability set** $\mathcal{W}(\tau, t_1, \mathcal{M}) = \mathcal{W}[\tau]$ *and a set-valued* **feedback control strategy** $u = \mathcal{U}(t, x)$, *such that all the solutions to the differential inclusion*

$$\dot{x} \in A(t)x + \mathcal{U}(t, x) + v(t), \; \tau \le t \le t_1, \tag{2.3}$$

that start from any given position $\{\tau, x_\tau\}$, $x_\tau \in \mathcal{W}(\tau, t_1, \mathcal{M})$, $\tau \in [t_0, t_1)$, *would reach the terminal set* $\mathcal{M}$ *at time* $t_1 : x(t_1) \in \mathcal{M}$, *whatever be the unknown disturbance* $v(t)$.

The set-valued function $\mathcal{W}[\cdot]$ is known as the *solvability tube.* The formulated problem again does not have any optimization criteria and requires finding just a feasible solution. Nevertheless, we shall seek such a solution by applying some *minimaximization* schemes through a DP interpretation of the problem. Introduce the value function

$$\mathcal{V}(t, x) \;=\; \min_{\mathcal{U}} \max_{x(\cdot)} \{\mathcal{I}(t, x) | \mathcal{U}(\cdot, \cdot) \in U_{\mathcal{P}}, x(\cdot) \in \mathcal{X}_{\mathcal{U}}(\cdot)\},$$

where

$$\mathcal{I}(t, x) = d^2(x[t_1], \mathcal{M}),$$

and $\mathcal{X}_{\mathcal{U}}(\cdot)$ is the variety of all trajectories $x(\cdot)$ of inclusion (2.3) generated by given strategy $\mathcal{U}$ and as in the above $U_{\mathcal{P}}$ stands for the class of strategies that ensure the existence and extendability of solutions to the differential inclusion (1.2) (this, for example, is the class of set-valued functions with values in $comp\mathbb{R}^n$, upper semicontinuous in x and continuous in t).

The formal H-J-B-I (Hamilton-Jacobi-Bellman-Isaacs) equation for the value $\mathcal{V}(t, x)$ looks as follows

$$\frac{\partial \mathcal{V}}{\partial t} + \min_u \max_v \left(\frac{\partial \mathcal{V}}{\partial x}, A(t)x + u + v \right) = 0 \tag{2.4}$$

with boundary condition

$$\mathcal{V}(t_1, x) \;=\; h_+^2(x, \mathcal{M}) \; . \tag{2.5}$$

The last equation may not have a classical solution though, and its treatment may therefore again require use of the generalized notion of viscosity solution [5] or its equivalent – the minmax solution [18]. However, in the specific problems of this paper, the value function $\mathcal{V}(t, x)$ shall turn out to be directionally differentiable for any direction $(1, h), h \in \mathbb{R}^n$ which ensures the respective requirements will be satisfied. We shall have this remark in mind while continuing to write the H-J-B-I equation in the formal "classical" transcription (2.4). (In this paper all the value functions shall turn out to be either convex or quasi-convex in x and all the solutions to the respective H-J-B or H-J-B-I equations shall not lead beyond the notions of viscosity or minmax solutions.)

We further deal with the following nondegenerate case.

Assumption 2.1 . The solvability tube $\mathcal{W}[\cdot]$ is such that there exists an $\epsilon > 0$ and a continuous function $w_*(t)$, that yield the inclusion $w_*(t) + \epsilon\mathcal{S} \in \mathcal{W}[t]$, $t_0 \le t \le t_1$, where $\mathcal{S} = \{x : (x, x) \le 1\}$.

Lemma 2 *Under Assumption 2.1 equation (2.4), with boundary condition (2.5), has a unique solution $\mathcal{V}(t,x)$ in the domain $[t_0,t_1] \times \mathbb{R}^n$.*

Denote $\mathcal{W}_*[t] = \{x : \mathcal{V}(t,x) \leq 0\}$. An important fact is given by

Lemma 3 *Under Assumption 2.1 the solvability set $\mathcal{W}[t]$ of Definition 2.1 may be represented as*

$$\mathcal{W}[t] = \mathcal{W}_*[t], \ t_0 \leq t \leq t_1, \tag{2.6}$$

If equation (2.4) ((2.5)) were solved, one could obtain the solution strategy as

$$\mathcal{U}_*(t,x) = \arg\min\{(\partial\mathcal{V}(t,x)/\partial x, u) | u \in \mathcal{P}(t)\} \tag{2.7}$$

(if the gradient $\partial\mathcal{V}(t,x)/\partial x$ does exist at $\{t,x\}$), or, more generally, as

$$\mathcal{U}_*(t,x) = \{u : \max\{dh_+^2(x, \mathcal{W}_*[t])/dt | v \in \mathcal{Q}(t)\} \leq 0\}. \tag{2.8}$$

To specify the control strategy $\mathcal{U}_*$ it therefore suffices to know only the cut $\mathcal{W}_*[t]$ of the value function $\mathcal{V}(t,x)$. [3] Since the direct integration of the H-J-B-I equation may require rather subtle numerical techniques, we shall further investigate and approximate the cuts of the value function rather than the value function itself. By dealing only with the cuts, we shall thus *avoid the integration* of the H-J-B-I equation for the class of problems considered in this paper. Particularly, we shall introduce an array of parametrized internal and external *ellipsoidal approximations* for these cuts. It is therefore important to be able to find the cuts $\mathcal{W}_*[t]$ without calculating $\mathcal{V}$. Thus, it is not uninteresting to observe that there exists a set-valued integral whose value happens to be precisely the set $\mathcal{W}[t]$ and therefore also the cut $\mathcal{W}_*[t]$ of the value function $\mathcal{V}(t,x)$. This is the "alternated integral" introduced by L.S. Pontryagin [16], (see [13], Section 1.7).

Lemma 4 *Under Assumption 2.1 set $\mathcal{W}_*[t]$ may be represented as*

$$\mathcal{W}_*[t] = I(t, t_1, \mathcal{M}), \ t_0 \leq t \leq t_1, \tag{2.9}$$

where $I(t,t_1,\mathcal{M})$ is the set-valued alternated integral of Pontryagin.

Finally, the set-valued function $\mathcal{W}[t]$ is a solution to the following evolution equation of the "funnel type":

Lemma 5 *Under Assumption 2.1 the set-valued function $\mathcal{W}[t]$ satisfies for all $t \in [t_0,t_1]$ the evolution equation*

$$\lim_{\sigma\to 0} \sigma^{-1} h_+(\mathcal{W}[t-\sigma] + \sigma\mathcal{Q}(t), \ (I - \sigma A(t))\mathcal{W}[t] - \sigma\mathcal{P}(t)) = 0 \ , \mathcal{W}[t_1] = \mathcal{M} \ . \tag{2.10}$$

The map $\mathcal{W}[t] = \mathcal{W}(t,t_1,\mathcal{M})$ satisfies the semigroup property

$$\mathcal{W}(t,t_1,\mathcal{M}) = \mathcal{W}(t,\tau,\mathcal{W}(\tau,t_1,\mathcal{M})), \ t \leq \tau \leq t_1. \tag{2.11}$$

[3] Here the properties of the value function $\mathcal{V}(t,x)$ are such that the respective strategy $\mathcal{U}(t,x) \in U_{\mathcal{P}}$ is always feasible.

The set-valued function $\mathcal{W}[\cdot]$ also turns out to be a "stable bridge", as defined in [8], [9]. The basic property of the solution is as follows:

Theorem 2.1 *Suppose Assumption 1.1 holds. Then, with* $x_\tau = x(\tau) \in \mathcal{W}_*[t]$, *all the solutions* $x[t] = x(t, \tau, x_\tau)$ *to the differential inclusion*

$$\dot{x} \in \mathcal{U}_*(t, x) + v(t), \quad x[\tau] = x_\tau, \tag{2.12}$$

satisfy the relation $x[t] \in \mathcal{W}_*[t]$, $\tau \leq t \leq t_1$, *and therefore reach the terminal set* $\mathcal{M} : x[t_1] \in \mathcal{M}$, *whatever be the unknown disturbance* $v(t)$.

Among the feasible set-valued strategies $\mathcal{U}(t, x)$ that satisfy the inclusion $\mathcal{U}(t, x) \subseteq \mathcal{U}_*(t, x)$ and therefore ensure the result of theorem 1.1, if substituted in (1.11) instead of $\mathcal{U}_*(t, x)$, is $\mathcal{U}_e(t, x)$ – the "extremal aiming strategy" ([8], [9]), given by relation

$$\mathcal{U}_e(t, x) = \partial_l k(t, -l^0) \;=\; \arg\min\{(l^0, u) | u \in \mathcal{P}(t)\}. \tag{2.13}$$

Here $\partial_l k$ stands for the subdifferential,of function $k(t, l)$ in the variable l, $k(t, l) = \rho(l|\mathcal{P}(t))$, and $l^0 = l^0(t, x) \neq 0$ is the maximizer for the problem

$$d_*[\tau, x] = \max\{(l, x) - \rho(l|\mathcal{W}_*[\tau])| \; \|l\| \leq 1\} > 0. \tag{2.14}$$

with $l^0(t, x) = 0$ when $d_*[\tau, x] = 0$. [4]

Lemma 6 *With* $x_\tau \in \mathcal{W}_*[\tau]$ *strategy* $\mathcal{U}_e(t, x)$ *ensures the inclusion* $X_{\mathcal{W}_*}(t, \tau, x_\tau) \subseteq \mathcal{W}_*[t], \quad \tau \leq t \leq t_1$.

Here $X_{\mathcal{W}_*}(\cdot, \tau, x_\tau)$ is the tube of all solutions to system (1.2), $x[\tau] = x_\tau$, with $\mathcal{U}(t, x) = \mathcal{U}_e(t, x)$. The results of the above may be summarized into

Theorem 2.2 *The synthesizing strategy* $\mathcal{U}_e(t, x)$ *resolves the problem of guaranteed control synthesis under uncertainty.*

Under Assumption 2.1 the last Theorems do not require any matching conditions for the constraints on the inputs u, v, (of type $\mathcal{P}(t) = \alpha \mathcal{Q}(t), \; 0 < \alpha < 1$).
Finally, we also mention that the cut $\mathcal{W}_*[t]$ may be obtained from the following value function

$$\mathcal{V}_*(\tau, x) = \min_{\mathcal{U}} \max_{v} \max_{x(\cdot)} \{\mathcal{I}_*(\tau, x) | \mathcal{U}(\cdot, \cdot), v(\cdot)\},$$

as the set $\mathcal{W}_*[\tau] = \{x : \mathcal{V}_*(\tau, x) \leq 0\}$, where

$$\mathcal{I}_*(\tau, x) = \int_\tau^{t_1} (h_+^2(u, \mathcal{P}(t)) - h_+^2(v, \mathcal{Q}(t)))dt + h_+^2(x(t_1), \mathcal{M}),$$

and $x(\cdot)$ varies over all the trajectories of system (1.11) generated by strategy $\mathcal{U}$ and function v. Value function $\mathcal{V}_*(t, x)$ reflects the so-called H_∞ approach to the problem of control synthesis (see [2]).
The application of the described techniques requires to calculate the tube $\mathcal{W}_*[\cdot]$ in one or another way, producing finally the desired strategy $\mathcal{U}_*$ or $\mathcal{U}_e$ *in the form of*

[4] A justification of the last formula, which reflects N.N. Krasovski's "extremal aiming" rule is given in [8], [9], see also [13].

an algorithm. The aim of this paper is further to emphasize an ellipsoidal technique that would allow production of a computational algorithm for the solvability tubes and further on, to generate the "guaranteed" control synthesis in the form of a design – an "analytical" controller.

We shall now pass to the description of ellipsoidal approximations for the solvability tubes $\mathcal{W}[\cdot]$ and their crossections $\mathcal{W}[t]$ - the cuts of the value function $\mathcal{V}(t,x)$.

3. Ellipsoid solvability tubes

In this section and in the next one we continue to assume all the sets $\mathcal{P}, \mathcal{Q}, X^0, \mathcal{M}$ to be ellipsoidal-valued as given by (2.1), (2.2), considering the problem of Definition 2.1 under Assumption 2.1. In this case the cut $\mathcal{W}_*[t]$ of the value function $\mathcal{V}(t,x)$ is described by the alternated integral of Pontryagin and its evolution in time is due to the funnel equation (2.10).

Take the differential equations

$$\dot{x}^* = A(t)x^*p(t) + q(t) \tag{3.1}$$

and also

$$\dot{X}_+ = A(t)X_+ + X_+A(t) - \pi(t)X_+ - \pi^{-1}(t)P(t) + \tag{3.2}$$
$$+H^{-1}(t)F(H(t), X_+, Q(t))H'^{-1}(t),$$
$$\dot{X}_- = A(t)X_- + X_-A(t) + \pi(t)X_- + \pi^{-1}(t)Q(t) - \tag{3.3}$$
$$-H^{-1}(t)F(H(t), X_-, P(t))H'^{-1}(t),$$

with boundary conditions

$$x^*(t_1) = m\ ,\ X_+(t_1) = M,\ X_-(t_1) = M. \tag{3.4}$$

Here

$$F(H,P,X) = (HXH')^{1/2}(HPH')^{1/2} + (HPH')^{1/2}(HXH')^{1/2}.$$

Denote the solutions to (3.2), (3.3) under boundary condition (3.4) as $X_+(t|\pi(\cdot), H(\cdot))$, $X_-(t|\pi(\cdot), H(\cdot))$. Following the techniques of [13], it is possible to prove the following assertion:

Theorem 3.1 *For every vector $l \in \mathbb{R}^n$, the following inclusions are true for any measurable functions $\pi(t) > 0$, and $H(t) = H'(t)$:*

$$\mathcal{E}_-[t] = \mathcal{E}(x^*(t), X_+(t|\pi(\cdot), H(\cdot)) \supseteq \mathcal{W}[t] \supseteq \mathcal{E}(x^*(t), X_+(t|\pi(\cdot), H(\cdot)) = \mathcal{E}_+[t]. \tag{3.5}$$

Moreover, the following relations hold:

$$\rho(l|\mathcal{W}[t]) \ = \ \inf\{\rho(l|\mathcal{E}(x^*(t), X_+(t|\pi(\cdot), H(\cdot))|\pi(\cdot), H(\cdot)\}, \tag{3.6}$$

and

$$\mathcal{W}[t] \ = \ \cap\{\mathcal{E}(x^*(t), X_+(t|\pi(\cdot), H(\cdot))\}, \tag{3.7}$$

as well as

$$\rho(l|\mathcal{W}[t]) \;=\; \sup\{\rho(l|\mathcal{E}(x(t), X_-(t|\pi(\cdot), H(\cdot)))|\pi(\cdot), H(\cdot)\}, \tag{3.8}$$

and

$$\mathcal{W}[t] = \overline{\cup\{\mathcal{E}(x^*(t), X_-(t|\pi(\cdot), H(\cdot))|\pi(\cdot), H(\cdot)\}} \tag{3.9}$$

where $\pi(t) > 0,\ H(t) = H'(t)$, *are measurable functions.*

Here the dash stands for the closure of the respective set.
For given pairs of functions $\pi(\cdot), H(\cdot)$ and boundary values m^*, M^* denote

$$E_+(t, \tau, \mathcal{E}(m^*, M^*)) = \mathcal{E}(x^*(t, \tau, m^*), X_+(t, \tau, M^*)), \tag{3.10}$$

$$E_-(t, \tau, \mathcal{E}(m^*, M^*)) = \mathcal{E}(x^*(t, \tau, m^*), X_-(t, \tau, M^*)). \tag{3.11}$$

Then, obviously, $E_-(t, \tau, \mathcal{E}(m^*, M^*)) \subseteq \mathcal{W}[t] \subseteq E_+(t, \tau, \mathcal{E}(m^*, M^*))$.

Lemma 7 *The following identities are true with* $t \le \tau \le t_1$:

$$E_-(t, \tau, E_-(\tau, t_1, \mathcal{E}(m, M))) \equiv E_-(t, t_1, \mathcal{E}(m, M)) \tag{3.12}$$

$$E_+(t, t_1, \mathcal{E}(m, M)) \equiv E_+(t, \tau, E_+(\tau, t_1, \mathcal{E}(m, M))) \tag{3.13}$$

Relations (2.11), (2.12) thus define, in backward time, the "lower" and "upper" semigroup properties of the corresponding mappings.
We emphasize once more, that $\mathcal{E}_-[t]$ is an internal approximation of Pontryagin's alternated integral $I(t, t_1, \mathcal{M})$ of (2.9). One may also observe from the previous Theorem that the upper and lower ellipsoidal estimates $\mathcal{E}_+[t] = \mathcal{E}(x(t), X_+(t|\pi(\cdot), H(\cdot)))$ and $\mathcal{E}_-[t] = \mathcal{E}(x(t), X_-(t, |\pi(\cdot), H(\cdot)))$ are respectively *inclusion-minimal and inclusion-maximal* as compared to any other external or internal ellipsoidal approximations of set $\mathcal{W}[t]$.
The property of being *inclusion-maximal* as well as *the semigroup property* of the internal approximation $E_-(t, \tau, \mathcal{E}(m^*, M^*))$ are crucial for using ellipsoidal techniques in the analytical design of the solution strategies for the problem of control synthesis. For the problem of control synthesis we need only the internal approximations of the solvability set. Nevertheless, we presented also the external approximations of this set, since, with a reversal of time, the results of this section give the external and internal approximations of domains of "attainability under uncertainty", see [13].

4. Synthesis under uncertainty through ellipsoid techniques

Let us return to the original problem of control synthesis. There the strategy $\mathcal{U}_e(t, x)$ was synthesized such that it ensured all solutions $x[t] = x(t, \tau, x_\tau)$ of the differential inclusion (2.3), $x_\tau = x[\tau] \in \mathcal{W}[\tau]$, to satisfy the relation $x[t] \in \mathcal{W}[t],\ \ \tau \le t \le t_1$, and therefore to ensure $x[t_1] \in \mathcal{M} = \mathcal{W}[t_1]$, *whatever be the disturbance* $v(t)$. Here $\mathcal{W}[t]$ is the *solvability set* described in Section 2. The exact solution requires, as indicated above, calculating the tube $\mathcal{W}[\cdot]$ and then, for each instant t, to solve an extremal problem of type (2.4) whose solution finally yields the desired strategy $\mathcal{U}_e(t, x)$, which is therefore actually defined *as an algorithm.*

In order to obtain a simpler scheme, we will now substitute $\mathcal{W}[t]$ by one of its *internal ellipsoidal approximations* $\mathcal{E}_-[t] = \mathcal{E}(x^*, X(t))$. The conjecture is that once $\mathcal{W}[t]$ is substituted by $\mathcal{E}_-[t]$, we should just copy the scheme of Section 2, constructing a strategy $\mathcal{U}_-(t,x)$ such that for every solution $x[t] = x(t,\tau,x_\tau)$ to equation

$$\dot{x}[t] \in A(t)x + \mathcal{U}_-(t,x[t]) + v(t), \quad \tau \le t \le t_1, \quad x[\tau] = x_\tau, \quad x_\tau \in \mathcal{E}_-[\tau], \tag{4.1}$$

the inclusion $x[t] \in \mathcal{E}_-[t], \quad \tau \le t \le t_1$, would be true and therefore $x[t_1] \in \mathcal{E}(m,M) = \mathcal{M} = \mathcal{E}_-[t]$, whatever be the disturbance $v(t)$.

In constructing the "ellipsoidal synthesis", we shall imitate the strategy $\mathcal{U}_e$ given by (2.13), applying the respective scheme to the internal approximation $\mathcal{E}_-[t]$ of $\mathcal{W}[t]$. It was indicated in [13] that once approximation $\mathcal{E}_-[t]$ is selected "appropriately", (that is due to (3.1), (3.3), (3.4), (3.8), (3.9)), the respective "ellipsoidal-based" extremal strategy $\mathcal{U}_-(t,x)$ does solve the problem. More precisely, we have the same relations (2.13), (2.14), except that $\mathcal{W}[t]$ will now be substituted by $\mathcal{E}_-[t]$. Namely, this gives

$$\mathcal{U}_-(t,x) = \begin{cases} \mathcal{E}(p(t),P(t)) & \text{if } x \in \mathcal{E}_-[t] \\ p(t) - P(t)l^0(l^0,P(t)l^0)^{-1/2} & \text{if } x \notin \mathcal{E}_-[t], \end{cases} \tag{4.2}$$

where $l^0 = l^0(t,x)$ is the unit vector that solves the problem

$$d[t,x] = (l^0,x) - \rho(l^0|\mathcal{E}_-[t]) = \max\{(l,x) - \rho(l|\mathcal{E}_-[t]) | \|l\| \le 1\} \tag{4.3}$$

and $d[t,x] = h_+(x,\mathcal{E}_-[t])$.

One may readily observe that relation (4.3) coincides with (2.14), if set $\mathcal{W}[t]$ is substituted for $\mathcal{E}_-[t]$ and $\mathcal{P}(t)$ for $\mathcal{E}(p(t),P(t))$; however, here the maximization problem (4.3) may be solved in more detail than its more general analogue (2.14), (since $\mathcal{E}_-[t]$ is an ellipsoid).

If s^0 is the solution to the minimization problem

$$s^0 = \arg\min\{\|(x-s)\| | s \in \mathcal{E}_-[t], \ x = x(t)\}, \tag{4.4}$$

then in (4.3) one may take $l^0 = k(x(t) - s^0), \quad k > 0$, so that l^0 will be *the gradient* of the distance $d[t,x] = h_+(x,\mathcal{E}_-[t])$ with t fixed. (This can be verified by differentiating either (4.3) or (4.4) in x).

Lemma 8 *Consider a nondegenerate ellipsoid $\mathcal{E} = \mathcal{E}(x^*,X)$ and a vector $x \notin \mathcal{E}(a,Q)$. Then the gradient*

$$l^0 = \partial h_+(x,\mathcal{E}(x^*,X))/\partial x$$

may be expressed as

$$l^0 = (x-s^0)/\|x-s^0\|, \ s^0 = (I+\lambda X^{-1})^{-1}(x-a) + a, \tag{4.5}$$

where the multiplier $\lambda > 0$ is the unique root of the equation $f(\lambda) = 0$, and where

$$f(\lambda) = \left((I+\lambda X^{-1})^{-1}(x-a), X^{-1}(I+\lambda X^{-1})^{-1}(x-a)\right)^{-1}$$

Corollary 4.1 *With parameters x^*, X given and x varying, the multiplier λ may be uniquely expressed as a function $\lambda = \lambda(x)$.*

Looking at relation (2.13) in view of the fact that $\mathcal{P}(t) = \mathcal{E}(p(t), P(t))$, we observe

$$\arg\min\{(l^0, u) | u \in \mathcal{E}(p(t), P(t))\} = \mathcal{U}_-(t, x). \quad (4.6)$$

The result (4.2) is then a consequence of the following fact:

Lemma 9 *Given ellipsoid $\mathcal{E}(p, P)$, the minimizer u^* for the problem*

$$\min\{(l, u) | u \in \mathcal{E}(p, P)\} = (l, u^*) \quad , l \neq 0,$$

is the vector $u^* = p \; - \; Pl(l, Pl)^{-\frac{1}{2}}$.

This lemma follows from the formula for the support function of an ellipsoid. The result may now be summarized in the assertion:

Theorem 4.1 *For the set $\mathcal{W}[t]$ define an internal approximation $\mathcal{E}_-[t] = \mathcal{E}_-(x^*(t), X_-(t))$ with fixed parametrizing functions $\pi(t), H(t)$ in (3.3). If $x[\tau] \in \mathcal{E}_-[\tau]$ and the synthesizing strategy is taken as $\mathcal{U}_-(t, x)$ of (4.2), then the following inclusion is true: $x[t] \in \mathcal{E}_-[t], \quad \tau \leq t \leq t_1$, and therefore $x[t_1] \in \mathcal{E}(m, M)$, whatever be the unknown disturbance $v(t)$.*

The "ellipsoidal" synthesis described in this section thus gives a solution strategy $\mathcal{U}_-(t, x)$ for any internal approximation $\mathcal{E}_-[t] = \mathcal{E}_-(x(t), X_-(t))$ of $\mathcal{W}[t]$. With $x \notin \mathcal{E}_-[t]$, the function $\mathcal{U}_-(t, x)$ is single-valued, whilst with $x \in \mathcal{E}_-[t]$ we have $\mathcal{U}_-(t, x) = \mathcal{E}_-[t]$. The overall solution $\mathcal{U}(t, x)$ turns out to be upper-semicontinuous in x and measurable in t, ensuring therefore the existence of a solution to the differential inclusion (4.1). Due to Theorem 3.1 (see (3.9)), each element $x \in \text{int}W[t]$ belongs to a certain ellipsoid $\mathcal{E}_-[t]$ and may therefore be steered to the terminal set $\mathcal{M}$ by means of a certain "ellipsoidal-based" strategy $\mathcal{U}_-(t, x)$, under Assumption 2.1.

Relations (4.2),(4.5) indicate that strategy $\mathcal{U}_-(t, x)$ is given explicitly, with the only unknown being the multiplier λ of Lemma 8, which can be calculated as the only root of equation $f(\lambda) = 0$. But in view of Corollary 4.1 the function $\lambda = \lambda(t, x)$ may be calculated *in advance*, depending on the parameters $x^*(t), X_-(t)$ of the internal approximation $\mathcal{E}_-[t]$ (which may also be calculated in advance and is such that it ensures the feasibility of $\mathcal{U}_-$). With this specificity, the suggested strategy $U_-(t, x)$ may be considered as an *analytical design*.

References

[1] Aubin, J.-P.: *Viability Theory*, Birkhäuser, Boston, 1991.

[2] Basar, T.; Bernhard, P.: *H^∞ Optimal Control and Related Minimax Design Problems* ser. SCFA, 2-nd ed., Birkhäuser, Boston, 1995.

[3] Capuzzo-Dolcetta, I.; Lions, P. L.: *Viscosity solutions of Hamilton-Jacobi-Bellman equations and and state constraints.* Trans. Amer. Math. Soc., v. 318, 1990, pp. 643–683.

[4] Chernousko, F. L.: *State Estimation for Dynamic Systems,* CRC Press, 1994.

[5] Crandall, M. G.; Lions, P. L.: *Viscosity Solutions of Hamilton-Jacobi Equations*, Trans. Amer. Math. Soc., 277, 1983, pp. 1–42.

[6] Fleming, W. H.; Soner, H. M.: *Controlled Markov Processes and Viscosity Solutions*, Springer-Verlag, 1993.

[7] Knobloch, H.; Isidori, A.; Flockerzi, D.: *Topics in Control Theory*, Birkhäuser, DMV-Seminar, Band 22, 1993.

[8] Krasovski, N. N.: *Game-Theoretic Problems on the Encounter of Motions* Nauka, Moscow, 1970, (in Russian), English Translation : *Rendezvous Game Problems* Nat. Tech. Inf. Serv., Springfield, VA, 1971.

[9] Krasovski, N. N.; Subbotin, A. N.: *Positional Differential Games*, Springer-Verlag, 1988.

[10] Kurzhanski, A. B.: *Control and Observation Under Uncertainty.* Nauka, Moscow, 1977.

[11] Kurzhanski, A. B.; Nikonov, O. I.: *On the Problem of Synthesizing Control Strategies.* Evolution Equations and Set-Valued Integration, Doklady Akad. Nauk SSSR, 311, 1990, pp. 788–793, Sov. Math. Doklady, v. 41, 1990.

[12] Kurzhanski, A.; Nikonov, O. I.: *Evolution Equations for Tubes of Trajectories of Synthesized Control Systems.* Russ. Acad. of Sci. Math. Doklady, v. 48, N3, 1994, pp. 606–611.

[13] Kurzhanski, A. B.; Vályi, I.: *Ellipsoidal Calculus for Estimation and Control.* Birkhäuser, Boston, ser. SCFA, 1996.

[14] Leitmann, G.: *One approach to control of uncertain dynamical systems.* Proc. 6-th. Workshop on Dynamics and Control, Vienna, 1993.

[15] Lions, P.-L.; Souganidis, P. E.: *Differential Games, Optimal Control and Directional Derivatives of Viscosity Solutions of Bellman's and Isaac's Equations*, SIAM J. Cont. Opt., v. 23, 1985, pp. 566–583.

[16] Pontryagin, L. S.: *Linear Differential Games of Pursuit.* Mat. Sbornik v. 112, (154):3(7), 1980.

[17] Rockafellar, R. T.: *Convex Analysis,* Princeton University Press, 1970.

[18] Subbotin, A. I.: *Generalized Solutions of First-Order PDE's. The Dynamic Optimization Perspective*, Ser. SC, Birkhäuser, Boston, 1995.

International Series of Numerical Mathematics
Vol. 124, © 1998 Birkhäuser Verlag, Basel

Strong Observability of Time-Dependent Linear Systems

Dirk Liebscher*

Abstract. There are considered time-dependent linear systems of the form

$$\dot{x} = Ax + Bu, \ y = Cx$$

with state $x \in I\!R^n$, control (input) $u \in I\!R^m$ and output $y \in I\!R^p$. We derive local characterizations of observability of (A, C) and strong observability of (A, B, C). These criteria are pointwise rank conditions on a certain matrix, which is explicitly built up from the first $n-2$ derivatives of A and B and the first $n-1$ derivatives of C. The results generalize well-known theorems for time-invariant systems.

1. Introduction

We consider linear systems of the form

$$\dot{x} = A(t)x + B(t)u, \ y = C(t)x \tag{1}$$

with state $x \in I\!R^n$, control $u \in I\!R^m$ and output $y \in I\!R^p$. By $C_s^j(\mathcal{I})$ we denote the set of functions, for which the jth derivative exists a.e. and is piecewise continuous on $\mathcal{I}$. We define the notions of observability, strong observability and controllability.

Definition 1 *(i) The pair $(A(t), C(t))$ is called (completely)* **observable** *(on $\mathcal{I}$), if $\dot{x} = A(t)x$, $C(t)x(t) \equiv 0$ on some non-degenerate interval $\mathcal{J} \subset \mathcal{I}$ always implies that $x(t) \equiv 0$ on $\mathcal{J}$;*

(ii) the triple $(A(t), B(t), C(t))$ is called **strongly observable** *(on $\mathcal{I}$), if $\dot{x} = A(t)x + B(t)u$, $C(t)x(t) \equiv 0$ on some non-degenerate interval $\mathcal{J} \subset \mathcal{I}$ for some control u with $Bu \in C_s(\mathcal{I})$ always implies that $x(t) \equiv 0$ on $\mathcal{J}$; and*

(iii) the pair $(A(t), B(t))$ is called (completely) **controllable** *(on $\mathcal{I}$) if $(-A^T(t), B^T(t))$ is observable.*

Note that our definition of controllability is equivalent to the usual one, that the pair (A, B) is controllable on $\mathcal{I}$, if for all $t_1, t_2 \in \mathcal{I}$, $t_1 < t_2$, $x_1, x_2 \in I\!R^n$ there exists a control u, such that the solution $x(t)$ of the initial value problem $\dot{x} = Ax + Bu$, $x(t_1) = x_1$ satisfies $x(t_2) = x_2$. Note also that observability of (A, C) is the same as strong observability of (A, B, C) with $B = 0$.

In the time-dependent case there are no direct characterizations of these notions via matrix valued functions A, B and C. There exist global characterizations via a fundamental matrix of $\dot{x} = A(t)x$ for observability (and controllability) (see [1], [2], [3]). In this paper we prove local characterizations of the central notions, which equal in time-invariant case the known results of R.E.Kalman [4] and W.Kratz [5]

*Abteilung Mathematik V, Universität Ulm, 89069 Ulm, Germany, dal@mathematik.uni-ulm.de

2. Auxiliary results

We start with some properties of the rank of continuous matrix-valued functions.

Lemma 1 *Assume that $D : \mathcal{I} \to \mathbb{R}^{k\times l}$ is piecewise continuous on the non-degenerate interval $\mathcal{I} \subset \mathbb{R}$. Then there exists a non-degenerate interval $\mathcal{J} \subset \mathcal{I}$ such that the following holds:*

(i) the ranks of all submatrices of $D(t)$ are constant on $\mathcal{J}$;

(ii) there exists $r < l$ such that

$$r \equiv \operatorname{rank} D(t) < l \text{ on } \mathcal{J},$$

provided that the set $\{t \in \mathcal{I} : \operatorname{rank} D(t) < l\}$ is not nowhere dense in $\mathcal{I}$.

Proof: The assertions are simple consequences of the following property of the piecewise continuous subdeterminants of $D(t)$: A piecewise continuous function equals zero, if its zeros are dense. ∎

For the proof of our main results in the next section we require the following lemma.

Lemma 2 *Let be given $D : \mathcal{I} \to \mathbb{R}^{k\times l}$ such that the ranks of all submatrices of $D(t)$ are constant on $\mathcal{I}$, and let $D \in C^j(\mathcal{I})$ for some $j \in \mathbb{N}$, $t_0 \in \mathcal{I}$, $z_0 \in \operatorname{Ker} D(t_0)$. Then there exists $z : \mathcal{I} \to \mathbb{R}^l$ with*

$$z \in C^j(\mathcal{I}),\ z(t_0) = z_0, \text{ and } z(t) \in \operatorname{Ker} D(t) \text{ for all } t \in \mathcal{I}\,.$$

Proof: The proof is simple, and we will skip it. ∎

The main tool for the proof in the next section is the following generalized product rule. Its proof is very technical and consists of many calculations, so we will skip it. One can find the complete proof in [6] and [7].

We consider the following situation:

Let be given matrix-valued functions $D_\mu : \mathcal{I} \to \mathbb{R}^{k\times l}$ with $D_\mu \in C^{s-\mu}(\mathcal{I})$ for $\mu = 1, \dots, s \in \mathbb{N}$, and some function $u : \mathcal{I} \to \mathbb{R}^l$ such that the recursively defined functions

$$f_0(t) \equiv 0\,,\ f_\mu(t) := D_\mu(t)u(t) + \frac{\mathrm{d}}{\mathrm{d}t} f_{\mu-1}(t) \text{ for } \mu = 1, \dots, s \tag{2}$$

exist and are continuous on $\mathcal{I}$.

Then, if $u \in C^{s-1}(\mathcal{I})$, it follows from the usual product rule that

$$f_\mu(t) = D_{\mu 0}(t)u(t) + \sum_{\nu=1}^{\mu-1} D_{\mu\nu}(t)u^{(\nu)}(t)$$

for $0 < \mu \le s$, $t \in \mathcal{I}$, where the matrices $D_{\mu\nu} = D_{\mu\nu}(t)$ are defined by the recursion

$$\begin{cases} D_{\mu+1,\mu} := D_1 & \text{for } 0 \le \mu \le s-1\,, \\ D_{\mu+1,0} := D_{\mu+1} + \dot{D}_{\mu 0} & \text{for } 1 \le \mu \le s-1\,, \\ D_{\mu+1,\nu} := D_{\mu,\nu-1} + \dot{D}_{\mu\nu} & \text{for } 1 \le \nu < \mu \le s-1\,. \end{cases} \tag{3}$$

Theorem 1 (generalized product rule) *Let be given matrix-valued functions $D_\mu : \mathcal{I} \to \mathbb{R}^{k\times l}$ with $D_\mu \in C^{s-\mu}(\mathcal{I})$ for $\mu = 1,\ldots,s \in \mathbb{N}$, and some function $u : \mathcal{I} \to \mathbb{R}^l$ on a non-degenerate interval $\mathcal{I} \subset \mathbb{R}$ such that the functions $f_\mu : \mathcal{I} \to \mathbb{R}^k$, defined by the recursion (2), exist and are continuous on $\mathcal{I}$ for $\mu = 1,\ldots,s$. Then there exist a non-degenerate interval $\mathcal{J} \subset \mathcal{I}$ and functions $v_\nu : \mathcal{J} \to \mathbb{R}^l$ for $1 \le \nu \le s-1$, such that*

$$f_\mu(t) = D_{\mu 0}(t)u(t) + \sum_{\nu=1}^{\mu-1} D_{\mu\nu}(t)v_\nu(t) \tag{4}$$

holds for all $t \in \mathcal{J}$, $1 \le \mu \le s$, where the matrices $D_{\mu\nu}(t)$ are defined by (3).

The simplest case of this theorem reads as follows

Corollary 1 *Let be given $D : \mathcal{I} \to \mathbb{R}^{k\times l}$ and $u : \mathcal{I} \to \mathbb{R}^l$ such that $D(t)$ and the product $D(t)u(t)$ are piecewise continuously differentiable on $\mathcal{I}$, i.e. $\in C^1_s(\mathcal{I})$. Then there exist a non-degenerate interval $\mathcal{J} \subset \mathcal{I}$ and a function $v : \mathcal{J} \to \mathbb{R}^l$ such that, for all $t \in \mathcal{J}$,*

$$\frac{\mathrm{d}}{\mathrm{d}t}\{D(t)u(t)\} = \dot{D}(t)u(t) + D(t)v(t)\,.$$

3. Characterization of strong observability

We always assume $n \ge 2$.

Theorem 2 *Let be given matrix-valued functions $A : \mathcal{I} \to \mathbb{R}^{n\times n}$ and $C : \mathcal{I} \to \mathbb{R}^{p\times n}$ on some interval $\mathcal{I} \subset \mathbb{R}$ such that*

$$A \in C^{n-2}_s(\mathcal{I})\,,\; C \in C^{n-1}_s(\mathcal{I})\,, \tag{5}$$

and define matrices $C_k = C_k(t)\,,\ t \in \mathcal{I}$, recursively by

$$C_1 := C\,,\; C_{\mu+1} := \dot{C}_\mu + C_\mu A \quad for \quad \mu = 1,\ldots,n-1\,. \tag{6}$$

Then the pair $(A\,,\,C)$ is observable on $\mathcal{I}$ if and only if

$$\operatorname{rank}\, Q(t) = n \tag{7}$$

for $t \in \mathcal{I}$ except on a nowhere dense subset of $\mathcal{I}$, where the **observability matrix** *$Q : \mathcal{I} \to \mathbb{R}^{pn\times n}$ is defined by*

$$Q := [C_1^T,\ldots,C_n^T]^T\,. \tag{8}$$

Note that the smoothness assumptions (5) are needed only to ensure the existence of $C_1,\ldots,C_n$, and Q. Moreover, observe that in the time-invariant case $C_\mu = CA^{\mu-1}$. Hence, Q is the usual Kalman matrix in that case, and our result reduces to Kalman's criterium. From theorem 2 follows immediately a characterization of controllability.

Corollary 2 *Let be given matrix-valued functions* $A : \mathcal{I} \to \mathbb{R}^{n\times n}$ *and* $B : \mathcal{I} \to \mathbb{R}^{n\times m}$ *on some interval* $\mathcal{I} \subset \mathbb{R}$ *such that*

$$A \in C_s^{n-2}(\mathcal{I})\,,\; B \in C_s^{n-1}(\mathcal{I})\,, \tag{9}$$

and define matrices $B_k = B_k(t)\,,\; t \in \mathcal{I}$, *recursively by*

$$B_1 := B\,,\; B_{\mu+1} := AB_\mu - \dot{B}_\mu \;\; \textit{for} \;\; \mu = 1, \ldots, n-1\,. \tag{10}$$

Then the pair $(A\,,\, B)$ *is controllable on* $\mathcal{I}$ *if and only if*

$$\operatorname{rank} \tilde{Q}(t) = n \tag{11}$$

for $t \in \mathcal{I}$ *except on a nowhere dense subset of* $\mathcal{I}$, *where the* **controllability matrix** $\tilde{Q} : \mathcal{I} \to \mathbb{R}^{n\times pn}$ *is defined by*

$$\tilde{Q} := [B_1\,, \ldots, B_n]\,. \tag{12}$$

The most general result is the characterization of strong observability.

Theorem 3 *Let be given matrix-valued functions* $A : \mathcal{I} \to \mathbb{R}^{n\times n}$, $B : \mathcal{I} \to \mathbb{R}^{n\times m}$ *and* $C : \mathcal{I} \to \mathbb{R}^{p\times n}$ *on some interval* $\mathcal{I} \subset \mathbb{R}$ *such that*

$$A \in C_s^{2n-3}(\mathcal{I})\,,\; B \in C_s^{2n-3}(\mathcal{I})\,,\; C \in C_s^{2n-2}(\mathcal{I})\,, \tag{13}$$

and define $p \times m$*-matrices* $B_{\mu\nu} = B_{\mu\nu}(t)\,,\; t \in \mathcal{I}$, *recursively by*

$$\begin{cases} B_{\mu+1,\mu} := CB & \textit{for} \;\; 0 \le \mu \le n-1\,, \\ B_{\mu+1,0} := C_{\mu+1}B + \dot{B}_{\mu 0} & \textit{for} \;\; 1 \le \mu \le n-1\,, \\ B_{\mu+1,\nu} := B_{\mu,\nu-1} + \dot{B}_{\mu\nu} & \textit{for} \;\; 1 \le \nu < \mu \le n-1\,. \end{cases} \tag{14}$$

Then the triple $(A\,,\, B\,,\, C)$ *is strongly observable on* $\mathcal{I}$ *if and only if*

$$\operatorname{rank} S(t) = \operatorname{rank} S^*(t) \tag{15}$$

for $t \in \mathcal{I}$ *except on a nowhere dense subset of* $\mathcal{I}$, *where the matrix-valued functions* $S : \mathcal{I} \to \mathbb{R}^{pn\times(n+(n-1)m)}$, *and* $S^* : \mathcal{I} \to \mathbb{R}^{(pn+n)\times(n+(n-1)m)}$ *are defined by (in blocked form)*

$$\begin{cases} S := [Q\,, T]\,,\; S^* := \begin{bmatrix} I & 0 \\ Q & T \end{bmatrix} \quad \textit{with} \\ T := \begin{bmatrix} 0 & 0 & \ldots & 0 \\ B_{10} & 0 & \ldots & 0 \\ B_{20} & B_{21} & \ldots & 0 \\ \vdots & \vdots & \ddots & \vdots \\ B_{n-1,0} & B_{n-1,1} & \ldots & B_{n-1,n-2} \end{bmatrix}, \end{cases} \tag{16}$$

$n \times n$*-identity matrix* I, *and where the matrices* C_k, Q *are defined as in Theorem 2 by the formulae* (6) *and* (8).

Note that Theorem 3 reduces to Theorem 2 if $B(t) \equiv 0$ except when the smoothness assumptions (13) are stronger than (5). Nevertheless we give here only a proof for Theorem 3, because this is the main result. The proof of Theorem 2 is much simpler (in particular it does not require Theorem 1) and can be found in [6] and [7]. In the time-invariant case Theorem 3 reduces to [[2], Th. 3.5.7] or [[5], Th. 2].

Proof: First suppose that (15) holds for $t \in \mathcal{I} \setminus \mathcal{E}$ with some exceptional set $\mathcal{E} \subset \mathcal{I}$. Let $K(t) \in I\!R^{pn \times pn}$ so that Ker $K(t) = $ Im $T(t)$, and define

$$H(t) = Q^T(t)K^T(t)K(t)Q(t)\,.$$

Then we have that

$$H(t) \text{ is invertible for all } t \in \mathcal{I} \setminus \mathcal{E}\,. \tag{17}$$

Otherwise, there would exist $x \neq 0$ with $KQx = 0$ (we omit the argument), so that $Qx \in \operatorname{Ker} K = \operatorname{Im} T$, i.e., $Qx + Tu = 0$ for some vector u. Hence, $z = \begin{pmatrix} x \\ u \end{pmatrix} \in \operatorname{Ker} S$, but $\notin \operatorname{Ker} S^*$, which contradicts (15). Note that our assumption (13) guarantees that the occuring matrices are well defined. Let $\dot{x} = A(t)x(t) + B(t)u(t)\,,\ y(t) = C(t)x(t)$ on a non-degenerate interval $\mathcal{J} \subset \mathcal{I}$ with $Bu \in C_s(\mathcal{J})\,,\ y \in C^{n-1}(\mathcal{J})$. Then our assumptions and notation imply that, on $\mathcal{J}$,

$$y^{(\mu)} = C_{\mu+1}x + f_\mu \quad \text{for } 0 \leq \mu \leq n-1\,, \tag{18}$$

where the f_μ satisfy (2), i.e., $f_0 \equiv 0\,,\ f_\mu = D_\mu u + \dot{f}_{\mu-1}$ for $1 \leq \mu \leq n$, with $D_\mu := C_\mu B \in C^{n-\mu}(\mathcal{J})$. Now, Theorem 1 implies that there exist $v_\nu : \mathcal{J} \to I\!R^m$, such that

$$f_\mu(t) = B_{\mu 0}(t)u(t) + \sum_{\nu=1}^{\mu-1} B_{\mu\nu}(t)v_\nu(t) \tag{19}$$

for $1 \leq \nu \leq n-1\,,\ t \in \mathcal{J} \setminus \mathcal{E}_1$, where $\mathcal{E}_1$ is nowhere dense. Note that the definitions (14) and (3) of the $B_{\mu\nu}$ and $D_{\mu\nu}$ coincide here. If we define $\hat{y}(t) := (y^T(t)\,, \ldots, y^{(n-1)T}(t))^T$, and if we put $\hat{v} = (v_0^T\,, \ldots,\ v_{n-2}^T)^T$ with $v_0 = u$, then (18) and (19) imply that $\hat{y} = Qx + T\hat{v}$. Hence, $K\hat{y} = KQx$, such that

$$H(t)x(t) = Q^T(t)K^T(t)K(t)\hat{y}(t) \quad \text{for all } t \in \mathcal{J} \setminus \mathcal{E}_1\,, \tag{20}$$

and, $y \equiv 0$ implies by continuity that $x \equiv 0$ for all $t \in \mathcal{J}$, because $\mathcal{E}_1$ is nowhere dense. Hence the triple $(A\,, B\,, C)$ is strongly observable on $\mathcal{I}$, if the exceptional $\mathcal{E}$ from the beginning is nowhere dense.

Now suppose that the exceptional set $\mathcal{E}$, where (15) is false, is not nowhere dense. Then, by Lemma 1 (ii), there exists a non-degenerate interval $\mathcal{J} \subset \mathcal{I}$ with rank $S(t) <$ rank $S^*(t)$ for all $t \in \mathcal{J}$. By assumption (13), $S^* \in C_s^{n-1}(\mathcal{I})$, and according to Lemma 1 we may choose $\mathcal{J} \subset \mathcal{I}$ such that, moreover, $S^* \in C^{n-1}(\mathcal{J})$ and the ranks of all submatrices of S^* are constant on $\mathcal{J}$. Then, for a fixed $\tau_0 \in \mathcal{J}$ there exists

$$z_0 = (x^{0T}\,, u_1^{0T}\,, \ldots,\ u_{n-1}^{0T})^T \in \operatorname{Ker} S(\tau_0) \text{ with } z_0 \notin \operatorname{Ker} S^*(\tau_0),$$

i.e., $x^0 \neq 0$. From Lemma 2 we can conclude that there exists a "path" $z(t)$, which satisfies

$$\begin{cases} z(t) = (x^T(t)\,,\, u_1^T(t)\,, \ldots,\, u_{n-1}^T(t)) \in C^{n-1}(\mathcal{J}) \ \text{ with} \\ z(\tau_0) = z_0\,,\; x(\tau_0) = x^0 \neq 0\,,\; z(t) \in \operatorname{Ker} S(t) \ \text{ for all } \ t \in \mathcal{J}\,. \end{cases} \tag{21}$$

Actually, this is the only place where we use the strong smoothness assumptions (13). Next, we define functions $u_{\nu\mu}\,,\ 1 \leq \nu \leq n-1\,,\ 1 \leq \mu \leq n-\nu$ and $x_\nu\,,\ 1 \leq \nu \leq n$ by the recursions:

$$\begin{cases} u_{1\mu} := u_\mu\,,\ 1 \leq \mu \leq n-1\,,\ u_{\nu+1,\mu} := \dot{u}_{\nu\mu} - u_{\nu,\mu+1}\,, \\ 1 \leq \nu \leq n-2\,,\ 1 \leq \mu \leq n-\nu-1\,; \text{ and} \\ x_1 := x\,,\ x_{\nu+1} = \dot{x}_\nu - Ax_\nu - Bu_{\nu 1}\,,\ 1 \leq \nu \leq n-1\,. \end{cases} \tag{22}$$

It follows from (21) and (13) that $u_{\nu\mu} \in C^{n-\nu}(\mathcal{J})\,,\ x_\nu \in C^{n-\nu}(\mathcal{J})$, $C_\mu \in C^{n-\mu}(\mathcal{J})$, and $B_{\mu\nu} \in C^{n-\mu-1}(\mathcal{J})$. Next, we prove by induction on ν that the formulae

$$(\nu,\mu) \qquad\qquad C_\mu x_\nu + \sum_{j=1}^{\mu-1} B_{\mu-1,j-1} u_{\nu j} \equiv 0 \ \text{ on } \ \mathcal{J}$$

hold for all $1 \leq \nu \leq n\,,\ 1 \leq \mu \leq n+1-\nu$. The formulae $(1,\mu)$ for $1 \leq \mu \leq n$ are equivalent with $S(t)z(t) = 0$ for $t \in \mathcal{J}$, which is true by (21). Hence, let $1 \leq \nu < n\,,\ 1 \leq \mu \leq n-1$. Then, we can conclude from (6), (14), (22), and the induction hypothesis for $(\nu\,,\,\mu)$ and $(\nu\,,\,\mu+1)$ that the following holds:

$$\begin{aligned} C_\mu x_{\nu+1} + \sum_{j=1}^{\mu-1} B_{\mu-1,j-1} u_{\nu+1,j} &= C_\mu(\dot{x}_\nu - Ax_\nu - Bu_{\nu 1}) + \sum_{j=1}^{\mu-1} B_{\mu-1,j-1}(\dot{u}_{\nu j} - u_{\nu,j+1}) \\ &= \frac{\mathrm{d}}{\mathrm{d}t}\Big\{C_\mu x_\nu + \sum_{j=1}^{\mu-1} B_{\mu-1,j-1} u_{\nu j}\Big\} - \dot{C}_\mu x_\nu - C_\mu(Ax_\nu + Bu_{\nu 1}) \\ &\qquad - \sum_{j=1}^{\mu-1}\{\dot{B}_{\mu-1,j-1} u_{\nu j} + B_{\mu-1,j-1} u_{\nu,j+1}\} \\ &= -C_{\mu+1} x_\nu + \{\dot{B}_{\mu-1,0} - B_{\mu 0}\} u_{\nu 1} - \dot{B}_{\mu-1,0} u_{\nu 1} \\ &\qquad - \sum_{j=2}^{\mu-1}\{B_{\mu,j-1} - B_{\mu-1,j-2}\} u_{\nu j} - \sum_{j=1}^{\mu-1} B_{\mu-1,j-1} u_{\nu,j+1} \\ &= -C_{\mu+1} x_\nu - \sum_{j=2}^{\mu-1} B_{\mu,j-1} u_{\nu j} - B_{\mu 0} u_{\nu 1} - CBu_{\nu\mu} \\ &= -C_{\mu+1} x_\nu - \sum_{j=1}^{\mu} B_{\mu,j-1} u_{\nu j} = 0\,, \end{aligned}$$

and this shows that the formula $(\nu+1,\mu)$ holds. For $\mu = 1$ we obtain that

$$Cx_\nu \equiv 0 \ \text{ on } \ \mathcal{J} \ \text{ for } \ \nu = 1\,, \ldots,\, n\,, \tag{23}$$

i.e., $CX \equiv 0$ on $\mathcal{J}$ for the $n \times n$-matrix-valued function $X(t) = (x_1(t),\, \ldots, x_n(t))$. Again, by Lemma 1, we may assume that the ranks of all submatrices of $X(t)$ are constant on $\mathcal{J}$ and that $x(t) = x_1(t) \neq 0$ for $t \in \mathcal{J}$ (since $x^0 = x(\tau_0) \neq 0$).

If $C(t) \equiv 0$ on $\mathcal{J}$, then the triple (A, B, C) is trivially not strongly observable. Hence, we may assume $C(t) \not\equiv 0$ on $\mathcal{J}$. It follows from (23) and the choice of $\mathcal{J}$ that there exist $1 \leq r < n$, and continuous functions $\alpha_\nu(t)$ on $\mathcal{J}$ such that

$$x_{r+1} = \sum_{\nu=1}^{r} \alpha_\nu x_\nu \quad \text{on} \quad \mathcal{J}. \tag{24}$$

Let $\beta_1, \ldots, \beta_r$ be defined as the unique solution of the following initial value problem:

$$\begin{cases} \dot\beta_1 = -\alpha_1\beta_r\,,\ \dot\beta_\nu = -\beta_{\nu-1} - \alpha_\nu\beta_r \ \text{ for } \ 2 \leq \nu \leq r\,, \ \text{with} \\ \beta_1(t_0) = 1\,,\ \beta_\nu(t_0) = 0 \ \text{ for } \ 2 \leq \nu \leq r\,, \end{cases} \tag{25}$$

where $t_0 \in \mathcal{J}$ is fixed. Then, we define

$$\tilde{x} := \sum_{\nu=1}^{r} \beta_\nu x_\nu\,,\ \tilde{u} := \sum_{\nu=1}^{r} \beta_\nu u_{\nu 1}\,. \tag{26}$$

It follows that $\tilde{x} \in C^1(\mathcal{J})$, $B\tilde{u} \in C(\mathcal{J})$, and that $C\tilde{x} \equiv 0$ on $\mathcal{J}$ by (23). Moreover, (24), (25), (26), and (22) imply that

$$\begin{aligned} \dot{\tilde{x}} &= \sum_{\nu=1}^{r} \dot\beta_\nu x_\nu + \sum_{\nu=1}^{r} \beta_\nu \{x_{\nu+1} + Ax_\nu + Bu_{\nu 1}\} \\ &= A\tilde{x} + B\tilde{u} + \sum_{\nu=2}^{r} \{\dot\beta_\nu + \beta_{\nu-1} + \alpha_\nu\beta_r\} x_\nu + \{\dot\beta_1 + \alpha_1\beta_r\} x_1 \\ &= A\tilde{x} + B\tilde{u}\,, \ \text{and} \ \ \tilde{x}(t_0) = x_1(t_0) = x(t_0) \neq 0\,. \end{aligned}$$

Thus $\tilde{x}$, $\tilde{u}$ satisfy on $\mathcal{J}$:

$$\tilde{x} \not\equiv 0\,,\ C\tilde{x} \equiv 0\,,\ \dot{\tilde{x}} = A\tilde{x} + B\tilde{u}\,.$$

Therefore, by definition, the triple (A, B, C) is not strongly observable on $\mathcal{I}$, which completes the proof. ∎

4. Remarks and comments

(i) Note that the criterion (15) for strong observability of (A, B, C) depends only on the first $n-2$ derivatives of A and B and the first $n-1$ derivatives of C, while the proof of Theorem 3 required the stronger smoothness assumptions (13). Moreover, one direction of Theorem 3 can be shown under the weaker conditions

$$A \in C_s^{n-2}(\mathcal{I})\,,\ B \in C_s^{n-2}(\mathcal{I})\,,\ C \in C_s^{n-1}(\mathcal{I})\,.$$

We expect that the other direction is also true under this weaker condition.

(ii) In [[6], Paragraph 2.3] there are constructed examples of linear systems (1) for every $\varepsilon > 0$ with the following properties:

(A, B, C) is strongly observable on $\mathcal{I} = [0,1]$,
$A, B, C \in C^\infty(\mathcal{I})$, and
the Lebesgue measure of the exceptional set in Theorem 3 (i.e., where (15) does not hold) is larger than $1 - \varepsilon$. The technique in [6] is quite similar to the construction of Cantor sets.

(iii) On the other hand it is clear (from the condition (15)) that the exceptional set in Theorem 3 is **finite** for a compact interval $\mathcal{I}$, if the data A, B, C are not only in $C^\infty(\mathcal{I})$ but even piecewise **holomorphic** (in particular piecewise polynomial, which is certainly the case in many applications).

(iv) We mention that in case of proper differentiability order of the output y (that means if there exists $\hat{y}$), one can construct an observer from (20) by multiplication with $H^{-1}(t)$ from the left.

References

[1] J. Klamka. *Controllability of dynamical systems.* Kluwer, Dordrecht, 1991.

[2] W. Kratz. *Quadratic Functionals in Variational Analysis and Control Theory.* Akademie Verlag, Berlin, 1995.

[3] H.E. Meadows L.M. Silverman. Controllability and observability in time-variable linear systems. *SIAM J. on Control*, 5:64–73, 1967.

[4] R.E. Kalman. Mathematical description of linear dynamical systems. *SIAM J. on Control*, 1:152–192, 1963.

[5] W. Kratz. Characterization of strong observability. *Linear Algebra Appl.*, 221:31–40, 1995.

[6] D. Liebscher. *Zeitabhängige lineare Systeme: Lokale Rangkriterien und Rayleigh-Prinzip.* PhD thesis, Universität Ulm, 1996.

[7] D. Liebscher W. Kratz. A local characterization of observability. submitted to Linear Algebra Appl.

Numerical Methods and their Application to Flight Path Optimization and Fluid Dynamics

International Series of Numerical Mathematics
Vol. 124,

Sensitivity Analysis and Real-Time Control of Nonlinear Optimal Control Systems via Nonlinear Programming Methods

Christof Büskens* Helmut Maurer*

Abstract. Parametric nonlinear optimal control problems subject to control and state constraints are studied. Based on recent stability results we propose a robust nonlinear programming method to compute the sensitivity derivatives of optimal solutions. Real-time control approximations of perturbed optimal solutions are obtained by evaluating a first order Taylor expansion of the perturbed solution. The numerical methods are illustrated by two examples. We consider the Rayleigh problem from electrical engineering and the maximum range flight of a hang glider.

1. Introduction

Stability and sensitivity analysis of parametric nonlinear optimal control problems has become an active area of research in recent years. Disturbances or perturbations of system data are modelled by parameters in the dynamics, in the boundary conditions or in the constraints on state and control. In *stability analysis*, differential properties of the optimal solutions with respect to perturbation parameters are studied. *Sensitivity analysis* is concerned with the computation of sensitivity differentials for optimal solutions. This sensitivity information enables the control engineer to estimate the changes in the modelling function and optimal solution due to small deviations of the design parameters from fixed nominal values.

In case that an actual deviation from nominal parameters occurs in the system, it may be too time-demanding to compute the optimal perturbed solution with high precision. In this situation, the control engineer wishes to design a fast and reliable *real-time control approximation* of the perturbed optimal solution. It will be shown that such a real-time control approximation can be developed on the basis of stability and sensitivity analysis.

Lipschitz stability results for optimal solutions to parametric control problems have been given in [6, 7, 14]. Fréchet differentiability of optimal solutions and associated adjoint multipliers are established in [15, 16, 18, 19]. These authors prove solution differentiability by using shooting methods for solving the boundary value problem derived from Pontryagin's minimum principle.

In Section 2, the property of Fréchet differentiability of optimal solutions is used to approximate the perturbed solution by its first order Taylor's expansion with respect

*Westfälische Wilhelms Universität Münster, Institut für Numerische Mathematik, Einsteinstrasse 62, 48149 Münster, Germany, E-mail: buskens@math.uni-muenster.de, maurer@math.uni-muenster.de

to the parameter. This approach leads to a numerical procedure of obtaining real-time approximations of perturbed solutions. Implementations of this approach using *boundary value methods* have been reported in [4, 13, 15, 18, 19, 21, 22]. In section 3, we outline how *nonlinear programming methods* can be employed in an efficient and robust way to compute both the nominal solution and the parameter sensitivities of perturbed solutions. Section 4 presents two numerical examples that illustrate the application of these methods.

2. Parametric nonlinear optimal control problems

We consider parametric nonlinear control problems subject to control and state constraints. Data perturbations are modelled by a parameter $p \in P = \mathbb{R}^k$. The following parametric control problem will be referred to as problem OC(p). Minimize the functional

$$g(x(t_f), t_f, p) + \int_0^{t_f} f_0(x(t), u(t), p)dt \tag{1}$$

subject to

$$\dot{x}(t) = f(x(t), u(t), p) \quad \text{for a.e. } t \in [0, t_f]\,, \tag{2}$$

$$x(0) = \varphi(p)\,, \quad \psi(x(t_f), p) = 0\,, \tag{3}$$

$$C(x(t), u(t), p) \leq 0 \quad \text{for } t \in [0, t_f]\,. \tag{4}$$

We assume that the functions $g : \mathbb{R}^{n+1} \times P \to \mathbb{R}$, $f_0 : \mathbb{R}^{n+m} \times P \to \mathbb{R}$, $f : \mathbb{R}^{n+m} \times P \to \mathbb{R}^n$, $\varphi : P \to \mathbb{R}^n$, $\psi : \mathbb{R}^n \times P \to \mathbb{R}^r$, $0 \leq r \leq n$, and $C : \mathbb{R}^{n+m} \times P \to \mathbb{R}^k$ are sufficiently smooth on appropriate open sets. The final time t_f is either fixed or free. The admissible class is that of piecewise continuous control functions. In stability analysis, conditions are imposed such that the optimal control is *continuous* and piecewise of class C^1.

Let us fix a *reference* or *nominal* parameter p_0 and consider problem $OC(p_0)$ as the *unperturbed* or *nominal* problem. We assume that there exists a local solution (x_0, u_0) of the reference problem OC(p_0) such that $u_0 \in C(0, t_f; \mathbb{R}^m)$. The reference solution $x_0(t)$, $u_0(t)$ and the associated adjoint function $\lambda_0(t)$, $0 \leq t \leq t_f$, satisfy a *boundary value problem* that is derived from Pontryagin's minimum principle.

The stability analysis in Malanowski, Maurer and Pesch [15, 16, 18, 19] provides conditions for the following property of *solution differentiability* : The unperturbed solution $x_0(t), \lambda_0(t), u_0(t)$ can be embedded into a family of optimal solutions $x(t,p), \lambda(t,p)$, $u(t,p)$ to the perturbed problem OC(p) which is C^1 for $x(t,p), \lambda(t,p)$ and piecewise C^1 for $u(t,p)$ in a neighborhood of the reference parameter p_0.

This type of strong C^1-stability where optimal solutions are of class C^1 with respect to *both* variables (t,p) allows us to develop an approximation of the perturbed solution $x(t,p), \lambda(t,p), u(t,p)$ by considering the following first order Taylor expansion; here the variable y represents any one of the variables x, u, λ:

$$y(t,p) \approx y_0(t) + \frac{\partial y}{\partial p}(t, p_0)\,(p - p_0)\,. \tag{5}$$

It is shown in [15, 16, 18, 19] that the sensitivity differentials

$$\frac{\partial y}{\partial p}(t, p_0)$$

satisfy a *linear* boundary value problem evaluated along the unperturbed solution. Both the nominal functions $y_0(t)$ and $\frac{\partial y}{\partial p}(t, p_0)$ are computed *off-line*. When an actual deviation p from the nominal parameter p_0 is detected, the expression (5) provides a fast approximation for the perturbed solution since it requires only matrix–multiplications. Further numerical investigations of this approach may be found in [4, 13, 21, 22].

A well-known drawback of solving the complete boundary value problem can be seen in the fact that one has to include adjoint equations. Usually it is rather difficult to find appropriate estimates for adjoint variables and to solve adjoint equations accurately. In the next section, we propose a direct optimization method for computing the real-time control approximation on the right-hand side of (5) without employing adjoint variables. The method proceeds by first discretizing the control problem over a grid and then applying nonlinear programming techniques.

3. Real-time control via nonlinear programming methods

Direct optimization methods provide a powerful tool for solving optimal control problems with control and state constraints; cf. e.g. [3, 5, 8, 23, 24]. We restrict the discussion to Euler's method applied to the control problem $OC(p)$ in (1)–(4). Let $N > 0$ be a positive integer and choose mesh points τ_i, $i = 0, 1, \ldots, N$, with $0 = \tau_0 < \tau_1 < \tau_2 < \ldots < \tau_{N-1} < \tau_N = t_f$. Denoting approximations of the values $x(\tau_i)$ and $u(\tau_i)$ by x^i and u^i, the control problem (1)–(4) is replaced by the following nonlinear programming problem with equality and inequality constraints:

$$\begin{array}{ll} \text{Minimize} & g(x^N, \tau_N, p) + \sum\limits_{i=0}^{N-1} (\tau_{i+1} - \tau_i)\, f_0(x^i, u^i, p)\,, \\ \text{subject to} & x^{i+1} = x^i + (\tau_{i+1} - \tau_i) \cdot f(x^i, u^i, p)\,, \quad i = 0, \ldots, N-1\,, \\ & x^0 = \varphi(p)\,,\ \psi(x^N, p) = 0\,, \\ & C(x^i, u^i, p) \le 0\,, \quad i = 0, \ldots, N\,. \end{array} \tag{6}$$

The dimension of this NLP-problem can be further reduced by treating the control variables as the only optimization variables. Introduce the optimization variables

$$z := (u^0, u^1, \ldots, u^{N-1}, u^N) \in \mathbb{R}^{m\cdot(N+1)}$$

and compute $x^i = x^i(u^0, \ldots, u^{i-1}, p) =: x^i(z, p)$ as the solution of the recursive state equation in (6) with initial conditions $x^0 = \varphi(p)$ given in (3). A *free* final t_f is handled as an additional optimization variable. Then we consider the following parametric

NLP-problem $NLP(p)$ with optimization variables z:

$$\begin{array}{ll} \text{Minimize} & F(z,p) := g(x^N(z,p),\tau_N,p) + \sum_{i=0}^{N-1}(\tau_{i+1}-\tau_i)\, f_0(x^i(z,p),u^i,p) \\ \multicolumn{2}{l}{\text{subject to equality and inequality constraints}} \\ & \psi_i(x^N(z,p),p) = 0\,, \quad i=0,\dots,r\,, \\ & C(x^i(z,p),u^i,p) \le 0\,, \quad i=0,\dots,N\,. \end{array} \tag{7}$$

Let $G(z,p) = (\,G_1(z,p),\dots,G_{r+k\cdot(N+1)}(z,p)\,)$ denote the collection of functions defining equality and inequality constraints in (7). Then the parametric problem $NLP(p)$ has the form:

$$\begin{array}{lll} \text{Minimize} & F(z,p) & \\ \text{subject to} & G_i(z,p) = 0\,, & i=1,\dots,r\,, \\ & G_i(z,p) \le 0\,, & i=r+1,\dots,r+k\cdot(N+1)\,. \end{array} \tag{8}$$

Several reliable optimization codes have been developed to solve NLP-problems of this type; cf., e.g., the sequential quadratic programming code E04UCF in the NAG-library. Instead of Euler's method incorporated into the definition of the NLP-problem (7), one can use any higher order single step method, e.g., a higher order Runge-Kutta scheme. Also, one can replace the piecewise constant control by higher order spline approximations of the control. Several of these options have been implemented in the code NUDOCCCS of Büskens [5].

An excellent reference for sensitivity analysis of parametric NLP-problems is Fiacco [9]. The Lagrangian function for problem NLP(p) is :

$$L(z,\mu,p) = F(z,p) + \mu^* G(z,p)\,, \quad \mu \in \mathbb{R}^{r+k\cdot(N+1)}\,.$$

Let z_0 be the unperturbed solution for the nominal parameter p_0 and let μ_0 be the associated Lagrange multiplier. Consider the set of active indices defined by $I_a := \{i \mid G_i(z_0,p_0) = 0\}$ and denote the active constraints by $G^a := (G_i)_{i\in I_a}$. Suppose that second order sufficient conditions hold for the nominal problem $NLP(p_0)$, i.e.,

$$L_{zz}(z_0,\mu_0,p_0) \quad \text{is positive definite on} \quad Ker(\,G_z^a\,)\,, \tag{9}$$

and, moreover, assume that the Lagrange multipliers μ_i for active inequality constraints in (8) are *positive*. Then the unperturbed solution z_0, μ_0 can be embedded into a C^1-family of perturbed solutions $z(p), \mu(p)$ to NLP(p) with $z(p_0) = z_0$, $\mu(p_0) = \mu_0$. The sensitivity differentials are given by the formula

$$\begin{pmatrix} \dfrac{dz}{dp}(p_0) \\[2mm] \dfrac{d\mu^a}{dp}(p_0) \end{pmatrix} = -\begin{pmatrix} L_{zz} & (G_z^a)^* \\ G_z^a & 0 \end{pmatrix}^{-1} \begin{pmatrix} L_{zp} \\ G_p^a \end{pmatrix}. \tag{10}$$

Here, the multiplier corresponding to active constraints is denoted by μ^a. The right-hand side is evaluated at the nominal solution. Note that the matrix on the right-hand side is non-singular since second order sufficient conditions are assumed to hold.

A numerical check of condition (9) consists in evaluating the projected Hessian on $Ker(G_z^a)$ and in testing if its eigenvalues are positive. We mention that this approach has been implemented in [1, 2, 10, 11] for nonlinear programming problems. Beltracchi and Gabriele [1, 2] have tested whether it is possible to substitute the Hessian in (10) by an SQP-update obtained in the process of computing the nominal solution. In our approach, the Hessian is evaluated after computing the nominal solution z_0. This explicit computation yields much more accurate results.

Now we apply this formula to the discretized control problem (6). The resulting expression $\frac{dz}{dp}(p_0)$ provides a good approximation for the sensitivity of the perturbed optimal control at the mesh points, i.e., for the quantities $\frac{du}{dp}(\tau_i, p_0)$, $i = 0, \ldots, N$. From this we obtain the state sensitivities $\frac{dx}{dp}(\tau_i, p_0)$ via the linearization of the state equation (6). Moreover, it can be shown that the Lagrange multipliers μ^a lead to good approximations of the adjoint variables $\lambda(\tau_i)$ for the control problem (1)–(4). Hence we also get the sensitivities $\frac{d\lambda}{dp}(\tau_i, p_0)$ from relation (10). As we have seen in (5), the sensitivity expressions lead to an easily implementable real-time control of perturbed solutions. In the next section, we are going to illustrate the quality of this approximation by two numerical examples.

4. Numerical examples

4.1 Rayleigh problem: Optimal control of an electric circuit

Sensitivity analysis of the following Rayleigh problem has been investigated in [17] using boundary value methods. There it is verified that all assumptions of solution differentiability in [15] hold. Hence, we can compare sensitivity results obtained via nonlinear programming with the high precision results derived from boundary value methods.

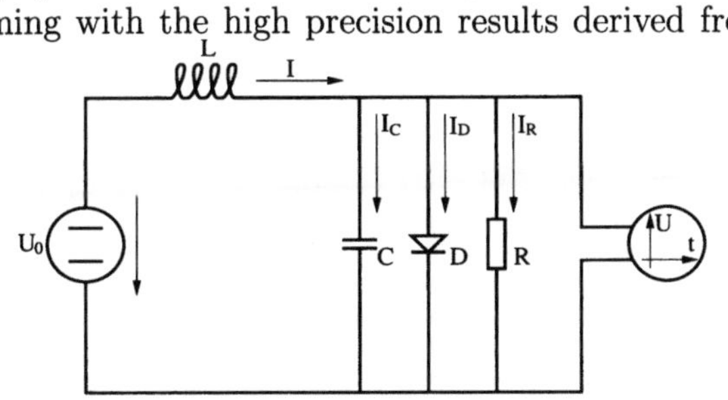

Figure 1: Tunneldiode oscillator

Consider the electric circuit (tunneldiode oscillator) shown in Figure 1 where L denotes inductivity, C capacity, R resistance, I electric current, U_0 is the voltage at the generator, and where D is the diode. The state variable $x(t)$ represents the electric current I at time t. The control function is the voltage $U_0(t)$ at the generator. After a suitable transformation of $U_0(t)$ we arrive at the following specific Rayleigh equation with a scalar control $u(t)$ and a parameter $p \in \mathbb{R}$; see [12, 25]:

$$\ddot{x}(t) = -x(t) + \dot{x}(t)\,(\,1.4 - p\,\dot{x}(t)^2\,) + 4\,u(t)\,. \tag{11}$$

The nominal parameter is chosen as $p_0 = 0.14$. It is easy to verify that the Rayleigh equation (11) with zero control $u(t) \equiv 0$ and nominal parameter has a *limit cycle* in the $(x, \dot{x})$–plane. Then the aim is to control oscillations of the electric current in the tunneldiode oscillator.

Introducing the state variables $x_1 = x$ and $x_2 = \dot{x}$, we obtain the following control problem: Minimize the functional

$$J(u) = \int_0^{t_f} (\, u(t)^2 + x_1(t)^2 \,)\, dt$$

subject to

$$\begin{aligned}
&\dot{x}_1(t) = x_2(t)\,,\\
&\dot{x}_2(t) = -x_1(t) + x_2(t)\,(\,1.4 - p\,x_2(t)^2\,) + 4\,u(t)\,,\\
&x_1(0) = x_2(0) = -5\,, \quad x_1(t_f) = x_2(t_f) = 0\,,\\
&|\,u(t)\,| \le 1 \quad \text{for } t \in [\,0, t_f\,] \text{ with } t_f = 4.5\,.
\end{aligned}$$

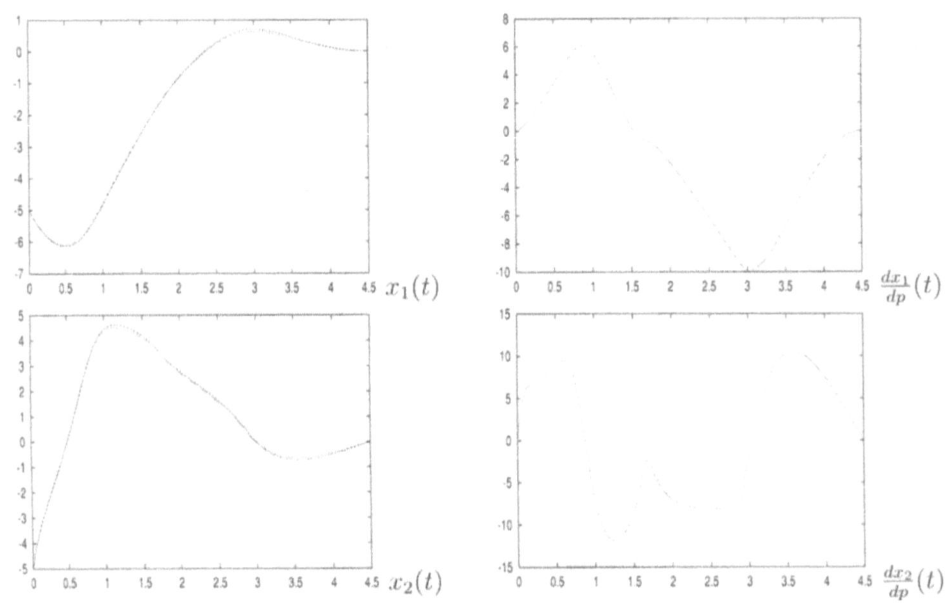

Figure 2: Solutions x_1, x_2, and sensitivities $\frac{dx_1}{dp}$, $\frac{dx_2}{dp}$, .

We choose the grid size $N = 50$ and the initial estimate $u^i = 0, i = 0, 1, \ldots, N$, for the control variables. The CPU-time on a SUN SPARC 20 workstation was 1.7 sec. The state variables resp. the control variable are displayed in the left column of Figure 2 resp. Figure 3 while the sensitivity differentials are shown in the resp. right columns. Figures 2 and 3 exhibit both the unperturbed (nominal) solutions for

$p_0 = 0.14$ which are marked by dashed lines (- - -) and the perturbed solutions for $p = 0.16$ marked by dotted lines $(\cdots)$. Due to space limitations we refrain from showing the adjoint variables and its sensitivities. Comparing the exact perturbed state with the approximation defined on the right side of (5) we get the relative error $6 \cdot 10^{-3}$.

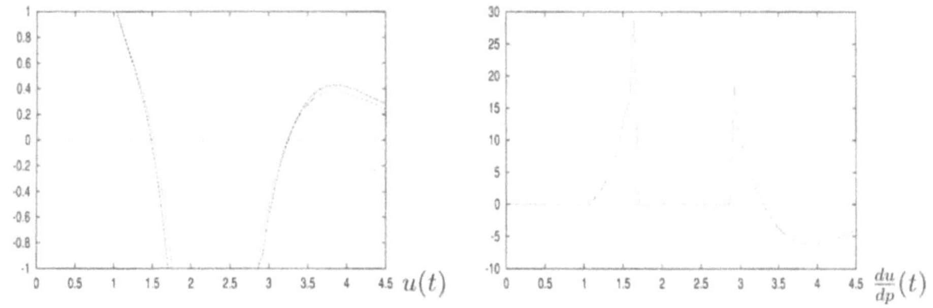

Figure 3: Optimal control u and sensitivity $\frac{du}{dp}$.

4.2 Maximum range flight of a hang glider

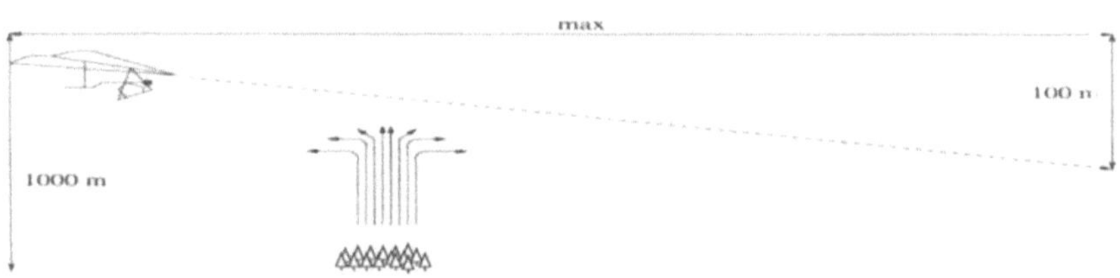

Figure 4: Maximum range flight of a hang glider

The two dimensional maximum range flight of a hang glider in a thermal upwind (Figure 4) has become a famous benchmark problem for testing direct optimization methods in optimal control. Under the simplified assumption that the glider can be treated as a point mass, the mathematical model can be formulated as the following optimal control problem (c.f. [20], [23]):

$$
\begin{array}{llll}
\text{Maximize} & F(u) = x_1(t_f) & & \\
\text{subject to} & \dot{x}_1 = x_3, & & \text{(horizontal distance)} \\
& \dot{x}_2 = x_4, & & \text{(altitude)} \\
& \dot{x}_3 = \frac{1}{m}\left(-r_l \frac{x_4 - u_a}{v_r} - d \frac{x_3}{v_r}\right), & & \text{(horizontal velocity)} \\
& \dot{x}_4 = \frac{1}{m}\left(r_l + \frac{x_3}{v_r} - d \frac{x_4 - u_a}{v_r} - mg\right), & & \text{(vertical velocity)}
\end{array}
$$

where

$$\begin{aligned} v_r &= \sqrt{x_3^2 + (x_4 - u_a)^2}, & c_D &= c_{D_0} + r_k u^2, \\ d &= \frac{1}{2}\, c_D\, \varrho\, s\, v_r^2, & r_l &= \frac{1}{2}\, u\, \varrho\, s\, v_r^2, \\ u_a &= u_{a_m}\, e^{-\left(\frac{x_1 - x_{a_0}}{r}\right)^2} \left(1 - \left(\frac{x_1 - x_{a_0}}{r}\right)^2\right) \end{aligned}$$

x_1 denotes the horizontal distance, x_2 the altitude, x_3 the horizontal velocity and x_4 the vertical velocity. The control function u describes the lift coefficient, r_l the lift force vertical to the velocity v_r relative to the air, d the drag force opposite to v_r, c_D the drag coefficient and u_a the upward wind velocity in the thermal depending on the horizontal distance. The values of the constants are $c_{D_0} = 0.034$, $r_k = 0.069662$, $s = 14$ (wing area), $\varrho = 1.13$ (air density corresponding to the standard pressure and temperature at a height of about 1000 m above sea level), $g = 9.80665$ (gravitational acceleration), $r = 100$ (factor describing the horizontal extension of the thermal upwind), $x_{a_0} = 350$, $u_{a_m} = 2.5$ (maximal upwind velocity). The boundary conditions are given by

$$\begin{aligned} x_1(0) &= 0, & & \\ x_2(0) &= 1000, & x_2(t_f) &= 900, \\ x_3(0) &= 13.2275675, & x_3(t_f) &= x_3(0), \\ x_4(0) &= -1.28750052, & x_4(t_f) &= x_4(0). \end{aligned}$$

The terminal time t_f is free and the control function u is restricted by

$$u(t) \le 1.4 \qquad \text{for all } t \in [0, t_f]\,.$$

We consider now the mass m of the pilot and the glider in the optimal control problem as the *perturbation parameter*. Its nominal value is choosen as $m_0 = 100$. To have a comparison we also calculate the solution for $m = m_p = 98$. In the following figures, the unperturbed (nominal) solutions are marked by dashed lines (- - -) while the perturbed solution are marked by dotted lines ($\cdots$).

The nominal solution is calculated with the direct method NUDOCCCS implemented in Büskens [5]. We use a Runge-Kutta integration method of order 4 and a continuous, piecewise linear approximation of the control. The initial guess for the control was $u(t) \equiv 0$ and for the terminal time $t_f = 100$. After applying an automatic grid generator to minimize the local and global discretization error we preserved a mesh of 131 non equidistant discrete points for the direct sensitivity analysis. The calculated value for the objective function $F(u) \approx 1247.9284$ has 5 correct significant decimal digits, while the terminal time $t_f \approx 98.765$ has 4 correct significant decimal digits.

Figure 5 shows the optimal solutions and the corresponding sensitivity differentials on the normalized time interval $[0, 1]$. The optimal control and its sensitivity are displayed in Figure 6. The control exhibits one boundary arc and, hence, the corresponding sensitivity is equal to zero in this region. The boundary arc is located around the area with maximal upwind. Only small differences between nominal and perturbed solution can be found in the beginning and at the end of the time interval.

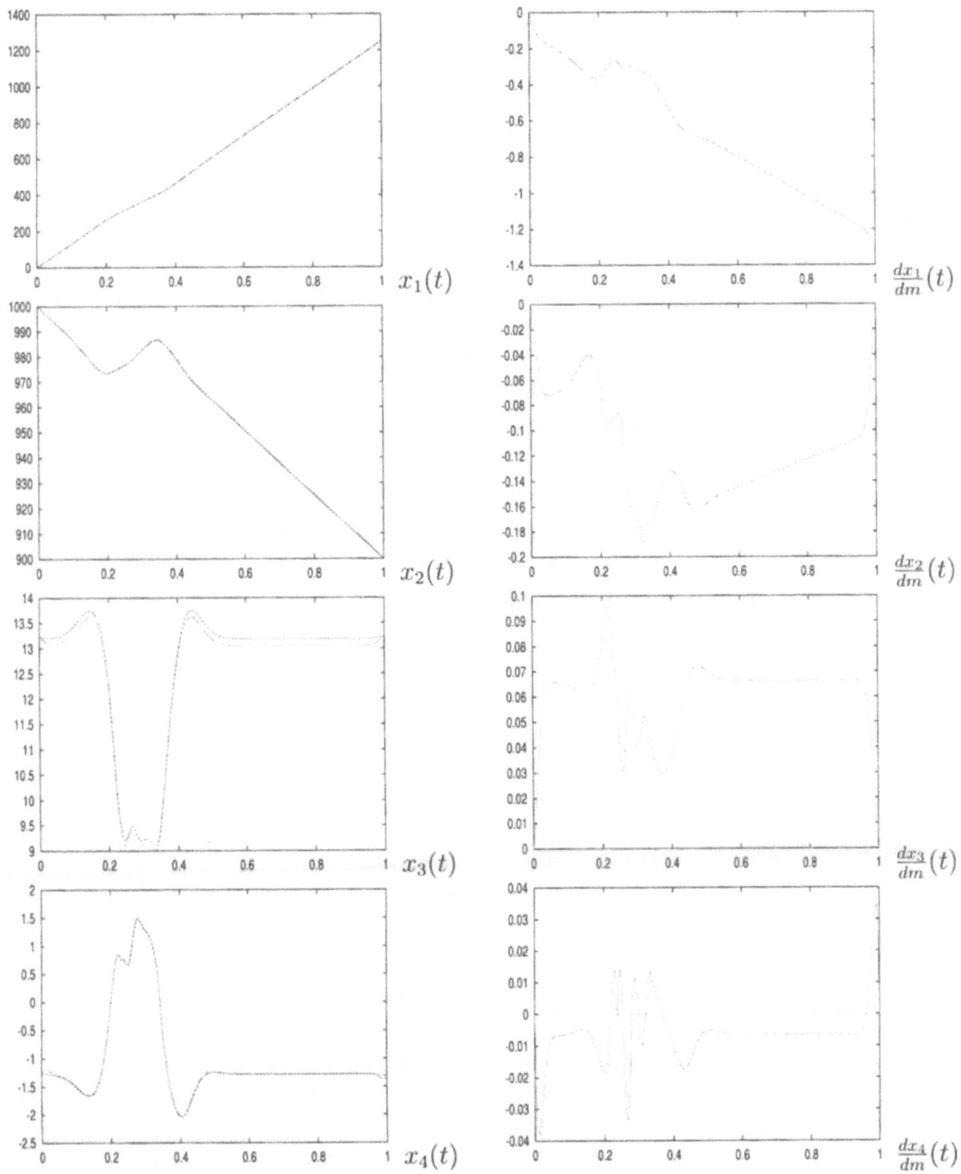

Figure 5: Optimal states x_i and sensitivities $\frac{dx_i}{dm}$, $i = 1, \ldots, 4$.

N	CPU	$x_1(t_f)$	ERR
70	3.9 sec.	1247.8535	$6.577 \cdot 10^{-5}$
84	6.1 sec.	1247.9072	$2.276 \cdot 10^{-5}$
131	9.5 sec.	1247.9284	$5.787 \cdot 10^{-6}$
257	39.2 sec.	1247.9351	$1.488 \cdot 10^{-7}$
398	97.4 sec.	1247.9356	-

Table 1: Automatic Grid Generation

Again, because of space limitations we refrain from presenting the adjoint variables and its sensitivity differentials.

The most critical phase of the flight through the thermal upwind leads to the largest oscillations in the optimal solutions. Figures 5 and 6 illustrate the fact that the proposed algorithm is able to capture the sensitivity structure in detail. Since the terminal time t_f is treated as additional optimization variable in the NLP-problem, we can use relations (10) again to calculate its sensitivity. For the perturbation $m_p = 98$ the optimal final time is $t_f \approx 99.938$ while the first order Taylor expansion yields $t_f \approx 99.923$. The value of the objective function in the perturbed case is $F(u) \approx 1250.660$. Here the second order Taylor expansion gives a value of $F(u) \approx 1250.660$.

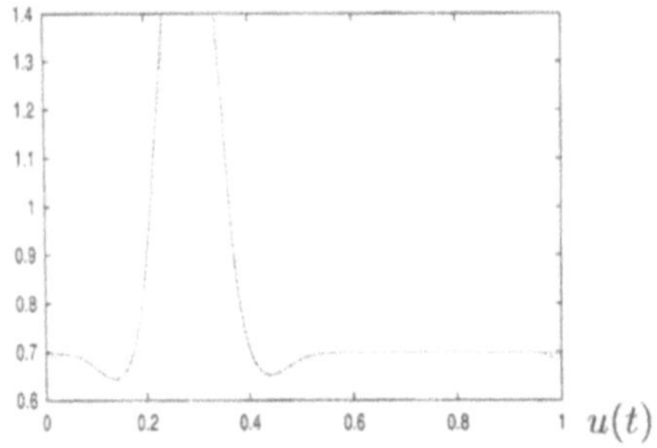

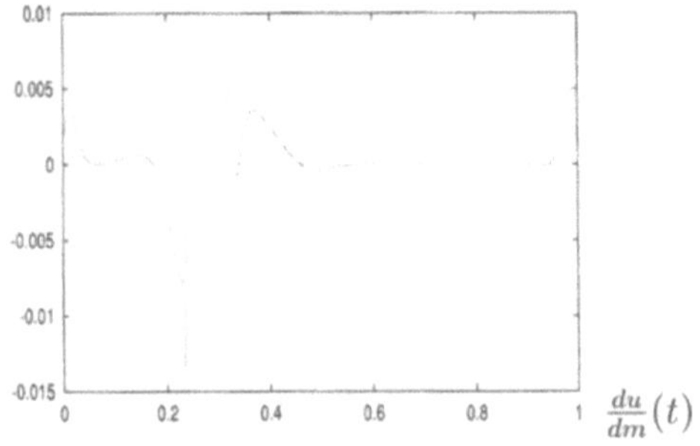

Figure 6: Optimal control u and sensitivity $\frac{du}{dm}$.

Table 1 reflects the behaviour of the automatic grid generation. Starting with a grid of 70 equidistant time intervals the algorithm yields a solution for 398 non equidistant time intervals which is exact on 8 digits. CPU specifies the computing time for the different step sizes on a SUN SPARC 20 workstation, ERR is the relative error of the solution.

References

[1] Beltracchi, T. J.; Gabriele, G. A.: *An investigation of using an RQP based method to calculate parameter sensitivity derivatives*, in "Recent Advances in Multidisciplinary Analysis and Optimization", NASA Conference Publication 3031, Part 2, Proceedings of a symposium held in Hampton, USA, September 28–30, 1988.

[2] Beltracchi, T. J.; Nguyen, H. N.: *Experience with post optimality parameter sensitivity analysis in FONSIZE*, American Institute of Aeronautics and Astronautics, Report AIAA-92-4749-CP, pp. 496–506, 1992.

[3] Betts, J. T.; Huffmann, W. P.: *Path constrained trajectory optimization using sparse sequential quadratic programming*, Applied Mathematics and Statistics Group, Boeing Computer Services, Seattle, USA, 1991.

[4] Bock, H. G.; Krämer-Eis, P.: *An efficient algorithm for approximate computation of feedback control laws in nonlinear processes*, ZAMM, **61** (1981), T 330–T 332.

[5] Büskens, C.: *Direkte Optimierungsmethoden zur numerischen Berechnung optimaler Steuerungen*, Diploma thesis, Institut für Numerische Mathematik, Universität Münster, Münster, Germany, 1993.

[6] Dontchev, A. L.; Hager, W. W.: *Lipschitz stability in nonlinear control and optimization*, SIAM J. Control and Optimization, **31** (1993), pp. 569–603.

[7] Dontchev, A. L.; Hager, W. W.; Poore, A. B.; and Yang, B.: *Optimality, stability and convergence in nonlinear control*, Applied Mathematics and Optimization, **31** (1995), pp. 297–326.

[8] Evtushenko, Y. G.: *Numerical Optimization Techniques*, Translation Series in Mathematics and Engineering, Optimisation Software Inc., Publications Division, New York, 1985.

[9] Fiacco, A. V.: *Introduction to Sensitivity and Stability Analysis in Nonlinear Programming*, Mathematics in Science and Engineering, Vol. 165, Academic Press, New York, 1983.

[10] Hallmann, W.: *Sensitivity analysis for trajectory optimization*, 28th Aerospace Sciences Meeting, AIAA 90-0471, Reno, USA, January 8–11, 1990.

[11] Hallmann, W.: *Optimal scaling techniques for the nonlinear programming problem*, 5th AIAA/NASA/USAF/ISSMO Symposium on Multidisciplinary Analysis and Optimization, Panama City Beach, USA, September 7–9, 1994.

[12] Jacobson, D. H.; Mayne, D. Q.: *Differential Dynamic Programming*, American Elsevier Publishing Company Inc., New York 1970.

[13] Krämer-Eis, P.: *Ein Mehrzielverfahren zur numerischen Berechnung optimaler Feedback-Steuerungen bei beschränkten nichtlinearen Steuerungsproblemen*, Bonner Mathematische Schriften, **164**, 1985.

[14] Malanowski, K.: *Stability and sensitivity of solutions to nonlinear optimal control problems*, Applied Mathematics and Optimization, **32** (1995), pp. 111–141.

[15] Malanowski, K.; Maurer, H.: *Sensitivity analysis for parametric control problems with control-state constraints*, Computational Optimization and Applications, **5** (1996), pp. 253–283.

[16] Malanowski, K.; Maurer, H.: *Sensitivity analysis for state constrained control problems,* Angewandte Mathematik und Informatik, Universität Münster, Preprint No. 21/96-N, 1996.

[17] Maurer H.; Augustin, D.: *Second order sufficient conditions and sensitivity analysis for the controlled Rayleigh problem*, to appear in Proceedings of the 4th Conference on Parametric Optimization and Related Topics, Enschede, June 6–9, 1995.

[18] Maurer, H.; Pesch, H. J.: *Solution differentiability for parametric nonlinear control problems*, SIAM Journal on Control and Optimization, **32** (1994), pp. 1542–1554.

[19] Maurer, H.; Pesch, H. J.: *Solution differentiability for parametric nonlinear control problems with control-state constraints*, Control and Cybernetics, **23** (1994), pp. 201–227.

[20] Nerz, E.: *Optimale Steuerung eines Hängegleiters*, Munich University of Technology, Department of Mathematics, Diploma Thesis, 1990.

[21] Pesch, H. J.: *Real-time computation of feedback controls for constrained optimal control problems, Part 1: Neighbouring extremals*, Optimal Control Applications & Methods, **10** (1989), pp. 129–145.

[22] Pesch, H. J.: *Real-time computation of feedback controls for constrained optimal control problems, Part 2: A correction method based on multiple shooting*, Optimal Control Applications & Methods, **10** (1989), pp. 147–171.

[23] von Stryk, O.: *Numerische Lösung optimaler Steuerungsprobleme: Diskretisierung, Parameteroptimierung und Berechnung der adjungierten Variablen.* Fortschritt-Berichte VDI, Reihe 8, Nr. 441, VDI Verlag, Germany (1995).

[24] Teo, K. L.; Goh C. J.; Wong, K. H.: *A Unified Computational Approach to Optimal Control Problems*, Longman Scientific and Technical, New York, 1991.

[25] Tun T.; Dillon, T. S.: *Extensions of the differential dynamic programming method to include systems with state dependent control constraints and state variable inequality constraints*, Journal of Applied Science and Engineering A, **3**, (1978), pp. 171–192.

International Series of Numerical Mathematics
Vol. 124, © 1998 Birkhäuser Verlag, Basel

Accelerating Multiple Shooting for State-Constrained Trajectory Optimization Problems

Kurt Chudej *

Abstract. Multiple shooting is an indirect solution method for complicated optimal control problems. It yields precise numerical solutions. The rate of convergence is usually quadratic due to the underlying Newton type method. However, it was recently noticed that if the general necessary conditions of optimal control are applied to state constrained problem formulations from aerospace engineering the method decelerates often to a slow linear rate of convergence. We identify the culprit, namely the appearance of a quadratic control component in the optimal control problem and certain boundary and/or interior point conditions. Finally we provide a reformulation of these boundary and/or interior point conditions in order to avoid this situation.

The reformulation of the boundary-value problem is especially helpful for more efficient solutions of state constrained trajectory optimization problems with a drag polar of parabolic shape from aerospace engineering.

1. Introduction

A reliable numerical solution method for complicated optimal control problems is multiple shooting. This was demonstrated for a large number of hard bench-mark problems, cf. the survey in [1]. Due to the Newton-type method used for solving the nonlinear shooting equations one expects quadratic convergence in a close neighborhood of the solution. However the observed rate of convergence is occasionally only linear when the general necessary optimality conditions are applied to optimal control problems with a quadratic control component without modification. We identify the responsible boundary and/or interior point conditions and give a remedy. This motivates the general advice, to try to factorize and simplify boundary conditions involving the Hamiltonian for optimal control problems of this kind. Finally we present two examples, a well-known state-constrained spline problem and a typical trajectory optimization problem with a drag polar of parabolic shape and a dynamic pressure limit.

2. Optimal control problem with quadratic control

For brevity we consider at first a state-constrained optimal control problem of determining a scalar[1], piecewise continuous control $u(t)$, $t \in [0,T]$, which minimizes the

*Lehrstuhl für Höhere Mathematik und Numerische Mathematik, Sonderforschungsbereich 255 Transatmosphärische Flugsysteme, Technische Universität München, D-80290 München, Germany, email: chudej@mathematik.tu-muenchen.de

[1]This assumption will be discarded later, as realistic modelled optimal control problems have vector-valued control functions.

functional

$$\varphi(x(T))$$

subject to

$$\dot{x} = f(x,u) = a(x)\,u^2 + b(x)\,u + c(x), \quad 0 \le t \le T,$$
$$x(0) = x_0, \quad \psi(x(T)) = 0$$

and the scalar state inequality constraint

$$S(x) \le 0, \quad 0 \le t \le T.$$

The state x is a n-dimensional vector, final time T is free or given. The functions $\varphi: I\!R^n \to I\!R$, $f: I\!R^n \times I\!R \to I\!R^n$, $a, b, c: I\!R^n \to I\!R^n$, $\psi: I\!R^n \to I\!R^k$ $(k < n)$, $S: I\!R^n \to I\!R$ are assumed to be C^∞ with respect to all variables, $a \ne 0$.

We define the functions S^i recursively by $S^i := (S^{i-1})_x f$ $(i = 1, \dots, p)$, $S^0 := S$. We assume that $(S^i)_u \equiv 0$ for $i = 0, \dots, p-1$ and $(S^p)_u \not\equiv 0$, i.e. the order of the scalar function $S(x)$ with respect to the ODE $\dot{x} = f(x,u)$ is p.

We assume that the extremal arc is composed by interior arcs $(S(x) < 0)$ and boundary arcs $(S(x) = 0)$ as follows

$$\begin{aligned} S(x) &< 0, && 0 \le t < t_1 \\ S(x) &= 0, && t_1 \le t \le t_2 \\ S(x) &< 0, && t_2 < t \le T. \end{aligned}$$

When certain regularity conditions are fulfilled one can apply the necessary conditions of optimal control theory (see e.g. [2–4] and the survey articles [1,5]). To simplify the presentation we assume that no jumps of the costate function $\lambda(t)$ appear for $t_1 < t < t_2$ (cf. [5]). The Hamiltonian $\mathcal{H}$ is defined as

$$\mathcal{H}(x, u, \lambda, \mu) := \lambda^T f(x,u) + \mu S^p(x,u).$$

The control $u(t) := u(x, \lambda)$ is determined by

$$\begin{cases} \mathcal{H}_u = 0 & \text{if} \quad 0 \le t < t_1 \text{ or } t_2 < t \le T \\ S^p(x,u) = 0 & \text{if} \quad t_1 \le t \le t_2 \end{cases}$$

and the Lagrange multiplier $\mu(t) := \mu(x, \lambda, u)$ satisfies

$$\begin{cases} \mu = 0 & \text{if} \quad 0 \le t < t_1 \text{ or } t_2 < t \le T \\ \mu = -\lambda^T f_u(x,u)/S^p_u(x,u) & \text{if} \quad t_1 \le t \le t_2. \end{cases}$$

This yields $H(x, \lambda) := \mathcal{H}(x, u(x,\lambda), \lambda, \mu(x, \lambda, u(x,\lambda)))$.

One derives a multipoint boundary-value problem with jump conditions for (x, λ): Find a state function $x(t)$, a costate function $\lambda(t)$, constants $\nu_0, \dots, \nu_{p-1}$, σ, junction times t_1 and t_2 and final time T, which satisfy the piecewise defined ODE system

$$\begin{cases} \dot{x} &= f(x, u(x,\lambda)) \\ -\dot{\lambda} &= \lambda^T f_x(x, u(x,\lambda)) + \mu(x, \lambda, u(x,\lambda)) S^p_x(x, u(x,\lambda)) \end{cases} \tag{1}$$

and the continuity, interior point and boundary conditions

$$x(0) = x_0, \quad \psi(x(T)) = 0, \quad \lambda(T) = \varphi_x(x(T)) + \sigma^T \psi_x(x(T)) \tag{2}$$
$$x(t_1^+) = x(t_1^-), \quad \lambda(t_1^+) = \lambda(t_1^-) - \sum_{i=0}^{p-1} \nu_i S_x^i(x(t_1)) \tag{3}$$
$$x(t_2^+) = x(t_2^-), \quad \lambda(t_2^+) = \lambda(t_2^-) \tag{4}$$
$$H(x(t_1^+), \lambda(t_1^+)) = H(x(t_1^-), \lambda(t_1^-)) \tag{5}$$
$$H(x(t_2^+), \lambda(t_2^+)) = H(x(t_2^-), \lambda(t_2^-)) \tag{6}$$
$$S(x(t_1)) = \ldots = S^{p-1}(x(t_1)) = 0 \tag{7}$$
$$T \text{ is given or } H(x(T), \lambda(T)) = 0. \tag{8}$$

We want to discuss the conditioning of the numerical solution of this multipoint boundary-value problem by shooting techniques.

3. Multiple shooting

The core of multiple shooting is the solution of the shooting equations (see e.g. the textbooks [6, 7] or [8–13]). To simplify the presentation we only use $0, t_1, t_2$ and T as multiple shooting nodes, see Fig. 1. The shooting equations can then be written as a set of nonlinear equations, $\mathcal{F}: \mathbb{R}^{12n+p+k+3} \to \mathbb{R}^{6n+p+k+3}$, $\mathcal{G}: \mathbb{R}^{12n+p+k+3} \to \mathbb{R}^{6n}$,

$$0 = \mathcal{F}(Z), \quad \text{given in detail by Eqs. (2-8),} \tag{9}$$
$$0 = \mathcal{G}(Z) = \begin{pmatrix} z(t_1^-) - \Phi(0\ , z(0)\ , t_1^-) \\ z(t_2^-) - \Phi(t_1^+, z(t_1^+), t_2^-) \\ z(T) \ - \Phi(t_2^+, z(t_2^+), T\ ) \end{pmatrix}, \tag{10}$$

for $Z := (z(0), z(t_1^-), z(t_1^+), z(t_2^-), z(t_2^+), z(T), \nu_0, \ldots, \nu_{p-1}, \sigma, t_1, t_2, T) \in \mathbb{R}^{12n+p+k+3}$, $z(t) := (x(t), \lambda(t))$. $\Phi(\tau, \tilde{z}, t)$ denotes the solution of the initial-value problem

$$\dot{z} = G(z), \quad \text{given in detail by Eq. (1),}$$
$$z(\tau) = \tilde{z}$$

at time t.

Motivated by the desired quadratic rate of convergence, the set of nonlinear equations (9,10) is solved by Newton type methods in the various implementations of multiple shooting (e.g. [8–13]). What causes the slow linear convergence and the large condition number of the Jacobian, which was observed for some examples of the considered class of optimal control problems with a quadratic control function?

4. Detailed examination of some boundary and interior point conditions

Motivated by a description of a nondifferentiable Hamiltonian, for a previously used problem formulation of a boundary-value problem with switching function and certain final conditions by Oberle [14], a detailed examination of the boundary and interior point conditions involving the Hamiltonian seems very promising.

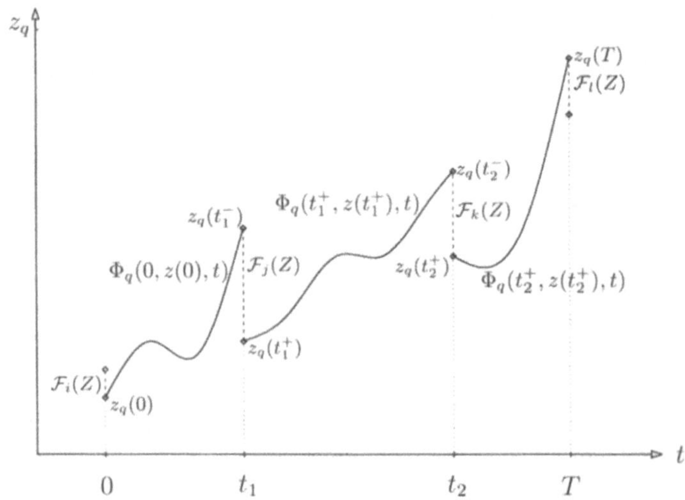

Figure 1: Iteration of Newton's method applied to the shooting equations

Some technical assuptions are required for the following calculations: $\lambda^T a \neq 0$ on $[0, t_1] \cup [t_2, T]$ and $S_x^{p-1} a \neq 0$ on $[t_1, t_2]$.

For free final time T one has the boundary condition $H(x(T), \lambda(T)) = 0$. By substituting the optimal control on free subarcs $u = \dfrac{-\lambda^T b}{2\,\lambda^T a}$ into the Hamiltonian one obtains

$$H(x(T), \lambda(T)) = -\frac{1}{4} \frac{[\lambda(T)^T b(x(T))]^2}{\lambda(T)^T a(x(T))} + \lambda(T)^T c(x(T)) = 0. \tag{11}$$

Lemma 1 *One observes that Eq. (11) as part of Eq. (9) has a root of multiplicity two with respect to the variables $x(T), \lambda(T)$, provided $\lambda(T)^T c(x(T)) = 0$.*

The assumption that $\lambda^T c$ vanishes at final time T is fulfilled e.g. for a circular target orbit of a space vehicle in high altitude, but that is a rather special case.

It motivates on the other hand a detailed examination of the continuity condition $H(x(t_i^+), \lambda(t_i^+)) - H(x(t_i^-), \lambda(t_i^-)) = 0$ at the junction times t_1, t_2 of a boundary arc. Both terms $\lambda^T c$ evaluated at t_2^- and t_2^+ obviously cancel each other. But one has to take into account that both the free optimal control law and a boundary optimal control have to be used appropriately. Moreover one has to consider the jump discontinuity of the costate variable λ at t_1.

Let $u_b(t) := u_b(x(t))$ be an optimal control on the state constrained subarc, which fulfills $0 = S^p(x, u_b) = S_x^{p-1}(x)[a(x)u_b^2 + b(x)u_b + c(x)]$. Some tedious manipulation yields at the junction times t_1, t_2:

$$\begin{aligned} H(x(t_1^+), \lambda(t_1^+)) \; &- \; H(x(t_1^-), \lambda(t_1^-)) = \\ &= \frac{\{[\lambda(t_1^-)^T a(x(t_1))]\, u_b(t_1^+) + [\lambda(t_1^-)^T b(x(t_1))]/2\}^2}{\lambda(t_1^-)^T a(x(t_1))} = 0 \end{aligned} \tag{12}$$

$$\begin{aligned} H(x(t_2^+),\lambda(t_2^+)) &- H(x(t_2^-),\lambda(t_2^-)) = \\ &= \frac{\{[\lambda(t_2)^T a(x(t_2))]\, u_b(t_2^-) + [\lambda(t_2)^T b(x(t_2))]/2\}^2}{\lambda(t_2)^T a(x(t_2))} = 0 \end{aligned} \tag{13}$$

Lemma 2 *One observes that Eqs. (12,13) as part of Eq. (9) have a root of multiplicity two.*

What are the *implications* of these two lemmas for the conditioning of the numerical solution of the shooting equations?

Corollary 1 *If we solve the shooting equations by Newton's method and at least one component of these equations has a root of multiplicity two, then the Jacobian (shooting matrix) will be singular at the solution point and the convergence rate will decrease to linear (cf. [15])in the domain of convergence.*

Therefore one observes a slow linear rate of convergence for the Newton type methods implemented in the various multiple shooting codes (e.g. [9–13]). The condition number of the (approximate) shooting matrix increases as the iteration gets closer to the solution point.

Remark 1 *The situation is even worse for* homotopies *applied to this kind of state-constrained optimal control problems. Due to construction the start of the next iteration with the new homotopy value coincides with the solution of the previous iteration with the old homotopy value. Therefore the shooting matrix is at least badly conditioned at the beginning and the end of each iteration. If the homotopy increments are very small, then during the whole iteration the shooting matrix is badly conditioned.*

Conclusion *In order to restore quadratic convergence, it is therefore required to factorize and reformulate the conditions involving the Hamiltonian.*

Example 1 We use a well-known state constrained spline problem (see e.g. [2,16]) to demonstrate the above mentioned undesirable effect [17]:

$$\begin{gathered} \min\ x_3(1) \\ \dot{x_1} = u, \quad \dot{x_2} = x_1, \quad \dot{x_3} = u^2/2 \\ S(x) = x_2 - k \le 0, \quad k \text{ is a given constant with } 0 < k < 1/6, \\ x_1(0) = 1,\ x_2(0) = 0,\ x_3(0) = 0,\ x_1(1) = -1,\ x_2(1) = 0 \end{gathered}$$

In this problem the conditions (12,13) simplify to

$$H(x(t_1^+),\lambda(t_1^+)) - H(x(t_1^-),\lambda(t_1^-)) = -\lambda_1(t_1^-)^2/(2\lambda_3(t_1^-)) = 0 \tag{14}$$
$$H(x(t_2^+),\lambda(t_2^+)) - H(x(t_2^-),\lambda(t_2^-)) = -\lambda_1(t_2^+)^2/(2\lambda_3(t_2^+)) = 0 \tag{15}$$

This yields only linear convergence due to two components with a root of multiplicity two. Quadratic convergence is restored if one uses $\lambda_1(t_1^-) = \lambda_1(t_2^+) = 0$ instead of (14,15).

Remark 2 *One observes that the Hamiltonian is* regular *for the considered optimal control problem with a scalar quadratic control function. This implies the continuity of the scalar control function at the junction times* t_1, t_2,

$$u(x(t_1^-), \lambda(t_1^-)) = u(x(t_1^+), \lambda(t_1^+)) \quad (16)$$
$$u(x(t_2^-), \lambda(t_2^-)) = u(x(t_2^+), \lambda(t_2^+)). \quad (17)$$

Usually one applies Eqs. (16,17) instead of Eqs. (12,13) in the multipoint boundary-value problem which avoides automatically the above described complications of Lemma 2.

This observation concealed for a long time the findings of Lemma 2. More important is the fact that Lemma 2 and its implications are applicable also for vector-valued optimal control problems which may be non-regular.

5. Generalization

The situation gets more interesting if the optimal control problem has a vector-valued control function where at least one component appears linearly in the r.h.s. of the ODE system and one component appears quadratically in the r.h.s. of the ODE system besides possible further control components. This is a typical problem setting in aerospace engineering.

Then the Hamiltonian might be *non-regular* for an optimal trajectory and several candidate control laws exist on a boundary arc of the state inequality constraint which compete to minimize the Hamiltonian.

Therefore several different transformations of the Eqs. (5,6) apply, depending on the actual switching structure of the candidate control laws on the boundary arc. Therefore one might be tempted to use the most general interior point condition at the junction times t_1, t_2, namely Eqs. (5,6). We examine the consequences of this approach.

Lets consider a more general state-constrained optimal control problem of determining a piecewise continuous control $u(t) := (u_1(t), u_2(t))^T$, $t \in [0, T]$, which minimizes the functional

$$\varphi(x(T))$$

subject to

$$\dot{x} = f(x, u) = a(x, u_2)\, u_1^2 + b(x, u_2)\, u_1 + c(x, u_2), \quad 0 \le t \le T,$$
$$x(0) = x_0, \quad \psi(x(T)) = 0$$
$$u_2 \in U_2$$

and the scalar state inequality constraint

$$S(x) \le 0, \quad 0 \le t \le T.$$

In contrast to Chap. 2 the control u consists of a scalar control component u_1 and a m-dimensional component u_2. The admissible control set for u_2 is described by U_2. The functions $a, b, c\colon I\!R^n \times I\!R^m \to I\!R^n$ are assumed to be C^∞ with respect to all variables, $a \neq 0$.

The calculation of the formulas for $(u(t), \mu(t))$ in dependence of $(x(t), \lambda(t))$ has to be done simultaneously and is much more complicated than in Chap. 2.

Lemma 3 *The findings of Lemma 1 carry over directly to the formula*

$$H(x(T),\lambda(T)) = -\frac{1}{4}\frac{[\lambda(T)^T b(x(T),u(T))]^2}{\lambda(T)^T a(x(T),u(T))} + \lambda(T)^T c(x(T),u(T)) = 0. \tag{18}$$

If the functions $a(x,u)$, $b(x,u)$, $c(x,u)$ are continuous at the junction points t_1 and t_2 then the same calculations which were performed to derive Eqs. (12, 13) yield

$$\begin{aligned} H(x(t_1^+),\lambda(t_1^+)) &- H(x(t_1^-),\lambda(t_1^-)) = \qquad (19)\\ &= \frac{\{[\lambda(t_1^-)^T a(x(t_1),u(t_1^-))]\, u_b(t_1^+) + [\lambda(t_1^-)^T b(x(t_1),u(t_1^-))]/2\}^2}{\lambda(t_1^-)^T a(x(t_1),u(t_1^-))} = 0\\ H(x(t_2^+),\lambda(t_2^+)) &- H(x(t_2^-),\lambda(t_2^-)) = \qquad (20)\\ &= \frac{\{[\lambda(t_2)^T a(x(t_2),u(t_2^+))]\, u_b(t_2^-) + [\lambda(t_2)^T b(x(t_2),u(t_2^+))]/2\}^2}{\lambda(t_2)^T a(x(t_2),u(t_2^+))} = 0, \end{aligned}$$

where $u(t)$, $u_b(t)$ are abbreviations for $u(x(t),\lambda(t))$, $u_b(x(t),\lambda(t))$.

Therefore the findings of Lemma 2 and Corollary 1 carry over to the more general problem setting with vector-valued controls.

The importance of Lemma 3 is demonstrated by the following typical optimal control problem from aerospace engineering:

Example 2 We consider a trajectory optimization problem of an aircraft or a space vehicle. We use a point mass model over a spherical and rotating Earth subject to a dynamic pressure constraint (state constraint of order one). The state vector x consists of velocity v, path inclination γ, azimuth inclination χ, altitude h, latitude Λ and mass m. The control vector u consists of the quadratic component C_{L} (lift coefficient), throttle setting δ (linear component), thrust angle ε and bank angle ζ (nonlinear components). Important quantities are thrust $\tilde{T}$, drag D and lift L of the considered aircraft or space vehicle. For more details with respect to the differential equations see [18] and for an application see e.g. [19].

Minimize an arbitrary Mayer type cost functional

$$\varphi(x(T))$$

subject to

$$\begin{aligned} \dot{v} &= [\tilde{T}(x)\delta\cos\varepsilon - D(x,C_{\mathrm{L}})]/m + g_1(x)\\ \dot{\gamma} &= [\tilde{T}(x)\delta\sin\varepsilon + L(x,C_{\mathrm{L}})]\frac{\cos\zeta}{mv} + g_2(x)\\ \dot{\chi} &= [\tilde{T}(x)\delta\sin\varepsilon + L(x,C_{\mathrm{L}})]\frac{\sin\zeta}{mv\cos\gamma} + g_3(x)\\ \dot{h} &= g_4(x), \quad \dot{\Lambda} = g_5(x), \quad \dot{m} = g_6(x)\,\delta \end{aligned}$$

with a drag polar of parabolic shape (with air density ϱ, minimum drag coefficient C_{D_0}, drag-due-to-lift factor k, reference area F)

$$L = \varrho(h)v^2FC_{\mathrm{L}}/2, \qquad D = \varrho(h)v^2F[C_{\mathrm{D}_0}(v,h) + k(v,h)\; C_{\mathrm{L}}^2]/2$$

and the dynamic pressure constraint

$$S(x) = \varrho(h)v^2/2 - q_{\max} \leq 0$$

and control inequality constraints

$$0 \leq \delta \leq 1, \quad |C_{\rm L}| \leq C_{\rm L,max}$$

and boundary conditions

$$x(0) = x_0, \quad \psi(x(T)) = 0.$$

Note that the Hamiltonian of this typical aerospace example is non-regular, and that there are more then twenty different candidate control laws on a boundary arc of the dynamic pressure limit [20].

Assume that the optimal controls $C_{\rm L}$ and δ satisfy $|C_{\rm L}| < C_{\rm L,max}$ and $\delta = 1$ on $[t_1 - \alpha, t_2 + \alpha]$ ($\alpha > 0$, small). This yields continuity of the controls $C_{\rm L}$, δ, ε, ζ at t_1 and t_2 and therefore also continuity of $a(x)$, $b(x,\zeta)$, $c(x,\delta,\varepsilon,\zeta)$ at t_1 and t_2. Assume that the optimal values for the control components ε, ζ have already been determined. Then the quadratic control component $C_{\rm L}$ plays the role of u_b and satisfies $0 = S^1(x, C_{\rm L}) = S_x(x)[a(x)C_{\rm L}^2 + c(x,\delta,\varepsilon,\zeta)]$ due to $S_x(x)b(x,\zeta) \equiv 0$.

Therefore Eqs. (19,20) have a root of multiplicity two due to Lemma 3 and should be avoided!

Remark 3 *Similar results hold for state-constrained trajectory optimization problems with a polar of parabolic shape, where either angle of attack or load factor are used as a quadratic control instead of the lift coefficient.*

This leads to the following
General Advice *If one has to solve an optimal control problem with a quadratic control component by multiple shooting, and the necessary conditions include boundary and/or interior point conditions involving the Hamiltonian, then one should try to simplify and factorize these conditions in order to eliminate possible roots of multiplicity two. If the assumptions of Lemma 2 or 3 are satisfied, then use* $(\lambda^-)^T a u_b + (\lambda^-)^T b/2|_{t_1} = 0$ *instead of Eqs. (12,19) and use* $\lambda^T a u_b + \lambda^T b/2|_{t_2} = 0$ *instead of Eqs. (13,20).*

Remark 4 *The advice is especially valuable for complicated optimal control problems, if the adjoint equations as well as the optimal control laws are derived by computer algebra systems or automatic differentiation. E.g. then applying the LaTeX to FORTRAN77 compiler [21] for the automatic generation of the application-dependent code for the multiple shooting code MUMUS [12]. Here the software engineer might be tempted to use the* simple *code with the Hamiltonian, which should be avoided.*

Remark 5 *The statements of Lemmas 1–3 and Corollary 1 hold also, if the necessary conditions with respect to the Hamiltonians* $\mathcal{H}^{(i)} := \lambda^{(i)T} f(x,u) + \mu_i S^i(x)$, $i = 0, \ldots, p-1$ *are used.*

Remark 6 *There may be even discontinuities in the shooting equations; then junction times* t_i *move across constant multiple shooting nodes. Therefore special modifications of the shooting equations are implemented in the multiple shooting codes MUMUS (Hiltmann [12,22]) and VBDSCO (Kiehl [23]).*

6. Conclusion

Certain boundary and interior point conditions are identified to destroy quadratic convergence of multiple shooting for an important class of optimal control problems. This class includes especially a large number of state constrained trajectory optimization problems, describing the flight of an aeroplane or space vehicle through the atmosphere of a planet using a quadratic drag polar. General advice is given on how to avoid the identified "bad" conditions and to improve the conditioning of the shooting equations.

Acknowledgement
This research was supported by the Deutsche Forschungsgemeinschaft (DFG) through the Sonderforschungsbereich 255 Transatmosphärische Flugsysteme.

The author is indebted to Prof. Dr. Dr.h.c. R. Bulirsch for the longstanding great support and encouragement. He also thanks Dr. P. Hiltmann, Dr. B. Kugelmann and Dr. K.-D. Reinsch for many helpful discussions and advice on this subject.

References

[1] Pesch, H. J.: *A Practical Guide to the Solution of Real-Life Optimal Control Problems*, Control and Cybernetics, 23, 1/2 (1994), 7–60.

[2] Bryson, A. E.; Denham, W. F.; Dreyfus, S. E.: *Optimal Programming Problems with Inequality Constraints I: Necessary Conditions for Extremal Solutions*, AIAA Journal, 1 (1963), 2544–2550.

[3] Jacobson, D. H.; Lele, M. M.; Speyer, J. L.: *New Necessary Conditions of Optimality for Control Problems with State-Variable Inequality Constraints*, Journal of Mathematical Analysis and Application, 35 (1971), 255–284.

[4] Maurer, H.; Gillessen, W.: *Application of Multiple Shooting to the Numerical Solution of Optimal Control Problems with Bounded State Variables*, Computing, 15 (1975), 105–126.

[5] Hartl, R. F.; Sethi, S. P.; Vickson, R. G.: *A Survey of the Maximum Principles for Optimal Control Problems with State Constraints*, SIAM Review, 37, 2 (1995), 181–218.

[6] Stoer, J.; Bulirsch, R.: *Introduction to Numerical Analysis,* Springer, New York, 1993^2.

[7] Ascher, U. M.; Mattheij, R. M. M.; Russel, R. D.: *Numerical Solution of Boundary Value Problems for Ordinary Differential Equations*, SIAM, Philadelphia, 1995^2.

[8] Bulirsch, R.: *Die Mehrzielmethode zur numerischen Lösung von nichtlinearen Randwertproblemen und Aufgaben der optimalen Steuerung*, Report der Carl-Cranz-Gesellschaft e.V., Oberpfaffenhofen, 1971. Reprint: Report des Sonderforschungsbereichs 255 Transatmosphärische Flugsysteme, Lehrstuhl für Höhere Mathematik und Numerische Mathematik, Techn. Univ. München, 1993.

[9] Oberle, H. J.: *Numerische Berechnung optimaler Steuerungen von Heizung und Kühlung für ein realistisches Sonnenhausmodell*, Habilitation, Institut für Mathematik, Techn. Univ. München, 1982. Reprint: Report TUM-Math-8310, Institut für Mathematik, Techn. Univ. München, 1983.

[10] Oberle, H. J.; Grimm, W.: *BNDSCO – A program for the numerical solution of optimal control problems, user guide*, Report DLR IB/515-89/22, Institut für Dynamik der Flugsysteme, Deutsche Forschungsanstalt für Luft- und Raumfahrt, Oberpfaffenhofen, 1989.

[11] Kiehl, M.: *Vektorisierung der Mehrzielmethode zur Lösung von Mehrpunkt-Randwertproblemen und Aufgaben der optimalen Steuerung*, Dissertation, Institut für Mathematik, Techn. Univ. München, 1989. Reprint: Report No. 115, DFG-Schwerpunktprogramm Anwendungsbezogene Optimierung und Steuerung, Mathematisches Institut, Techn. Univ. München, 1989.

[12] Hiltmann, P.: *Numerische Lösung von Mehrpunkt-Randwertproblemen und Aufgaben der optimalen Steuerung mit Steuerfunktionen über endlichdimensionalen Räumen*, Dissertation, Mathematisches Institut, Techn. Univ. München, 1990. Reprint: Report No. 448, DFG-Schwerpunktprogramm Anwendungsbezogene Optimierung und Steuerung, Mathematisches Institut, Techn. Univ. München, 1993.

[13] Hiltmann, P.; Chudej, K.; Breitner, M.: *Eine modifizierte Mehrzielmethode zur Lösung von Mehrpunkt-Randwertproblemen, Benutzeranleitung*, Report No. 14, Sonderforschungsbereich 255 Transatmosphärische Flugsysteme, Lehrstuhl für Höhere Mathematik und Numerische Mathematik, Techn. Univ. München, 1993.

[14] Oberle, H. J.: *On the Numerical Computation of Minimum-Fuel, Earth-Mars Transfer*, Journal of Optimization Theory and Applications, 22, 3 (1977), 447–453.

[15] Decker, D. W.; Kelley, C. T.: *Newton's Method at Singular Points I,II.* SIAM J. Numer. Anal., 17 (1980), 66-70, 465-471.

[16] Bryson, A. E.; Ho, Y.-C.: *Applied Optimal Control,* Hemisphere Publ. Corp., Washington, D.C., 1975^2.

[17] Hiltmann, P.: *Private communication*, 1993.

[18] Miele, A.: *Flight Mechanics I, Theory of Flight Paths*, Addison-Wesley, Reading, Massachusetts, 1962.

[19] Bulirsch, R.; Chudej, K.: *Combined Optimization of Trajectory and Stage Separation of a Hypersonic Two-Stage Space Vehicle*, Zeitschrift für Flugwissenschaften und Weltraumforschung, 19, 1 (1995), 55–60.

[20] Chudej, K.: *Optimale Steuerung des Aufstiegs eines zweistufigen Hyperschall-Raumtransporters*, Dissertation, Fakultät für Mathematik, Techn. Univ. München, 1994.

[21] Mehlhorn, R.: *Integrierte Arbeitsumgebung zur numerischen Berechnung von Problemen der optimalen Steuerung*, Dissertation, Fakultät für Maschinenwesen, Techn. Univ. München, 1996.

[22] Hiltmann, P.: *A Multiple Shooting Algorithm for Multipoint Boundary Value Problems,* Report No. 17, Sonderforschungsbereich 255 Transatmosphärische Flugsysteme, Lehrstuhl für Höhere Mathematik und Numerische Mathematik, Techn. Univ. München, 1994.

[23] Kiehl, M.: *Smoothing the Function of the Multiple-Shooting Equation*, Applied Mathematics and Optimization, 24 (1991), 171–181.

International Series of Numerical Mathematics
Vol. 124,

SQP Methods and their Application to Numerical Optimal Control*

Alex Barclay† Philip E. Gill† J. Ben Rosen‡

Abstract. In recent years, sequential quadratic programming (SQP) methods have been developed that can reliably solve constrained optimization problems with many hundreds of variables and constraints. These methods require remarkably few evaluations of the problem functions and can be shown to converge to a solution under very mild conditions on the problem.

Some practical and theoretical aspects of applying SQP methods to optimal control problems are discussed, including the influence of the problem discretization and the zero/nonzero structure of the problem derivatives. We conclude with some recent approaches that tailor the SQP method to the control problem.

1. Introduction

Recently there has been considerable progress in the development of general-purpose sequential quadratic programming (SQP) methods for large-scale nonlinear optimization (see, e.g., Eldersveld [12], Tjoa and Biegler [34], Betts and Frank [4] and Gill, Murray and Saunders [16]). These methods are efficient and reliable, and can be applied to large sparse problems with a mixture of linear and nonlinear constraints. The methods require remarkably few evaluations of the problem functions and converge to a solution under very mild conditions on the problem. An important application for these methods is the class of problems derived by discretizing optimal control problems. These problems have several important characteristics: (i) many variables and constraints; (ii) sparse and structured constraint and objective derivatives; (iii) objective and constraint functions (and their first derivatives) that are expensive to evaluate; and (iv) many constraints active (i.e., exactly satisfied) at the solution.

In §2. we give a brief discussion of general-purpose SQP methods, with an emphasis on those aspects that most affect performance. In §3. we consider the application of these general-purpose methods to optimal control problems. In §4. we discuss the most common forms of discretization and consider their effect on the efficiency of general-purpose SQP methods. Finally, in §5. we briefly consider some SQP methods that are specially tailored to problems derived from optimal control problems.

---

* This research was partially supported by National Science Foundation grants CCR-95-27151 and DMI-9424639, Office of Naval Research grants N00014-90-J-1242 and N00014-96-1-0274.

† Department of Mathematics, University of California, San Diego, La Jolla, California 92093-0112, e-mail: abarclay@ucsd.edu, pgill@ucsd.edu

‡ Department of Computer Science and Engineering, University of California, San Diego, La Jolla, California 92093-0114, e-mail: jbrosen@cs.ucsd.edu.

2. SQP methods

The general nonlinearly constrained optimization problem can be written in the form

$$\begin{array}{ll} \underset{x\in\mathbb{R}^n}{\text{minimize}} & F(x) \\ \text{subject to} & b_l \le \left\{ \begin{array}{c} x \\ Ax \\ c(x) \end{array} \right\} \le b_u, \end{array} \tag{1}$$

where c is a vector of nonlinear functions, A is a constant matrix that defines the linear constraints, and b_l and b_u are constant upper and lower bounds. It is assumed that first derivatives of the problem are known explicitly, i.e., at any point x it is possible to compute the gradient $\nabla F(x)$ of the objective F and the Jacobian $J(x)$ of the nonlinear constraints c. Our principal concern is with large problems, although the precise definition of the size of a problem depends on several factors, including the zero/nonzero structure of the problem derivatives and the number of degrees of freedom at a solution (i.e., the number of variables less the number of active constraints at a solution)[1].

At a constrained minimizer x^*, the objective gradient ∇F can be written as a linear combination of the constraint gradients. The multipliers in this linear combination are known as the *Lagrange multipliers.* The Lagrange multipliers for an upper bound constraint are nonpositive, the multipliers for a lower bound constraint are nonnegative. The vector of Lagrange multipliers associated with the *nonlinear* constraints of (1) is denoted by π^*.

SQP methods are a class of optimization methods that solve a quadratic programming subproblem at each iteration. Each QP subproblem minimizes a quadratic model of a certain modified Lagrangian function subject to linearized constraints. A merit function is reduced along each search direction to ensure convergence from any starting point. The basic structure of an SQP method involves *major* and *minor* iterations. The major iterations generate a sequence of iterates (x_k, π_k) that converge to (x^*, π^*). At each iterate a QP subproblem is used to generate a search direction towards the next iterate (x_{k+1}, π_{k+1}). Solving such a subproblem is itself an iterative procedure, with the *minor* iterations of an SQP method being the iterations of the QP method. (For an overview of SQP methods, see, for example, Gill, Murray and Wright [18].)

In the SQP formulation considered here, the objective and constraint derivatives ∇F and J are required once each major iteration in order to define the objective and constraints of the QP subproblem. SQP methods have two important properties. First, they are most robust when the derivatives of the objective and constraint functions are computed exactly. Second, the SQP iterates do not usually satisfy the nonlinear equality constraints except as the solution is approached. (However, it is possible to ensure that the iterates always satisfy the *linear* constraints.)

Each QP subproblem minimizes a quadratic model of the *modified Lagrangian* $\mathcal{L}(x, x_k, \pi_k) = F(x) - \pi_k^T d_L(x, x_k)$, which is defined in terms of the *constraint lineariza-*

[1] Some authors define the number of degrees of freedom as the number of variables less the number of equality constraints (i.e., constraints with $(b_l)_i = (b_u)_i$).

tion, $c_L(x, x_k) = c(x_k) + J(x_k)(x - x_k)$, and the *departure from linearity*, $d_L(x, x_k) = c(x) - c_L(x, x_k)$. Given estimates (x_k, π_k) of (x^*, π^*), an improved estimate is found from $(\widehat{x}_k, \widehat{\pi}_k)$, the solution of the following QP subproblem:

$$\begin{array}{ll} \underset{x \in \mathbb{R}^n}{\text{minimize}} & F(x_k) + \nabla F(x_k)^T(x - x_k) + \frac{1}{2}(x - x_k)^T H_k (x - x_k) \\ \text{subject to} & b_l \leq \left\{ \begin{array}{c} x \\ Ax \\ c(x_k) + J(x_k)(x - x_k) \end{array} \right\} \leq b_u, \end{array}$$

where H_k is a positive-definite approximation to $\nabla^2_x \mathcal{L}(x_k, x_k, \pi_k)$. Once the QP solution $(\widehat{x}_k, \widehat{\pi}_k)$ has been determined, the major iteration proceeds by determining new variables (x_{k+1}, π_{k+1}) as $x_{k+1} = x_k + \alpha_k(\widehat{x}_k - x_k)$ and $\pi_{k+1} = \pi_k + \beta_k(\widehat{\pi}_k - \pi_k)$, where the scalars α_k and β_k are chosen to yield a sufficient decrease in a *merit function* (a suitable combination of the objective and constraint functions). Some methods determine α_k and β_k simultaneously using a linesearch, others use a linesearch for α_k but keep β_k fixed at 1. The choice of merit function will not be considered here, but choices that have proved effective in practice are the l_1 penalty function (see e.g., Fletcher [14]) and an augmented Lagrangian function (see Schittkowski [30] and Gill, Murray and Saunders [16]).

The definition of the QP Hessian H_k is crucial to the success of an SQP method. In the method of SNOPT (Gill, Murray and Saunders [16]), H_k is a limited-memory quasi-Newton approximation to $G = \nabla^2_x \mathcal{L}(x_k, x_k, \pi_k)$, the Hessian of the modified Lagrangian. Another possibility is to define H_k as a positive-definite matrix related to a finite-difference approximation to G. Exact second derivatives have also been proposed in this context (see, e.g., Fletcher [14], Eldersveld and Gill [13]).

The QP solver must repeatedly solve linear systems formed from rows and columns of the QP constraint matrices. The class of active-set QP methods solve a system of the form

$$\begin{pmatrix} H & W^T \\ W & 0 \end{pmatrix} \begin{pmatrix} p \\ q \end{pmatrix} = \begin{pmatrix} g \\ h \end{pmatrix} \tag{2}$$

each minor iteration. The matrices H and W consists of a subset of the rows and columns of H_k, $J(x_k)$ and A specified by the *working-set*, which is an estimate of the constraints active at a solution. Eventually, the working sets of the QP subproblem usually "settle down" to the active set of the nonlinear problem. For this reason, the final working set from one QP subproblem is used to guide the initial working set in the next. Once the active set has been identified, the QP subproblems reach optimality in one iteration, and hence only a single system (2) need be solved in later subproblems.

The method used to solve (2) largely determines the characteristics of the QP method. When the system is large and sparse, we know of only two viable QP methods—the reduced-Hessian method and the Schur-complement method.

Reduced-Hessian QP methods. These methods solve (2) using a full-rank matrix Z that spans the null space of W (i.e., $WZ = 0$). The matrix Z is used to transform (2) into two smaller systems, one of which involves the *reduced Hessian* Z^THZ. The reduced Hessian is stored and updated in dense form, which is efficient as long as the number of degrees of freedom is small compared to the number of variables. In particular, reduced Hessian methods are efficient if W is nearly square and products $H_k x$ can be formed efficiently.

If W is large and sparse, Z is usually maintained in "reduced-gradient" form, using sparse LU factors of a square matrix B that alters as the working set changes. The defining equations are

$$WP = \begin{pmatrix} B & S \end{pmatrix}, \qquad Z = P\begin{pmatrix} -B^{-1}S \\ I \end{pmatrix}, \tag{3}$$

where P is a permutation that ensures B is nonsingular. The number of degrees of freedom is the number of columns of S. Products of the form Zv and $Z^T g$ are obtained by solving with B or B^T. For more details, see Gill, Murray and Saunders [16].

Schur-complement QP methods. After each change to W, the system (2) is equivalent to a system of the form

$$\begin{pmatrix} H_0 & W_0^T & U \\ W_0 & 0 & 0 \\ U^T & 0 & V \end{pmatrix}\begin{pmatrix} u \\ v \\ w \end{pmatrix} = \begin{pmatrix} a \\ b \\ c \end{pmatrix},$$

where the number of columns of U is equal to the number of QP iterations. Given the system

$$K_0 = \begin{pmatrix} H_0 & W_0^T \\ W_0 & 0 \end{pmatrix} \tag{4}$$

at the start of the QP, factorizations of K_0 and the Schur complement $C = V - U^T K_0^{-1} U$ can be used to solve (2). It is efficient to work with a sparse factorization of K_0 and dense factors of C. The factorization of K_0 may be computed using any suitable code. If the number of updates is small enough, C may be maintained as a dense factorization (based on either orthogonal or elementary transformations), and updated to reflect changes in W. As the dimension of C grows, it is eventually necessary to refactorize the system (2) from scratch. For more details, see Gill, Murray, Saunders and Wright [17]. The Schur-complement QP method is implemented in the code SOCS of Betts and Frank [4].

3. SQP methods for optimal control

In this section we give a brief review of the application of general-purpose SQP methods to optimal control problems in which the dynamics are determined by a system of ordinary differential equations (ODEs). The problem is assumed to be of the form

$$\underset{u,y}{\text{minimize}} \quad y_1(t_f) \tag{5}$$

$$\text{subject to} \quad y(t_0) = y_0, \tag{6}$$

$$\dot{y}(t) = f(y,u,t), \quad t \in [0,t_f], \tag{7}$$

$$g(y,u,t) \geq 0, \quad t \in [0,t_f], \tag{8}$$

where $y(t)$ is an n_y-vector of state variables and $u(t)$ is an n_u-vector of controls. The inequalities (8) often include upper and lower bounds on control variables $u(t)$ and state variables $y(t)$. It is assumed that given the initial condition y_0 and the control function $u = u(t)$, $t \in [0, t_f]$, the state vector function $y = y(t)$ is uniquely determined by the differential system (7). We also assume that the control $u(t)$ is continuous and satisfies some standard conditions needed for the existence of an optimal control (see, e.g., Leitmann [25]).

4. Discretizing the control problem

Although the size of the ODE system (7) can be large, the dimension of the control vector $u(t)$ is typically much smaller. The usual approach is to represent $u(t)$ as a low-order spline or piecewise polynomial function on $[0, t_f]$. The coefficients of this spline or piecewise polynomial are then adjusted during the optimization.

Given the initial conditions y_0 and values of the spline coefficients, the controls can be evaluated at any point in $[0, t_f]$ and the state vector function $y = y(t)$ is uniquely determined by the differential system (7). However, it is neither necessary (nor advisable) to explicitly solve the differential system at each step. Instead, finite-dimensional nonlinear equations associated with certain well-known ODE methods are imposed as nonlinear constraints during the optimization. A feature of this approach is that, since SQP methods do not generally satisfy nonlinear equality constraints until the solution is approached, the differential system is only satisfied near the end of the computation.

4.1 Discretizing the control functions

The total time interval $[0, t_f]$ is divided into N equal subintervals of length Δt each. These subintervals define $N + 1$ *nodes*:

$$t_k = k\Delta t, \quad k = 0, 1, \ldots, N, \quad \text{with} \quad t_N = t_f. \tag{9}$$

A piecewise polynomial $u^k(t)$ is used to represent $u(t)$ for $t \in [t_k, t_{k+1}]$. For example, if $u^k(t)$ is a cubic polynomial, then $u^k(t)$ can be represented by $4n_u$ coefficients, which are then determined by the optimization. The continuity of the $u^k(t)$ and their derivatives at the nodes is enforced by means of appropriate linear equality constraints. Any bounds on the $u^k(t)$ at the nodes imply additional linear inequalities on the coefficients of the polynomial.

A convenient alternative representation can be defined using Hermite interpolation. In the case of a cubic approximation, optimization variables u_k and w_k are introduced to represent the values of $u(t)$ and $\dot{u}(t)$ at each node t_k. The unique piecewise cubic $u^k(t)$ is defined that has values (u_k, u_{k+1}) and derivatives (w_k, w_{k+1}) at the end points of $[t_k, t_{k+1}]$. This cubic allows the control at any point $t = t_k + \rho\Delta t$ in $[t_k, t_{k+1}]$ to be written as

$$u^k(t) = u_\rho + \rho(\rho - 1)\left[(u_{k+1} - u_k)(1 - 2\rho) + w_\rho \Delta t\right], \tag{10}$$

where $u_\rho = (1 - \rho)u_k + \rho u_{k+1}$ and $w_\rho = (\rho - 1)w_k + \rho w_{k+1}$ (see, e.g., Hairer, Nørsett and Wanner [20, p.190]). This scheme automatically gives continuity of $u^k(t)$ and its derivative at the nodes. Moreover, simple upper and lower bounds on the controls $u(t)$ lead to simple upper and lower bounds on the discretized variables $\{u_k\}$.

4.2 Discretizing the differential equations

Three methods for discretizing the ODEs (7) will be considered: single shooting, multiple shooting and collocation. Each of these methods gives a finite-dimensional problem with a different character.

Single shooting. Given the initial conditions y_0 and a piecewise polynomial control approximation $\{u^k(t)\}$, the state vector $y = y(t)$ is uniquely determined by the differential system (7). The method of single shooting minimizes with respect to the control variables (say, $\{u_k, w_k\}$) and solves (7) over $[0, t_f]$ at each new point generated by the optimization algorithm. It follows that the differential equations are being used to *eliminate* the state variables from the problem. It is common for an adaptive ODE solver to be used for this process.

The inequality constraints (8) can be imposed explicitly at each control subinterval boundary as requirements on the u_k. These become

$$g(y_k, u_k, t_k) \geq 0, \qquad k = 0, 1, \ldots, N, \tag{11}$$

where y_k denotes the state value at t_k. The objective value is the first component of y_N.

The discretized problem is a nonlinear optimization problem with variables $x = (u_0, w_0, u_1, w_1, \ldots, u_N, w_N)^T$. In order to apply an SQP method, it is necessary to be able to compute the derivatives of the inequality constraints (11). These calculations involve the partial derivatives $\partial y_i / \partial u_j$ and $\partial y_i / \partial w_j$, where y_i denotes the ODE solution evaluated at t_i. Each of these derivatives is itself the solution of a differential equation whose right-hand side involves the derivatives f'_y and f'_u. This system must be solved in conjunction with the original ODEs. Since the state value y_k is independent of (u_i, w_i) for $i > k$, the Jacobian of y with respect to u and w is block lower triangular. Packages are available that allow the efficient and reliable computation of these Jacobians (see, e.g., Buchauer, Hiltmann and Kiehl [7], Maly and Petzold [26]). These packages allow a constraint derivative to be computed with approximately the same effort and the same accuracy as the constraint value itself.

Multiple shooting. In this case, the interval $[0, t_f]$ is divided into subintervals and the differential equations are solved over each subinterval. Continuity of the solutions between the subintervals is achieved by enforcing the continuity conditions as nonlinear equations (see Ascher, Mattheij and Russell [1]). In the optimal control context, the multiple shooting equations are imposed as constraints in the optimization. When the constraints need to be evaluated, the differential equation (7) is regarded as an independent initial-value problem over each of the subintervals $[t_k, t_{k+1}]$. A continuous solution over $[0, t_f]$ is obtained by matching the initial conditions at t_k with the final values obtained from the previous subinterval $[t_{k-1}, t_k]$. The initial values of y for each subinterval are treated as variables of the optimization and they are adjusted to achieve continuity over $[0, t_f]$.

Let $y^k(t)$ denote the solution of the differential equations (7) on the interval $[t_k, t_{k+1}]$ with initial conditions $y^k(t_k) = y_k$. The value of y_0 is given, and the vectors y_k, $k = 1$, $2, \ldots, N-1$, are variables to be adjusted by the optimizer. For simplicity, we assume that $u(t)$ is represented by the Hermite piecewise cubic polynomial $u^k(t)$ defined above. Given y_k and the control values u_k and w_k at the nodes, the differential system (7) can be solved by a "black-box" solver to give $y^k(t_{k+1})$. In general, $y^k(t_{k+1})$ will not equal the initial value y_{k+1} for the next interval, and it is necessary to adjust y_k, u_k and w_k so that the continuity conditions

$$c_k = y^k(t_{k+1}) - y_{k+1} = 0, \qquad k = 0, 1, \ldots, N-1, \tag{12}$$

are satisfied at the subinterval boundaries. The last of these conditions involves y_N, which does not specify an initial value for any differential equation. This vector can be used to impose a condition on $y^{N-1}(t_N)$ arising from either an explicit condition on $y(t_f)$ or a condition on y from the inequality constraints $g \geq 0$, which are imposed pointwise at the nodes, as in single shooting. If there are no inequalities, y_N can be a free variable in the optimization. The initial-value problems for each subinterval are independent and can be solved in parallel (see, e.g., Kiehl [22], and Petzold *et al.* [29]).

As in single shooting, the calculation of the problem derivatives requires the solution of additional ODEs. However, in this case the Jacobian and Hessian are sparse and structured. If Hermite interpolation is used to define each $u^k(t)$, the vector of variables x for the finite-dimensional problem can be written in the form $x = (u_0, w_0, y_1, u_1, w_1, \ldots, y_N, u_N, w_N)^T$, with a grand total of $n = n_y N + 2n_u(N+1)$ variables. With this ordering, the Jacobian of the continuity conditions (12) has the form

$$J = \begin{pmatrix} U_0 & -I & & & & & & \\ & Y_1 & U_1 & -I & & & & \\ & & & Y_2 & U_2 & -I & & \\ & & & & \ddots & \ddots & \ddots & \\ & & & & & Y_{N-1} & U_{N-1} & -I \end{pmatrix},$$

where $Y_k = \partial c_k / \partial y_k$ and U_k contains the partial derivatives $\partial c_k / \partial u_k$ and $\partial c_k / \partial w_k$ in appropriate order. The Jacobian for the inequality constraints $g \geq 0$ (11) will depend upon the particular application. The Lagrangian Hessian is also highly structured. If there are no nonlinear constraints other than the continuity conditions, then $\nabla^2_x \mathcal{L}(x, x_k, \pi_k)$ is a block-diagonal matrix with the same structure as $J^T J$.

Methods based upon multiple shooting have been proposed, amongst others, by Pantelides, Sargent and Vassiliadis [27], Gritsis, Pantelides and Sargent [19], and Petzold *et al.* [29]).

Collocation. Methods for discretizing the differential constraints using collocation have been proposed by Dickmanns and Well [11], Kraft [23], Hargraves and Paris [21], and Betts and Huffman [5]. The method of collocation is closely related to multiple shooting. Our discussion follows that of Brenan [6].

Discretization by collocation may be regarded as a form of multiple shooting in which an appropriate implicit Runge Kutta (IRK) formula is used to solve the initial-value problem over $[t_k, t_{k+1}]$. An s-stage IRK formula involves constants $\{\alpha_{ij}\}$, $\{\beta_j\}$

$$\begin{array}{c|ccc} \rho_1 & \alpha_{11} & \cdots & \alpha_{1s} \\ \vdots & \vdots & \ddots & \vdots \\ \rho_s & \alpha_{s1} & \cdots & \alpha_{ss} \\ \hline & \beta_1 & \cdots & \beta_s \end{array} \qquad \begin{array}{c|ccc} 0 & 0 & 0 & 0 \\ \frac{1}{2} & \frac{5}{24} & \frac{1}{3} & -\frac{1}{24} \\ 1 & \frac{1}{6} & \frac{2}{3} & \frac{1}{6} \\ \hline & \frac{1}{6} & \frac{2}{3} & \frac{1}{6} \end{array}$$

Figure 1: The Butcher tableau and its definition for Lobatto IIIA, $s = 3$.

and $\{\rho_i\}$ specified by a Butcher tableau (see Figure 1) and involves the evaluation of the right-hand side f at the points $t_{kj} = t_k + \rho_j \Delta t$, for $j = 1, 2, \ldots, s$. The solution of the initial-value problem is the state vector $y^k(t_{k+1})$ such that

$$y^k(t_{k+1}) = y_k + \Delta t \sum_{j=1}^{s} \beta_j f(y_{kj}, u_{kj}, t_{kj}), \tag{13}$$

where the quantities y_{kj} satisfy the s collocation conditions

$$y_{kj} - y_k - \Delta t \sum_{l=1}^{s} \alpha_{jl} f(y_{kl}, u_{kl}, t_{kl}) = 0, \qquad j = 1, \ldots, s, \tag{14}$$

with the suffix ij denoting a quantity defined at t_{ij}. As in multiple shooting, the vector $y^k(t_{k+1})$ is matched with y_{k+1} by imposing the continuity condition $c_k = 0$ as an optimization constraint, where $c_k = y^k(t_{k+1}) - y_{k+1}$. The collocation conditions are also imposed as constraints, giving additional s nonlinear equality constraints and s unknowns $\{y_{kl}\}$ at each node.

The number of variables and constraints is reduced if the IRK scheme includes $\rho_1 = 0$ or $\rho_s = 1$, since one or two redundant collocation equations can be eliminated. For example, with the Lobatto IIIA scheme with $s = 3$, the Butcher tableau of Figure 1 indicates that only three collocation points are required: $t_{k1} = t_k$, $t_{k2} = \frac{1}{2}(t_k + t_{k+1})$, and $t_{k3} = t_{k+1}$. The continuity conditions (12) are then defined with

$$y^k(t_{k+1}) = y_k + \frac{\Delta t}{6}(f(y_k, u_k, t_k) + 4f(y_{k2}, u_{k2}, t_{k2}) + f(y_{k+1}, u_{k+1}, t_{k+1})), \tag{15}$$

where the state value y_{k2} satisfies the single collocation condition

$$y_k + \frac{\Delta t}{24}(5f(y_k, u_k, t_k) + 8f(y_{k2}, u_{k2}, t_{k2}) - f(y_{k+1}, u_{k+1}, t_{k+1})) - y_{k2} = 0, \tag{16}$$

with $u_{k2} = \frac{1}{2}(u_k + u_{k+1}) + \frac{1}{8}\Delta t(w_k - w_{k+1})$. A similar scheme using a 2-stage Lobatto IRK is proposed by Betts and Huffman [5]. See also, Pesch [28], Lamour [24], von Stryk and Bulirsch [36], Bulirsch *et al.* [8], von Stryk [35], Betts [3], and Schulz, Bock and Steinbach [32].

An important property of collocation is that the partial derivatives of problem functions are relatively simple to compute. For the moment, suppose that the relation $u_{k2} = \frac{1}{2}(u_k + u_{k+1}) + \frac{1}{8}\Delta t(w_k - w_{k+1})$ is imposed as the *linear constraint*

$$\frac{1}{2}(u_k + u_{k+1}) + \gamma(w_k - w_{k+1}) - u_{k2} = 0, \tag{17}$$

where for brevity, we have denoted $\Delta t/8$ by γ. It follows that at each subinterval, the differential constraints give three derivatives, c_k', c_{k2}' and A_k, defined by the continuity conditions, the collocation conditions and the linear constraints above. These derivatives may be listed schematically as follows:

	y_k	u_k	w_k	y_{k2}	u_{k2}	y_{k+1}	u_{k+1}	w_{k+1}
c_k'	$I+\bar\beta_1 Y_k$	$\bar\beta_1 U_k$		$\bar\beta_2 Y_{k2}$	$\bar\beta_2 U_{k2}$	$\bar\beta_3 Y_{k+1}-I$	$\bar\beta_3 U_{k+1}$	
c_{k2}'	$I+\bar\alpha_{21} Y_k$	$\bar\alpha_{21} U_k$		$\bar\alpha_{22} Y_{k2}-I$	$\bar\alpha_{22} U_{k2}$	$\bar\alpha_{23} Y_{k+1}$	$\bar\alpha_{23} U_{k+1}$	
A_k		$\frac{1}{2}I$	γI		$-I$		$\frac{1}{2}I$	$-\gamma I$

where Y_k and U_k denote the Jacobians f_y' and f_u' evaluated at t_k and $\bar\alpha$ and $\bar\beta$ denote the Lobatto coefficients scaled by Δt. It follows that the full Jacobian has a block diagonal structure, with N overlapping diagonal blocks. Except for the first block (which must account for the ODE initial conditions) each diagonal block has sparsity and structure identical to that defined above.

If the derivatives f_y' and f_u' are sparse, as they are in many applications, this constraint Jacobian will be sparse and structured. This structure will not be altered significantly if the linear constraints (17) are used to eliminate the variables u_{k2}. In this case the derivatives are

	y_k	u_k	w_k	y_{k2}	y_{k+1}	u_{k+1}	w_{k+1}
c_k'	—	$\bar\beta_1 U_k+\frac{1}{2}\bar\beta_2 U_{k2}$	$\gamma\bar\beta_2 U_{k2}$	—	—	$\bar\beta_3 U_{k+1}+\frac{1}{2}\bar\beta_2 U_{k2}$	$-\gamma\bar\beta_2 U_{k2}$
c_{k2}'	—	$\bar\alpha_{21} U_k+\frac{1}{2}\bar\alpha_{22} U_{k2}$	$\gamma\bar\alpha_{22} U_{k2}$	—	—	$\bar\alpha_{23} U_{k+1}+\frac{1}{2}\bar\alpha_{22} U_{k2}$	$-\gamma\bar\alpha_{22} U_{k2}$

where an entry "—" implies that the derivative is unchanged by the elimination.

The constraints can be simplified further if Hermite cubic interpolation is used to represent $y(t)$ on the interval $[t_k, t_{k+1}]$. It follows from the identity $\dot y_k = f(y_k, u_k, t_k)$, that the state variable at the collocation point can be represented in terms of the y_k's, u_k's and w_k's (as in (10)). Hence we are left with just one implicit nonlinear equation

$$y_{k+1} - y_k - \frac{\Delta t}{6}\big(f(y_k,u_k,t_k) + 4f(y_{k2},u_{k2},t_{k2}) + f(y_{k+1},u_{k+1},t_{k+1})\big) = 0,$$

where $y_{k2} = \frac{1}{2}(y_k + y_{k+1}) + \frac{\Delta t}{8}\big(f(y_k,u_k,t_k) - f(y_{k+1},u_{k+1},t_{k+1})\big)$. This is the basis of the Hermite-Simpson method.

Although the variable y_{k2} can be eliminated from the problem, there is a strong argument from the optimization viewpoint for not performing the elimination explicitly. If the explicit relation for y_{k2} is kept as an explicit constraint, the associated derivatives are given by

	y_k	u_k	w_k	y_{k2}	u_{k2}	y_{k+1}	u_{k+1}	w_{k+1}
c_k'	$I+\bar\beta_1 Y_k$	$\bar\beta_1 U_k$		$\bar\beta_2 Y_{k2}$	$\bar\beta_2 U_{k2}$	$\bar\beta_3 Y_{k+1}-I$	$\bar\beta_3 U_{k+1}$	
c_{k2}'	$\frac{1}{2}I+\gamma Y_k$	γU_k		$-I$		$\frac{1}{2}I-\gamma Y_{k+1}$	$-\gamma U_{k+1}$	
A_k		$\frac{1}{2}I$	γI		$-I$		$\frac{1}{2}I$	$-\gamma I$

As before, if f_y' and f_u' are sparse, this scheme will give a sparse Jacobian for the solver. Alternatively, if y_{k2} is not an explicit variable, the derivatives become significantly more complicated (and hence, less sparse). For example, the partial derivatives of c_k' with respect to y_k and u_k are $-(I + \bar{\beta}_1 Y_k + \bar{\beta}_2 Y_{k2}(\frac{1}{2}I + \gamma Y_k))$ and $-(\bar{\beta}_1 U_k + \bar{\beta}_2 \gamma Y_{k2} U_k)$ respectively.

When modeling large control problems, there is always the temptation to eliminate as many constraints and variables as possible. In some cases this is justified–even essential, but the elimination is often done at the expense of creating a constraint Jacobian that is less sparse than the original. In large-scale optimization, it is not the size of the problem that is important as much as the nonzero structure of the Jacobian and Hessian. To be more precise, it is the interaction of this structure with the matrix factorizations used to solve the linear systems (3) and (4) that is crucial. In general, there is little to be lost, and much to be gained by hesitating before eliminating variables and constraints by hand. This observation is based on the assumption that the more sparse the Jacobian, the better the sparse solver is able to exploit the sparsity during the factorizations.

4.3 Properties of the discretizations

If there are few control variables, the number of degrees of freedom will be small compared to the number of variables, and methods based on the explicit estimation of the reduced Hessian $Z^T H Z$ will work efficiently.

Single shooting gives a discretized problem with the smallest number of variables and constraints. Moreover, the constraint Jacobian is essentially dense, and it is not necessary to use a sparse solver. Single shooting is usually used in conjunction with adaptive ODE software, and when it converges, it can be very efficient. In many cases, good results can be obtained from single shooting (see, e.g., Büskens and Maurer [9]). However, it is well known that single shooting can suffer from a lack of robustness and stability (see, e.g., Ascher, Mattheij and Russell [1]). For some nonlinear problems it can generate intermediate iterates that are nonphysical and/or not computable. For some well-conditioned boundary-value problems, it can generate unstable initial-value problems.

If any one of the active inequality constraints (11) involves state variables, both single and multiple shooting require the calculation of the partial derivatives $\partial y_i/\partial u_j$ and $\partial y_i/\partial w_j$ at each node. There seems little justification for choosing single shooting in this case, given the superior stability properties of multiple shooting (see below). The decision is less straightforward if g does not involve the state variables, since no additional ODEs need to be solved in this case.

Multiple shooting with an appropriate ODE solver does not have the potential stability problems of single shooting. However, both single and multiple shooting have a weakness that is not present in collocation. In a "black-box" shooting scheme, it is common for the constraints and their derivatives to be evaluated using an adaptive ODE package in which the step sizes are chosen adaptively based on local error estimates. Current SQP methods require that the Jacobian elements be computed to high accuracy, which implies a relatively tight tolerance for the ODE solver. This can make the problem derivatives extremely expensive, with as much as 95% of the computation time being spent in the derivative calculation. By contrast, collocation derivatives can be obtained to (essentially) full precision at little cost if f'_y and f'_u are available. Moreover, if second derivatives of f can be computed (either by hand or with the use of a symbolic differentiation package), then second-derivative SQP methods can be applied.

Another advantage of collocation is that it is very convenient to implement a multigrid-type approach in which a solution on a coarse grid of nodes is used as an initial estimate for the larger problem on a finer grid. In this context, second derivative methods are likely to be significantly more efficient than first-derivative methods. Since quasi-Newton methods estimate second derivative information over a sequence of iterations, a poor initial Hessian may cause the algorithm to wander away from a good estimate of the solution. It may be seen that the reduced Hessian Z^THZ associated with different discretizations are not related in a simple way, even when one mesh is the refinement of another. This implies that the optimal reduced Hessian from one problem is not easily used to start an optimization on a finer grid. This difficulty does not occur with exact second derivatives since the reduced Hessian for the refined system is computed from scratch.

A disadvantage of collocation is that the number of variables and constraints is significantly increased compared to multiple shooting. However, as we have already indicated, this need not be a serious disadvantage if the problem derivatives are sufficiently sparse. A potentially more serious difficulty is that, for a given complexity of optimization problem, the state variables obtained by collocation may not be as accurate as those obtained by multiple shooting. In this case, a compromise is to use collocation to obtain a good starting estimate for a multiple shooting approach.

5. SQP methods that exploit structure

A general-purpose SQP algorithm treats all variables and constraints equally, and does not distinguish between state and control variables. Many methods have been devised that exploit the structure of the problem with the aim of increasing efficiency. The formulation of such "modified" or "structured" SQP methods are considered next. For clarity, the notation will be changed so that y and u denote the components of the *discretized* versions of the state and control vectors, with y an n_1-vector and u an n_2-vector. The structured problem has n $(n = n_1 + n_2)$ variables and m $(m = m_1 + m_2)$ constraints, and is written as

$$\begin{array}{ll} \underset{y\in \mathbb{R}^{n_1}, u\in \mathbb{R}^{n_2}}{\text{minimize}} & F(y,u) \\ \text{subject to} & c_1(y,u) = 0, \quad c_2(y,u) \geq 0. \end{array} \tag{18}$$

The m_1 constraints $c_1(y,u) = 0$ represent the continuity and collocation conditions associated with the discretized ODEs. The m_2 inequalities $c_2(y,u) \geq 0$ define all the inequality constraints, including simple upper and lower bounds. We assume that $n_1 = m_1$, so that the number of equality constraints is equal to the number of state variables. It follows that the constraint Jacobian for this problem has the form

$$J(x) = \begin{pmatrix} J_1 \\ J_2 \end{pmatrix} = \begin{pmatrix} J_1^y & J_1^u \\ J_2^y & J_2^u \end{pmatrix},$$

where J_1 and J_2 are $n_1 \times (n_1+n_2)$ and $m_2 \times (n_1+n_2)$, and the matrices J_1^y, J_1^u, J_2^y and J_2^u reflect the (y,u) partition of the variables. The QP subproblem associated with the structured nonliner problem (18) can be written as

$$\begin{array}{ll} \underset{\Delta y\in\mathbb{R}^{n_1},\Delta u\in\mathbb{R}^{n_2}}{\text{minimize}} & g_y^T\Delta y + g_u^T\Delta u + \frac{1}{2}(\ \Delta y^T \quad \Delta u^T\ )H\begin{pmatrix}\Delta y\\ \Delta u\end{pmatrix} \\ \text{subject to} & J_1^y\Delta y + J_1^u\Delta u = -c_1, \\ & J_2^y\Delta y + J_2^u\Delta u \geq -c_2, \end{array} \tag{19}$$

where Δy and Δu denote the predicted change in the state and control parameters respectively.

5.1 Structured Reduced-Hessian methods

Under the assumption that J_1^y is invertible, the equality constraints $J_1^y\Delta y + J_1^u\Delta u = -c_1$ can be used to express Δy as a function of Δu, with

$$\Delta y = -(J_1^y)^{-1}c_1 - (J_1^y)^{-1}J_1^u\Delta u. \tag{20}$$

This expression may be used to establish the identity

$$\begin{pmatrix}\Delta y\\ \Delta u\end{pmatrix} = Y_1c_1 + Z_1\Delta u,$$

where Z_1 and Y_1 are $n\times n_2$ and $n\times n_1$ matrices such that

$$Y_1 = \begin{pmatrix}-(J_1^y)^{-1}\\ 0\end{pmatrix}, \quad\text{and}\quad Z_1 = \begin{pmatrix}-(J_1^y)^{-1}J_1^u\\ I\end{pmatrix}.$$

Substituting for Δy in the definition of the QP subproblem gives

$$\begin{array}{ll} \underset{\Delta u\in\mathbb{R}^{n_2}}{\text{minimize}} & (g+HY_1c_1)^TZ_1\Delta u + \frac{1}{2}\Delta u^TZ_1^THZ_1\Delta u \\ \text{subject to} & J_2Z_1\Delta u \geq -J_2Y_1c_1 - c_2, \end{array} \tag{21}$$

which is a problem involving only the control variables Δu. Once Δu has been calculated, the state increment is determined from (20). The dimensions of the Hessian and constraints for this problem are $n_2\times n_2$ and $m_2\times n_2$, respectively. In the common situation where the number of control variables is significantly smaller than the number of state variables, then $n_2 \ll n$, and the dimension of this QP is considerably smaller than that of (19). Tanartkit and Biegler [33] propose forming the matrices $Z_1^THZ_1$ and J_2Z_1 explicitly and solving the subproblem (21) using a dense QP algorithm. A feature of this method is that any additional structure in the derivatives matrix J_2 is lost (the extreme case of this is when the inequalities $c_2(y,u)$ consist of only simple bounds. In this case the bounds are transformed into dense general inequalities.

Other approaches involve solving (21) with different choices for $g + HY_1c_1$ (see, e.g., Beltracchi and Gabriele [2] and Schultz [31]). A structured problem of the form (21) lies at the heart of many methods for optimal control. Dennis, Heinkenschloss and Vicente [10] propose a trust-region SQP method for the case where the only inequalities are simple bounds.

5.2 Using approximate Jacobians

A disadvantage of the straightforward implementation of multiple shooting is that it is necessary to compute the n_y^2 Jacobian elements at each major iteration. Gill, Jay and Petzold [15] propose an algorithm that has the numerical complexity of single shooting and the stability and robustness of multiple shooting. The first step is to transform the nonlinear constraints of (18) and define the problem

$$\begin{array}{ll} \underset{y\in\mathbb{R}^{n_1},u\in\mathbb{R}^{n_2}}{\text{minimize}} & F(y,u) \\ \text{subject to} & c_1^M(y,u)=0, \quad c_2^M(y,u)\geq 0, \end{array} \tag{22}$$

where the m_1-vector $c_1^M(y,u)$ and m_2-vector $c_2^M(y,u)$ form elements of the vector c such that

$$c^M(x)=M(x)c(x), \quad \text{with} \quad M(x)=\begin{pmatrix} (J_1^y)^{-1} & 0 \\ -J_2^y(J_1^y)^{-1} & I \end{pmatrix}.$$

The idea is now to use an SQP algorithm to solve the problem (22). The only modification to the standard SQP strategy is that an *approximate* Jacobian $\widetilde{J}$ is used in place of $J(x)$. The approximation is given by

$$\widetilde{J}(x)=M(x)J(x)=\begin{pmatrix} I & (J_1^y)^{-1}J_1^u \\ 0 & J_2^u-J_2^y(J_1^y)^{-1}J_1^u \end{pmatrix}.$$

The entries for the unit columns of this Jacobian are available with no work. Moreover, the products $(J_1^y)^{-1}J_1^u$ used to define the other columns can be found with single-shooting complexity (but multiple shooting stability) using directional sensitivity techniques (see Maly and Petzold [26]).

The original variables of (18) are not subject to the transformation, which implies that only minor modifications need to be made to general-purpose SQP methods in order to accommodate the approximate Jacobian (one implementation is based on the SQP code SNOPT of Gill, Murray and Saunders [16]). In addition, linear constraints and selected nonlinear inequalities can be left untransformed, allowing their structure to be exploited by the solver in the usual way. For more details, see Gill, Jay and Petzold [15].

Acknowledgment

We extend sincere thanks to Laurent Jay and Linda Petzold for many discussions on ordinary differential equations.

References

[1] Ascher, U. M.; Mattheij, R. M. M.; Russell, R. D.: *Numerical Solution of Boundary Value Problems for Ordinary Differential Equations.* Classics in Applied Mathematics, 13. Society for Industrial and Applied Mathematics (SIAM) Publications, Philadelphia, PA, 1995. ISBN 0-89871-354-4.

[2] Beltracchi, J. T.; Gabriele, G. A.: *An investigation of using a RQP based method to calculate parameter sensitivity derivatives.* In: Recent advances in multidisciplinary analysis and optimization. NASA conference publication 3031, Part 2, proceedings of a symposium held in Hampton, Virginia, September 28–30, 1988, 1988.

[3] Betts, J. T.: *Issues in the direct transcription of optimal control problems to sparse nonlinear programs.* In: Bulirsch, R.; Kraft, D. (editors): *Control Applications of Optimization*, volume 115 of Internat. Ser. Numer. Math., pages 3–17, Basel, 1994. Birkhäuser.

[4] Betts, J. T.; Frank, P. D.: *A sparse nonlinear optimization algorithm.* J. Optim. Theory and Applics., 82, 519–541, 1994.

[5] Betts, J. T.; Huffman, W. P.: *The application of sparse nonlinear programming to trajectory optimization.* J. of Guidance, Control, and Dynamics, 14, 338–348, 1991.

[6] Brenan, K. E.: *Differential-algebraic equations issues in the direct transcription of path constrained optimal control problems.* Ann. Numer. Math., 1(1–4), 247–263, 1994.

[7] Buchauer, O.; Hiltmann, P.; Kiehl, M.: *Sensitivity analysis of initial value problems with application to shooting techniques.* Numer. Math., 67, 151–159, 1994.

[8] Bulirsch, R.; Nerz, E.; Pesch, H. J.; von Stryk, O.: *Combining direct and indirect methods in optimal control: Range maximization of a hang glider.* In: Bulirsch, R.; Miele, A.; Stoer, J.; Well, K. H. (editors): *Calculus of Variations, Optimal Control Theory and Numerical Methods.*, volume 111 of *Internat. Ser. Numer. Math.*, pages 273–288, Basel, 1993. Birkhäuser.

[9] Büskens, C.; Maurer, H.: *Sensitivity analysis and real-time control of nonlinear optimal control systems via nonlinear programming.* In this volume

[10] Dennis, J. E. (Jr.); Heinkenschloss, M.; Vicente, L. N.: *Trust-region interior SQP methods for a class of nonlinear programming problems.* Technical Report 94-45 (revised 1996), Dept of Mathematical Sciences, Rice University, Houston, TX, 1994.

[11] Dickmanns, E.D.; Well, K. H.: *Approximate solution of optimal control problems using third-order Hermite polynomial functions.* In: Proc. 6th Technical Conference on Optimization Techniques. Berlin, Heidelberg, New York, London, Paris and Tokyo, 1975. Springer Verlag.

[12] Eldersveld, S. K.: *Large-scale sequential quadratic programming algorithms.* PhD thesis, Department of Operations Research, Stanford University, Stanford, CA, 1991.

[13] Eldersveld, S. K.; Gill, P. E.: *Preconditioned limited-storage quasi-Newton methods for large-scale constrained optimization.* Presented at the Fifth SIAM Conference on Optimization, Victoria, British Columbia, May 20–22, 1996.

[14] Fletcher, R.: *An ℓ_1 penalty method for nonlinear constraints.* In: Boggs, P. T.; Byrd, R.H.; Schnabel, R. B. (editors): *Numerical Optimization 1984*, pages 26–40, Philadelphia, 1985. SIAM.

[15] Gill, P. E.; Jay, L. O.; Petzold, L.: *An SQP method for the optimization of dynamical systems.* To appear.

[16] Gill, P. E.; Murray, W.; Saunders, M. A.: SNOPT: *An SQP algorithm for large-scale constrained optimization.* Numerical Analysis Report 97-1, Department of Mathematics, University of California, San Diego, La Jolla, CA, 1997.

[17] Gill, P. E.; Murray, W.; Saunders, M. A.; Wright, M. H.: *Sparse matrix methods in optimization.* SIAM J. on Scientific and Statistical Computing, 5, 562–589, 1984.

[18] Gill, P. E.; Murray, W.; Wright, M. H.: *Practical Optimization.* Academic Press, London and New York, 1981. ISBN 0-12-283952-8.

[19] Gritsis, D. M.; Pantelides, C. C.; Sargent, R. W. H.: *Optimal control of systems described by index two differential-algebraic equations.* SIAM J. Sci. Comput., 16, 1349–1366, 1995.

[20] Hairer, E.; Nørsett, S. P.; Wanner, G.: *Solving Ordinary Differential Equations I.* Springer Verlag, New York, Berlin and Heidelberg, second revised edition, 1993. ISBN 0387-56670-8.

[21] Hargraves, C. R.; Paris, S. W.: *Direct trajectory optimization using nonlinear programming and collocation.* J. of Guidance, Control, and Dynamics, 10, 338–348, 1987.

[22] Kiehl, M.: *Parallel multiple shooting for the solution of initial value problems.* Parallel Comput., 20, 275–295, 1994.

[23] Kraft, D.: *On converting optimal control problems into nonlinear programming problems.* In: Schittkowski, K. (editor): *Computational Mathematical Programming*, NATO ASI Series F: Computer and Systems Sciences 15, pages 261–280. Springer Verlag, Berlin and Heidelberg, 1985.

[24] Lamour, R.: *A well-posed shooting method for transferable DAEs.* Numer. Math., 59, 815–829, 1991.

[25] Leitmann, G.: *The Calculus of Variations and Optimal Control.* Mathematical Concepts and Methods in Science and Engineering, 24. Plenum Press, New York and London, 1981. ISBN 0-306-40707-8.

[26] Maly, T.; Petzold, L. R.: *Numerical methods and software for sensitivity analysis of differential-algebraic systems.* to appear in Applied Numerical Mathematics, 1996.

[27] Pantelides, C. C.; Sargent, R. W. H.; Vassiliadis, V. S.: *Optimal control of multistage systems described by high-index differential-algebraic equations.* Internat. Ser. Numer. Math., 115, 177–191, 1994.

[28] Pesch, H. J.: *Real-time computation of feedback controls for constrained optimal control problems.* Optimal Control Appl. Methods, 10, 129–171, 1989.

[29] Petzold, L.; Rosen, J. B.; Gill, P. E.; Jay, L.O.; Park, K.: *Numerical optimal control of parabolic PDEs using DASOPT.* In: Biegler, L. T.; Coleman, T. F.; Conn, A. R.; Santosa, F. N. (editors): *Large Scale Optimization with Applications, Part II: Optimal Design and Control*, volume 93 of *IMA Volumes in Mathematics and its Applications.* Springer Verlag, Berlin, Heidelberg and New York, 1997.

[30] Schittkowski, K.: NLPQL: *A Fortran subroutine for solving constrained nonlinear programming problems.* Ann. Oper. Res., 11, 485–500, 1985/1986.

[31] Schultz, V. H.: *Reduced SQP methods for large-scale optimal control problems in DAE with application to path planning problems for satellite mounted robots.* PhD thesis, University of Heidelberg, 1996.

[32] Schulz, V. H.; Bock, H.-G.; Steinbach, M. C.: *Exploiting invariants in the numerical solution of multipoint boundary value problems for DAE.* submitted to SIAM J. Sci. Comput., 1996.

[33] Tanartkit, P.; Biegler, L. T.: *Stable decomposition for dynamic optimization.* Ind. Eng. Chem. Res., pages 1253–1266, 1995.

[34] Tjoa, I.-B.; Biegler, L. T.: *A reduced SQP strategy for errors-in-variables estimation.* Comput. Chem. Eng., 16, 523–533, 1992.

[35] von Stryk, O.: *Numerical solution of optimal control problems by direct collocation.* In: Bulirsch, R.; Miele, A.; Stoer, J.; Well, K. H. (editors): *Control Applications of Optimization*, volume 111 of *Internat. Ser. Numer. Math.*, pages 129–143, Basel, 1993. Birkhäuser.

[36] von Stryk O.; Bulirsch, R.: *Direct and indirect methods for trajectory optimization.* Ann. Oper. Res., 37(1–4), 357–373, 1992.

International Series of Numerical Mathematics
Vol. 124, © 1998 Birkhäuser Verlag, Basel

Predictor-Corrector Continuation Method for Optimal Control Problems

Ernst Grigat* Gottfried Sachs*

Abstract. A new Predictor-Corrector homotopy method is proposed based on a variable order predictor and a stability oriented steplength control. The task of the corrector is performed by an adapted Multiple Shooting implementation including a variable grid strategy. Predictors of different orders are combined by two alternative order controls based on minimizing computational effort. A relaxed homotopy stepsize formula based on the *Newton-Kantorovich* theorem is developed that gives way to a significant speedup in the solution of optimal control problems by continuation methods. The proposed method is applied to an example from flight mechanics. The ascent of a Sänger-type hypersonic vehicle is optimized with respect to fuel consumption. Defining the maximum admissible dynamic pressure $\bar{q}_{\max}$ to be a natural homotopy parameter a transition from $\bar{q}_{\max} =$ 70000 Pa to $\bar{q}_{\max} = 60000$ Pa is efficiently performed by the described continuation method.

1. Introduction

Many optimization problems arising from scientific research or industrial applications can be formulated as *Optimal Control Problems. Variational Calculus* and *Optimal Control Theory* provide the theoretical framework to treat these problems numerically by transforming them into *Multipoint-Boundary-Value Problems.* Efficient techniques like the *Multiple Shooting Method* can be applied to generate a solution of the optimization problem. Difficulties for generating a solution may exist where there are no initial data for the underlying Newton-Method to ensure convergence. Therefore by introducing a *homotopy parameter* the original problem is embedded into a family of problems where a solution for a certain member of that family exists or is easily attainable. The solution of the original problem is then defined as the endpoint of a homotopy starting from the problem already solved. In practice so-called *Predictor-Corrector* methods are commonly used since they combine the advantages of two different theoretical approaches and avoid their disadvantages.

For classifying possible predictors an order definition for predictor methods introduced by *Deuflhard* can be used. Based on this definition it is possible to develop a class of predictors of different order related to the well-known *Runge-Kutta* integration formulas. Combining these by a suitable order control results in a variable order predictor.

Applying the *Newton-Kantorovich* theorem is a common approach to derive a steplength control for the homotopy parameter with the aim of ensuring stability of the corrector by assuring the prediction guess to lie within the local domain of convergence

*TU München, Lehrstuhl für Flugmechanik und Flugregelung, Boltzmannstr. 15, D-85747 Garching, email: grigat@lfm.mw.tu-muenchen.de, sachs@lfm.mw.tu-muenchen.de

of the underlying Newton-Method. Relaxing that postulation leads to significantly larger homotopy steps (in practical tests by several orders of magnitude) while the (relaxed) Newton method converges nevertheless in most cases. In case of divergence reduction strategies for the steplength can be applied.

2. Solution of optimal control problems by multiple shooting

The problem under consideration is to mimimize a given function $J[u]$ subject to a set of boundary conditions and state constraints described in the following.

$J[u] = \Phi(t_f, x(t_f)) \longrightarrow min$		Performance criterion
$\dot{x} = f(x, u)$	$0 \leq t \leq t_f$	State equations
$u(t)$		Control vector
$r(t_f, x(0), x(t_f)) = 0$		Boundary conditions
$S(s_k, x(s_k^-), x(s_k^+)) = 0$		Switching conditions
$C(x) \leq 0$	$0 \leq t \leq t_f$	State constraints

It may be noted that problems with more general state constraints can be handled with the calculus derived from the definition given above. This also applies to cases of state inequality constraints containing mixed constraints $C(x, u)$ ($u(t)$ and $x(t)$ appear explicitly in C) as well as cases of pure control constraints $C(u)$.

The addressed optimal control problem can be reformulated as Multipoint-Boundary-Value Problem (MPBVP) which can be treated numerically by the Multiple Shooting Method [3, 11] schematically shown in Fig.1. The implementation used in this case [7]

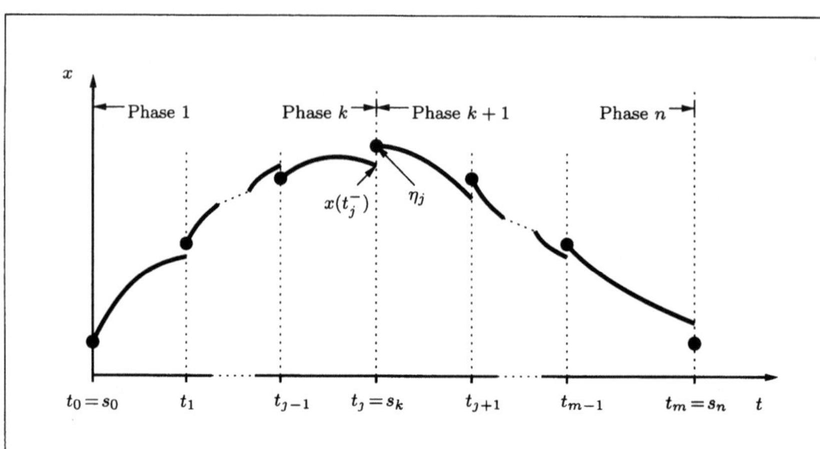

Figure 1: Multiple shooting method

is based on a modified Newton method applied to a problem of the type

$$F(\eta_0, \ldots, \eta_m, s_0, \ldots, s_n) =: F(Z) = 0$$

where $0 = t_0 < t_1 < \cdots < t_{m-1} < t_m = t_f$ denote the multiple shooting nodes and the η_i are guesses for the associated $x(t_i)$. As a special feature of this algorithm the switching points $0 = s_0 < s_1 < \cdots < s_{n-1} < s_n = t_f$ defining the n so called *integration phases* are inserted into the multiple shooting grid simplifying their computational treatment. This can be done by formally defining a mesh consisting only of switching points where the originally ordinary nodes are specified by continuity conditions for all dependent variables and $t_j = t_j^* =$ const. for the time variable.

3. Variable multiple shooting grid

One problem often occuring during Newton iterations is the situation when a switching point is moved over a gridpoint. In many cases this leads to oscillations of the switching point around the gridpoint and ends up with a program breakdown. A graphical approach to this problem [5] turned out to be successful in many cases but the necessary interactive manipulations are not satisfying with respect to an intended high level of user comfort. Using an appropriately constructed variable grid one can avoid the problems mentioned above without stopping the program run for some modifications.

After choosing an initial grid π_i

$$0 = t_0 < t_1 < t_2 < \quad . \quad . \quad . \quad < t_{m-2} < t_{m-1} < t_m = t_f$$

which can be done e.g. by a strategy based on a perturbation approach [2] this mesh is transformed into a so-called *reference grid* π_r

$$0 = \tau_0 < \tau_1 < \tau_2 < \quad . \quad . \quad . \quad < \tau_{m-2} < \tau_{m-1} < \tau_m = m$$

remaining invariant during Newton iterations (Fig. 2). So the relative distances be-

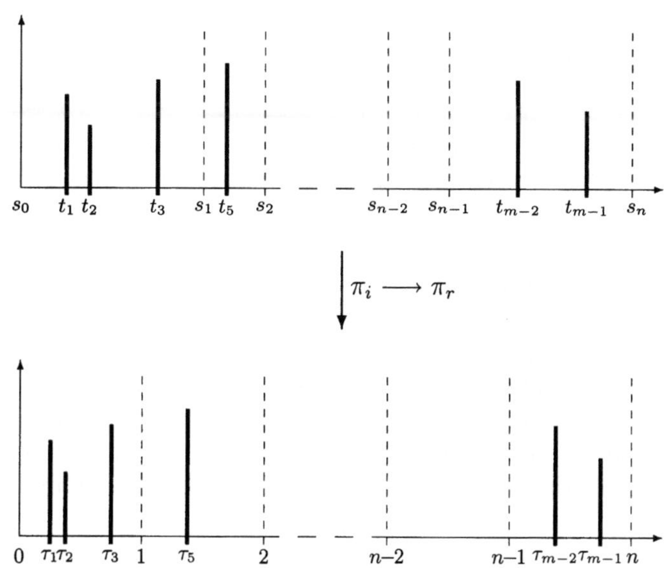

Figure 2: Transformation of initial grid

tween ordinary gridpoints are constant within an integration phase. With the notation introduced above, the transformation $t_j \longrightarrow \tau_j$ can be written formally

$$\tau_j = (k-1) + \frac{t_j - s_{k-1}}{s_k - s_{k-1}} \qquad\qquad s_{k-1} < t_j \le s_k$$

The formal switching condition for a constant node then has to be replaced by

$$s_{k-1} + (\tau_j - (k-1))(s_k - s_{k-1}) - t_j =: s_{k-1} + (\tau_j - (k-1))\Delta s_k - t_j \stackrel{!}{=} 0$$

After each Newton iteration step a new multiple shooting grid π_i^{new} can be defined by re-transforming the reference grid with new values Δs_k^{new}. Fig. 3 sums up the transition $\pi_i \longrightarrow \pi_i^{new}$ in the course of a Newton iteration step. It may be noted that

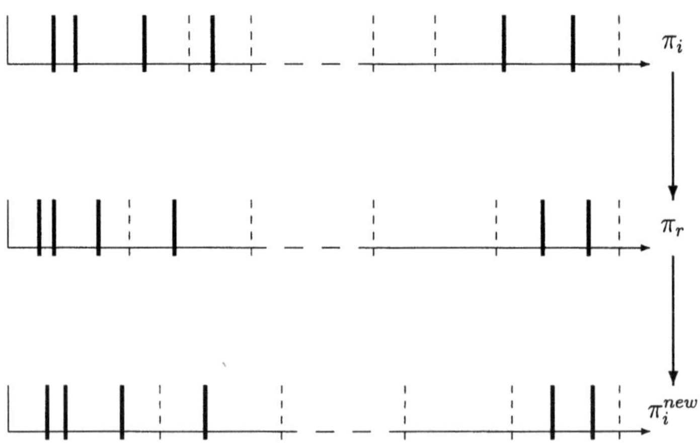

Figure 3: Grid transformation during Newton step

this grid strategy is just an auxiliary means of avoiding some special difficulties but cannot guarantee stability or even convergence of multiple shooting. These tasks will be discussed in the following.

4. Variable order continuation method

As mentioned before, Multiple Shooting needs adequate initial data to ensure convergence. Besides direct approaches [12], [13] the so-called continuation or homotopy method turned out to be a promising way to construct a sufficiently accurate starting trajectory. By introducing a so-called *homotopy parameter* (eventually normed to lie within the unit interval) the original problem is embedded into a family of problems.

$$F(Z) \stackrel{!}{=} 0$$
$$\downarrow$$
$$H(Z(\tau);\tau) \stackrel{!}{=} 0$$

with $H(Z(1);1) \equiv F(Z)$. With an already known or easily attainable solution for $H(Z(0);0)$ the original problem can be solved by a transition from $\tau = 0$ to $\tau = 1$ which can be basically done in two different ways.

The first approach is called *classical continuation method* where some intermediate points are defined for the homotopy parameter. Having produced a solution for a certain value of τ this serves as initial guess for the problem belonging to the following value of τ. This procedure can suffer from a huge amount of intermediate points needed to provide stability.

The *continuous approach* uses the *Davidenko Differential Equation* arising from the differentiated homotopy operator H.

$$\frac{dZ}{d\tau} = -\left[\frac{\partial H(Z(\tau);\tau)}{\partial Z}\right]^{-1} \cdot \frac{\partial H(Z(\tau);\tau)}{\partial \tau}$$

Integrating this differential equation generates the desired solution, but for sensitive problems (and realistic problems are very sensitive in many cases) severe numerical obstacles occur.

A promising compromise is the *hybrid approach*, a combination of the two methods described above. Starting from an already known solution for a certain value of τ one forward integration step produces a guess or predictor which is then projected onto the solution manifold H in a correcting step (Fig. 4). Therefore this class of methods

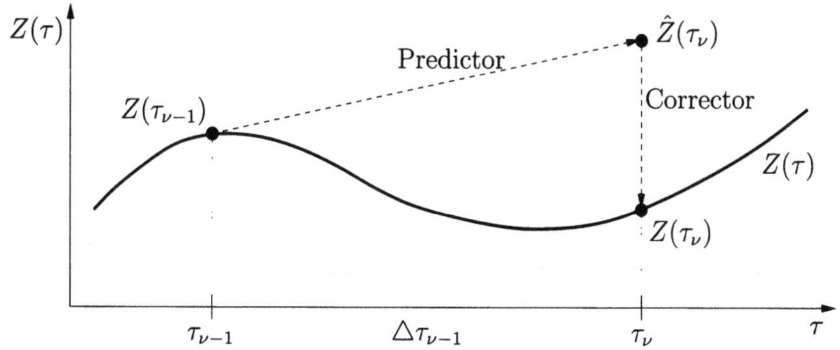

Figure 4: Predictor-Corrector continuation method

is called *Predictor-Corrector Continuation Methods.* In our context Multiple Shooting serves as corrector. In order to construct a suitable method of that kind one has to provide efficient algorithms for two more components.

- Predictor Method
- Stepsize Control

In preparation of these tasks we repeat an order definition by *Deuflhard*

Definition: Order of a Homotopy Method. *A homotopy method is called to be* of order *p, if $\exists\ \bar{\tau} > 0$ and a constant $\eta_p \geq 0$ such that for the predictor generated guess $\hat{Z}(\tau)$ holds*

$$\|\hat{Z}(\tau) - Z(\tau)\| \leq \eta_p \cdot \tau^p \qquad \forall\ \tau \in [0, \bar{\tau}]$$

Taking into consideration the well-known *Runge-Kutta* integration formulas [6] one can easily prove the following

Lemma. *Let there be a Runge-Kutta integration scheme of order p taken as predictor in a continuation method described above. Then the homotopy method is of order p+1.*

With these statements a huge set of possible predictors is available. Combining several predictors of different order by an appropriately constructed order control results in a variable order continuation method which allows better adaption of the solution algorithm to the problem under consideration.

5. Order control for variable order homotopy method

When designing a continuation method it is essential to ensure *stability* and *efficiency* for the algorithm. While the first requirement can be fulfilled by developing a stability oriented stepsize control based on the well-known *Newton-Kantorovich* theorem [8] the second problem can be addressed by a suitably designed order control. So the primary task of an order control developped in this context is to simply reduce (minimize) computational complexity. To this end two different strategies are presented.

In order to evaluate the complexity of a homotopy step, one has to define an appropriate measure independent from computer architecture. Analysing the structure of the Multiple Shooting method used here it can be shown that all computations except linear algebra can be significantly measured by the number of evaluations N_f of the right-hand side f describing the underlying system dynamics. In a sequence of numerical experiments [9] the solution of the linear system in the Newton method took about five percent of the time needed for computing the multiple shooting matrix. A close look at the set of predictor routines proposed (basically Runge-Kutta formulas) shows that the computational effort can be measured analogously by N_f. Therefore N_f can be stated to be a reasonable measure for the complexity of a homotopy step.

An obvious possibility to guess the computational effort needed for a following homotopy step consists in extrapolation of former steps. To this end we distinguish $N_f^{pre,p}(\nu)$ as measure for the complexity of an predictor of order p and $N_f^{cor}(\nu)$ for the corrector in the ν-th homotopy step and store all these values. For each p the complexities (normed with regard to the stepsize) of all former steps with that order can be extrapolated to get a good guess for the effort of the following step with that order. Comparison of the several estimates gives hint what order selection might be best. This approach can be called *extrapolation order control.*

In opposition to that (with respect to the homotopy path) global point of view in some situations it might be favourable to judge the effect of an order selection more locally. For these cases order selection rules should be available that can be summarized under the name *trend order control.* The last two homotopy steps are compared with regard to their complexity and by a dedicated selection scheme the order for the following step is determined. Due to space a detailed description of that scheme is omitted here. Roughly speaking, in most cases a decrease of computational effort leads to a constant or even decreased order whereas an increase effects an increase of order mostly. A general recommendation of what order control is best for a certain problem cannot be given here.

6. Relaxed homotopy stepsize control

As already mentioned, stability in Predictor-Corrector continuation methods is often tried to be assured by an appropriate stepsize control. When Newton method is used as corrector the well-known *Newton-Kantorovich* theorem can be applied to calculate the actual stepsize $\Delta\tau$ such that the new prediction value lies within the local radius

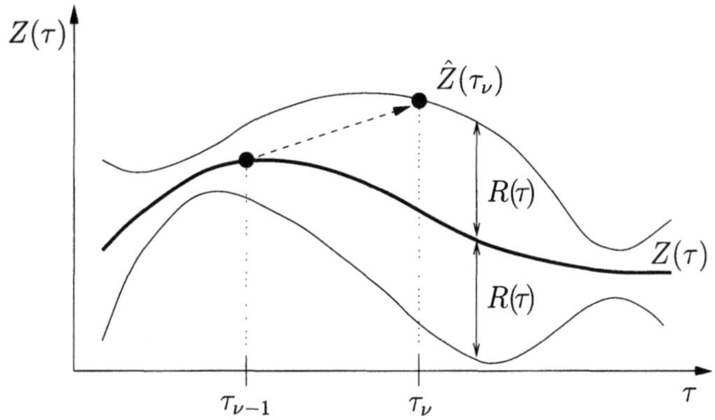

Figure 5: Predictor into local radius of convergence

of convergence $R(\tau)$ as shown in Fig.5. With local estimates $\hat{\omega}$ (Lipschitz constant for H) and $\hat{\eta}_p$ [4] the stepsize $\Delta\tau$ for the ν-th homotopy step is calculated by

$$\Delta\tau_\nu = \left(\frac{\sqrt{2}-1}{\hat{\eta}_p \cdot \hat{\omega}}\right)^{\frac{1}{p}} = \left(\frac{\sqrt{2}-1}{\|Z^{*(\nu-1)} - Z_0^{(\nu-1)}\| \cdot \hat{\omega}}\right)^{\frac{1}{p}} \cdot \Delta\tau_{\nu-1}$$

with prediction guess $Z_0^{(\nu-1)}$ and solution $Z^{*(\nu-1)}$. In practice the resulting stepsizes very often turn out to be quite small due to the small radius of convergence. Relaxing the postulation for the prediction guess to lie within a wider tube of bandwidth *tol* instead of $R(\tau)$ shows to be a reasonable approach for obtaining larger stepsizies (Fig. 6). It can be proved that with this relaxed condition the optimal stepsize can be calculated by the new formula

$$\Delta\tau_\nu = \left(\frac{\sqrt{2 \cdot tol \cdot \hat{\omega} + 1} - 1}{\|Z^{*(\nu-1)} - Z_0^{(\nu-1)}\| \cdot \hat{\omega}}\right)^{\frac{1}{p}} \cdot \Delta\tau_{\nu-1}.$$

In numerical experiments the improvement of this new formula showed up to be of several orders of magnitude for solving problems with a computer. A further possiblity for a significant speedup consists in relaxing the stop criterion in a related manner. The prediction guess is not projected exactly on the solution manifold $H(\tau)$ (which is not possible anyway using digital calculators) but just up to an accuracy *tol* replacing H by a tube of bandwidth *tol* around H.

The drawback of this relaxation strategy is a possibly not converging corrector, a disadvantage that exists anyway as the original formula uses only guesses for $\hat{\omega}$ and $\hat{\eta}_p$. So as a remedy for a possibly diverging corrector, a suitable stepsize reduction

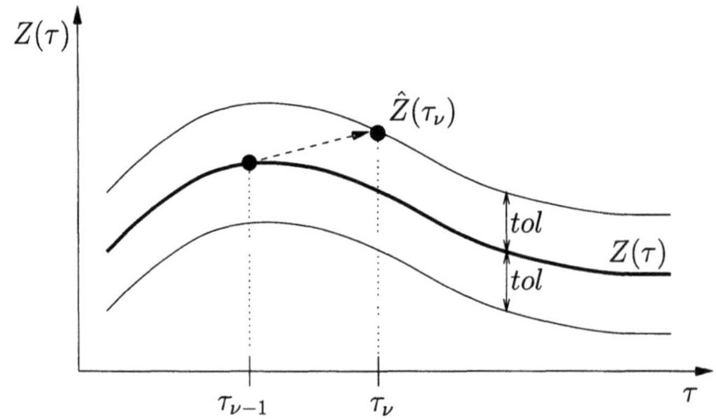

Figure 6: Predictor into tube of bandwidth *tol*

strategy has to be developed. As the value of *tol* in most cases determines some special order of magnitude the first reduction approach consists of dividing *tol* by 10. This will be continued until convergence of the corrector is attained or a bound $tol = 1/2\hat{\omega}$ is passed. From this point a reduction strategy due to *Deuflhard* [4] can be applied. It may be stated that considering the analogy between integration methods for ordinary differential equations and the class of predictors used here a suitable starting step size for the homotopy parameter τ can be easily derived from corresponding algorithms [14].

7. Application: ASCENT of a hypersonic aircraft

The effect of the continuation method proposed is demonstrated solving an optimal control problem from flight mechanics. The ascent of a hypersonic aicraft in the vertical plane is to be minimized with respect to fuel consumption. Modelling of the aircraft under consideration is taken from [1], [10]. With control variables α (angle of attack) and δ (throttle setting) dynamics are described by

$$\begin{aligned}
\dot{h} &= V \sin\gamma \\
\dot{V} &= \frac{T\cos\alpha - D}{m} - g\sin\gamma \\
\dot{\gamma} &= \frac{T\sin\alpha + L}{Vm} + \cos\gamma\left(\frac{V}{r} - \frac{g}{V}\right) \\
\dot{m} &= -T\dot{b}_T
\end{aligned}$$

where the aerodynamic forces D and L, the thrust T and fuel consumption characteristics $\dot{b}_T$ are described by intricate dependencies on state and control variables. The optimization objective of a minimium fuel consumption can be formulated as minimization of weight loss defining the performance criterion to

$$\Phi := m|_{t_0} - m|_{t_f}.$$

For technical reasons a physical constraint is imposed limiting the maximum allowed dynamic pressure to $\bar{q}_{\max} = 60000$ Pa describing a first order state constraint for the optimal control problem. As a trajectory for $\bar{q}_{\max} = 70000$ Pa exists from former computations, a way is given to generate a solution of the original problem by continuation methods. Both order controls (extrapolation and trend) proposed are applied testing extrapolation order control for two bandwidths *tol*. Runtime analysis shows superiority of extrapolation over trend control in this case mainly caused by a significantly lower number n_{hom} of homotopy steps. Moreover Tab. 1 gives an impression of possible

	n_{hom}	T_{pre} [s]	T_{cor} [s]	T_{hom} [s]
trend ($tol = 10^1$)	1028	470965	206605	677570
extrapolation ($tol = 10^1$)	299	12637	79267	91904
extrapolation ($tol = 10^0$)	3551	342614	2421135	2763749

Table 1: Runtime Analysis

speedup factors by relaxation of bandwidth *tol*.

It may be stated that the present implementation offers predictor routines of order 1-5, a set easily enlarged by Runge-Kutta formulas of higher order. Fig. 7 shows the selection of the two order strategies during homotopy run. Using trend order control

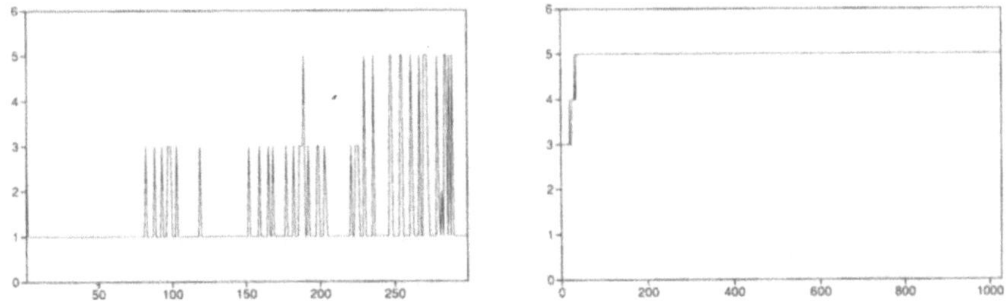

Figure 7: Extrapolation (left) and trend (right) order control

the highest possible order is selected in most cases, another reason for the relatively high complexity displayed in Tab. 1. Extrapolation Order Control can be interpreted as some kind of sensitivity indicator for that problem. The more the dynamical system is restricted the higher the order selection, which means that better prediction guesses are needed. Nevertheless it may be stressed that the results obtained for this special example cannot be generalized. Other systems may prefer other order controls.

8. Conclusion

A new continuation method of Predictor-Corrector type has been presented. The corrector method consists of a dedicated Multiple Shooting implementation improved with regard to stability by an adaptive grid. According to an order definition for predictor routines by *Deuflhard* several predictors of different order are presented. These are combined by two alternative order controls to a predictor of *variable* order. For the development of an efficient homotopy stepsize control a stability condition for the

Newton method has been relaxed leading to speedups of orders of magnitude. The new method has been applied to an example from flight mechanics, i.e. the ascent of a hypersonic aircraft.

References

[1] Bayer, R.: *Optimierung von Flugbahnen der Unterstufe eines luftatmenden Raumtransporters*, Dissertation, Technische Universität München, 1993.

[2] Bock, H. G.: *Randwertproblemmethoden zur Parameteridentifizierung in Systemen nichtlinearer Differentialgleichungen*, Dissertation, Universität Bonn, 1985.

[3] Bulirsch, R.: *Die Mehrzielmethode zur numerischen Lösung von nichtlinearen Randwertproblemen und Aufgaben der optimalen Steuerung.* Report der Carl-Cranz-Gesellschaft e.V, Oberpfaffenhofen, 1971. Nachdruck: Report des Sonderforschungsbereichs 255, Lehrstuhl für Höhere Mathematik und Numerische Mathematik, Technische Universität München (1993).

[4] Deuflhard, P.: *A Stepsize Control for Continuation Methods and its Special Application to Multiple Shooting Techniques*, Numerische Mathematik, 33 (1979), pp. 115–146.

[5] Grigat, E.; Sachs, G.; Kiehl, M.; Bulirsch, R.: *Graphic Interface for the Solution of Optimal Control Problems.* In: XI. Herbstschule Variationsrechnung, optimale Prozesse und Anwendungen, 1994, pp. 22–25. Preprint-Reihe Mathematik – Nr. 1 – 1994, Greifswald.

[6] Hairer, E.; Nørsett, S. P.; Wanner, G.: *Solving Ordinary Differential Equations I: Nonstiff Problems*, Springer-Verlag, Berlin, Heidelberg, New York, 1987.

[7] Hiltmann, P.: *Numerische Lösung von Mehrpunkt-Randwertaufgaben und Aufgaben der optimalen Steuerung mit Steuerfunktionen über endlichdimensionalen Räumen*, Dissertation, Technische Universität München, 1990.

[8] Kantorovich, L. V.; Akilov, G. P.: *Functional Analysis*, Pergamon Press, Oxford, New York, Toronto, Sydney, Paris, Frankfurt, 1982.

[9] Mehlhorn, R.: *Parallelisierung der Mehrzielmethode und Implementierung auf einem iPSC-Hypercube mit Anwendungen in der Flugbahnoptimierung*, Diplomarbeit, Lehrstuhl für Flugmechanik und Flugregelung, Technische Universität München, 1993.

[10] Sachs, G,; Dinkelmann, M.: *Optimal Three-Dimensional Cruise and Ascent of a Two-Stage Hypersonic Vehicle*, in 46th IAF Congress, Oslo, October 1995. IAF-95-A.2.03.

[11] Stoer, J.; Bulirsch, R.: *Einführung in die Numerische Mathematik 2*, Springer-Verlag, Berlin, Heidelberg, New York, 1990.

[12] von Stryk, O.: *Numerische Lösung optimaler Steuerungsprobleme: Diskretisierung, Parameteroptimierung und Berechnung der adjungierten Variablen*, Dissertation, Technische Universität München, 1994.

[13] von Stryk, O.; Bulirsch, R.: *Direct and Indirect Methods for Trajectory Optimization*, Annals of Operations Research, 37 (1992), pp. 357–373.

[14] Watts, H. A.: *Starting Step Size for an ODE Solver*, Journal of Computational and Applied Mathematics, 9 (1983), pp. 177–191.

Applications to Mechanical and Aerospace Systems

International Series of Numerical Mathematics
Vol. 124,

Time Invariant Global Stabilization of a Mobile Robot

Harald Abeßer Michael Katzschmann Joachim Steigenberger *

Abstract. In this paper we present time invariant controllers which globally asymptotically stabilize a simple mobile robot to an equilibrium posture. For both kinematics and dynamics the construction follows a strategy which in fact allows for a family of controllers parameterized by the initial state and ensures smoothness of each motion.

1. Introduction

From the vast literature on wheeled mobile robots (see [4] for references) it is well known that both kinematical and dynamical models of such robots appear as nonholonomic control systems. Due to a theorem by Brockett [2] they cannot be stabilized to an equilibrium point by time invariant state feedback which is smooth on a neighborhood of that point. Among the diverse proposals which have been given in order to manage this unpleasant feature mainly two different types of strategies can be observed:

(i) Take the smooth system as it is (up to feasible transformations), accept Brockett's conclusion and, thus, above all search for time variant or/and nonsmooth feedback. Most of these investigations are based on kinematical models in chained form. Pomet [6] proposes periodic time varying controllers while Canudas de Wit et al. [3], [5] introduce and experimentally check a hybrid controller.

(ii) In order to succeed with smooth time invariant feedback nevertheless one assumes the premise in Brockett's theorem to be false. Astolfi [1] first transforms by some $\sigma-$process the given to-be-stabilized smooth system to a nonsmooth system for which he constructs a smooth feedback that "stabilizes a singular point". The feedback finally transforms back to a nonsmooth stabilizing controller.

The basic idea of the strategy we are going to describe is to restrict the system to some proper subset M of the state space in such a way that

(a) the interesting equilibrium point p becomes a boundary point of M (so there is no full neighborhood of p and Brockett's premise is not hit anymore) and

(b) one can find on M a smooth state feedback which makes the positive orbits remain in M and have $\{p\}$ as their $\omega-$limit set.

If one had several (hopefully few) such subsets M which form a covering of the state space then, finally, the smooth (on the subsets) feedback laws combine to a "piecewise smooth" feedback on the state space. Practically, each initial state selects one of the

*Technical University of Ilmenau, Institute of Mathematics, PSF 327, D-98684 Ilmenau, Germany, e-mail: jstei@mathematik.tu-ilmenau.de

feedback branches and then, following the respective feedback modified dynamics, the system moves (monotonically) to the equilibrium point p as t tends to infinity.

Formally, our procedure is as follows.

Let $\sum : \dot{x} = f(x,u),\ x \in \mathbb{R}^n,\ u \in \mathbb{R}^m$, be a smooth control system with equilibrium $x_1,\ f(x_1,0) = 0$. Suppose that f does not allow a smooth feedback $u = \alpha(x)$ stabilizing $\sum$ to x_1. Then look for feedback laws $\alpha(\cdot, x_o)$ which depend on the initial state x_o in such a way that

$$\alpha(x, x_o) = \sum_{\varkappa} a_\varkappa(x)\chi_{A_\varkappa}(x_o) \tag{1}$$

where the sets $\operatorname{dom} a_\varkappa = B_\varkappa$ have x_1 as a common boundary point (belonging to $B_\varkappa$ or not), the $a_\varkappa(\cdot)$ are smooth on $B_\varkappa$, $\chi_{A_\varkappa}$ is the characteristic function of a set $A_\varkappa \subset B_\varkappa$, the sets $A_\varkappa$ are disjoint and cover $\mathbb{R}^n$. Thus

$$\alpha(\cdot,\cdot)|\cup_\varkappa (B_\varkappa \times A_\varkappa) \to \mathbb{R}^m$$

and the feedback modified system is

$$\begin{aligned}\tilde{\sum} : \dot{x} &= f(x, \alpha(x, x_o)) \\ &= f(x, a_\varkappa(x)) =: f_\varkappa(x) \ \text{ iff } \ x_o \in A_\varkappa.\end{aligned}$$

An appropriate choice of $A_\varkappa, B_\varkappa, a_\varkappa(\cdot)$ has to guarantee

(i) that the positive orbit through any $x_o \in A_\varkappa$ lies in $B_\varkappa = \operatorname{dom} f_\varkappa$ (this implies that every motion of the closed-loop system is smooth) and

(ii) that x_o is driven to $x_1 \in \partial B_\varkappa$ as t tends to infinity.

In the following we consider a mobile robot which consists of a (possibly heavy) trunk, two separately motor-driven (rear) wheels and a massless supporting (front) castor. The modeling control systems can be stabilized following the above strategy with $\varkappa \in \{1,2,3,4\}$. The sets $B_\varkappa$ are inspired by the domains (slit regions) of smooth transformations (cartesian to polar) which act on the control systems in such a way that linear feedbacks are very promising for the transformed systems.

2. Kinematics

From [4] we adopt the 3–dimensional control system

$$\dot{x} = v\cos\vartheta, \quad \dot{y} = v\sin\vartheta, \quad \dot{\vartheta} = \omega \tag{2}$$

as (the core of) a kinematical model. $(x, y, \vartheta) \in \mathbb{R}^2 \times S^1$ describes the position and orientation of the robot w.r.t. a frame fixed in the plane, v and ω are translational and angular velocity, respectively, and considered as independent inputs. We want to find a controller of type (1) which steers the robot from any initial posture (x_o, y_o, ϑ_o) to $(x_1, y_1, \vartheta_1) = (0, 0, 0)$.

For the sake of a simple formal description we introduce different charts on $\mathbb{R}^2 \times S^1$ (polar coordinates r, φ in $\mathbb{R}^2$ and a relative heading angle ψ):

$$
\begin{array}{llll}
\text{Transformation:} & x = r\cos\varphi, & y = r\sin\varphi, & \vartheta = \varphi + \psi \bmod 2\pi \\
\text{Domain } 1: & r > 0, & \varphi \in \left(-\frac{3\pi}{2}, \frac{\pi}{2}\right), & \psi \in (-\pi, \pi) \\
2: & r > 0, & \varphi \in \left(-\frac{3\pi}{2}, \frac{\pi}{2}\right), & \psi \in (0, 2\pi) \\
3: & r > 0, & \varphi \in \left(-\frac{\pi}{2}, \frac{3\pi}{2}\right), & \psi \in (-\pi, \pi) \\
4: & r > 0, & \varphi \in \left(-\frac{\pi}{2}, \frac{3\pi}{2}\right), & \psi \in (-2\pi, 0)
\end{array}
$$

In each chart the system (2) writes

$$
\dot{r} = v\cos\psi, \quad \dot{\varphi} = v\frac{1}{r}\sin\psi, \quad \dot{\psi} = -v\frac{1}{r}\sin\psi + \omega. \tag{3}
$$

Mind that the chart domains are open slit regions of $\mathbb{R}^2 \times S^1$ on which (3) is a smooth system.

Now we introduce the following subsets $A_\varkappa, B_\varkappa$ of respective chart domains and attached feedback laws $a_\varkappa \mid B_\varkappa \to \mathbb{R}^2$:

$$
\begin{array}{ll}
A_1 := (0, +\infty) \times (-\pi, 0) \times \left[-\frac{\pi}{2}, \frac{\pi}{2}\right] & \\
B_1 := (0, +\infty) \times \left(-\frac{5\pi}{4}, \frac{\pi}{4}\right) \times \left[-\frac{\pi}{2}, \frac{\pi}{2}\right] & a_1 : v = -r, \quad \omega = -a\psi + b\varphi \\
A_2 := (0, +\infty) \times (-\pi, 0) \times \left(\frac{\pi}{2}, \frac{3\pi}{2}\right) & \\
B_2 := (0, +\infty) \times \left(-\frac{5\pi}{4}, \frac{\pi}{4}\right) \times \left(\frac{\pi}{2}, \frac{3\pi}{2}\right) & a_2 : v = r, \quad \omega = -a(\psi - \pi) + b(\varphi + \pi) \\
A_3 := (0, +\infty) \times [0, \pi] \times \left[\frac{\pi}{2}, \frac{\pi}{2}\right] & \\
B_3 := (0, +\infty) \times \left(-\frac{\pi}{4}, \frac{5\pi}{4}\right) \times \left[-\frac{\pi}{2}, \frac{\pi}{2}\right] & a_3 : v = -r, \quad \omega = -a\psi + b\varphi \\
A_4 := (0, +\infty) \times [0, \pi] \times \left(-\frac{3\pi}{2}, -\frac{\pi}{2}\right) & \\
B_4 := (0, +\infty) \times \left(-\frac{\pi}{4}, \frac{5\pi}{4}\right) \times \left(-\frac{3\pi}{2}, -\frac{\pi}{2}\right) & a_4 : v = r, \quad \omega = -a(\psi + \pi) + b(\varphi - \pi).
\end{array}
$$

Then we can prove

Proposition 1 *Let the parameters a, b be chosen such that*

$$
a > 3, \ b > 0, \ 2a - 5b > \frac{4}{\pi}.
$$

Then the following holds: If the initial point (r_o, φ_o, ψ_o) belongs to $A_\varkappa$, $\varkappa \in \{1, 2, 3, 4\}$, then the solution $(r, \varphi, \psi)(\cdot)$ of the respective closed-loop system (with feedback $a_\varkappa$) is $B_\varkappa$–valued for $t > 0$ and, for $t \to +\infty$, tends to $(0, 0, 0)$ or $(-1)^{\varkappa/2}(0, \pi, -\pi)$ if $\varkappa = 1, 3$ or $\varkappa = 2, 4$, respectively.

Remarks (i) The sets $A_\varkappa$ describe a covering of $\mathbb{R}^2 \times S^1$ up to the trivial states with $r = 0$. Therefore the mobile robot is steered to the desired final position starting from arbitrary initial position: the asymptotic stability is global; in fact it is even exponential, and r tends strongly monotonically to zero.

(ii) Any initial position determines a unique $\varkappa \in \{1, 2, 3, 4\}$. Since the orbit does not leave the set $B_\varkappa$ the controller and thus the motion is smooth.

(iii) The proof deals in fact with certain subsets of $B_\varkappa$ which are shown, mainly by inspection of the direction field, to be positively invariant and do not contain closed orbits [7].

3. Dynamics

The dynamics of the robot, modeled, e.g., via Lagrange's equations and using motor torques as controls, shows up, after a feedback transformation, by two integrators added to the system (2):

$$\dot{x} = v\cos\vartheta, \quad \dot{y} = v\sin\vartheta, \quad \dot{\vartheta} = \omega, \quad \dot{v} = w_1, \quad \dot{\omega} = w_2. \tag{4}$$

Remark The feedback transformation which connects the driving torques with the acceleration controls is entered by inertia terms square in v and ω and by frictional terms. Modeling the latter ones could require incorporating a detailed castor kinematics [7].

The construction of a stabilizing feedback for (4) is based on the following philosophy. Again using local coordinates (r, φ, ψ) let $v = h_1(r, \varphi, \psi)$, $\omega = h_2(r, \varphi, \psi)$ be one of the kinematical feedback laws used above. Despite the unpleasant fact of nonzero initial values of the speeds v and ω we consider h_1 and h_2 as "good" feedback laws during the motion (for $t > 0$). Therefore we want to have a feedback $w_{1,2}(r, \varphi, \psi, v, \omega)$ for (4) which, starting with initial velocities $v(0) = \omega(0) = 0$, makes

$$e_v := v - h_1(r, \varphi, \psi) \quad \text{and} \quad e_\omega := \omega - h_2(r, \varphi, \psi)$$

tend to zero as fast as possible, expecting the robot then to behave like the closed-loop kinematical model.
To be specific, let us consider the above case $\varkappa = 1$. Then we have $v = e_v - r$, $\omega = e_\omega - a\psi + b\varphi$. Together with the feedback

$$\begin{aligned} w_1 &= -v\cos\psi - \lambda(v + r) \\ w_2 &= (a+b)\tfrac{v}{r}\sin\psi + \mu(-a\psi + b\varphi) - (a+\mu)\omega, \end{aligned} \tag{5}$$

where λ and μ are further parameters, we arrive at the smooth closed-loop system

$$\begin{aligned} \dot{r} &= (e_v - r)\cos\psi \\ \dot{\varphi} &= \tfrac{1}{r}(e_v - r)\sin\psi \\ \dot{\psi} &= -\tfrac{1}{r}(e_v - r)\sin\psi - a\psi + b\varphi + e_\omega \\ \dot{e}_v &= -\lambda e_v \\ \dot{e}_\omega &= -\mu e_\omega. \end{aligned}$$

Proposition 2 *If the parameters a, b are as in Proposition 1 and if $\lambda > 1$ and $\mu > \lambda + a + 1$ then the following holds: Any initial state $(r_o, \varphi_o, \psi_o, v_o, \omega_o) \in A_1 \times \{(0,0)\}$ is steered to the origin without leaving the set $B_1 \times \mathbb{R}^2$ and keeping $\frac{v}{r}$ bounded.*

Remarks (i) Analogue statements hold for the cases $\varkappa = 2, 3, 4$.
(ii) The mobile robot is steered smoothly and with bounded controls to the desired final position and zero final velocity when starting from rest at arbitrary initial position: the asymptotic stability is global on the configuration space (still open question for arbitrary initial velocities).
(iii) The proof follows the lines of that of Proposition 1 but is a bit more involved [8].

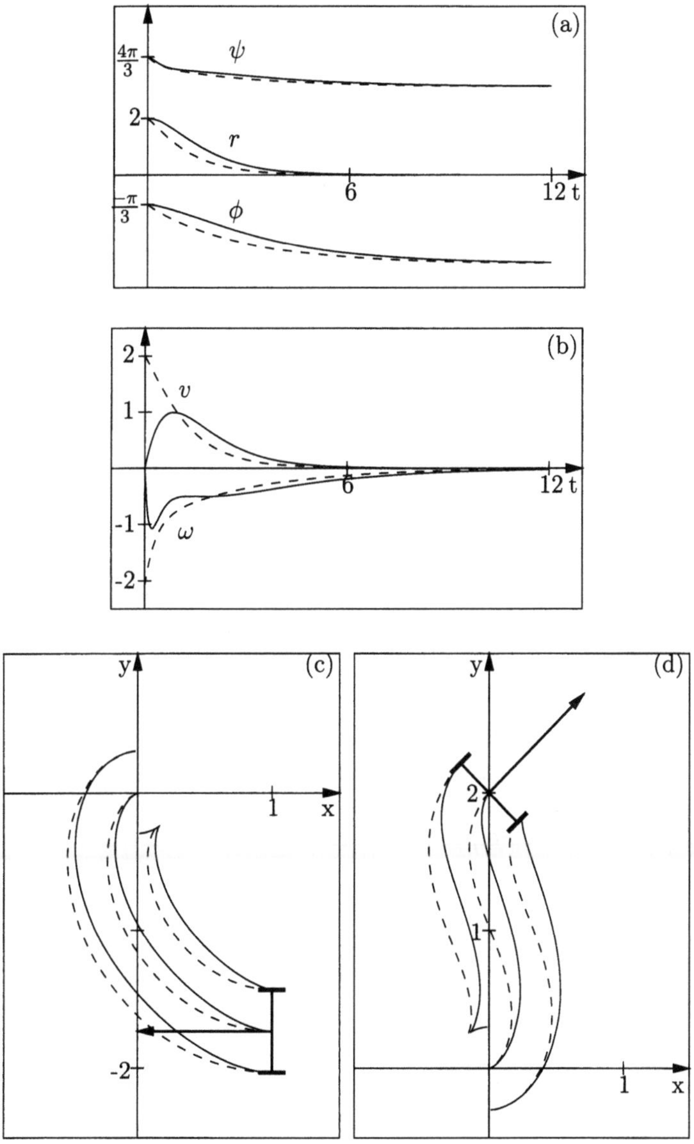

Figure 1: (a)–(c) robot starting in A_2, kinematical: dashed lines, dynamical: solid lines, (d) robot starting in A_3

4. Simulations

Using fixed parameter values $a = 4$, $b = 1$, $\lambda = 1.5$, $\mu = 7$ we compare the behavior of the robot governed by the feedback modified kinematical or dynamical model (see figure 1). Mind the lag of rotational motion.

5. Conclusion

In this paper we proposed a time-invariant globally stabilizing controller for a mobile robot. The controller is made up of several branches each of which is uniquely selected by the initial position and ensures smoothness of the closed-loop system. The only criterion of choice for the kinematic feedbacks was linearity; one should search for other feedback laws which render the motions optimal in some sense.
For proofs we refer to [7] and [8] (available on request) where also some more details about the kinematical model and dynamical effects can be found.

References

[1] Astolfi, A.: *A unifying approach to the asymptotic stabilization of nonholonomic systems via discontinuous feedback*, Report 94-01, Autom. Control Lab. ETH, Zürich.

[2] Brockett, R. W.: *Asymptotic stability and feedback stabilization*, in: Brockett, R. W.; Millmann, R. S.; Sussmann, H. J. (Eds.): *Differential Geometric Control Theory*, (Birkhäuser, 1983) 181–208.

[3] Canudas de Wit, C.; Berghuis, H.; Nijmeijer, H.: *Practical stabilization of nonlinear systems in chained form*, Preprint, Twente University, 1993.

[4] Canudas de Wit, C.; Khennouf, H.; Samson, C.; Sordalen, O. J.: *Nonlinear control design for mobile robots*, in: F. Yuan, Zheng, (Eds.): *Recent Trends in Mobile Robots*, World Scientific Series in Robotics and Automated Systems, Vol. 11, 121–156.

[5] Oelen, W.; Berghuis, H.; Nijmeijer, H.; Canudas de Wit, C.: *Implementation of a hybrid stabilizing controller on a mobile robot with two degrees of freedom*, IEEE Conf. on Robotics and Automation, San Diego, California, May, 1994, 1196–1201.

[6] Pomet, J.-B.: *Explicit design of time-varying stabilizing control laws for a class of controllable systems without drift*, Systems and Control Letters 18 (1992), 147–158.

[7] Steigenberger, J.; Abeßer, H.; Katzschmann, M.: *On mobile robots and their feedback stabilization, Part 1: Kinematics*, Preprint No. M 7/95, TU Ilmenau, Fak. Mathematik u. Naturwissenschaften, 1995.

[8] Steigenberger, J.; Abeßer, H.; Katzschmann, M.: *On mobile robots and their feedback stabilization, Part 2: Dynamics*, Preprint No. M 22/95, TU Ilmenau, Fak. Mathematik u. Naturwissenschaften, 1995.

International Series of Numerical Mathematics
Vol. 124, © 1998 Birkhäuser Verlag, Basel

Competitive Running on a Hilly Track

Elena Andreeva* Horst Behncke†

In [4] J.B. Keller treated the problem of competitive running by means of variational calculus. This simple model was improved and put on a more realistic basis in [1]. Both results, however, predict a slowing down in the last phase of a run, which is particularly pronounced in the model of Keller. While this is true for sprints and on the average for longer distance runs, the final sprint in many races seems to belie these conclusions. Two factors are most likely responsible for this effect. The first are psychological influences. Since this is difficult to model mathematically, we shall concentrate here on the physiological factor, glycolysis.

In longer distance runs glycolysis exerts a negative effect, because it uses the muscle glycogen rather inefficiently and because it may lead to a severe acidosis in the muscle. Hence it is necessary to use glycolysis at most in the final part of a longer distance run. Nature takes this into account by means of the Pasteur effect, which inhibits glycolysis when the oxygen uptake has reached its maximal level. In order to study these effects and to exhibit the final sprint, the model in [1] will be extended by introducing separate glycogen compartment. In addition, the resistance term may depend on the slope of the track. As in [1] the analysis of the optimal running strategy is based on the qualitative study of the switching function of the optimal control problem. We show that a long distance run on an even track consists essentially of three phases, the initial acceleration phase, an interval where v is almost constant, but slightly increasing and a power-constraint phase with the final spurt. On hilly tracks there may be several such phases, but in all cases such runs are at most power constrained. The negative influence of glycolysis, however, moves the power constraint phase further to the end of the run. It is obvious that this model can be extended also to skating, swimming, rowing or cycling.

This paper is divided into two sections:

1. The model
2. Running on a hilly track.

The notation in this paper is largely standard and that of [1]. Thus the derivative of a function of a single variable will be denoted by f', while partial derivatives will be denoted by subscripts, e.g. σ_E. All functions are assumed to be sufficiently smooth, i.e. twice continuously differentiable. For easier presentation and numerical calculations,

*Tver State University, Department of Mathematics, Mozhajskaya d 73 kv 43, Tver 170 043, Russia, elena@icc.tversu.ac.ru

†Universität Osnabrück, FB Mathematik/Informatik, 49069 Osnabrück, Germany, behncke@chryseis.mathematik.uni-osnabrueck.de

however, some functions will be given as continuous piecewise linear expressions however. This case is then considered as an approximant to an otherwise smooth function. Some constants of the model are taken over from [1]. But also these values should be considered only as approximations to be used in the numerical analysis. All this reflects the fact that constants in physiology and biomechanics will have an error and a variation of about 10 to 20 per cent. These approximations and ambiguities do not invalidate our conclusions, because the analysis is largely qualitative.

Whenever we refer to actual data and give estimates of parameters the m, kg, sec system is used. It should be noted, however, that as in [1] some quantities like force, energy, breathing rate, ... are mass specific, i.e. they are defined per kilogram of body weight.

1. The model

As in [1] and [2] we describe a run as a one dimensional motion where $x(t)$ respectively $v(t)$ denotes the position respectively velocity of the center of mass of the athlete. Newton's law of motion in the mass rescaled form is then

$$v' = f - r(x,v) \qquad \text{along with } x' = v. \tag{1}$$

Here f is the mass-specific force of the runner averaged over a step cycle. The resistance term r describes the internal dissipation of energy as well as the air resistance. For simplicity we write

$$r(x,v) = r_0(x) + r_1(v) + cv^2. \tag{2}$$

In [1] the air resistance coefficient c was found to be $c \approx .0037\text{m}^{-1}$. For a track with slope $\gamma(x)$ one would expect

$$r_0(x) = a + g \sin\gamma(x)$$

with g the gravitational acceleration. The situation, however, is more complex because of the negative work in the braking phase of the step cycle. The approximate form of r_0 as a function of the inclination is given by Margaria [5].

For the slope gradient $\gamma(x)$ we assume $|\gamma(x)| \leq 12^0$ and that it takes more than 40 meters to change from one extreme to the other. Thus

$$1.8 \leq r_0(x) \leq 5\ ;\ |r_0'(x)| \leq 0.13\ ;\ |r_0''(x)| \leq .0033.$$

In addition we will assume that the track has only finitely many smooth hills, i.e. $r_0'(x)$ and $r_0''(x)$ are piecewise monotonic.

In [1] it was shown that r_1 is approximately given by

$$r_1(v) = \begin{cases} 0 & v \leq 6\text{m/sec} \\ b(v-6) & v \geq 6\text{m/sec} \end{cases}$$

with $b \approx .7$. Here we will use a smooth approximant of this with a transition region in the interval [5,7]. Equation (1) is not valid for the start, small and negative velocities.

Since we are mainly interested in longer distance runs, we will ignore the starting phase and assume $v(0) = 6\text{m/sec}$. In addition we set

$$r(x,v) = r_0(x)v \quad \text{for } v \leq 1.$$

This is a purely technical assumption, because, as we will see later, the state $v(t) = 1$ is never reached in a competitive run. In addition we use this as an approximation to $r = 0$ once $v = 0$. This assumption assures that the "unphysical" state $v(t) \leq 0$ is never reached.

Muscles and tendons have limited strength. Thus the force is bounded

$$0 \leq f \leq F \tag{3}$$

The energy for running and other physical activities is mainly derived from the following four sources listed in their order of recruitment and power

i) ATP and creatine phosphate, the anaerobic alactic energy store
ii) Glycolysis, the anaerobic fermentation of glucose to lactic or pyruvic acid
iii) the oxydation of glucose or glycogen
iv) the oxydation of lipids.

ATP is the only immediate source of energy and all other processes operate via ATP production from ADP or AMP. Glycolysis is triggered by higher ADP concentrations. Its main function together with ATP and phosphocreatine is to provide sufficient power until breathing has attained its full capacity. Both sources provide enough power for heavy workloads of short duration, e.g. a 400 m run. Glycolysis is an extremely inefficient source of mechanical power, because it uses only about 5.26% of the chemical energy of glucose. In addition its metabolites lactic and pyruvic acid can lead to an acidosis which impairs a proper functioning of the muscle. These metabolites are transported by the blood to other muscles. The associated rate constant is about 900 sec. In the working muscle it will be much smaller though. Thus even endurance work will lead to a very low equilibrium level of lactic acid.

The anaerobic reserves at time t in the muscle will be denoted by $E(t)$, while $G(t)$ stands for the reserves of glucose and lipids in the muscle and blood. $E(t)$ consist of the alactic reserves, ATP and phosphocreatine, and the glycolytic reserves, i.e. the ability to employ the glycolytic pathway. Since the first comprises about 35% of the anaerobic reserves, which is close to the oxygen deficit, $E_{00} - E(t)$ with $E_{00} \approx .65E_0$ can be taken as a measure of the lactic acid concentration in the muscle. But $E(t)$ is also a measure for the ATP concentration. Conservation of energy leads to

$$E'(t) = -fv/\eta + \sigma(t) + d_1(E). \tag{4}$$

Here fv is the mechanical power used for running and the efficiency factor η relates this to chemical power. η is known to increase slightly with the running speed v and the inclination γ. Since longer distance runs are mostly run with velocities around 6 m/sec, we will assume η to be constant with an approximate value of $\eta \approx .6$. d_1 in equation (4) describes the removal of lactic acid by the blood and its transport to other muscles, where it is oxydized. Thus $d_1(E) \approx 0$ for large E and $d_1(E) \approx \gamma_1(E_{00} - E)$ with $\gamma_1^{-1} \approx 300$ sec for $E \leq E_{00} \approx .65E_0$, because this removal process is only active when glycolysis operates. Since the muscles cannot store an unlimited amount of energy one has

$$0 \leq E(t) \leq E_0, \quad 0 \leq G(t) \leq G_0. \tag{5}$$

The initial conditions $E(0) = E_0$ and $G(0) = G_0$ for a race then mean that the runners are rested. Since we will neglect the starting phase we will use $E(0) \approx E_0 - 40$ along with $v(0) = 6$, however.

Equation (4) is not valid if $E(t) = E_0$ and if the right-hand side of (4) is positive. In this case (4) has to be replaced by $E' = 0$, because the muscles cannot store any more anaerobic energy. As we will see, this constraint is never active in a competitive run.

The breathing rate σ has approximately the form

$$\sigma(t) = \sigma_0(E, G)e^{-t/\tau}(1 - e^{-t/30}) \tag{6}$$

The first factor describes the dependence of the breathing rate on the available glycogen and the acidotic state of the muscle. Thus $\sigma_E = 0$ if $E \geq E_{00}$. In fact we will assume $\sigma_E = 0$ for $E \geq .55E_0$ because the use of ATP and phosphocreatine counters the acidity of the lactic acid to some extent. The second factor models long time fatigue. Since $\tau \approx 30.000$ sec it can be neglected in most cases. Similarly we will neglect the last factor for times $t \geq 120$ sec, since it describes the initial increase of the breathing rate.

The oxydation of glucose is approximately a first order process. Thus $\sigma_0(E, G) \approx \sigma_{00}(E)G/G_0$ and we expect $\sigma_G \approx 3 \cdot 10^{-4}$. Since however glycogen is delivered by the bloods from its depots on demand, we expect $\sigma_G(E, G) = 0$ for large G, e.g. for $G \geq .8G_0$. As a function of E σ_0 will have a sigmoidal shape. We approximate it by a piecewise linear function with $\sigma_0(E) = \sigma_{0\,\max}$ for $E \leq .55E_0$ and $\sigma_0(0) = 2/3\sigma_{0,\mathrm{max}}$.

With these considerations we can write

$$G' = -\sigma - \alpha(fv/\eta - \sigma) + d_2(G) \tag{7}$$

The first term stands for the glucose oxydized in the muscle by breathing, while the second term gives the amount of glucose used in the glycolytic process. Thus $\alpha \approx 19$ if $E \leq E_{00}$, because one molecule of glucose leads to the formation of 38 molecules of ATP from ADP, while glycolysis produces only 2. For $E \geq E_{00}$ we will have $\alpha = 0$ if glycolysis is not used at all. Thus α will depend on E and it is a measure to what extent the glycolytic pathway is used. Equation ((7)) is correct only if $fv/\eta > \sigma$. If $fv/\eta \leq \sigma$, it should be replaced by

$$G' = -\sigma + d_2.$$

i.e. $\alpha = 0$ in this situation. This case however will occur in competitive running on hilly tracks only rarely. If it occurs the equations for the covariables will have to be modified. The general setup of this system is summarized in Figure 1.

The ability to run or perform other kinds of work decreases as E decreases. More precisely it is the ATP concentration, which determines the power output, because the

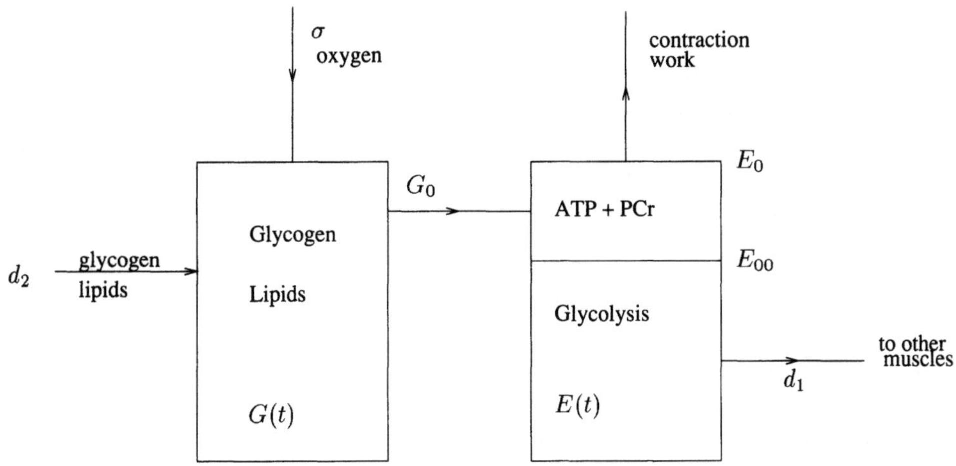

Figure 1: The compartments of the energetics and the flow of substrates.

hydrolysis of ATP drives the sliding muscle filaments. Since for low E, $E \le E_{00} = 0.65E_0$ is also a measure for the ATP concentration, we can write as in [1]

$$fv \cdot \eta^{-1} \le P(E) + \sigma. \tag{8}$$

From physiology one would expect P to have a sigmoidal shape with $P(0) = 0$. As in [1] we will approximate it by

$$P(E) = \min(P_0, E/\rho) \quad \text{with } \rho \approx 22. \tag{9}$$

This constraint will be used in two ways. In the initial phase of the run, $t \le 100$ sec, we will use a limit of $P_0 \approx v_\infty F/\eta$ as in [1] where v_∞ is the maximal velocity on an even track, i.e. $F = r(v_\infty, x)$, where $\gamma(x) = 0$.

For longer distance runs, P_0 will be much smaller, because of the Pasteur effect. Thus we will assume

$$P_0 \le 20 \; W/\text{kg}$$

once the oxygen uptake is maximal or once $t \ge 120$ sec. The time interval [100, 120] can then be used to interpolate P_0 linearly. A small P_0 is thus the mathematical formulation of the Pasteur effect. Additionally this effect could result in a reduction of the glycolytic capacity, i.e. a reduction of E_0.

The aim of the athlete is to cover a given distance D in the shortest time T, possible or equivalently to maximize.

$$D = \int_0^T v \, dt = x(T) \to \max$$

subject to (1), (4), (7) and the constraints (3), (5) and (9), T being given. This is a problem of optimal control with phase inequality constraints and f as the control variable. In contrast to [1], where we studied competitive running of short and middle distances only, the emphasis is on longer distance runs here. Thus the starting phase and initial increase of σ will be neglected. Even though the above model is quite general, we have in mind the concrete set of parameters for world record athletes of [1]. The conclusions however will be valid in much more general circumstances, because some parameters are almost universal, while in other cases the ratios are almost constant.

As reference values we will use:

Men:	$F \approx 6.7\ \mathrm{m/sec}^2$, $E_0 \approx 2500\ \mathrm{m}^2/\mathrm{sec}^2, \sigma_0(E_0, G_0) \approx 27, G_0 \approx 10^5\ \mathrm{m}^2/\mathrm{sec}^2$
Women:	$F \approx 6\ \mathrm{m/sec}^2$, $E_0 \approx 2000\ \mathrm{m}^2/\mathrm{sec}^2, \sigma_0(E_0, G_0) \approx 20, G_0 \approx .9 \cdot 10^5\ \mathrm{m}^2/\mathrm{sec}^2$

2. Running on a hilly track

Our analysis of this model is largely qualitative. Essential use will be made of the fact that the biomechanical processes are much faster than the energetic processes. The problem (1), (4), (7) with initial conditions $x(0) = 0, v(0) \approx 6, E(0) = E_0 - 40$ and $G(0) = G_0$ and constraints (3), (5) and (8) has a measurable solution f by Filippov's Theorem. The Hamiltonian for this problem is given by [6, ch. VI 3]

$$\begin{aligned}\mathcal{H} &= \lambda_0 v + \lambda_1(f - r) + \lambda_2(\sigma - fv/\eta + d_1) + \lambda_3(-\sigma - \alpha(fv/\eta - \sigma) + d_2) \\ &+\nu_1(fv/\eta - P - \sigma) + \nu_2(-f) + \nu_3(f - F)\end{aligned} \tag{10}$$

where $\lambda_0, \lambda_1, \lambda_2$ and λ_3 are the covariables for x, v, E and G respectively. The constraint $E \leq E_0$ will not be considered here, because the oxygen deficit accumulated in the first 120 seconds almost corresponds to the alactic reserves.

The covariables are absolutely continuous and satisfy

$$\begin{aligned}&\lambda_0' = \lambda_1 r_x, \lambda_1' = -\lambda_0 + r_v\lambda_1 + f\left(\frac{v}{\eta}\right)'(\lambda_2 + \alpha\lambda_3) - f\left(\frac{v}{\eta}\right)'\nu_1 \\ &\lambda_2' = -(\sigma_E + d_1')\lambda_2 - (\alpha - 1)\sigma_E\lambda_3 + \alpha_E(f\frac{v}{\eta} - \sigma)\lambda_3 + (P' + \sigma_E)\nu_1 \\ &\lambda_3' = -\sigma_G\lambda_2 - (\sigma_G(\alpha - 1) + d_2')\lambda_3 + \sigma_G\nu_1\end{aligned} \tag{11}$$

Here ν_1 is the multiplier for the power constraint (8). Thus $\nu_1 \leq 0$ and $\nu_1 = 0$ whenever strict inequality holds in (8). Since v, E and G are free variables at T the terminal constraints are $\lambda_1(T) = \lambda_2(T) = \lambda_3(T) = 0$ and $\lambda_0(T) \geq 0$. It should be noted that the equations for λ_2 and λ_3 do not involve λ_0 and λ_1. Since $\mathcal{H}$ is linear in the control f,

$$\varphi = \lambda_1 - (\lambda_2 + \alpha\lambda_3) \cdot \frac{v}{\eta} = \lambda_1 - \psi\frac{v}{\eta} \tag{12}$$

is a switching function. It follows from [6, ch. VI (118)] that

$$\nu_1 = -\frac{\eta}{v}\varphi \tag{13}$$

whenever $0 < f < F$ and when (8) is active. Since the constraint $E \leq E_0$ will not be active near T the terminal constraints imply $\lambda_0(T) = 1$, because otherwise we would have $\lambda_i(t) = 0$ $i = 0, 1, 2, 3$ near T, which is impossible. The key to the remaining analysis is the properties of φ and the size of $E(t)$ compared to E_{00}, which marks the threshold of glycolysis. The constraints $E(t) = 0$ and $G(t) \leq G_0$ cannot be active in this setting and need not concern us here. The constraint $G(t) \geq 0$, is likewise strict, because there is an almost unlimited supply of lipids and because we will consider only runs which last less than 10 hours or so. In the analysis of φ essential use will be made of the fact that the equations (11) for λ_2 and λ_3 are decoupled from those of λ_1 and λ_0 and the corresponding coefficients are much smaller.

2.1 The terminal phase

Since $\lambda_0(T) = 1$ we see $\lambda_1(t) \approx (T - t)$ and $\lambda_2(t), \lambda_3(t) = o(T - t)$ near T. Thus φ is positive near T. If φ were positive for all $t < T$, the run would be an all out sprint with $f = F$ initially and a subsequent power constraint [1]. Thus we may assume that there is a maximal $t_2 < T$ with $\varphi(t_2) = 0$. The interval $[t_2, T]$ defines the final spurt. Since $f\frac{v}{\eta} - \sigma = P_0(E) > 0$, it follows from (11) that $\lambda_2(t), \lambda_3(t) > 0$ for $t \in [t_2, T]$. Actually one would expect λ_2 and λ_3 to be positive for all $t < T$ as the shadow price interpretation with $x(T)$ as the value function suggests. This is indeed the case.

Lemma 1 *λ_2 and λ_3 are positive on* $[0, T)$.

This follows from the monotonicity of the value function $x(T)$ with respect to $E(0)$ and $G(0)$. It can also be shown directly, but since this involves a careful analysis of various cases when integrating (11) backwards from T, we will not pursue this here. It is likewise obvious that $\varphi(t), \lambda_2(t), \lambda_3(t) > 0$ for $t < T$ near T as long as T is large enough.

Similarly one would expect λ_0 and λ_1 to be positive. Though this is plausible and true, the above proof cannot be adapted easily to this situation.

The shadow price interpretation of λ_2 and λ_3 also suggests $\lambda_2 \geq (\sigma_E + d_1')/\sigma_G\lambda_3$. This is indeed the case on $[t_2, T]$ and can be shown by considering the function $\phi = \sigma_G\lambda_2 - (\sigma_E + d_1')\lambda_3$ and its derivative on this interval. However this equation is useful only if $E(t) \leq .5E_0$ The form of our equations and general physiological principles suggest that this corresponds closely to the terminal phase or in other words $T - t_2 \approx .5E_0/P_0$.

Arguing as in [1] it is obvious that there is a short initial maximal acceleration phase $[0, t_1]$ with $\varphi(t) > 0$ on $t < t_1$ and $\varphi(t_1) = 0$. For long distance runs, which we are considering here, $t_1 \leq 1.5sec$.

For the remaining analysis we need to know more about the properties of φ. We have

$$\varphi' = -\lambda_0 + \varphi r_v + \psi\left(r\frac{v}{\eta}\right)_v + \psi\frac{v}{\eta}A - \lambda_3\frac{v}{\eta}B - \nu_1\left(f\left(\frac{v}{\eta}\right)' - \nu_1\left(\frac{v}{\eta}\right)(P' + \sigma_E + \alpha\sigma_G)\right) \tag{14}$$

with $A = (\sigma_E + d_1' + \alpha\sigma_G)$ and $B = (\sigma_E + \alpha d_1' + (\sigma_G - d_2')\alpha + \alpha_E d_1)$. The first three summands on the right-hand side of (14) are the dominant terms, while A and B are very small.

In addition we need the complete expression of φ'', even though it is rather lengthy. From (14) one gets

$$\begin{aligned}\varphi'' &= \varphi(-r_x + r_{vv}(f-r)) + r_v\varphi' + \psi'\left(\left(r\frac{v}{\eta}\right)_v + A\frac{v}{\eta}\right) \\ &+ \psi\left(\left(r\frac{v}{\eta}\right)_{vv,} + A/\eta\right)(f-r) + \dot{A}v/\eta + \psi\sigma_G Bv/\eta \\ &+ \lambda_3\left(-\left(\frac{v}{\eta}\right)'(f-r)B - \frac{v}{\eta}\dot{B} - \frac{v}{\eta}(\sigma_G - d_2')B\right) \\ &+ \chi\varphi'(f/v + P' + \sigma_E + \alpha\sigma_G) + \chi\varphi((f/v + P' + \sigma_E + \alpha\sigma_G)^\bullet + \sigma_G B)\end{aligned} \tag{15}$$

Here χ denotes the characteristic function of the power constraint domain and $^\bullet$ means derivative with respect to t. The leading term in (14) is the fourth and third summands. Since $\dot{A}$ and $\dot{B}$ are small the Legendre Clebsch condition $(\varphi'')_f \leq 0$ for singular arcs is clearly satisfied.

Now rewrite (15) as

$$\varphi'' + a\varphi + b\varphi' = C.$$

Remove the first order term by the transformation $y = (\exp D) \cdot \varphi$, with a suitable D and apply the well-known Sturmian oscillation theory [7] to the resulting equation. Since the oscillation of the corresponding homogeneous equation and of C is bounded below, our control problem is piecewise continuous.

We will speak of positive (negative) (t_1, t_2) arcs of φ if $\varphi(t_1) = \varphi(t_2) = 0$ and $\varphi(t) > 0(\varphi(t) < 0)$ in between.

The following result is central for the remainder.

Lemma 2 *φ has no negative arcs.*

Proof: It should be noted that now $\alpha = 0$. Even though this change of equation would be associated with a jump of λ_3, the continuity of $\mathcal{H}$ implies the continuity of λ_3 in this case. Now apply the oscillation theory to (15) for various r_x configurations and use the fact that v decreases to 0 rapidly. This in turn shows that negative arcs are not feasible.

2.2 Singular arcs

It is now possible to determine $(f-r)$ for singular arcs. Since λ_3 is generally rather small, we obtain

$$(f-r) \approx -\left(\left(r\frac{v}{\eta}\right)_{vv} + A/\eta\right)^{-1}\left(\left(r\frac{v}{\eta}\right)_v + \frac{v}{\eta}A\right)\psi'/\psi \tag{16}$$

where

$$\psi' = -\psi(\sigma_E + d_1' + \alpha\sigma_G) + \lambda_3(\sigma_E + \alpha d_1' + \alpha_E(f\frac{v}{\eta} - \sigma + E'))$$

$$-\chi\varphi\eta/v(P' + \sigma_E + \lambda\sigma_G)$$

This shows $f - r = 0$ if $E(t) \geq .65E_0 = E_{00}$, because then $A = B = \psi' = 0$. It is expected that the largest part of the race is run below the glycolytic threshold, i.e. $E(t) \geq E_{00}$. This in turn implies $r\eta/y \approx \sigma$.

For $E_{00} \geq E(t) > .5E_0$ one has $\sigma_E + d_1' \approx 0, A = \alpha\sigma_G$ and $B \approx 0$. Thus

$$f - r \approx \left(\left(r\frac{v}{\eta}\right)_{vv} + \alpha\sigma_{G/\eta}\right)^{-1} (\alpha\sigma_G)^2$$

In this case v would increase slightly, depending on the decrease of the local glycogen concentration.

Below the glycolytic threshold $E(t) \leq 0.5E_0$ f will increase more strongly in order to compensate the negative effects of glycolysis. Because of this increase this phase can not last very long. In general this phase is followed by the final sprint. Generally one can say that it becomes increasingly more difficult to leave a positive semi arc once E decreases below E_{00}. These considerations remain valid also if $f \cdot v/\eta \leq \sigma$, where one has to replace α by 0. With these preparations we can now state our main result.

Theorem 1 *The optimization problem* (1), (4), (7) *with* (3), (5) *and* (8) *is piecewise continuous. For longer distance runs all positive arcs are power constrained. Intermediate power constraints are entered only on uphill parts $r_x > 0$ and left on downhill parts $r_x \leq 0$. A run on a straight track will consist of at most three phases, an initial phase with maximal acceleration $f = F$, a phase where v is almost constant $f\frac{v}{\eta} \sim \sigma$ and a final power constraint phase.*

Proof: The proof is based on oscillation theory applied to (15). First one notes that pure $f = F$ positive arcs would last too long and thus violate the power constraint condition. ∎

Moreover the oscillation half period of the homogeneous version of (15) is clearly bounded below, 8 sec with our choice of parameters. Thus accumulation of positive φ-arcs can only result from high oscillations of the inhomogeneous part in (15). This in turn can be reduced to oscillations of r_x, if one analyzes various E configurations. It remains to analyze positive power constrained arcs. This is done for various E and r_x configurations. One finds that v' converges to 0 rapidly and becomes negative even if $r_x > 0$. Then however φ'' becomes negative so that the power constrained arc cannot be left. This also shows that positive φ arcs are entered only in the terminal interval or if $r_x > 0$. In particular power constraint intervals will not occur in intermediate phases if $r_x \leq 0$.

Remark 1 *This result is clearly in close agreement with our every day experience. In addition this result extends to the weakly time dependent situation, where σ decays very slowly in time. This will occur if one models fatigue via time dependent coefficients, e.g. the factor $e^{-t/\tau}$ in σ.*

References

[1] Behncke, H.: *Optimization models for the force and energy in competitive sports.* Math. Meth. Appl. Sci. **9** 298–311 (1987).

[2] Behncke, H.: *A mathematical model for the force and energetics in competitive running.* J. Math. Biol. **31** 853–878 (1993).

[3] Bryson, A.E; Ho, Y.C.: *Applied Optimal Control.* Hemisphere Publishing Co. Washington, New York London, 1975.

[4] Keller, J.B.: *A Theory of Competitive running.* Phys. Today **26** 43–47 (1973).

[5] Margaria, R.: *Biomechanics and energetics of muscular exercise.* Oxford Univ. Press., Oxford (1976).

[6] Neustadt, L.: *Optimization.* Princeton Univ. Press, Princeton (1976).

[7] Reid, W. T.: *Sturmian Theory for ordinary differential equations.* Springer Verlag New York, Heidelberg, Berlin 1980.

International Series of Numerical Mathematics
Vol. 124, © 1998 Birkhäuser Verlag, Basel

Convex Domains of Given Diameter with Greatest Volume

Uwe Klemt*

Abstract. In this paper we study the following geometric optimization problem: What are the convex domains of given diameter with greatest volume in the n-dimensional Euclidean space ($n \geq 2$)? By using the plane of support of a convex domain we formulate the above mentioned question analytically. This leads to a multidimensional variational problem in parametric form with respect to a class of state functions and associated control vectors of Grassmann coordinates fulfilling certain state restrictions, control restrictions and boundary conditions. In order to prove the conjecture that circle and ball, respectively, are domains solving the problem we apply a generalized duality theory in the sense of R. Klötzler. On the basis of this theory first-order necessary conditions and second-order sufficiency conditions for an auxiliary finite-dimensional parametric optimization problem are verified.

1. Introduction

Basing on the work of H. Minkowski [12] and W. Blaschke [2] in the first decades of the 20th century many authors dealt with geometrical optimization problems. The book of T. Bonnesen and W. Fenchel [3] summarizes most of the results obtained in these years. Among these problems is the following:

Problem 1 *What are the convex domains of given diameter D with greatest volume in the n-dimensional Euclidean space $\mathbb{E}^n$ ($n \geq 2$)?*

Depending on the dimension of space there are several ways to prove the

Conjecture 1 *Circle and ball of diameter D are solutions to Problem 1, respectively.*

For $n = 2$ J.M. Jaglom and W.G. Boltjanski used geometric properties of the circle [6]. To deal with the general case T. Kubota [10] employed sophisticated symmetrization methods. It is the object of this paper to give a new proof using modern methods of the calculus of variations connected with finite-dimensional optimization methods[1]. To this end we make use of the plane of support of a convex domain $\mathcal{B}$ and introduce a parametric representation of its boundary $\partial\mathcal{B}$. The main tool in our proof of Conjecture 1 is a generalized duality theory in the sense of R. Klötzler [8]. It seems that this theory enables us not only to deal with geometrical optimization problems in a relatively simple manner, but also to tackle more complicated problems arising in physics and technics (e.g. in the design of surfaces).

*BTU Cottbus, Lehrstuhl Optimierung, P. O. Box 101344 03013 Cottbus, Germany, email: klemt@math-tu-cottbus.de

[1]As far as we know this connection of multidimensional variational problems in parametric form and finite-dimensional optimization methods does not appear in the literature before.

2. Preliminaries

2.1 Some general remarks on variational problems in parametric form

In this section we state some facts on variational problems in parametric form. Let $\Omega \subset \mathbb{E}^m$, $G \subset \mathbb{E}^r$ be Lipschitz-domains, Ω bounded. We deal with problem (P)

$$J(x,p) = \int_\Omega f(x(t), p(t))\, dt \to \inf \tag{1}$$

s.t. $x \in \mathcal{K} \subset W^{1,r}_\infty(\Omega)$,
associated $\nu \left(= \binom{r}{m}\right)$-dimensional controls p of Grassmann coordinates

$$p_{i_1 \ldots i_m} = \frac{\partial(x^{i_1}, \ldots, x^{i_m})}{\partial(t_1, \ldots, t_m)} \tag{2}$$

$(i_1 < i_2 < \cdots < i_m)$
control constraints

$$p(t) \in V(x(t)) \quad \text{for a.e. } t \in \Omega \tag{3}$$

state constraints

$$x(t) \in G \quad \text{for all } t \in \overline{\Omega} \tag{4}$$

boundary conditions

$$\psi(x) = 0 \text{ on } \partial\Omega \quad . \tag{5}$$

Here $V : \overline{G} \to \mathbb{E}^\nu$ and the sets $V(\xi)$ for all $\xi \in G$ are cones of $\mathbb{E}^\nu$ with vertices in 0. Furthermore f is assumed to be continuous and $f(\xi, \cdot)$ positively homogeneous of the first degree. This ensures that the variational integral (1) is invariant with respect to transformations of the parameters with positive Jacobian. Since the coordinates $p_{i_1 \ldots i_m}$ are minors of

$$(x^i_\alpha) = \begin{pmatrix} \frac{\partial x^1}{\partial t_1} & \cdots & \frac{\partial x^r}{\partial t_1} \\ \cdots & \cdots & \cdots \\ \frac{\partial x^1}{\partial t_m} & \cdots & \frac{\partial x^r}{\partial t_m} \end{pmatrix} \tag{6}$$

they are antisymmetric in the indices and for $2 \leq m \leq r-2$ not independent by themselves. They have to lie on the Grassmann manifold Γ, i.e. they must satisfy the system of equations (see [1]):

$$p_{i_1 \ldots i_l i_{l+1} \ldots i_m} p_{j_1 \ldots j_l i_{l+1} \ldots i_m} - \sum_{\gamma=1}^{l} p_{j_\gamma i_2 \ldots i_l i_{l+1} \ldots i_m} p_{j_1 \ldots j_{\gamma-1} i_1 j_{\gamma+1} \ldots j_l i_{l+1} \ldots i_m} = 0 \tag{7}$$

where $i_\alpha, j_\beta \in \{1, 2, \ldots, m\}$, $l = 2, 3, \ldots, \mathrm{Min}\{m, r-m\}$. It suffices to take into account exactly $R = \nu - [1 + m(r-m)]$ of the equations (7) (see [5] for details). Let

$p^0_{(i)} = p_{i^0_1 \dots i^0_m} \neq 0$ (from now on we abbreviate $i_1 \dots i_m$ by (i) if $i_1 < i_2 \dots < i_m$), denote by $h_1, h_2, \dots, h_l$ columns of (x^i_α) appearing in $p^0_{(i)}$, by $k_1, k_2, \dots, k_l$ columns of (x^i_α) which do not appear in $p^0_{(i)}$ ($h_1 < h_2 < \dots < h_l$, $k_1 < k_2 < \dots < k_l$) and by $p^{k_1 k_2 \dots k_l}_{h_1 h_2 \dots h_l}$ the minor of (x^i_α) which is derived from $p^0_{(i)}$ by replacing the columns $h_1, h_2, \dots, h_l$ by $k_1, k_2, \dots, k_l$. Furthermore let $j^0_1 < j^0_2 < \dots < j^0_{m-l}$ be those elements from $\{i^0_1, i^0_2, \dots, i^0_m\}$ which remain after deleting $h_1, h_2, \dots, h_l$. Then we can choose for instance the equations

$$b_\rho(p) := p^0_{(i)} p^{k_1 k_2 \dots k_l}_{h_1 h_2 \dots h_l} - \sum_{\gamma=1}^{l} p_{k_\gamma h_2 \dots h_l j^0_1 \dots j^0_{m-l}} p_{k_1 \dots k_{\gamma-1} h_1 k_{\gamma+1} \dots k_l j^0_1 \dots j^0_{m-l}} = 0 \tag{8}$$

($l = 2, 3, \dots, \mathrm{Min}\{m, r-m\}$, $\rho = 1, \dots, R$). We especially include the claims $b_\rho(p) = 0$ in the control restrictions (3). Note that the gradients with respect to p of the functions b_ρ are linearly independent[2].

2.2 Variational problems in parametric form and duality

Let $v \in \mathbb{E}^\nu$ with components $v_{(i)}$. Denoting by Q_0 the set of functions

$$Q_0 = \{\zeta \in \mathcal{K} | \zeta(t) \in \overline{G}, \psi(\zeta) = 0 \text{ on } \partial\Omega\} \tag{9}$$

and by s the vector of all Jacobian determinants

$$s_{(i)} = \frac{\partial(S^1, \dots, S^m)}{\partial(\xi^{i_1}, \dots, \xi^{i_m})} \tag{10}$$

for a given $S = (S^1, \dots, S^m) \in W^{1,m}_\infty(G)$ the following theorem obtained by R. Klötzler [8] holds.

Theorem 1 (Klötzler)

$$L_0(S) := \inf_{\zeta \in Q_0} \int_\Omega dS^1(\zeta) \wedge \dots \wedge dS^m(\zeta) \to \sup \tag{11}$$

is a dual problem to (P) (i.e. $J(x,p) \geq L_0(S)$) s.t. all $S \in W^{1,m}_\infty(G)$, fulfilling the inequality[3]

$$s_{(i)}(\xi) v_{(i)} \leq f(\xi, v) \quad \textit{for all } v \in V(\xi) \textit{ and a.e. } \xi \in G \quad . \tag{12}$$

For totally differentiable S strong duality (i.e. $J(x,p) = L_0(S)$) holds if and only if

$$s_{(i)}(x(t)) p_{(i)}(t) = f(x(t), p(t)) \quad \textit{for a.e. } t \in \Omega \tag{13a}$$

and

$$L_0(S) = \int_\Omega dS^1(x(t)) \wedge \dots \wedge dS^m(x(t)) \quad . \tag{13b}$$

[2] In contrast to the classical field theory in the calculus of variations we do not divide the equations (8) by $\sqrt{p_{(i)} p_{(i)}}$ (with the summation convention being operative) to ensure the positive homogeneity of $b_\rho(p)$. To do this would complicate the calculations in Section 4.

[3] From now on the summation convention is operative with respect to all index combinations (i).

Corollary 1 *The conditions (13a) and (13b) together with (12) are sufficiency conditions for the optimality of* (x, p) *to (P).*

Proof: The proof of Theorem 1 is obtained by an estimation. This estimation is sharp if and only if conditions (12) and (13) hold. By means of duality this proves the optimality of (x, p) to (P). ∎

2.3 Finite-dimensional optimization problems

We now define Pontrjagin's function[4]:

$$H(\xi, v, s) := -f(\xi, v) + s_{(i)}(\xi)v_{(i)} \quad . \tag{14}$$

Using this function we derive from Theorem 1 and Corollary 1 a finite-dimensional optimization problem (F_W) for the verification of the weak local optimality of (x^*, p^*) to (P):

$$-H(\xi, v, s) \to \min_{\xi, v} \tag{15}$$

s.t. $\xi \in \overline{G}, v \in V$. Furthermore to prove (13a) we have to show

$$H(x^*(t), p^*(t), s(x^*(t))) = 0 \tag{16}$$

and for the validity of (12) we must show

$$(x^*(t)^T, p^*(t)^T)^T \in \text{ arg min } \{-H(\xi, v, s) | \xi \in \overline{G}, v \in V\} \quad . \tag{17}$$

In Section 4 we will aim to verify first-order necessary and second-order sufficiency conditions for (F_W). Because of the positive homogeneity of $f(\xi, \cdot)$ we cannot expect the latter to hold even if we assume that f and s are twice continuously differentiable with respect to the appropriate variables. We will fix this problem later by imposing a regularity condition on p as an additional control constraint.

3. Analytical formulation of problem 1

Following [9] we let $e(t) = (e^1, \dots, e^n)^T$ be an arbitrary basis vector of $\mathbb{E}^n$ with the coordinates

$$\begin{aligned} e^1 &= \cos t_1 \\ e^\alpha &= \cos t_\alpha \prod_{j=1}^{\alpha-1} \sin t_j \quad \alpha = 2, \dots, n-1 \\ &\vdots \\ e^n &= \prod_{j=1}^{n-1} \sin t_j \quad . \end{aligned} \tag{18}$$

[4] In a part of the literature this function is also called Hamiltonian.

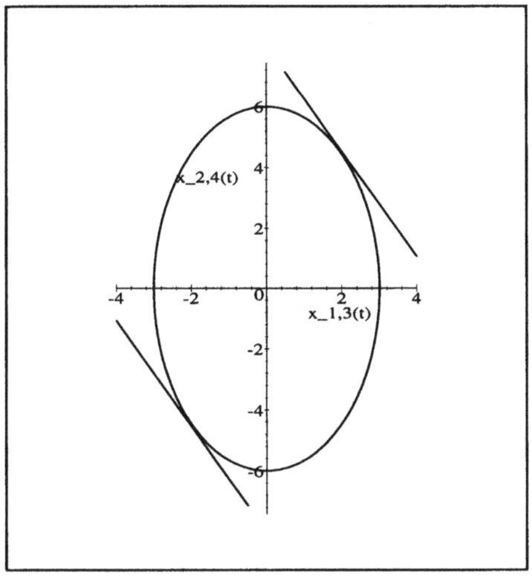

Figure 1: A convex domain and its planes of support.

The coordinates $t_1, \dots, t_{n-1}$ are spherical coordinates of the $(n-1)$-dimensional unit sphere (see [11]). We restrict these coordinates on the domain $0 \leq t_j \leq \pi$ $(j = 1, \dots, n-1)$.

If $\mathcal{B}$ is an arbitrary strongly convex domain in $\mathbb{E}^n$ the oriented plain of support E with normal vector $e(t)$ and the property that all points of $\mathcal{B}$ lie in the negative half space with respect to E is uniquely determined. The point of tangency of E and $\mathcal{B}$ is $x(t) = (x^1(t), \dots, x^n(t))^T$. The plane of support being parallel to E and having the normal vector $-e(t)$ touches $\mathcal{B}$ exactly in a point $\bar{x}(t) = -(x^{n+1}(t), \dots, x^{2n}(t))^T$ (see Fig. 1).

Hence for $j = 1, \dots, n$ and $i = 2, \dots, n-1$

$$\begin{aligned}
&x^{j+n}(0, t_2, \dots, t_{n-1}) = -x^j(\pi, t_2, \dots, t_{n-1}), \\
&x^{j+n}(\pi, t_2, \dots, t_{n-1}) = -x^j(0, t_2, \dots, t_{n-1}), \\
&x^{j+n}(t_1, \dots, t_{i-1}, \pi, t_{i+1}, \dots, t_{n-1}) \\
&= -x^j(\pi - t_1, \dots, \pi - t_{i-1}, 0, \pi - t_{i+1}, \dots, \pi - t_{n-1}), \\
&x^{j+n}(t_1, \dots, t_{i-1}, 0, t_{i+1}, \dots, t_{n-1}) \\
&= -x^j(\pi - t_1, \dots, \pi - t_{i-1}, \pi, \pi - t_{i+1}, \dots, \pi - t_{n-1}) \quad .
\end{aligned} \tag{19}$$

We now restrict ourselves to convex domains of diameter D. By a suitable rotation and translation we then get the conditions

$$\begin{aligned}
x^1(0, t_2, \dots, t_{n-1}) &= x^{1+n}(0, t_2, \dots, t_{n-1}) &= \tfrac{D}{2}, \\
x^1(\pi, t_2, \dots, t_{n-1}) &= x^{1+n}(\pi, t_2, \dots, t_{n-1}) &= -\tfrac{D}{2}, \\
x^\alpha(0, t_2, \dots, t_{n-1}) &= x^{\alpha+n}(0, t_2, \dots, t_{n-1}) &= 0, \\
x^\alpha(\pi, t_2, \dots, t_{n-1}) &= x^{\alpha+n}(\pi, t_2, \dots, t_{n-1}) &= 0
\end{aligned} \tag{20}$$

for $\alpha = 2, \dots, n$. Thus a state function $x(t) = (x^1(t), \dots, x^{2n}(t))^T$ on $\Omega = [0, \pi]^{n-1}$ which satisfies the boundary conditions (19) and (20) is assigned to every strongly convex domain $\mathcal{B} \subset \mathbb{E}^n$. In this paper we only deal with $\mathcal{B}$ which are regular so that the associated function x is an element of the class $C^{1,2n}(\Omega)$ and at the same time satisfies the condition

$$p_{(i)}(t)p_{(i)}(t) > 0 \quad \text{for a.e. } t \in \Omega \quad . \tag{21}$$

Without loss of generality we can assume

$$p_{(i)}(t)p_{(i)}(t) = 1 \quad \text{for a.e. } t \in \Omega \quad . \tag{22}$$

This is the additional regularity condition mentioned at the end of Section 2.3.

Let be $\eta^i = \eta^i(\sigma_1, \dots, \sigma_{n-1})$, $i = 1, \dots, n$, $\sigma \in \Sigma \subset \mathbb{E}^{n-1}$. Applying Gauss' Theorem we find for the volume of a domain $\mathcal{B} \subset \mathbb{E}^n$

$$\begin{aligned} |\mathcal{B}| &= \int_{\mathcal{B}} d\eta^1 \dots d\eta^n \\ &= \frac{1}{n} \int_{\Sigma} \left(\sum_{i=1}^{n} \eta^i \right) \frac{\partial(\eta^{i+1}, \dots, \eta^n, \eta^1, \dots, \eta^{i-1})}{\partial(\sigma_1, \dots, \sigma_{n-1})} \, d\sigma_1 \dots d\sigma_{n-1} \quad . \end{aligned} \tag{23}$$

We assume the parameters enumerated so that the last integrand in (23) remains nonnegative. Using our representation of $\mathcal{B}$ this leads to

$$\begin{aligned} |\mathcal{B}| = \frac{1}{n} \int_{\Omega} \sum_{i=1}^{n} (-1)^{i+1} \big(& x^i(t) p_{1\dots(i-1)(i+1)\dots n}(t) \\ & + x^{i+n}(t) p_{(1+n)\dots(i+n-1)(i+n+1)\dots 2n}(t) \big) \, dt_1 \dots dt_{n-1} \quad . \end{aligned} \tag{24}$$

Finally we skip the factor $\frac{1}{n}$ and rewrite our problem as a minimizing one getting $(\mathrm{P}_{D,V})$:

$$\begin{aligned} J(x,p) = -\int_{\Omega} \sum_{i=1}^{n} (-1)^{i+1} \big(& x^i(t) p_{1\dots(i-1)(i+1)\dots n}(t) \\ & + x^{i+n}(t) p_{(1+n)\dots(i+n-1)(i+n+1)\dots 2n}(t) \big) \, dt_1 \dots dt_{n-1} \to \inf \end{aligned} \tag{25}$$

s.t.
$x \in C^{1,2n}(\Omega)$,
Grassmann controls p fulfilling the state equations (2),
state constraints

$$\sum_{i=1}^{n} \left(x^i(t) + x^{i+n}(t) \right)^2 \leq D^2 \tag{26}$$

control constraints (8), (22),
boundary conditions (19), (20)
as an analytical formulation of Problem 1.

By Conjecture 1 in this setting the candidate for optimality is

$$x^*(t) = -\bar{x}^*(t) = \frac{D}{2} e(t), \tag{27}$$

i.e. the representative of a circle and a ball, respectively. We denote the associated Grassmann coordinates to x^* and $\bar{x}^*$ by p^*.

4. Proof of conjecture 1

The key problem now is to find an appropriate $S \in C^{3,n-1}(G)$. Generally one would start with a linear or quadratic setting ($a^j \in \mathbb{E}^{2n}, Q^j \in \mathbb{E}^{(2n)\times(2n)}$):

$$S^j(\xi^1, \ldots, \xi^{2n}) = (a^j)^T \xi \tag{28a}$$

$$S^j(\xi^1, \ldots, \xi^{2n}) = \frac{1}{2}\xi^T Q^j \xi + (a^j)^T \xi \tag{28b}$$

($j = 1, \ldots, n-1$). In view of the special structure of our problem we here use a modification of the formulation suggested in [9] with the correction given in [7]. With $C > 0$, $z = (z^1, \ldots, z^n)$, $z^i = \xi^i + \xi^{i+n}$,

$$\theta_j(z) = \operatorname{arccot}\left(\frac{z^j}{\sqrt{\sum\limits_{i=j+1}^{n} (z^i)^2}}\right) \tag{29}$$

and

$$\Psi^j(z) = C \int\limits_0^{\theta_j(z)} (\sin\theta)^{n-j-1}\, d\theta \quad (j = 1, \ldots, n-1) \tag{30}$$

we define

$$S^j(\xi) = \Psi^j(z(\xi)) \quad (j = 2, \ldots, n-1) \tag{31a}$$

and

$$S^1(\xi) = -\Psi^1(z(\xi)) \quad . \tag{31b}$$

We start the proof of Conjecture 1 with the verification of condition (13b) of Theorem 1. To this end we observe

$$I = \int\limits_\Omega dS^1(x(t)) \wedge \cdots \wedge dS^m(x(t)) = -C^{n-1} \int\limits_{\Omega_x} \prod_{j=1}^{n-1} (\sin\theta_j)^{n-j-1}\, d\theta_j \quad . \tag{32}$$

Here Ω_x denotes the image of Ω under the injective mapping $t \leftrightarrow \theta$. Since by definition the coordinates $\theta_j(z)$, ($z^i = x^i(t) + x^{i+n}(t)$), together with $\sqrt{\sum_{i=1}^n (z^i)^2}$ form a system of spherical coordinates of the point $z \in \mathbb{E}^n$ the value of the integral on the right-hand side of (32) is up to the factor $-C^{n-1}$ the measure of the n-dimensional unit sphere. Together with the boundary conditions (19) and (20) this implies the independency of I of the special state function x (see [9, p. 380] for details). Thus we have $L_0(S) = I$ and condition (13b) is satisfied.

To verify the remaining conditions stated in Klötzler's Theorem for arbitrary $n \geq 2$ would result in many technical difficulties. That is why we only deal with the case $n = 4$, although a generalization is no real problem. (Due to the fact that for $n = 2, 3$ our problem can be solved by elementarily geometric methods this is the easiest interesting case.) From (31) we calculate $s_{i_1 i_2 i_3}$ (omitting the dependency of $(\xi^1, \dots, \xi^8)$):

$$s_{123} = C^3 \frac{\xi^4 + \xi^8}{((\xi^1+\xi^5)^2 + (\xi^2+\xi^6)^2 + (\xi^3+\xi^7)^2 + (\xi^4+\xi^8)^2)^2} \tag{33a}$$

$$s_{124} = -C^3 \frac{\xi^3 + \xi^7}{((\xi^1+\xi^5)^2 + (\xi^2+\xi^6)^2 + (\xi^3+\xi^7)^2 + (\xi^4+\xi^8)^2)^2} \tag{33b}$$

$$s_{134} = C^3 \frac{\xi^2 + \xi^6}{((\xi^1+\xi^5)^2 + (\xi^2+\xi^6)^2 + (\xi^3+\xi^7)^2 + (\xi^4+\xi^8)^2)^2} \tag{33c}$$

$$s_{234} = -C^3 \frac{\xi^1 + \xi^5}{((\xi^1+\xi^5)^2 + (\xi^2+\xi^6)^2 + (\xi^3+\xi^7)^2 + (\xi^4+\xi^8)^2)^2} \tag{33d}$$

$$s_{125} = 0 \tag{33e}$$

and

$$\begin{array}{lllllllll}
s_{123} & = s_{127} & = -s_{136} & = s_{167} & = s_{235} & = -s_{257} & = s_{356} & = s_{567} \\
s_{124} & = s_{128} & = -s_{146} & = s_{168} & = s_{245} & = -s_{258} & = s_{456} & = s_{568} \\
s_{134} & = s_{138} & = -s_{147} & = s_{178} & = s_{345} & = -s_{358} & = s_{457} & = s_{578} \\
s_{234} & = s_{238} & = -s_{247} & = s_{278} & = s_{346} & = -s_{368} & = s_{467} & = s_{678} \\
s_{125} & = s_{126} & = s_{135} & = s_{137} & = s_{145} & = s_{148} & = s_{156} & = s_{157} \\
 & = s_{158} & = s_{236} & = s_{237} & = s_{246} & = s_{248} & = s_{256} & = s_{267} \\
 & = s_{268} & = s_{347} & = s_{348} & = s_{357} & = s_{367} & = s_{378} & = s_{458} \\
 & = s_{468} & = s_{478} & & & & &
\end{array} \tag{34}$$

We then calculate the constant C so that (16) holds:

$$C = \frac{1}{2} D^{\frac{4}{3}} \quad . \tag{35}$$

(By induction this result can be easily extended to the general case: $C = \frac{1}{2} D^{\frac{n}{n-1}}$.) In (8) we choose, without loss of generality, $p_{i_1^0 i_2^0 i_3^0} = p_{123}$. This is suggested by the fact that $p^*_{123}(t) = -\frac{1}{8} D^3 \sin^3 t_1 \sin^2 t_2 \sin t_3 \neq 0$ for a.e. $t \in \Omega$. Thus we can derive the $\binom{8}{3} - [1 + 3(8-3)] = 40$ equations $b_\rho(p) = 0$. Furthermore from (26) we obtain

$$G = \left\{ \xi \in \mathbb{E}^8 \,\middle|\, \sum_{i=1}^{4} (\xi^i + \xi^{i+4})^2 \leq D^2 \right\} \quad . \tag{36}$$

Next we consider (F_W). With the definitions given in Section 3 the Lagrangian for this finite-dimensional optimization problem reads as

$$\begin{aligned}\Lambda^W(\xi, v) = &-(\xi^1 v_{234} - \xi^2 v_{134} + \xi^3 v_{124} - \xi^4 v_{123} + \xi^5 v_{678} - \xi^6 v_{578} + \xi^7 v_{568} - \xi^8 v_{567}) \\ &- \sum_{i_1<i_2<i_3} s_{i_1 i_2 i_3}(\xi) v_{i_1 i_2 i_3} \\ &+ \sum_{\rho=1}^{40} \lambda_\rho b_\rho(v) + \lambda_0 \left(\sum_{i_1<i_2<i_3} v_{i_1 i_2 i_3} v_{i_1 i_2 i_3} - 1 \right) \\ &- \mu \left(D^2 - \sum_{i=1}^{4} (\xi^i + \xi^{i+4})^2 \right) \quad . \end{aligned} \tag{37}$$

Using Λ^W we now verify the well-known first-order necessary and second-order sufficiency conditions for the optimality of $(x^*(t)^T, \bar{x}^*(t)^T, p^*(t)^T)^T$ to (F_W) (see e.g. [4]). This was done by the help of the *Symbolic Computation System* Maple V. We find the Lagrange multipliers

$$\mu(t) = \frac{1}{4} D^2 \sin^2 t_1 \sin t_2 \tag{38}$$

and

$$\lambda_0(t) = 0 \tag{39}$$

(Together with the results for $n = 2, 3$ not mentioned in this paper this implies the generalization $\mu(t) = \frac{n}{2^n} D^{n-2} \prod_{j=1}^{n-2} (\sin t_j)^{n-j-1}$ and $\lambda_0(t) = 0$ for arbitrary $n \geq 2$.) Clearly the 40 multipliers $\lambda_\rho(t)$ will depend on the order of the equations $b_\rho(v) = 0$. Due to a lack of space we cannot quote them explicitly[5]. Finally we have shown a second-order sufficiency condition to hold. In summary, we can formulate the following theorem:

Theorem 2 *For $n = 4$ the triple $(x^*, \bar{x}^*, p^*)$ is a weak local minimizer for $(P_{D,V})$.*

Theorem 2 proves Conjecture 1 subject to all domains having a continuously differentiable representation.

5. Concluding remarks

Beside the generalization of Theorem 2 for arbitrary $n > 4$ a couple of questions remain open. First, one should be interested in proving that $(x^*, \bar{x}^*, p^*)$ is even a strong local minimizer for $(P_{D,V})$. It is possible to do this by calculating

$$\mathcal{H}(\xi, s) := \sup_{v \in V} H(\xi, v, s) \quad . \tag{40}$$

[5] Since the results produced by Maple V usually consume much space they cannot be given in detail here.

Similar to (F_W) we can formulate a finite-dimensional optimization problem (F_S)

$$-\mathcal{H}(\xi, s) \to \min_{\xi \in \overline{G}} \tag{41}$$

and verify conditions for the optimality of $(x^*(t)^T, \bar{x}^*(t)^T)^T$ to (F_S). Furthermore it would be interesting to explicitly compute the neighborhoods of $(x^*, \bar{x}^*, p^*)$ for which local optimality holds true. Finally we should mention again that we did not deal with nonregular convex domains of given diameter.

References

[1] Bertini, E.: *Einführung in die projektive Geometrie mehrdimensionaler Räume.* L. W. Seidel & Sohn, Wien, 1924.

[2] Blaschke, W.: *Kreis und Kugel.* Teubner-Verlag, Leipzig, 1916.

[3] Bonnesen, T.; Fenchel, W.: *Theorie der konvexen Körper.* Springer-Verlag, Berlin-Heidelberg-New York, 1974.

[4] Fiacco, A. V.; McCormic, G. P.: *Nonlinear Programming: Sequential Unconstrained Minimization Techniques.* SIAM, Philadelphia, 1990.

[5] Fuchs, G.: *Lagrangesche Variationsprobleme in Parameterdarstellung.* PhD thesis, Universität Halle-Wittenberg, 1969.

[6] Jaglom, J. M.; Boltyanski, W. G.: *Konvexe Figuren.* Deutscher Verlag der Wissenschaften, Berlin, 1956.

[7] Klemt, U.: A *Remark on the Paper "Convex Domains of Given Thickness with Smallest Surface Area" by R. Klötzler.* Reihe Mathematik M-08/1995, Technische Universität Cottbus, 1995.

[8] Klötzler, R.: Models and Applications of Duality in Optimal Control. In *Proc. IFIP Working Conference on Recent Advances in System Modeling and Optimization*, Hanoi, 1983.

[9] Klötzler, R.: Konvexe Bereiche kleinster Oberfläche bei gegebener Dicke. *Z. Anal. Anw.*, 4(4):373–383, 1984.

[10] Kubota, T.: Über konvex-geschlossene Mannigfaltigkeiten im n-dimensionalen Raume. *Sci. Rep. Tôhoku Univ.*, 14:85–99, 1925.

[11] Madelung, E.: *Die mathematischen Hilfsmittel des Physikers.* Springer-Verlag, Berlin-Göttingen-Heidelberg, 1953.

[12] Minkowski. H.: *Theorie der konvexen Körper.* In *Gesammelte Abhandlungen*, pages 131–229. Teubner-Verlag, Leipzig, 1911.

International Series of Numerical Mathematics
Vol. 124, © 1998 Birkhäuser Verlag, Basel

Isoperimetric and Isodiametric Area-minimal Plane Convex Figures

Anita Kripfganz *

Abstract. In the present paper, area-minimal plane convex figures with prescribed diameter and perimeter are studied. This geometrical extremal problem is a concave maximum problem. For figures with maximal circumradius it is associated with that of area-minimal sector-indomains. A corresponding perimeter partition problem is solved using methods of optimal control and nonlinear optimization. In dependence on the diameter and the perimeter a certain non-regular inpolyeder of the Reuleaux-triangle is area-minimal.

1. Introduction

Plane convex figures are involved with some characteristic parameters: area, diameter, perimeter, circumradius etc. Studying connections between some of them we get a special kind of geometrical extremal problem described in the survey [2] of Bonnesen and Fenchel and in the problem book [8] of Jaglom and Boltjanski. Behind them, isoperimetric and isodiametric plane convex figures of least area are yet unknown (see ref. [4]) for certain relations between the diameter and the perimeter. Using the notations

$\mathcal{A}_2$ for the set of plane convex figures $\mathcal{B} \subseteq \mathbb{E}^2$
$\mathcal{F}$ for the area,
$\mathcal{L}$ for the perimeter length,
$\mathcal{D}$ for the diameter

we get the problem formulation

$$(\mathrm{P_{L,D}}): \qquad \mathcal{F}(\mathcal{B}) \rightarrow \text{Min!} \qquad \mathcal{L}(\mathcal{B}) = L,\ \mathcal{D}(\mathcal{B}) = D\,.$$

For certain parameter values the optimal solution $\mathcal{B}^*$ is known. If $L \in [2D, 3D]$ then $\mathcal{B}^*$ is an isosceles triangle, if $L = \pi D$ then $\mathcal{B}^*$ is a Reuleaux triangle. Inequalities between area, diameter and perimeter were established by Kubota [14,15] and Favard [5]. But they are not sharp for $L \in (3D, \pi D)$. Using geometrical methods Sholander [18] has proved that the optimal figure is a polyeder. Furthermore, he has supposed that it is an inpolyeder of the Reuleaux triangle for $L \in (3D, \pi D)$.

In this paper, we suppose that $\mathcal{B}^*$ has maximal circumradius. The known solutions of problem $\mathrm{P_{L,D}}$ fulfill this property. For $L \in (3D, \pi D)$ this supposition implies that three boundary points of $\mathcal{B}^*$ are vertices of an equilateral triangle with sides of length

*Institut für Mathematik, Universität Leipzig, Augustusplatz 10, D-04109 Leipzig,
e-mail: kripfganz@mathematik.uni-leipzig.de

D. Then, the problem of area-minimal m-tuple of sector indomains is solved starting with $m = 1$ and $m = 2$. An area-minimal triple forms an inpolyeder of the Reuleaux-triangle. Under the formulated supposition this polyeder is a solution of problem $P_{L,D}$.

2. The generalized Favard problem

In 1929, Favard [5] has solved the problem of area-minimal plane convex sets where perimeter and circumradius are fixed. We generalize this problem substituting the circumcircle by a circular sector S_ω with central angle $\omega \in [0, 2\pi]$, radius $R = 1$ (without loss of generality). We denote the corresponding isosceles triangle by Δ_ω.

An ω-sector indomain $\mathcal{B}_\omega$ is a plane convex set with $S_\omega \cap \Delta_\omega \subseteq \mathcal{B}_\omega \subseteq S_\omega$ and $\Delta_\omega \cup \mathcal{B}_\omega$ convex. The perimeter of $\mathcal{B}_\omega$ is formed by the legs of Δ_ω and a convex curve. Fixing the length L of this variable boundary curve we ask for an area-minimal ω-sector indomain. This problem is solved in three steps (see ref. [9])

2.1 Structural analysis

Using the support function and the sector formula for the area we formulate the problem as an optimal control problem with distributional control, state and control restrictions:

$$(\mathrm{CP}^{\omega,\mathrm{L}}): \qquad I(x,u) = \frac{1}{2}\int_0^\omega (x_1(t)^2 - x_2(t)^2)dt \to \text{Min!}$$

$$\begin{array}{ll} \dot{x}_1 = x_2 & \\ "\dot{x}_2 = u - x_1" & x_1(0) = x_1(\omega) = 1 \\ \dot{x}_3 = x_1 & x_2(0) = x_2(\omega) = 0 \\ x_1 \leq 1 & x_3(0) = 0 \\ "u \geq 0" & x_3(\omega) = L. \end{array}$$

If the variable boundary curve does not contain any straight line then the second derivative of the support function x_1 is a summable function. In general, the conditions in quotation marks are to understand only in the sense of distributions.

An important tool for solving such problems is Klötzler's distributional version of Pontryagin's maximum principle [9]. Applying this principle to problem $(\mathrm{CP}^{\omega,\mathrm{L}})$ we have to evaluate
the maximum condition

$$(\mathrm{M}): \qquad \max_{v\geq 0,\ v\in L_\infty(0,\omega)} \langle v \cdot y_2,\ \psi\rangle = \langle u \cdot y_2,\ \psi\rangle \quad \forall\, \psi \in K,\ \psi \geq 0,$$

the canonical equations

$$\begin{aligned} (\mathrm{C}): \qquad y_1(t) &= -l_1^1 + \int_t^\omega (-x_1(\tau) + y_2(\tau) + y_3(\tau))\,d\tau - \int_t^\omega \mu(d\tau) \\ y_2(t) &= -l_2^1 + \int_t^\omega (x_2(\tau) + y_1(\tau))\,d\tau \\ y_3(t) &= -l_3^1, \end{aligned}$$

and the transversality conditions

$$(\mathrm{T}): \qquad y_1(0) = l_1^0, \;\; y_2(0) = l_2^0, \;\; y_3(0) = l_3^0.$$

In this, (x, u) is an optimal process of problem $(\mathrm{CP}^{\omega,\mathrm{L}})$, l^0, l^1 are vectors from $\mathbb{E}^3$, $y : [0, \omega] \to \mathbb{E}^3$ is a left-continuous vector valued function of bounded variation and μ is a nonnegative regular measure concentrated on the set $\{t \in [0, \omega] \mid x_1(t) = 1\}$. A distribution is a continuously linear functional over the basic space $K = W^1_{\infty,loc}$.

The maximum condition leads to $y_2 \leq 0$ and the complementarity condition

$$\langle u \cdot y_2, \psi \rangle = 0 \quad \forall \psi \in K, \; \psi \geq 0 \tag{1}$$

which is equivalent to

$$\operatorname{supp} u \subseteq \{t \mid y_2(t) = 0\}. \tag{2}$$

With respect to the state restriction $x_1 \leq 1$ we distinguish *active* ($x_1(t) = 1$) and *nonactive* parts of the solution. The canonical equations yield

$$y_2(t) = A_i \cos(t - \alpha_i) + y_3 \tag{3}$$

for $t \in (a_i, b_i) \subseteq [0, \omega] \setminus \operatorname{supp}\mu$ with $y_2(t) < 0$,
and

$$\mu([t, \omega]) = -l_1^1 + (\omega - t)(y_3 - 1) \tag{4}$$

for $t \in (\omega', \omega] \subseteq \operatorname{supp}\mu$. The analytical and geometrical discussion of the arising cases proves the following necessary condition for $\mathcal{B}^*_\omega$.

Lemma 1 *The optimal ω-sector indomain $\mathcal{B}^*_\omega$ is a composition of at most a generalized polyhedral partial sector indomain $\mathcal{P}_{\omega'} \subseteq S_{\omega'}$ and a remaining partial sector $S_{\omega''}$.*

A generalized polyeder may have accumulation points of vertices. The ω'-sector indomain $\mathcal{P}_{\omega'}$ has to be area-minimal with respect to its partial perimeter length L'. This implies that all vertices of $\mathcal{P}_{\omega'}$ except the vertex of S'_ω belong to the sector arc. If $\mathcal{P}_{\omega'}$ is a polyhedral sector indomain (without accumulation points of vertices) then it is called inpolyeder of $S_{\omega'}$. The corresponding boundary curve joining the legs of Δ'_ω is called inpolygon of S'_ω.

2.2 Area-minimal inpolyeder

We investigate the structure of area-minimal sector inpolyeders having fixed perimeter length. Describing an inpolyeder by means of his central angles φ_i we obtain the nonlinear optimization problem

$$(\mathrm{P}^{\omega,\mathrm{L}}): \qquad \begin{aligned} F_\omega(\varphi, n) &= \frac{1}{2}\sum_{i=1}^{n} \sin\varphi_i \to \text{Min!} \\ & 2\sum_{i=1}^{n} \sin\frac{\varphi_i}{2} = L \\ & \sum_{i=1}^{n} \varphi_i = \omega \\ & 0 < \varphi_i < \omega \quad \text{for } i = 1, \ldots, n. \end{aligned}$$

If we fix the optimal but unknown number n and evaluate necessary and sufficient optimality conditions we get

$$\alpha := \varphi_1 = \cdots = \varphi_{n-1} \geq \varphi_n = \omega - (n-1)\alpha =: \delta(\alpha) > 0. \tag{5}$$

The corresponding inpolyeder is called almost regular and regular for $\delta = \alpha$, resp. Putting these central angles into the isoperimetric condition we have to solve the non-linear equation

$$l_\omega(\alpha) = L. \tag{6}$$

In this, the function

$$l_\omega(\alpha) = \begin{cases} 2\sin\frac{\alpha}{2} + 2\sin\frac{\delta(\alpha)}{2} & \text{for } \alpha \in [\frac{\omega}{2}, Min\{\omega,\pi\}] \\ 2(k-1)\sin\frac{\alpha}{2} + 2\sin\frac{\delta(\alpha)}{2} & \text{for } \alpha \in [\frac{\omega}{k}, \frac{\omega}{k-1}), k \geq 3 \end{cases}$$

is a piece-wise smooth and strictly monotonously increasing function. Hence, equation (6) is uniquely solvable, and the optimal number n of central angles is uniquely determined from

$$L \in (l_{n-1}^\omega, l_n^\omega] \tag{7}$$

where $l_n^\omega = 2n\sin\frac{\omega}{2n}$ denotes the length of an n-sided regular inpolygon of S_ω.

Lemma 2 *An area-minimal inpolyeder of S_ω is uniquely determined and almost regular.*

The area of an almost regular inpolyeder depends on L. This optimal value function of problem ($\mathrm{P}^{\omega,\mathrm{L}}$) is implicitly given by

$$f_\omega(L) := \frac{n-1}{2}\sin\alpha_\omega(L) + \frac{1}{2}\sin\delta_\omega(L) \tag{8}$$

where $\alpha_\omega(L)$ is determined from

$$L = 2(n-1)\sin\frac{\alpha}{2} + 2\sin\frac{\omega-(n-1)\alpha}{2} \tag{9}$$

and $\delta_\omega(L) = \delta(\alpha_\omega(L))$. Recalling Favard we denote the function f_ω by *'fonction penetrante'*. It is a continuous and strictly monotonously increasing function which is piece-wise differentiable on each interval $I_n^0 := (l_{n-1}^\omega, l_n^\omega)$. For $n \geq 3$, f_ω' is strictly concave and positive, f_ω''' and $f_\omega^{(4)}$ are negative on each I_n^0. The second derivative has a unique zero λ_n^ω on I_n^0. Hence, f_ω is piece-wise strictly convex-concave with turning points λ_n^ω. It is a roughly convex function in the sense of Phu [16] (see refs. [7,13]). For $n = 2$, f_ω is strictly convex on I_2^0.

2.3 Area-minimal indomains

From Lemma 1 we have the representation $\mathcal{B}_\omega^* = \mathcal{P}_{\omega'} \cup S_{\omega''}$ with $\omega = \omega' + \omega''$. Now, we exclude polygonal accumulation points and circular arcs from the boundary structure of $\mathcal{B}^*$.

Because of Lemma 2 each partial polyeder of $\mathcal{P}_{\omega'}$ without polygonal accumulation points has to be almost regular. The corresponding central angles fulfill condition (5).

This excludes the existence of an accumulation point of vertices. Hence, $\mathcal{P}_{\omega'}$ is almost regular. The area of $\mathcal{B}^*_\omega$ is now given by the continuous function

$$\hat{f}(L') = \frac{1}{2}(L - L') + f_{\omega'}(L') \tag{10}$$

with $L = \omega'' + L'$ and $\omega' = \omega - L + L'$. Because $\mathcal{B}^*_\omega$ is area-minimal for given L the parameter L' minimizes $\hat{f}$ over $[\tilde{L}, L]$ where $\tilde{L}$ denotes the minimal possible value of L'. The function $\hat{f}$ is implicitly differentiable a.e. with

$$\frac{d\hat{f}}{dL'} = -\frac{1}{2} + \cos\frac{\alpha}{2} + \cos\frac{\delta}{2}(1 - \cos\frac{\alpha}{2}) - \frac{1}{2} \leq 0 \tag{11}$$

where $\alpha = \alpha(L', \omega')$ is given by (9). This implies $L' = L$ and $\omega'' = 0$.

Theorem 1 *For given L, the area-minimal ω-sector indomain is a uniquely determined almost regular inpolyeder of S_ω.*

3. The m-tuple partition problem

We consider m-tuple of plane figures $\mathcal{B}_i$ with perimeter length $L_i + 2$. Each of them is inscriped in a unit circular sector S_ω. Its total perimeter length is fixed. We study the problem of finding m-tuple of least total area:

$$(\mathrm{PP}): \qquad \sum_{i=1}^{m} \mathcal{F}(\mathcal{B}_i) \rightarrow \quad \text{Min!}$$
$$\sum_{i=1}^{m} \mathcal{L}(\mathcal{B}_i) = m \cdot \Lambda\,.$$

The parameter Λ is a suitable choice.

This extremal problem shows a hierarchic structure. It can be divided into m generalized Favard problems by fixing the perimeter length of each sector indomain, parametrically. Through the given total perimeter length these problems are linked together. All indomains have to be area-minimal. Theorem 1 implies that only m-tuple of almost regular inpolyeders come into question as optimal solutions. Using the *'fonction penetrante'* as optimal value function of the generalized Favard problem we obtain the isoperimetric partition problem

$$({}^{\mathrm{m}}\mathrm{P}^{\omega,\Lambda}): \qquad {}^{m}f_\omega(\mathbf{L}) = \sum_{i=1}^{m} f_\omega(L_i) \rightarrow \text{Min!}$$
$$\sum_{i=1}^{m} L_i = m \cdot \Lambda$$
$$\underline{L}_\omega \leq L_1 \leq \cdots \leq L_m \leq \omega.$$

The hierarchic structure of this problem is expressed in the following necessary optimality condition: If $\mathbf{L}^*$ is an optimal perimeter partition then each selection ${}^k\mathbf{L}^* = (L^*_{i_1}, \ldots, L^*_{i_k})$ with $\Lambda_k = \frac{1}{k}\sum_{j=1}^{k} L^*_{i_j}$ has to be an optimal solution of the partition problem $({}^{\mathrm{k}}\mathrm{P}^{\omega,\Lambda_k})$. This property suggests applying the method of complete induction for solving problem $({}^{\mathrm{m}}\mathrm{P}^{\omega,\Lambda})$.

3.1 The pair partition problem

For $m = 2$, we fix the number n of central angles by $\Lambda \in (l^\omega_{n-1}, l^\omega_n]$, study stationary points of ${}^2f_\omega$ for feasible $\mathbf{L} \in Q_n = (l^\omega_{n-1}, l^\omega_n) \times (l^\omega_{n-1}, l^\omega_n)$ as well as $\mathbf{L} \notin Q_n$, and compare their objective values (see refs. [6,11]).

It could be shown that there is no local minimal solution of ${}^2f_\omega$ over Q_n with $L_1 \neq L_2$. For this, second order necessary optimality conditions are studied. The properties of the *'fonction penetrante'* up to its fourth derivative are used in the proof. The minimal solution over Q_n could be attained for the symmetric pair $\mathbf{L} = (\Lambda, \Lambda)$, only. We denote it by twin solution.

Outside Q_n the global minimal solution is attained at an unsymmetric partition

$$\mathbf{L} = \begin{cases} (l^\omega_{n-1}, 2\Lambda - l^\omega_{n-1}) & \text{for } \Lambda \leq \frac{1}{2}(l^\omega_{n-1} + l^\omega_n) \\ (2\Lambda - l^\omega_n, l^\omega_n) & \text{for } \Lambda \geq \frac{1}{2}(l^\omega_{n-1} + l^\omega_n). \end{cases} \tag{12}$$

In these partitions, a regular inpolyeder of S_ω is involved. To show (12) a pair of almost regular inpolyeders of S_ω is considered as one inpolyeder of $S_{2\omega}$. The pair partition problem $({}^2\mathrm{P}^{\omega,\Lambda})$ is embedded into a generalized Favard problem $(\mathrm{P}^{2\omega,2\Lambda})$. An additional restriction ensures that $L \in Q_n$ is excluded from the feasible domain. It turns out that an optimal inpolyeder of $S_{2\omega}$ is a composition of two almost regular ω-sector inpolyeders with perimeter partitions of type (12).

Now, we compare the objective values of the founded solutions by means of the defect function

$${}^2D^n_\omega(\Lambda) = \begin{cases} f_\omega(l^\omega_{n-1}) + f_\omega(2\Lambda - l^\omega_{n-1}) - 2f_\omega(\Lambda) & \text{for } \Lambda \leq \frac{1}{2}(l^\omega_{n-1} + l^\omega_n) \\ f_\omega(2\Lambda - l^\omega_n) + f_\omega(l^\omega_n) - 2f_\omega(\Lambda) & \text{for } \Lambda \geq \frac{1}{2}(l^\omega_{n-1} + l^\omega_n) \end{cases}.$$

This function describes Focke's mid-point convexity defect [6] of f_ω. The function ${}^2D^n_\omega$ has a unique zero ${}^2\lambda^\omega_n$ over $(l^\omega_{n-1}, l^\omega_n)$ which causes a solution branching between the twin solution and the unsymmetric partitions (12).

Theorem 2 *Let $\Lambda \in (l^\omega_{n-1}, l^\omega_n]$. Then, the twin solution is an optimal partition of problem $({}^2\mathrm{P}^{\omega,\Lambda})$ for $n = 2$ and for $\Lambda \leq {}^2\lambda^\omega_n$ if $n \geq 3$. In the other cases, the nonsymmetric partitions are optimal. The branch points ${}^2\lambda^\omega_n$ satisfy the relation ${}^2\lambda^\omega_n < \frac{1}{2}(l^\omega_{n-1} + l^\omega_n)$, asymptotically.*

3.2 Area-minimal m-tuple

The induction proof is based on the hierarchic structure of problem $({}^m\mathrm{P}^{\omega,\Lambda})$ (see ref. [12]). The set of partitions satisfying corresponding optimality conditions contains at most two elements. The transition from symmetric to nonsymmetric partitions is described by an $\frac{1}{m}$-point convexity defect of the *'fonction penetrante'*. The unique existence of the zero ${}^m\lambda^\omega_n$ of the defect function

$${}^mD^n_\omega(\Lambda) = \begin{cases} (m-1)f_\omega(l^\omega_{n-1}) + f_\omega(m\Lambda - (m-1)l^\omega_{n-1}) - mf_\omega(\Lambda) & \text{for } \Lambda \leq \frac{1}{m}((m-1)l^\omega_{n-1} + l^\omega_n) \\ (m-1)f_\omega(\frac{m\Lambda - l^\omega_n}{m-1}) + f_\omega(l^\omega_n) - mf_\omega(\Lambda) & \text{else} \end{cases}$$

which is defined on $[l^\omega_{n-1}, l^\omega_n]$ follows from the properties of f_ω and its convexity defect.

Theorem 3 *Let $\Lambda \in (l^{\omega}_{n-1}, l^{\omega}_{n}]$. Then, there exists a sequence $\{^{k}\lambda^{\omega}_{n}\}^{m}_{k=1}$ of branch points satisfying the recursive conditions*

$$^{k}\lambda^{\omega}_{n} \in \left(\frac{1}{k}(l^{\omega}_{n-1} + (k-1)^{k-1}\lambda^{\omega}_{n}),^{k-1}\lambda^{\omega}_{n}\right)$$

and describing the branching structure of the optimal solution $\mathbf{L}^*$ of the partition problem $(^{m}\mathrm{P}^{\omega,\Lambda})$ in the following manner:

1)

$$\mathbf{L}^* = (\Lambda, \ldots, \Lambda)$$

for $\Lambda \in [l^{\omega}_{n-1},^{m}\lambda^{\omega}_{n}]$,

2)

$$\mathbf{L}^* = (\underbrace{l^{\omega}_{n-1}, \ldots, l^{\omega}_{n-1}}_{m-j+1},\ m \cdot \Lambda - j \cdot l^{\omega}_{n} - (m-j-1)l^{\omega}_{n-1}, \underbrace{l^{\omega}_{n}, \ldots, l^{\omega}_{n})}_{j}$$

for $\Lambda \in \frac{1}{m}[(m-j)^{m-j}\lambda^{\omega}_{n} + jl^{\omega}_{n},\ (m-(j+1))l^{\omega}_{n-1} + (j+1)l^{\omega}_{n}]$ with $j \in \{0, \ldots, m-1\}$,

3)

$$\mathbf{L}^* = (\frac{m\Lambda - (j+1)l^{\omega}_{n}}{m-(j+1)}, \ldots, \frac{m\Lambda - (j+1)l^{\omega}_{n}}{m-(j+1)},\ \underbrace{l^{\omega}_{n}, \ldots, l^{\omega}_{n})}_{j+1}$$

for $\Lambda \in \frac{1}{m}[Max\{(m-j) \cdot {}^{m-j}\lambda^{\omega}_{n} + j \cdot l^{\omega}_{n},\ (m-(j+1))l^{\omega}_{n-1} + (j+1)l^{\omega}_{n}\},\ (j+1)l^{\omega}_{n} + (m-(j+1))^{m-(j+1)\lambda^{\omega}_{n}}]$ with $j \in \{0, \ldots, m-2\}$.

The branch points $^{m}\lambda^{\omega}_{n}$ are numerically calculated by C. Bröcker [3] using a Newton method (c.f. table 1).
Theorem 3 shows that regular $(n-1)$-sided, regular n-sided and almost regular n-sided inpolygons can participate in the total solution. The solution quality changes from symmetric to nonsymmetric partitions at the branch point $^{m}\lambda^{\omega}_{n}$. A further solution branching appears at the endpoints of the third type of parameter intervals. These intervals are nonempty for all n. A parameter interval of the second type could be empty for small n.

n	λ^{ω}_{n}	$^{2}\lambda^{\omega}_{n}$	$^{3}\lambda^{\omega}_{n}$
3	1,040688511114	1.040450413358	1,040192062429
4	1,043274926103	1.043114785701	1,042914996012
5	1,044695465118	1.044661940328	1,044538632024
6	1,045493875409	1.045485013601	1,045432384556
7	1,045972672932	1.045969679516	1,045943653450
8	1,046278280656	1.046277086069	1,046262813669
9	1,046483837905	1.046483299367	1,046474844637
10	1,046628166088	1.046627899577	1,046622582538

Table 1: Branch points for $\omega = \frac{\pi}{3}$ and small n

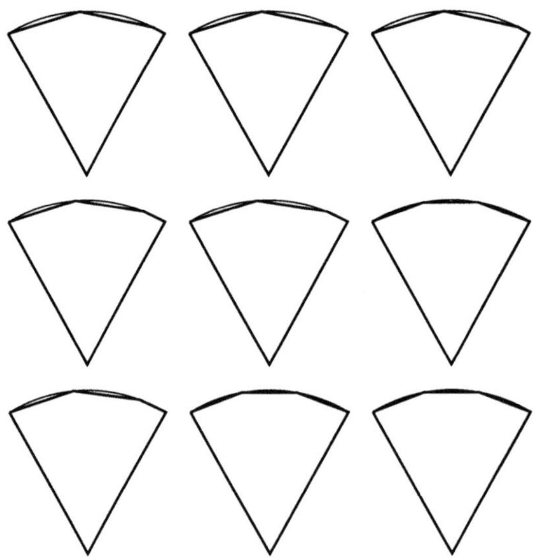

Figure 1: Area-minimal triples of $\frac{\pi}{3}$-sector indomains for $n = 4$

4. The least area problem ($\mathrm{P_{L,D}}$)

The isoperimetric and isodiametric least area problem is now discussed. First, we study this problem as an extremal problem in a complete metric space.

The feasible domain of problem ($\mathrm{P_{L,D}}$) is the set $\mathcal{A}_2(L, D) \subseteq \mathcal{A}_2$ of all isoperimetric and isodiametric congruent representatives of plane convex and compact figures. The set $\mathcal{A}_2$ is a closed convex cone with respect to the Minkowski addition and the multiplication with a non-negative real scalar. Equiped with the Hausdorff metric δ, $(\mathcal{A}_2, \delta)$ is a complete metric space (see ref. [17]). The set $\mathcal{A}_2(L, D)$ is a closed and bounded subset of $\mathcal{A}_2$ because $\mathcal{L}$ and $\mathcal{D}$ depend continuously on $\mathcal{B}$. Hence, $\mathcal{A}_2(L, D)$ is compact but without interior points. Furthermore, $\mathcal{A}_2(L, D)$ is a convex set if the diameter is suitable fixed. These statements contain the Blaschke selection theorem [1] which says that from each bounded sequence of convex bodies one can select a subsequence converging to a convex body. The root of the objective functional of problem ($\mathrm{P_{L,D}}$) is continuous and convex on $\mathcal{A}_2$. This statement follows from the plane case of the Brunn-Minkowski Theorem which describes the behaviour of mixed volumes of convex bodies (see refs. [2,17]).

We have shown that ($\mathrm{P_{L,D}}$) is a concave minimum problem, optimal solutions exist. Now, we take into consideration the circumradius of the optimal figure $\mathcal{B}^*$. The circumcircle is the least circle covering a convex figure $\mathcal{B}$. It is generated either by two diametral opposite boundary points of $\mathcal{B}$ or by three boundary points forming an acute-angled triangle. For given diameter D the circumradius $\mathcal{R}(\mathcal{B})$ satisfies the inequality $D/2 \leq \mathcal{R}(\mathcal{B}) \leq D/\sqrt{3}$. The functional $\mathcal{R}$ is continuous and convex.

For values $L \in [2D, 3D] \cup \{\pi D\}$ it is known that $\mathcal{B}^*$ has maximal circumradius. For $L \in (3D, \pi D)$ we study figures $\mathcal{B} \in \mathcal{A}_2(L, D)$ with $R = \frac{D}{\sqrt{3}}$. Area-minimal ones are composed of three almost regular $\frac{\pi}{3}$-sector inpolyeders. This follows from the above mentioned properties of the circumcircle as well as from the solution of a suitable perimeter partition problem discussed before.

Theorem 4 *The area-minimal figure of given perimeter $L \in (3D, \pi D)$, diameter D and circumradius $R = \frac{D}{\sqrt{3}}$ is an inpolyeder of the Reuleaux triangle composed of an optimal triple of $\frac{\pi}{3}$-sector inpolyeders.*

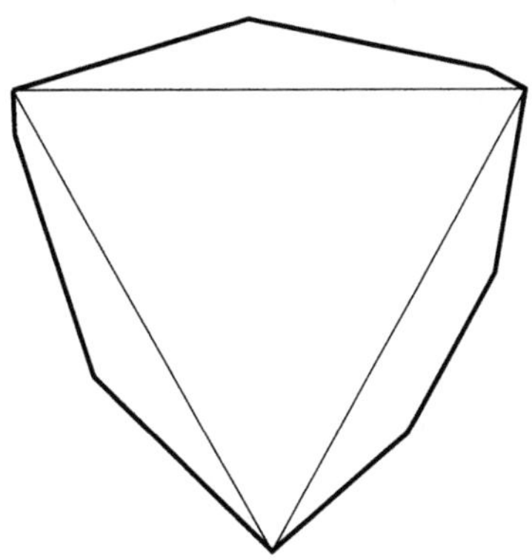

Figure 2: Area-minimal figure for $D = 1,\ L = 3,1176$

Figures having minimal circumradius are not optimal. This is realized as follows. A figure $\mathcal{B} \in \mathcal{A}_2(L, D)$ with $R = \frac{D}{2}$ is composed of two almost regular π-sector inpolyeders. If $\mathcal{B}$ is a regular inpolyeder of the circumcircle then it is not area-minimal for given L and D. To show this we vary two opposite vertices on an ellipse such that D and L stay unchanged. In the other case, we consider a non-centrosymmetric composition of the two almost regular π-sector inpolyeders. There are two vertices with distance less than D. But geometrically it is easy to see that each vertex of $\mathcal{B}^*$ has to be the endpoint of a diameter. Hence, the circumradius of $\mathcal{B}^*$ is greater than $\frac{D}{2}$. It remains to show that $\mathcal{R}(\mathcal{B}^*)$ is always maximal.

Acknowledgement
I would like to thank Joachim Focke and Rolf Klötzler for drawing my attention to the problems solved in this paper, and for stimulating discussions.

References

[1] Blaschke, W.: *Kreis und Kugel.* Leipzig: Teubner-Verlag 1916.

[2] Bonnesen, T.; Fenchel, W.: *Theorie der konvexen Körper.* Berlin: Springer-Verlag 1934.

[3] Bröcker, C.: *Über ein ungeklärtes Optimierungsproblem zu ebenen konvexen Bereichen.* Diplomarbeit. Universität Leipzig 1994.

[4] Croft, H.T.; Falconer, K.J.; Guy, R.K.: *Unsolved Problems in Geometry.* Berlin: Springer-Verlag 1991.

[5] Favard, J.M.: *Problèmes d'Extremums Relatifs aux Courbes Convexes* (I). Annales scientifiques de l' Ecole Normale Supérieure 46(1929), 345–369.

[6] Focke, J.; Klötzler, R.: *Flächenminimale Paare von Inpolygonen des Einheitskreises.* Manuskripte. Leipzig 1989.

[7] Hartwig, H.: *A Note on Roughly Convex Functions.* Optimization 38(1996), 319–327.

[8] Jaglom, I.M.; Boltjanski, W.G.: *Konvexe Figuren.* Berlin: Deutscher Verlag der Wissenschaften 1956.

[9] Klötzler, R.: *On a Distributional Version of Pontryagin's Maximum Principle.* Optimization 19(1988), 413–419.

[10] Kripfganz, A.: *The Generalized Favard Problem.* Contributions to Algebra and Geometry 36(1995)2, 185–202.

[11] Kripfganz, A.: *An Isoperimetric Partition Problem.* To appear in: Contributions to Algebra and Geometry.

[12] Kripfganz, A.: *Solution Branching of a General Perimeter Partition Problem.* Submitted to: Contributions to Algebra and Geometry.

[13] Kripfganz, A.: *Favard's 'Fonction Pénétrante' - A Roughly Convex Function.* Optimization 38(1996), 329–342.

[14] Kubota, T.: *Einige Ungleichheitsbeziehungen über Eilinien und Eiflächen.* Sci. Rep. Tôhoku Univ. 12(1923), 45–65.

[15] Kubota, T.: *Eine Ungleichheit für Eilinien.* Mathematische Zeitschrift 20(1924), 264–266.

[16] Phu, H.X.: *Some Analytical Properties of γ-convex Functions on the Real Line.* JOTA, 91(1996)3, 671–694.

[17] Schneider, R.: *Convex Bodies: The Brunn-Minkowski Theory.* Cambridge University Press 1993.

[18] Sholander, M.: *On Certain Minimum Problems in the Theory of Convex Curves.* Trans. Am. Math. Soc. 73(1952)1, 139–173.

International Series of Numerical Mathematics
Vol. 124,

Minimizing The Noise of an Aircraft During Landing Approach

Bernd Kugelmann*

Abstract. In the past, aircraft performance optimization studies have been mainly associated with time or energy criteria, i. e. minimizing the time of climb, maximizing range, minimizing fuel consumption and so on. There have been fewer attempts to restrict or minimize the aircraft noise during takeoff and/or landing. This noise can be reduced not only by improving the noise characteristics of the engines and the aircraft, but also by developing suitable guidance and control systems. In this paper the choice of reasonable optimality criteria will be discussed in order to minimize the amount of noise inconvenience encountered by people living in the community surrounding the airport. The resulting optimal control problems will be solved for a special aircraft model by the indirect approach using the maximum principle. Several state as well as control constraints will be taken into account. The numerical algorithms used for the computation of the noise minimal trajectories will be presented and some results will be shown.

1. Problem formulation

The dynamics of the aircraft are described by the following two-dimensional point-mass model:

$$\begin{aligned} m\dot{v} &= T\cos\alpha - D - W\sin\gamma &&=: mV(v,\gamma,h,x,T,\alpha),\\ mv\dot{\gamma} &= T\sin\alpha + L - W\cos\gamma &&=: mv\Gamma(v,\gamma,h,x,T,\alpha),\\ \dot{h} &= v\sin\gamma &&=: H(v,\gamma,h,x,T,\alpha),\\ \dot{x} &= v\cos\gamma &&=: X(v,\gamma,h,x,T,\alpha), \end{aligned} \tag{1}$$

for the state variables

$v(t)$	velocity [m/s],
$\gamma(t)$	flight path angle [rad],
$h(t)$	altitude [m],
$x(t)$	distance over ground [m].

The right-hand side depends on the controls

$T(t)$	thrust [N],
$\alpha(t)$	thrust-angle [rad],

and the aerodynamical forces

$D(v)$	drag [N],
L	lift [N],
W	weight [N].

*Technische Universität München, D-80290 München, Germany,
email: kugelm@mathematik.tu-muenchen.de

For simplicity lift L and weight W are assumed to be constant, satisfying the equality

$$L = W. \tag{2}$$

Assumption (2) together with the standard model of a quadratic polar

$$\begin{aligned} D &= \frac{1}{2}\rho v^2 (C_{D_0} + kC_L^2) \\ L &= \frac{1}{2}\rho v^2 C_L \end{aligned}$$

yields a formula for the drag D of the form

$$D(v) = c_1 v^2 + \frac{c_2}{v^2} \tag{3}$$

with coefficients c_1, c_2 depending on the air-density ρ and some parameters C_{D_0} and k. As a consequence formula (3) is only valid as long as (2) is satisfied and it is not valid for example if v is close or equal to 0. The data used for the numerical analysis are taken from [8]

$$c_1 = 0.226 \left[\frac{\text{kg}}{\text{m}}\right] \qquad c_2 = 5.2 \times 10^6 \left[\frac{\text{kg m}^3}{\text{s}^6}\right],$$

as well as the boundary conditions

$$\begin{array}{llll} v(0) = 124, & \gamma(0) = 0, & h(0) = 1197, & x(0) = 0, \\ v(t_f) = 77.5, & \gamma(t_f) = 0, & h(t_f) = 0, & x(t_f) = 15000. \end{array} \tag{4}$$

where the final time t_f is free and therefore part of the optimization process.

The performance criterion is an estimation of the noise inconvenience for the community surrounding an airport: The noise level N_p in decibel (dB) perceived by a person at a distance r from the aircraft with maximum thrust $T_{\max}$ is expressed as

$$N_p = N_0 + 25 \log \frac{r_{\text{ref}}}{r} + 52 \log \frac{T}{T_{\max}}$$

where N_0 is the perceived noise level in decibel at a reference distance $r_{\text{ref}} = 152\,[\text{m}]$ (compare [5], [8]). At any point along the flight trajectory one considers the molestation of that person who is affected most by the noise, i.e. that person who is closest to the aircraft. Therefore, in the performance index, the distance r has to be chosen as a function of the altitude h, this function representing the zoning-plan of the surrounding community. As an approximation for this function

$$r = h + 50$$

is used in the sequel. Finally the effect of the noise duration is added to the criterion by defining the effective perceived noise E_p

$$E_p = N_p + 20 \log \frac{v_{\text{ref}}}{v}$$

to yield the integral effective perceived noise I_1:

$$I_1 = \int_{x(0)}^{x(t_f)} E_p \, \mathrm{d}x. \tag{5}$$

However this noise criterion (5) is somewhat inadequate because of the logarithmic nature of the decibel scale. Doubling the decibel value results in a quadratic increase of the noise level. Therefore a peak value in the integrand of I_1 may occur on short intervals of the optimal solution, which results in short-term noise levels that are not acceptable for the community. There is another way of looking at the deficiency of I_1: The criterion contains penalties for flying with maximum thrust and for flying at low altitudes. But it includes no additional penalty if these two conditions come together. Therefore the worst configuration for the people in the neighborhood of the airport is not included in the performance index I_1, namely trajectories at low altitude with maximum thrust. Because of the conceptional disadvantage of (5) there are also great difficulties for the numerical treatment of this optimal control problem (see [10]).

In order to avoid these difficulties, (5) is replaced by

$$I_2 = \int_{x(0)}^{x(t_f)} 10^{E_p/10} \, \mathrm{d}x = \int_{x(0)}^{x(t_f)} \frac{c_3 T^{5.2}}{v^2 (h+50)^{2.5}} \, \mathrm{d}x \tag{6}$$

with the constant value $c_3 = 18.73$.

This model is a generalization of the formulation in [8], where the solutions with piecewisely constant flight path angle γ are investigated. In this paper the cases with arbitrary γ and arbitrary thrust angle α are to be discussed.

2. Indirect approach I

The aim of this section is to find the necessary conditions for a solution of (1) and (4) which minimizes (6). Additionally different state and/or control constraints are to be investigated, some of them being necessary for the existence of an optimal solution, some of them resulting from practical demands for a realistic flight trajectory.

Rewriting (6) with respect to the independent variable t yields

$$I_2 = \int_0^{t_f} \frac{c_3 T^{5.2}}{v (h+50)^{2.5}} \cos\gamma \, \mathrm{d}t. \tag{7}$$

According to the maximum principle [9] necessary conditions can be formulated by means of the Hamiltonian function $\mathcal{H}$:

$$\mathcal{H}(v,\gamma,h,x,\lambda_v,\lambda_\gamma,\lambda_h,\lambda_x,T,\alpha) = \frac{c_3 T^{5.2}}{v(h+50)^{2.5}} \cos\gamma + \lambda_v V + \lambda_\gamma \Gamma + \lambda_h H + \lambda_x X$$

where $V = V(v,\gamma,h,x,T,\alpha)$, $\Gamma = \Gamma(v,\gamma,h,x,T,\alpha)$, $H = H(v,\gamma,h,x,T,\alpha)$, $X = X(v,\gamma,h,x,T,\alpha)$ represent the right-hand side of the equations of motion (1) and

$\lambda_v(t)$, $\lambda_\gamma(t)$, $\lambda_h(t)$, $\lambda_x(t)$ are the so-called adjoint functions. The adjoint functions corresponding to the optimal solution satisfy the adjoint differential equations:

$$\dot{\lambda}_y(t) = -\frac{\partial \mathcal{H}}{\partial y} \tag{8}$$

where y stands for any of the state variables v, γ, h or x. Moreover the optimal controls T_0, α_0 minimize the Hamiltonian $\mathcal{H}$ along the optimal trajectory:

$$\mathcal{H}(v,\gamma,h,x,\lambda_v,\lambda_\gamma,\lambda_h,\lambda_x,T_0,\alpha_0) = \min_{T,\alpha} \mathcal{H}(v,\gamma,h,x,\lambda_v,\lambda_\gamma,\lambda_h,\lambda_x,T,\alpha) \tag{9}$$

where the minimum is sought in the appropriate, i.e. feasible control sets. Since there is no constraint for the controls, an explicit and unique expression for T_0, $\sin\alpha_0$ and $\cos\alpha_0$ in terms of the state and adjoint variables can be found for performance index (6)[1] by using (9):

$$\begin{aligned} \sin\alpha_0 &= -\frac{\lambda_\gamma/v}{\sqrt{(\lambda_\gamma/v)^2+\lambda_v^2}}, \\ \cos\alpha_0 &= -\frac{\lambda_v}{\sqrt{(\lambda_\gamma/v)^2+\lambda_v^2}}, \\ T_0 &= \left(\frac{v(h+50)^{2.5}\sqrt{(\lambda_\gamma/v)^2+\lambda_v^2}}{5.2c_3 m\cos\gamma}\right)^{\frac{1}{4.2}}. \end{aligned} \tag{10}$$

Substituting (10) into (1) and (8) yields a two-point boundary value problem for the state and adjoint variables with boundary conditions (4). In order to fix the free final time t_f the so-called transversality condition is used:

$$\mathcal{H}(v(t_f),\gamma(t_f),h(t_f),x(t_f),\lambda_v(t_f),\lambda_\gamma(t_f),\lambda_h(t_f),\lambda_x(t_f),T_0(t_f),\alpha_0(t_f)) =: \mathcal{H}(t_f) = 0.$$

3. Numerical solution of unconstrained problem

The two-point boundary value problem was solved numerically by means of the code MUMUS [4]. This is a multiple shooting algorithm (see [1]) that can be applied to multi-point boundary value problems too. Given a *good* initial guess for the solution, the method produces highly accurate results. If no *good* starting trajectory is available, the direct method DIRCOL [11] can be used to get a good approximation for the state and also for the adjoint variables. The numerical results for the unconstrained problem (performance index (6)) are shown in figures 1 and 2. The optimal thrust

[1]For performance index (5) the Hamiltonian has the form

$$\mathcal{H} =: 52\log\left(\frac{T}{T_{\max}}\right) v\cos\gamma + T\left[\lambda_v\frac{\cos\alpha}{m} + \lambda_\gamma\frac{\sin\alpha}{mv}\right] + \mathcal{H}_2(v,\gamma,h,x,\lambda_v,\lambda_\gamma,\lambda_h,\lambda_x).$$

This is a concave function with respect to the thrust T and therefore no solution T_0 of (9) can be found for $T \in \mathbb{R}$ and there is no solution of the corresponding optimal control problem. If an additional control constraint $T \in [T_{\min}, T_{\max}]$ is included in the problem formulation, the concavity of $\mathcal{H}$ results in a bang-bang structure of the thrust control. Of course, a constraint like this is also reasonable from a technical point of view.

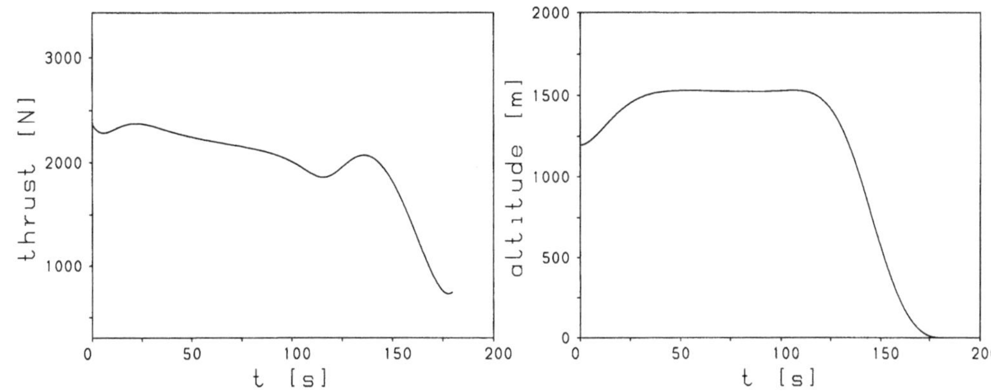

Figure 1: Thrust and altitude versus time for the unconstrained problem

history stays well inside the realistic interval [300, 3420], which is prescribed for the investigated type of aircraft (see [8]). Therefore there is no need to introduce an additional control constraint for the thrust. Constraints for the second control variable α will be investigated in section 6. The history for the altitude and for the flight path angle reflect the aim of the performance criterion, i.e. to minimize the noise inconvenience for the community on the ground: There is an ascent phase at the beginning of the trajectory in order to maximize the distance from the ground, followed by an almost horizontal flight, which ends in a steep descent to the ground in order to satisfy the end conditions. The optimal flight trajectory lasts 179.67 [s] and has a peak noise value of 81.75 [dB] and an average noise level of 71.67 [dB] (compare figure 2).

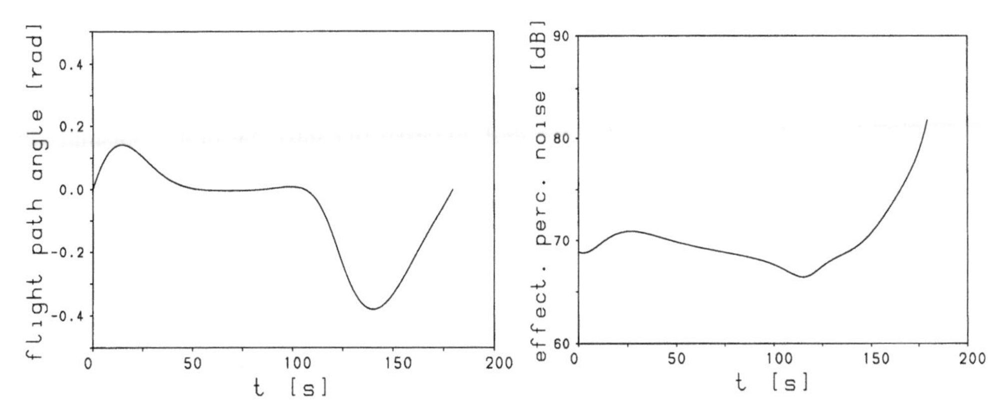

Figure 2: Flight path angle and effective perceived noise E_p versus time for the unconstrained problem

In order to apply such an optimal solution to a real-life landing approach, some additional restrictions have to be taken into account. This will be discussed in the following sections. For example, ascent phases during the landing process are not welcome because of technical reasons. In the mathematical model this can be formulated by means of a state constraint for the altitude

$$h(t) \leq h(0)$$

or even better for the flight path angle γ.

4. Indirect approach II

If an additional state constraint

$$\gamma(t) \leq 0 \tag{11}$$

is to be considered for an optimal solution of (6), (1), (4), the Hamiltonian has to be modified (see [3], [7]):

$$\mathcal{H} = \mathcal{H}_{\text{old}} + \mu\Gamma,$$

where $\mathcal{H}_{\text{old}}$ is the Hamiltonian from section 2 and $\mu = \mu(t)$ is another adjoint function satisfying

$$\mu(t) \begin{cases} = 0 & \text{if } \gamma(t) < 0, \\ \geq 0 & \text{if } \gamma(t) = 0. \end{cases}$$

Again the maximum principle (9) can be applied which yields the following equations for constrained subintervals $[t_{\text{entry}}, t_{\text{exit}}]$, where the state constraint (11) is active, i.e. $\gamma(t) \equiv 0, \quad t \in (t_{\text{entry}}, t_{\text{exit}})$:

$$\begin{aligned}
0 &= \mathcal{H}_T &&= \frac{5.2c_3T^{4.2}}{v(h+50)^{2.5}}\cos\gamma + \frac{\lambda_v\cos\alpha}{m} + [\lambda_\gamma + \mu]\frac{\sin\alpha}{mv} \\
0 &= \mathcal{H}_\alpha &&= \frac{T}{m}\left\{-\lambda_v\sin\alpha + [\lambda_\gamma + \mu]\frac{\cos\alpha}{v}\right\}, \\
0 &= \dot{\gamma} = \Gamma &&= \frac{T\sin\alpha + L - W\cos\gamma}{mv}.
\end{aligned}$$

Solving the second equation for $\lambda_\gamma + \mu$ and substituting into the first equation, the following system is obtained:

$$\begin{aligned}
&\frac{5.2c_3mT^{4.2}\cos\gamma\cos\alpha}{v(h+50)^{2.5}} + \lambda_v = 0 \\
&T\sin\alpha + L - W\cos\gamma = 0
\end{aligned} \tag{12}$$

Now $\gamma \equiv 0$ is used to conclude $\alpha = 0$ (notice assumption (2)) and to obtain explicit[2] formulas for the optimal controls and the adjoint function μ:

$$\begin{aligned}
\alpha_0 &= 0, \\
T_0 &= \left(-\frac{v\lambda_v(h+50)^{2.5}}{5.2c_3m}\right)^{\frac{1}{4.2}}, \\
\mu &= -\lambda_\gamma
\end{aligned} \tag{13}$$

[2]This can only be done because of (11). For any other box-constraint for γ, like $\gamma \leq C \neq 0$, system (12) cannot be solved explicitly and therefore has to be treated numerically.

Again these equations are substituted into (1) and (8) to yield a system of differential equations for the state and adjoint variables. Notice that the adjoint equations (8) are different from those of the unconstrained problem because of the modification of the Hamiltonian. Equations (12) and (13) are valid on constrained subarcs, while on unconstrained arcs ($\gamma < 0$) the system of differential equations of section 2 (unconstrained problem) is applicable. There is no theoretical result for the number and sequence of constrained and unconstrained subarcs. Therefore a reasonable switching structure has to be detected during the numerical solution process. The position of possible entry- or exit-points is determined by transversality conditions, which in this case (11) are equivalent to a continuity condition for the Hamiltonian:

$$\mathcal{H}(t_{\text{entry}}^{-}) = \mathcal{H}(t_{\text{entry}}^{+}) \qquad \text{and} \qquad \mathcal{H}(t_{\text{exit}}^{-}) = \mathcal{H}(t_{\text{exit}}^{+})$$

The adjoint function λ_γ is discontinuous at the entry point and continuous across the exit point, or vice versa, while all the other variables are continuous at both switching points (compare [3]). Putting all this together and choosing an appropriate switching structure produces a multi-point boundary value problem, which again can be solved by numerical algorithms.

5. A monotonously decreasing trajectory

A solution for the multi-point boundary value problem satisfying the flight path angle constraint (11) could be found by MUMUS and DIRCOL for the two-phase-switching structure:

constrained — unconstrained

The results are shown in figures 3 and 4. As for the unconstrained problem, there is

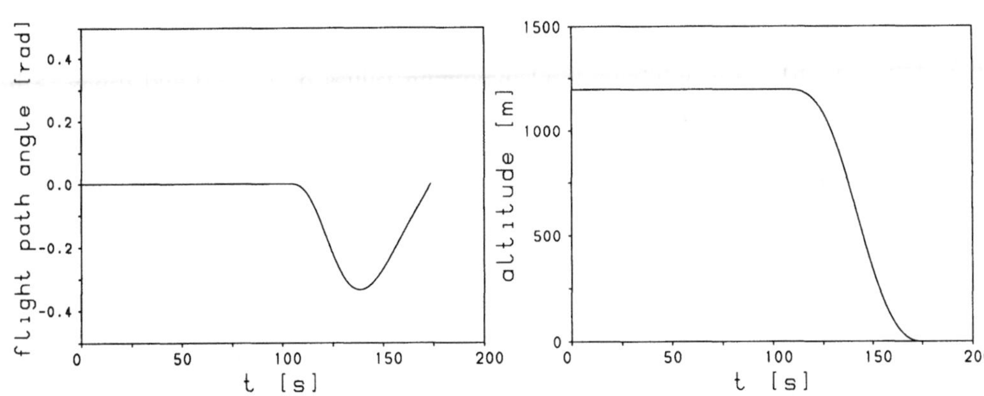

Figure 3: Flight path angle and altitude versus time for constraint (11)

a steep descent at the end accompanied by a peak in the velocity history. The thrust again stays within reasonable bounds. The duration 173.10 [s] of the flight is shorter than before, while the peak and average values of the effective perceived noise increase to 82.62 [dB] and 72.27 [dB] respectively.

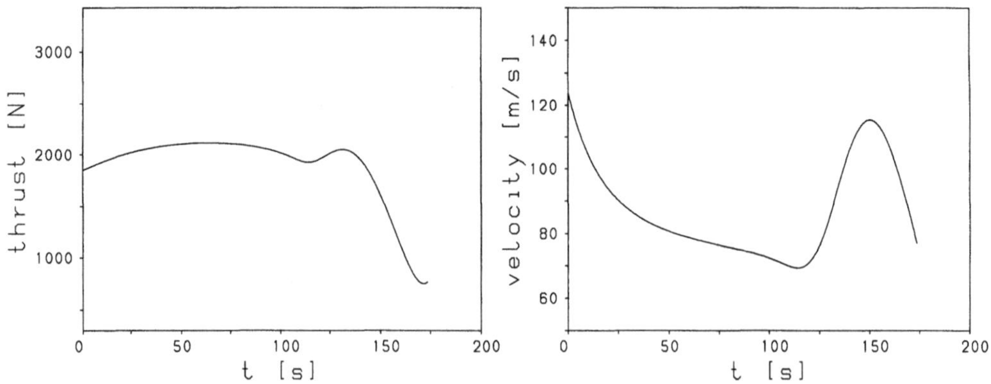

Figure 4: Thrust and velocity versus time for constraint (11)

In order to avoid the steep descent at the end, it would be possible to introduce another state constraint for the velocity, for example

$$v(t) \leq 90, \qquad \text{for} \quad t > 50.$$

While this problem will not be discussed in this paper, numerical solutions for this kind of velocity constraints can be found in [10].

6. A constraint for the thrust angle

Another step towards a more realistic modelling of the landing approach is to restrict the thrust angle α. For most aircraft this control can only be varied within a relatively small range. Therefore a control constraint

$$\alpha_{\min} \leq \alpha \leq \alpha_{\max} \tag{14}$$

is introduced with given minimum and maximum values $\alpha_{\min} < 0$ and $\alpha_{\max} > 0$. The derivation of the corresponding multi-point boundary value problem is simpler than it was for the state-constrained problem in section 4. The controls on constrained subarcs are computed by using a combination of the maximum principle and of the active constraint (14). This yields

$$\begin{aligned} \alpha_0 &= \begin{cases} \alpha_{\min} & \text{or} \\ \alpha_{\max} \end{cases} \\ T_0 &= \left[\left(-\frac{\lambda_v \cos\alpha_0}{m} - \frac{\lambda_\gamma \sin\alpha_0}{mv} \right) \frac{v(h+50)^{2.5}}{5.2 c_3 \cos\gamma} \right]^{\frac{1}{4.2}} \end{aligned} \tag{15}$$

For subarcs where (14) is not active, the controls have to be computed according to section 2 (see (10)). At all entry- and exit-points of the constraint (14) the state and adjoint variables as well as the controls are continuous.

The determination of a suitable switching structure again has to be done numerically. For this purpose the constraint (14) is introduced gradually, beginning with large values for $|\alpha_{\min}|$ and $|\alpha_{\max}|$, so that the corresponding switching structure can be guessed reliably from the α-history of the unconstrained problem. Then $|\alpha_{\max}|$ and $|\alpha_{\min}|$ are decreased step by step, while simultaneously the switching structure is modified in an appropriate way (see [2], [6] or [10] for a more detailed description of this homotopy procedure). When this homotopy chain is traced, the switching structure has to be changed repeatedly until a solution for $\alpha_{\max} = 0.16 = -\alpha_{\min}$ is obtained having the structure:

$$\alpha_{\max} \text{ — unconstrained subarc — } \alpha_{\min} \text{ — unconstrained subarc — } \alpha_{\max}$$

The results are shown in figures 5 and 6. It can be seen that the thrust angle has almost

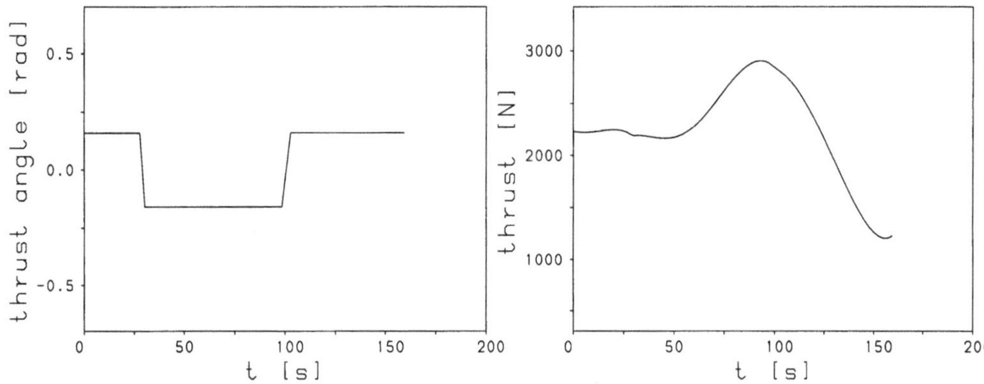

Figure 5: Thrust angle and thrust versus time for constraint (14)

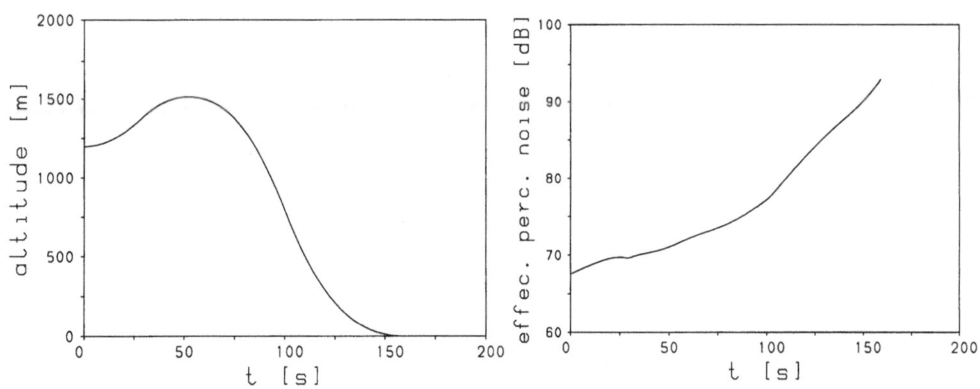

Figure 6: Altitude and effective perceived noise E_p versus time for constraint (14)

a bang-bang structure with only two very short intervals, where the constraint (14) is inactive. There is again an ascent phase at the beginning, but with a much smaller gradient compared to the unconstrained problem and also the descent phase at the end is much smoother than before. The whole landing procedure now lasts 159.18 [s] with a maximum noise of 93.08 [dB] and an average noise of 83.12 [dB].

7. Conclusion

This paper describes the application of the indirect method and of the multiple shooting algorithm to the optimal control problem of minimizing the noise emission during the landing approach of an aircraft. Starting with the unconstrained problem, the solutions have been adapted step-by-step to more realistic scenarios with respect to practical and technical demands. The effect of these additional requirements to the performance index has been documented. For future investigations, assumption (2) will be got rid of by using a more sophisticated model for the aerodynamic forces and the thrust angle will be regarded as a state variable by introducing its derivative as a new control.

References

[1] Bulirsch, R.: *Die Mehrzielmethode zur numerischen Lösung von nichtlinearen Randwertproblemen und Aufgaben der optimalen Steuerung*, Report der Carl-Cranz Gesellschaft, 1971.

[2] Bulirsch, R.; Montrone, F.; Pesch, H. J.: *Abort Landing in the Presence of a Windshear as a Minimax Optimal Control Problem. Part 2: Multiple Shooting and Homotopy* , Journal of Optimization Theory & Applications 70, 1991, pp. 223-254.

[3] Hartl, R. F.; Sethi, S. P.; Vickson, R. G.: *A Survey of the Maximum Principles for Optimal Control Problems with State Constraints*, SIAM Review, Vol. 37, Nr. 2, June 1995, pp. 181-218.

[4] Hiltmann, P.; Chudej, K.; Breitner, M. H.: *Eine modifizierte Mehrzielmethode zur Lösung von Mehrpunkt-Randwertproblemen — Benutzeranleitung*, Sonderforschungsbereich 255 der Deutschen Forschungsgemeinschaft, Report 14, München, 1993.

[5] Jacob, H. G.: *An Engineering Optimization Method with Application to STOL-Aircraft Approach and Landing Trajectories*, NASA TN D-6978, September 1972.

[6] Kugelmann, B.; Pesch, H. J.: *New General Guidance Method in Constrained Optimal Control, Part 2: Application to Space Shuttle Guidance*, Journal of Optimization Theory and Applications 67, S. 437-446, 1990.

[7] Maurer, H.: *Optimale Steuerprozesse mit Zustandsbeschränkungen*, Habilitationsschrift, Naturwissenschaftlicher Fachbereich IV der Universität Würzburg, 1976.

[8] Ohta, H.: *Analysis of Minimum Noise Landing Approach Trajectory*, J. Guidance, Vol. 5, No. 3, May-June 1982, pp. 263-269.

[9] Pontrjagin, L. S.; Boltjanskij, V. G.; Gamkrelidze, R. V.; Miscenko, E. F.: *Mathematische Theorie optimaler Prozesse*, Oldenbourg, 1964.

[10] Schmidt-Rheindt, E.: *Minimierung der Lärmbelastung bei der Landung von Flugzeugen*, Diploma Thesis, Department of Mathematics, University of Technology, February, 1997.

[11] von Stryk, O.: *Numerische Lösung optimaler Steuerungsprobleme: Diskretisierung, Parameteroptimierung und Berechnung der adjungierten Variablen*, Fortschritt-Berichte VDI, Reihe 8, Meß-, Steuerungs- und Regelungstechnik, Nr. 441, VDI-Verlag, Düsseldorf, 1995.

International Series of Numerical Mathematics
Vol. 124, © 1998 Birkhäuser Verlag, Basel

Real-Time Computation of Strategies of Differential Games with Applications to Collision Avoidance

Rainer Lachner* Michael H. Breitner* Hans J. Pesch*

Abstract. Contemporary developments of on-board systems for automatic or semi-automatic driving include car collision avoidance. For this purpose a worst case approach based on pursuit-evasion differential games is investigated. On a freeway a correct driver (evader) is faced with a wrong driver (pursuer) ahead. The correct driver tries to avoid collision against all possible maneuvers of the wrong driver and additionally tries to stay on the freeway. The representation of an optimal collision avoidance strategy along a lot of optimal paths is used to synthesize implementations with neural networks. Examples of simulations which proved satisfactory performance of the on-board collision avoidance system against various typical maneuvers of wrong drivers are presented.

1. Introduction

More than 3000 wrong drivers are registered officially on German freeways every year, but the actual number is estimated three to five times higher. On the average 75 people are injured and 16 killed for 100 wrong drives, compare [14], [15] and [21]. Therefore contemporary developments of on-board systems for automatic or semi-automatic driving include car collision avoidance. Current position and velocity vector of the car and all neighboring cars on the freeway can be measured by on-board sensors with radar or ultrasonic waves. Alternatively, freeway mounted systems can be used in the near future which transmit their measurements to on-board systems of cars around, compare [5], [9] and [13]. Due to drunkenness, tiredness of life or panic, future maneuvers of wrong drivers often are unpredictable. Faced with this uncertainty, on-board collision avoidance systems must

(1) detect reliably and as early as possible other cars on collision course,
(2) warn drivers immediately,
(3) calculate optimal steering and velocity control strategies in real-time against all possible maneuvers of all neighboring cars and
(4) advise drivers on monitors or windscreen head-up displays or adopt cars' steering and velocity control,

compare *PROMETHEUS* (Programme for European traffic flow with highest efficiency and unprecedented safety) [13] and *ACASD-TRP* (Automotive Collision Avoidance Systems Development – Technology Reinvestment Project) [9].

*Technische Universität Clausthal, Institut für Mathematik, Erzstr. 1, D–38678 Clausthal-Zellerfeld, Germany. e-mail: lachner@math.tu-clausthal.de , breitner@math.tu-clausthal.de , pesch@math.tu-clausthal.de .

2. Hierarchical formulation based on the game of two cars

We limit to a collision avoidance problem for two cars; one is driven by the correct driver C, the other by the wrong driver W. For this collision avoidance problem the kinematic equations can be modeled realistically with point mass models

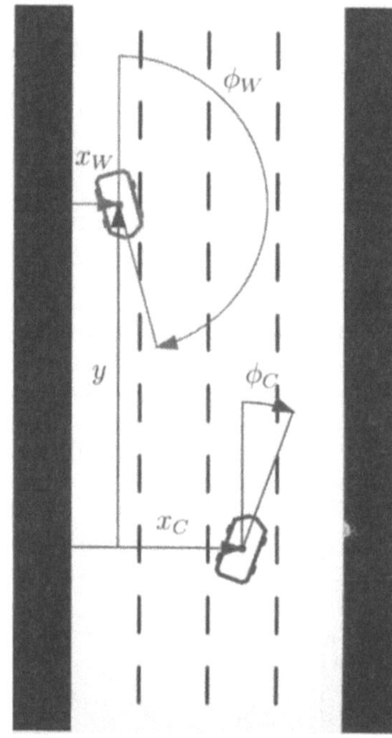

$$\dot{x}_W = v_W \sin\phi_W, \tag{1}$$

$$\dot{x}_C = v_C \sin\phi_C, \tag{2}$$

$$\dot{y} = v_W \cos\phi_W - v_C \cos\phi_C, \tag{3}$$

$$\dot{v}_C = b_C\, \eta_C, \tag{4}$$

$$\dot{\phi}_W = w_W\, u_W, \tag{5}$$

$$\dot{\phi}_C = w_C(v_C)\, u_C, \tag{6}$$

with independent variable time t, with state variables x_W, x_C (distance of W and C, respectively, from the left freeway side) and y (distance between W and C along the freeway) for the positions and with state variables ϕ_W, ϕ_C (driving direction of W and C, respectively), v_W and v_C (velocity of W and C, respectively) for the velocity vectors.

Control variables are u_W, u_C (turn rate of W and C, respectively) and η_C (velocity change rate). Without oversimplification v_W is taken as constant and the maximum angular velocities $w_W(v_W)$ and $w_C(v_C)$ are given car dependent. The kinematic constraints of bounded radius of curvature are included by the control variable inequality constraints

$$-1 \leq u_W \leq +1 \tag{7}$$

and

$$-1 \leq u_C \leq +1 \tag{8}$$

for the wrong driver W and the correct driver C, respectively. Steering $u_W = \pm 1$ and $u_C = \pm 1$ executes the sharpest possible right/left turn for W and C, respectively. The kinematic constraint of acceleration and deceleration is included by the control variable inequality constraint

$$-0.1 \leq \eta_C \leq +1 \tag{9}$$

for C, where $\eta_C = 1$ executes maximum braking and $\eta_C = -0.1$ executes maximum acceleration. In order to stay on the freeway W has to obey the state constraint

$$0\,\mathrm{m} \leq x_W \leq 15\,\mathrm{m}, \tag{10}$$

whereas C has to obey the state constraint

$$0\,\mathrm{m} \leq x_C \leq 15\,\mathrm{m}. \tag{11}$$

Correct driver C's goal to avoid collision against all possible maneuvers of wrong driver W can be formulated mathematically as

$$\min_{\gamma_W \in \Gamma_W} \max_{\gamma_C \in \Gamma_C} \min_{t_0 \leq t < \infty} d(\boldsymbol{z}(t)) \quad \text{w. r. t.} \quad \boldsymbol{z}(t_0) = \boldsymbol{z}_0 \tag{12}$$

with vector $\boldsymbol{z} := (x_W, x_C, y, v_C, \phi_W, \phi_C)^\top$ of state variables, with (feedback) strategies $\gamma_W(\boldsymbol{z})$ and $\boldsymbol{\gamma}_C(\boldsymbol{z})$, with sets Γ_W and Γ_C of all admissible strategies and with Euclidean distance $d(\boldsymbol{z})$ between W and C. The control variables are determined by $u_W(\boldsymbol{z}) := \gamma_W(\boldsymbol{z})$ and $(u_C(\boldsymbol{z}), \eta_C(\boldsymbol{z}))^\top := \boldsymbol{\gamma}_C(\boldsymbol{z})$. For the collision avoidance problem investigated here γ_W is admissible, if constraints (7) and (10) are obeyed for all $t \in [t_0, t_f]$, and $\boldsymbol{\gamma}_C$ is admissible, if constraints (8), (9) and (11) are obeyed for all $t \in [t_0, t_f]$. The collision avoidance maneuver starts at the moment t_0 at $\boldsymbol{z}(t_0) = \boldsymbol{z}_0$ when the correct driver C notices the wrong driver W. The collision avoidance maneuver ends at the terminal time $t_f < \infty$ when the minimum distance between W and C is reached.

The idea to analyze collision avoidance problems in a differential game framework dates back to RUFUS PHILIP ISAACS' work [20] and [19] in the early fifties, unclassified and published in the mid sixties [18]. The theory of collision avoidance, e. g. between cars, aircraft or ships, can be embedded in the theory of pursuit and evasion. Omitting state constraints (10) and (11) and assuming constant velocity v_C the collision avoidance problem now under consideration is the famous *game of two cars* introduced in [19] and [18]. If additionally the correct driver C does not have to obey the constraint of bounded curvature, i. e. $u_C \in]-\infty, \infty[$, the also famous *homicidal chauffeur game* arises. Even though the essential number of state variables is only 3 for the game of two cars and even only 2 for the homicidal chauffeur game their full solution is extraordinarily baffling. The state space is cut by diverse singular manifolds, e. g. barrier, universal and dispersal manifolds, which subdivide the state space into several component regions. Neither the game of two cars nor the homicidal chauffeur game can be solved fully, i. e. optimal strategies ${\gamma_W}^*(\boldsymbol{z})$ and ${\boldsymbol{\gamma}_C}^*(\boldsymbol{z})$ cannot be calculated explicitly for all $\boldsymbol{z}$, for all velocities v_W and v_C and for all maximum angular velocities w_W and w_C. Nevertheless various planar collision avoidance problems have been investigated as game of two cars or homicidal chauffeur game in the last three decades, see, e. g., [28], [26], [27], [36], [35], [3], [12], [31] and [34]. Note that state space constraints comparable to (10) and (11) have not been taken into account in these publications.

3. Computation of optimal paths

The approach described in the section before uses a hierarchical formulation based on the game of two cars. However, for this sophisticated modeling, optimal paths can be obtained only numerically by the method of characteristics and the numerical solution of the related multi-point boundary value problems.

3.1 Method of characteristics

It is well known that optimal strategies theoretically could be obtained by integrating the Isaacs' equation

$$V_z \cdot \dot{\boldsymbol{z}}\Big(\boldsymbol{z}, {u_W}^*(\boldsymbol{z}, V_z), {u_C}^*(\boldsymbol{z}, V_z), {\eta_C}^*(\boldsymbol{z}, V_z)\Big) \equiv 0 \tag{13}$$

directly, see [18], [3], [6] and [31]. The Isaacs' equation in general is a nonlinear first order partial differential equation for the value function $V(\boldsymbol{z})$ with

$$V(\boldsymbol{z}) := \min_{\gamma_W \in \Gamma_W} \max_{\gamma_C \in \Gamma_C} \|x_W(t_f) - x_C(t_f)\| \,. \tag{14}$$

for the approach under consideration. The terminal time t_f is fixed when there holds $y(t) = 0$ for the first time. Unfortunately for almost all complex differential games direct integration of the Isaacs' equation is impossible due to the state space dimension and various singular surfaces. Alternatively, the method of characteristics enables the derivation of the ordinary differential equation system along characteristics of the Isaacs' equation. Assuming V exists and is twice piecewise continuously differentiable, the gradient $V_z(\boldsymbol{z})$ of the value function $V(\boldsymbol{z})$ satisfies

$$\dot{V}_z = -V_z \cdot \frac{\partial}{\partial \boldsymbol{z}} \dot{\boldsymbol{z}}\Big(\boldsymbol{z}, u_W{}^*(\boldsymbol{z}, V_z), u_C{}^*(\boldsymbol{z}, V_z), \eta_C{}^*(\boldsymbol{z}, V_z)\Big) \tag{15}$$

along a characteristic. Terminal conditions for $V_z(t_f)$ are obtained by parameterizing the end condition $y(t_f) = 0$ and differentiating the value function $V_z(\boldsymbol{z}(t_f))$ which is known as a function of the parameters, compare [18], [31], [7], [8] and [6]. Finally a boundary value problem is obtained with the kinematic equations (1) – (6) and the differential equations (15) for the gradient $V_z(\boldsymbol{z})$. Boundary conditions are initial conditions $\boldsymbol{z}(t_0) = \boldsymbol{z}_0$ and the aforementioned final conditions. Interior point conditions are related to intersection points of the characteristic to singular surface and correspond to zeros of the switching functions or transitions to active constraints. For many characteristics these interior point conditions give rise to multi-point boundary value problems.

The zone of usable initial scenarios is now meshed by a six dimensional grid. Each grid point serves as initial state and is related to a multi-point boundary value problem. The solution of the multi-point boundary value problem can be obtained numerically by the highly accurate multiple shooting method, compare [8], [23], [7], [6] and [22]. Some necessary optimality conditions and sign conditions must be checked a posteriori. Even for a crude discretization with a large mesh size the number of potential initial states $\boldsymbol{z}_0$ is considerably high.

3.2 Numerical solution of the multi-point boundary value problems

Due to the large number of grid points it would be very time consuming and tedious to compute the optimal solutions starting at the grid points one by one manually. Thus one has to provide a semi-automatic procedure which automatically computes at least all optimal paths sharing the same switching structure, i. e. having in common the same structure of the multi-point boundary value problem. Since the heart of the multiple shooting method consists of an implementation of Newton's method, a sufficiently accurate initial guess for the desired solution at the multiple shooting nodes is needed. If at least one optimal path for one of the grid points is already known, this path can be used for an initial guess for a solution trajectory starting at a neighboring grid point. By checking all the neighbors and saving the solution in the case of convergence and optimality, one can recursively proceed and cover connected regions of the state space with the same switching structure. Unfortunately, it turned out that the pursuit evasion

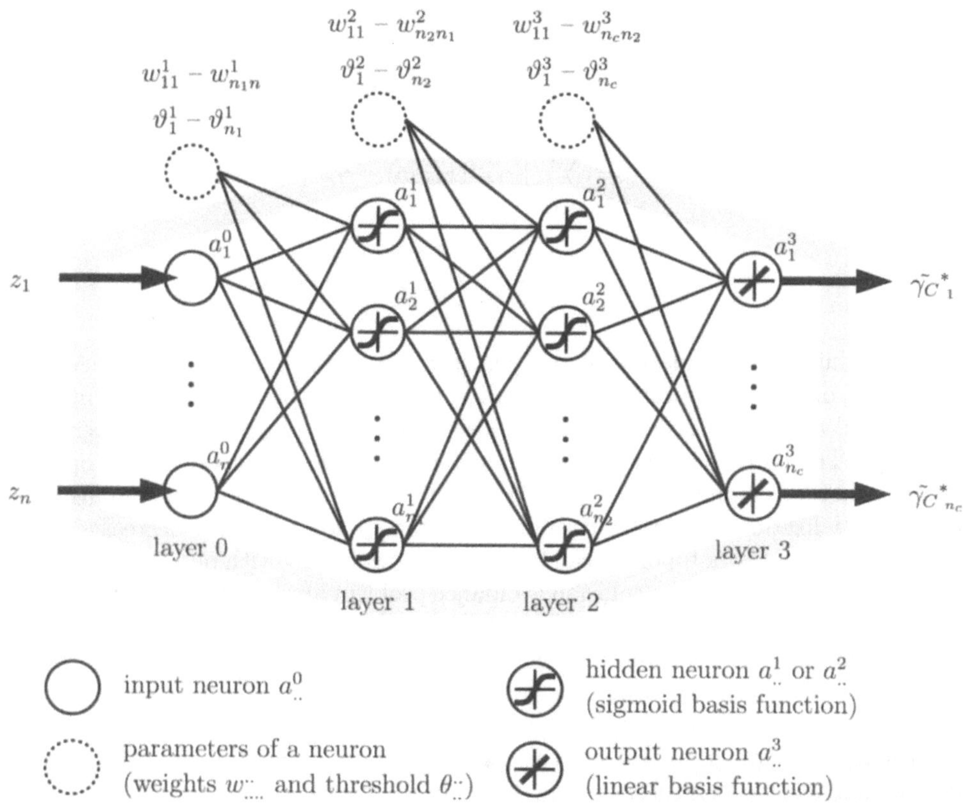

Figure 1: Synthesis of an optimal strategy with a four layer perceptron.

game under investigation possesses a variety of more than sixty different switching structures.

4. Approximation of optimal strategies by neural networks

Along optimal paths the realization of optimal strategies is known in open-loop form after the numerical solution of the multi-point boundary value problems. In order to gain an accurate approximation of optimal strategies, as many optimal paths as possible must be computed and the information yielded must be used for a synthesis.

Optimal strategies are only known at points along optimal trajectories and therefore $\boldsymbol{\gamma}_C{}^*(\boldsymbol{z}) = (u_C{}^*(\boldsymbol{z}), \eta_C{}^*(\boldsymbol{z}))^\top$ has to be synthesized in the relevant parts of the state space. For this purpose the approximation properties of trained artificial neural networks are exploited, see [17], [4], [16], [37], [24] and Fig. 1. and compare [30], [10] and [6]. For the artificial neural network we use the approved feedforward multilayered perceptrons, see [29], [39], [33], [1], [2], [32], [11], [38] and Fig. 1. Neural network topology and weights specify the neural mapping. A priori the topology, i. e. type and number of basis functions, must be fixed suitably. The data set $(\boldsymbol{z}_k, {\gamma_C}^*(\boldsymbol{z}_k))$, $k = 1, 2, \ldots$, at points along many optimal paths is split into training data set with $k \in I_T$ and cross-validation data set with $k \in I_V$ $(I_T \cap I_V = \emptyset)$. With a differentiable norm $\|\,.\,\|$ the approximation errors

$$\varepsilon_T(\boldsymbol{w}) := \sum_{k \in I_T} \| \tilde{\gamma_C}^*(\boldsymbol{z}_k; \boldsymbol{w}) - {\gamma_C}^*(\boldsymbol{z}_k) \| \tag{16}$$

and

$$\varepsilon_V(\boldsymbol{w}) := \sum_{k \in I_V} \| \tilde{\gamma_C}^*(\boldsymbol{z}_k; \boldsymbol{w}) - {\gamma_C}^*(\boldsymbol{z}_k) \| \tag{17}$$

enable training and validation of the perceptron's approximation quality. After initialization of $\boldsymbol{w}$ the approximation quality of the perceptron has to be improved, i. e. $\varepsilon_T(\boldsymbol{w})$ has to be decreased, by choosing $\boldsymbol{w}_{\text{new}} := \boldsymbol{w}_{\text{old}} + \Delta\boldsymbol{w}$ for every learning step. The correction $\Delta\boldsymbol{w}$ usually is determined by supervised back-propagation learning methods, compare [39], [33], [1], [2], [32], [11] and [38]. The training for the chosen initialization terminates if validation error $\varepsilon_V(\boldsymbol{w})$ no longer decreases.

The neural network topology can be designed graphically with public domain neural network simulators. For the collision avoidance problem investigated here the Stuttgart Neural Network Simulator (SNNS) has been used, see [40] and [39]. Three layered perceptrons have been proven convenient and successful. The inherent symmetry of the problem has been exploited to lessen the relevant part of the state space. Concept and implementation have been applied for German patent [25].

The following diagram sketches the cycle for the approximation of optimal steering ${u_C}^*(\boldsymbol{z})$ and velocity control ${\eta_C}^*(\boldsymbol{z})$ by the neural network.

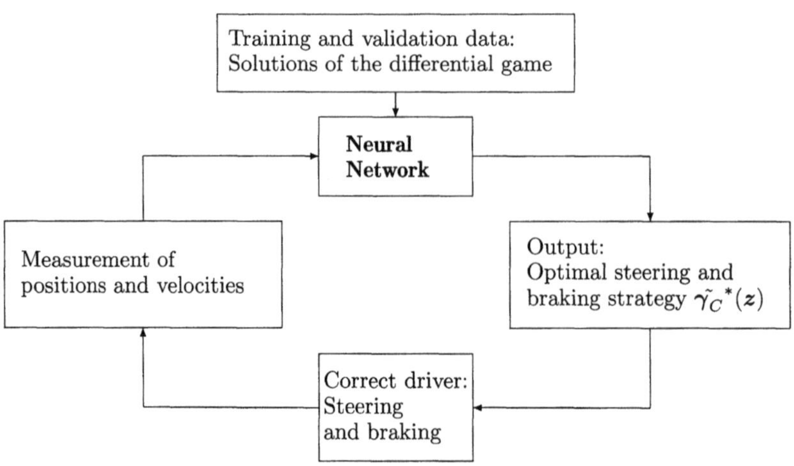

4.1 Simulations

Many simulations against various typical maneuvers of wrong drivers have been performed to check real-time capability and reliability of the neural collision avoidance system for real world scenarios. These simulations proved the satisfactory performance of the on-board collision avoidance system, see Fig. 2.

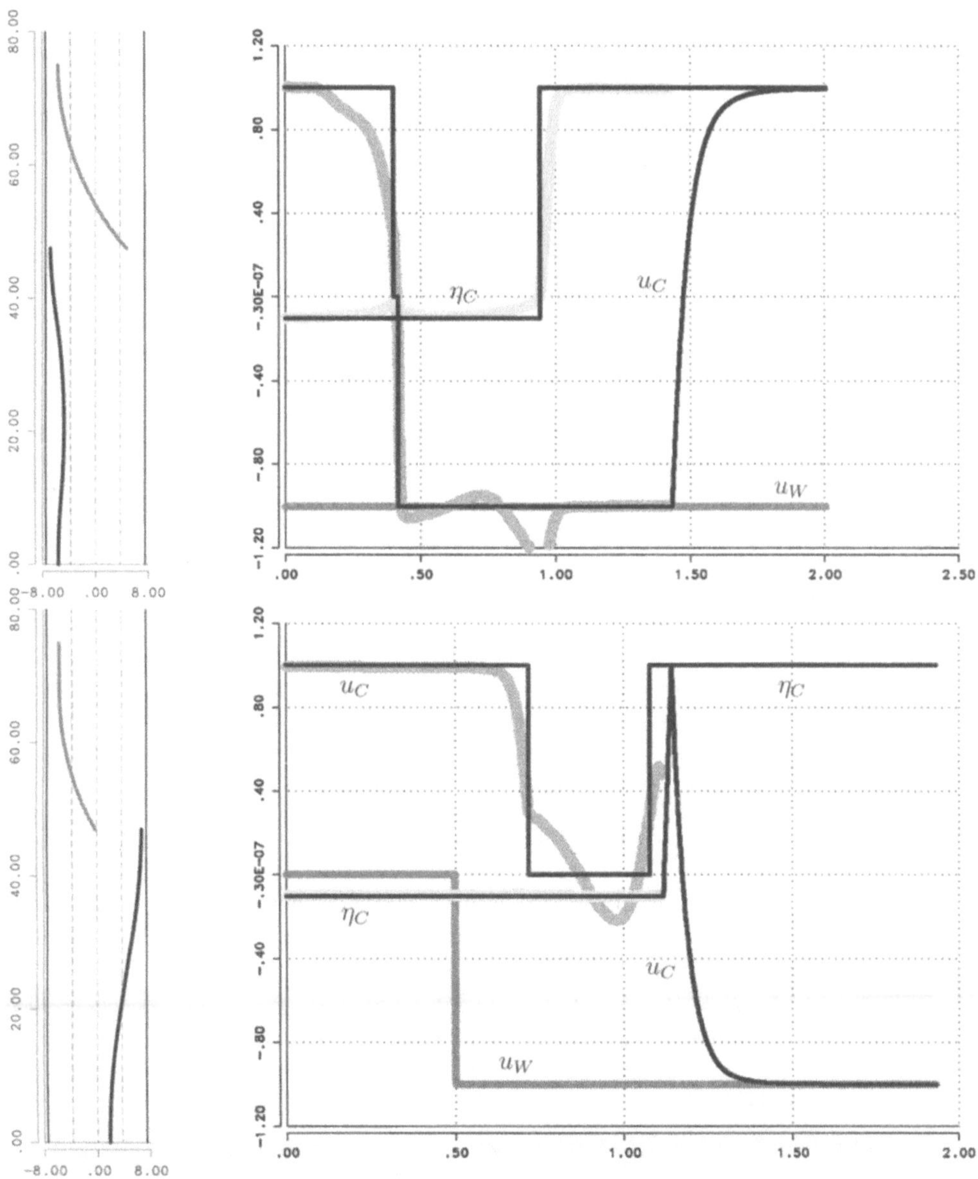

Figure 2: Two simulations with the real-time collision avoidance system for an initial distance of 75 m. Right: Path of W (grey) and prescribed path and C (black) on the freeway. Left: Corresponding time history of the controls. Prescribed control u_W of W (grey) and network output $\tilde{u}_C$ and $\tilde{\eta}_C$ (thick light grey lines). Since in the differential game under consideration the control variables appear linearly in the system equations, the optimal controls are of bang-bang type or singular. Therefore, instead of the actual network output a modification of it is used which has the bang-bang property. Both simulations end up with active constraint, the second one has a short singular subarc.

5. Conclusion

The successful synthesis of optimal collision avoidance strategies with quite easy to use feedforward perceptrons gives rise to high hopes. Since 6 state and 3 control variables have been involved in the collision avoidance problem investigated, we think that other complex high dimensional differential games can be solved following the procedure presented here. However, the unavoidable analytical or numerical calculation of hundreds, thousands or even ten thousands of optimal trajectories to obtain the optimal strategies' representation generally is a difficult task.

Acknowledgement

The authors thank C. CHINI, formerly TU München, and S. MIESBACH, Siemens AG, for joint research in the initial phase.

References

[1] Anderson, J. A.: *An Introduction to Neural Networks.* MIT Press, Cambridge, Massachusetts, 1995.

[2] Annema, A. J.: *Feedforward Neural Networks: Vector Decomposition Analysis, Modelling and Analog Implementation.* Kluwer, Boston, Massachusetts, 1995.

[3] Başar, T.; Olsder, G. J.: *Dynamic Noncooperative Game Theory.* Academic Press, London 1982 and 1995 (2. edition).

[4] Blum, E. K.; Li, L. K.: *Approximation Theory and Feedforward Networks,* Neural Networks **4** (1990).

[5] Boulter, J.; Maher, D.; Aburdene, M. F.: *Collision Avoidance Using Ultrasonic Waves.* Report, Bucknell University, Department of Electrical Engineering, Lewisburg, Pennsylvania 17837, 1994.

[6] Breitner, M. H.: *Robust optimale Rückkopplungssteuerungen gegen unvorhersehbare Einflüsse: Differentialspielansatz, numerische Berechnung und Echtzeitapproximation.* Ph. D. thesis, Technische Universität Clausthal, Institut für Mathematik, Nr. 596, VDI-Verlag, VDI Fortschritt-Bericht, Reihe **8** "Meß-, Steuerungs- und Regelungstechnik", Düsseldorf 1996.

[7] Breitner, M. H.; Pesch, H. J.: *Reentry Trajectory Optimization under Atmospheric Uncertainty as a Differential Game.* Annals of the International Society of Dynamic Games **1** (1994). URL: http://www.hut.fi/HUT/Systems.Analysis/isdg

[8] Breitner, M. H.; Pesch, H. J.; Grimm, W.: *Complex Differential Games of Pursuit-Evasion Type with State Constraints, Necessary Conditions for Optimal Open-Loop Strategies (Part 1) and Numerical Computation of Optimal Open-Loop Strategies (Part 2).* Journal of Optimization Theory and Applications **78** (1993).

[9] Chachich, A. C.: *Collision Avoidance and Automated Traffic Management Sensors.* Proceedings **2592** of a Conference in Philadelphia, October 22th – 26th, 1995. International Society for Optical Engineering (SPIE), P. O. B. 10, Bellingham, Washington 98227-0010. URL: http://www.spie.org/spie_home.html .

[10] Chini, C.: *Numerische Berechnung optimaler Ausweichmanöver mit einem Differentialspielansatz und Neuronalen Netzen.* Diploma thesis, Technische Universität München, Mathematisches Institut, München 1994.

[11] Gallant, S. I.: *Neural Network Learning and Expert Systems.* MIT Press, Cambridge, Massachusetts, 1993.

[12] Galperin, E. A.: *The Cubic Algorithm for Global Games with Application to Pursuit-evasion Games.* Computers and Mathematics with Applications **26** (1993).

[13] Harris, C. J.; An, P. E.; Matthews, N. D.; Roberts, J. M.: *Intelligent Collision Avoidance, Control and Guidance.* Report, Prometheus (Programme for European traffic flow with highest efficiency and unprecedented safety), University of Southampton, Department of Electronics and Computer Sciences, Southampton SO17 1BJ, 1995. To appear in IMechE Autotech '95.

[14] Hautzinger, H.; Heidemann, D.; Krämer, B.: *Fahrleistungserhebung 1990 – Inlandsfahrleistung und Kfz-Unfallrisiko in der Bundesrepublik Deutschland (alte Bundesländer).* Bericht der Bundesanstalt für Straßenwesen **M20**, Wirtschaftsverlag NW/Verlag für neue Wissenschaft, Bremerhaven 1993.

[15] Hautzinger, H.; Tassaux-Becker, B.; Hamacher, R.: *Verkehrsunfallrisiko in Deutschland.* Bericht der Bundesanstalt für Straßenwesen **M58**, Wirtschaftsverlag NW/Verlag für neue Wissenschaft, Bremerhaven 1996.

[16] Hornik, K.: *Approximation Capabilities of Multilayer Feedforward Networks.* Neural Networks **4** (1991).

[17] Hornik, K.; Stinchcombe, M.; White, H.: *Multi-Layer Feedforward Networks are Universal Approximators.* Neural Networks **2** (1989).

[18] Isaacs, R. P.: *Differential Games – A Mathematical Theory with Applications to Warfare and Pursuit, Control and Optimization.* John Wiley & Sons, New York 1965. Robert E. Krieger, Huntington, New York, 1975 (3. edition).

[19] Isaacs, R. P.: *Differential Games I: Introduction. Differential Games II: The Definition and Formulation. Differential Games III: The Basic Principles of the Solution Process. Differential Games IV: Mainly Examples.* Memoranda **RM-1391**, **RM-1399**,**RM-1411** and **RM-1486**. The RAND Corporation, P. O. B. 2138, Santa Monica, California 90407-2138, 1954/55. URL: http://www.rand.org .

[20] Isaacs, R. P.: *Games of Pursuit.* Paper **P-257**. The RAND Corporation, Santa Monica 1951.

[21] Kühnen, M. A.; Brühning, E.; Schepers, A.: *Unfallgeschehen auf Autobahnen – Strukturuntersuchung.* Bericht der Bundesanstalt für Straßenwesen **M50**, Wirtschaftsverlag NW/Verlag für neue Wissenschaft, Bremerhaven 1995.

[22] Lachner, R.; Breitner, M. H.; Pesch, H. J.: *Three-Dimensional Air Combat Analysis – An Example for the Numerical Solution of Complex Differential Games.* Annals of the International Society of Dynamic Games **3** (1996).

[23] Lachner, R.; Breitner, M. H.; Pesch, H. J.: *Optimal Strategies of a Complex Pursuit-Evasion Game.* Journal of Computing and Information **4** (1994).

[24] Leshno, M.; Lin, V. Y.; Pinkus, A.; Schocken, S.: *Multilayer Feedforward Networks with a Nonpolynomial Activation Function Can Approximate Any Function.* Neural Networks **6** (1993).

[25] Miesbach, S.; Breitner, M. H.; Chini, C.; Pesch, H. J.: *Neuronale Kennfeldsteuerung zur Kollisionsvermeidung von entgegenkommenden Fahrzeugen.* German patent application **95 P 8066 DE**, Europäisches Patentamt, Erhardtstrasse 27, D-80331 München, 1995. URL: http://www.epo.co.at/epo/.

[26] Miloh, T.: *The Game of Two Elliptical Ships.* Optimal Control Applications & Methods **4** (1983).

[27] Miloh, T.; Pachter, M.: *Ship Collision Avoidance and Pursuit-evasion Differential Games with Speed-loss in a Turn.* Computers and Mathematics with Applications **18** (1989).

[28] Miloh, T.: Sharma, S. D.: *Maritime Collision Avoidance as a Differential Game.* Schiffstechnik **24** (1977).

[29] Neural Information Processing Systems Foundation (San Diego), *Proceedings of NIPSF Conferences NIPS1 – NIPS9 (1988 – 1996).* MIT Press, Cambridge, Massachusetts. URL: http://www.cs.cmu.edu/Web/Groups/NIPS/NIPS.html .

[30] Pesch, H. J.; Gabler, I.; Miesbach, S.; Breitner, M. H.: *Synthesis of Optimal Strategies for Differential Games by Neural Networks.* Annals of the International Society of Dynamic Games **3** (1996).

[31] Petrosjan, L. A.: *Differential Games of Pursuit.* World Scientific, River Edge, New York, 1993.

[32] Rigoll, G.: *Neuronale Netze – Eine Einführung für Ingenieure, Informatiker und Naturwissenschaftler.* Expert Verlag, Renningen-Malmsheim 1994.

[33] Scherer, A.: *Neuronale Netze – Grundlagen und Anwendungen.* Vieweg, Wiesbaden 1996.

[34] Shinar, J.; Tabak, R.: *New Results in Optimal Missile Avoidance Analysis.* Journal of Guidance, Control & Dynamics **17** (1994).

[35] Vincent, T. L.: *Collision Avoidance at Sea.* In: Hagedorn, P.; Knobloch, H. W.; Olsder, G. J. (editors): Differential Games and Applications, Lecture Notes in Control and Information Sciences **3**, Springer, Berlin 1977.

[36] Vincent, T. L.; Cliff, E. M.; Grantham, W. J.; Peng, W. Y.: *Some Aspects of Collision Avoidance.* AIAA Journal **12** (1974).

[37] White, H.; Gallant, A. R.; Hornik, K.; Stinchcombe, M.; Wooldridge, J.: *Artificial Neural Networks – Approximation and Learning Theory.* Blackwell, Oxford 1992.

[38] White, D. A.; Sofge, D. A.: *Handbook of Intelligent Control — Neural, Fuzzy and Adaptive Approaches.* Van Nostrand Reinhold, New York 1992.

[39] Zell, A.: *Simulation Neuronaler Netze.* Addison-Wesley, Bonn 1996.

[40] Zell, A. et al.: *SNNS – Stuttgart Neural Network Simulator (Version 4.0): User Manual.* Universität Stuttgart, Institut für paralleles und verteiltes Hochleistungsrechnen, Breitwiesenstraße 20 – 22, D-70565 Stuttgart, 1995. URL: http://www.informatik.uni-stuttgart.de/SNNSinfo/snns.html .

The Use of Screening for the Control of an Endemic Disease

Georg Leitmann *

Abstract. We employ recent results on robust control of uncertain dynamical systems to deduce a screening policy for use by a public health authority in the control of an infectious disease such as gonorrhea.

1. Introduction

Until recently, relatively little effort has been devoted to the design of constrained feedback controllers which assure stability of an uncertain dynamical system. Employing recent results of Corless and Leitmann [1,2] on constrained controllers for robust exponential convergence to a specified region of state space, one can now address many systems in which the control inputs are subject to *a priori* bounds. E.g., Lee and Leitmann [3,4] utilized these results to derive robust harvesting strategies for some ecological systems based on uncertain models. Here, we endeavor to demonstrate the above mentioned results for control of endemic diseases such as gonorrhea.

As pointed out by Cromer [5], gonorrhea is the most reported communicable disease in the U.S.A.; e.g., see [6-12]. These researchers have been quite successful in modelling seasonal fluctuations in the infected population. As for control strategies, Cromer [5] examined the effect of a seasonal screening program and noted a resultant decrease in the infected population.

In this paper we consider a single population model similar to that of Cromer [5] with the exception of allowing the system and input parameters to be uncertain, that is, to range in bounded sets. We seek a constrained control strategy, implementation of which results in stabilizing the infected population at a prescribed tolerable level.

2. Problem formulation

Let $I(t)$ and $S(t)$, respectively, denote the fractions of infected and susceptible individuals at time t. Since gonorrhea does not confer immunity, immediately upon recovery a cured infected individual becomes susceptible again. On the other hand, as the incubation period of the disease as well as the time period of curing it are relatively very short, we assume that the total population consists only of susceptible and infected individuals; namely, $I(t) + S(t) = 1$.

In the model used here it is assumed that the fraction of infected individuals increases at a rate proportional to the product of infected and susceptible fractions, and decreases

*College of Engineering, University of California, Berkeley, CA 94720, USA, e-mail: gleit@coe.Berkeley.edu

at a rate proportional to the fraction of infected individuals as a result of treatment sought by some portion of the infected population in the absence of intervention by the public health authority. Thus

$$\frac{dI(t)}{dt} = a(t)I(t)[1 - I(t)] - b(t)I(t) \tag{1}$$

where $a(t)$ and $b(t)$, respectively, are the contact and cure rates. To allow for temporal variation and uncertainty, we assume them to be of the form

$$a(t) = a^* + \Delta a(t) \ , \quad b(t) = b^* + \Delta b(t) \tag{2}$$

subject to

$$|\Delta a(t)| \leq \overline{\Delta a} < a^* \ , \ |\Delta b(t)| \leq \overline{\Delta b} < b^*$$

where $a^*, b^*, \overline{\Delta a}$ and $\overline{\Delta b}$ are known positive constants.

At present, no vaccine is available for the prevention of the disease, so that a control program of the public health authority might be aimed at identifying infected individuals and curing them. Let $C(t)$ denote the fractional rate of screening the total population at time t, and let $I'(t)$ denote the fraction of infected in the screened population at time t. To allow for the possibility of $I'(t) \neq I(t)$, we introduce the uncertain coefficient of screening efficiency

$$m(t) := I'(t)/I(t) \tag{3}$$

where it is assumed that

$$m(t) = m^* + \Delta m(t) \tag{4}$$

subject to

$$|\Delta m(t)| \leq \overline{\Delta m} < m^*$$

with m^* and $\overline{\Delta m}$ known positive constants.

Again assuming that the time period of curing the disease is negligible, so that infected individuals identified by screening are cured immediately, intervention by screening the total population results in a modified model.

$$\frac{dI(t)}{dt} = a(t)I(t)[1 - I(t)] - b(t)I(t) - C(t)m(t)I(t) \quad . \tag{5}$$

Ideally, one would like to choose $C(t)$ so as to wipe out the disease. However, the cost of screening is roughly proportional to the screening rate, for, as pointed out by Cromer [5], testing is more expensive than treating; furthermore, many screened individuals are not infected and in need of treatment. Thus, a more realistic control program might be one that stabilizes the fraction of infected individuals at a given tolerably low level, say I^*.

Also, for reasons such as shortage of facilities and personnel, the fraction of the population that can be screened at time t is limited. Thus, the screening rate is constrained by

$$0 \leq C(t) \leq \bar{C}, \text{ where } \bar{C} \leq 1 \text{ is prescribed.}$$

Finally, we introduce a change of variables which permits us to cast the system into the form treated in [1,2], namely,

$$x := ln\frac{I}{I^*} \quad , \quad u := C - \frac{1}{2}\bar{C}, \tag{6}$$ [1]

so that (5) becomes

$$\frac{dx}{dt} = (a^* + \Delta a)[1 - I^* exp(x)] - (b^* + \Delta b) - (m^* + \Delta m)(u + \frac{1}{2}\bar{C}) \tag{7}$$

In order to recognize the applicability of the results of [1,2], we rewrite (7) as

$$\frac{dx}{dt} = Ax + B(u + e) \tag{8}$$

where here $A = const. < 0, B = -m^*$ and

$$|e| \leq k_0 + k_1(|x|)|x| + k_2|u| \tag{9}$$

with

$$k_0 := |b^* + \overline{\Delta b} - a^* + \overline{\Delta a} + \frac{1}{2}(m^* + \overline{\Delta m})\bar{C} + I^*(a^* + \overline{\Delta a})|(m^*)^{-1}$$

$$k_1(|x|) := [\frac{I^*(a^* + \overline{\Delta a})[exp(|x|) - 1]}{|x|} - A](m^*)^{-1}$$

$$k_2 := \overline{\Delta m}(m^*)^{-1} \quad .$$

Now we seek a feedback control $p : R \to R$, that is $u = p(x)$, such that, for all e satisfying (9), system (8) is uniformly exponentially convergent to $\{x| \ \ |x| \leq r\}$ with given r and rate α from a computable set of attraction $\mathcal{A}$. In terms of the original variables, this means a mapping $I(t) \longmapsto C(t)$ which assures the corresponding convergence of $I(t)$ to a neighborhood I^*. This question was addressed for a somewhat less general case in [14].

3. Stabilizing controls

According to [1,2], the control sought in sec. 2 is

$$u = p(x) = \rho s(\epsilon^{-1} m^* Px) + \tilde{\rho} sat[\tilde{\rho}^{-1}\gamma(|x|)m^* Px]$$

where

$$\rho := \frac{k_0}{1 - k_2}, \ \tilde{\rho} := \frac{1}{2}\bar{C} - \rho \, , \qquad \gamma(|x|) := \frac{\sigma + \mu^{-1}k_1^2(|x|)}{2(1 - k_2)} \quad ,$$

and $P > 0$ is a solution of

$$\sigma(m^*)^2P^2 - 2(A + \alpha)P - \mu = 0$$

[1]Henceforth, we omit the argument t whenever the meaning is clear.

where ϵ, σ, μ and α are positive constants at the control designer's discretion (with α as the prescribed convergence rate). The design parameter ϵ must be chosen sufficiently small so that

$$\epsilon < \frac{\alpha c^2}{k_0}$$

with the constant $c > 0$ satisfying

$$\frac{m^* \sqrt{P}[\sigma\mu + k_1^2(\frac{c}{\sqrt{P}}]c}{2\mu(1-k_2)} \leq \tilde{\rho} \ .$$

The functions s$(\cdot)$ and $sat(\cdot)$ are defined by

$$s(z) := \frac{z}{1+|z|} \ , \quad sat(z) := \begin{cases} z & if \ \ |z| \leq 1 \\ \frac{z}{|z|} & if \ \ |z| > 1 \ . \end{cases}$$

To assure uniform exponential convergence of system (8) at rate α to $\{x| \ \ |x| \leq r_\epsilon\}$, where

$$r_\epsilon := [\frac{\epsilon k_0}{\alpha P}]^{\frac{1}{2}} \ ,$$

it suffices that the initial state $x(t_0)$ belong to

$$\mathcal{A} := \{x | P x^2 \leq c^2\} \ .$$

Finally, in view of the scalar state x, the constant $\beta = 1$ (see Appendix), and hence

$$|x(t)| \leq r_\epsilon + |x(t_0)| exp[-\alpha(t-t_0)] \ .$$

Some observations are in order:

1. Here we address only the determination of a fractional screening rate which assures stabilization (in the sense defined here) of the infected population at a prescribed level. The implementation of the proposed screening strategy, that is, how to induce the appropriate fraction of the population to undergo screening is the task of the public health authority; its success is, of course, highly dependent on local circumstances.
2. The efficacy of the proposed screening policy depends on the validity of the mathematical model, including the nominal values of the system parameters as well as of the assumed bounds imposed on the uncertain elements. The determination of these model parameters and, indeed, the verification of the model are yet another task of the public health authority.
3. Finally it should be noted that the information on which the control is based, namely, the fraction of infected in the total population, $I(t)$, is not available. What is available is the fraction of infected in the screened population, $I'(t)$. In the model it is assumed that $I(t) = m^{-1}(t)I'(t)$ where $m(t)$ is uncertain but has known bounds. The theory, as currently developed, does not account for uncertain state information. However, it can be generalized to include uncertain state information; e.g., see [15]. For the moment, one might take $I(t) = (m^*)^{-1}I'(t)$. However, if $I(t)$ is estimated on the basis of $I'(t)$, one must impose $C(t) \in [\underline{C}, \bar{C}]$ with $\underline{C} > 0$.

4. Simulation results

To illustrate the efficacy of the proposed scheme we present the result of a simulation utilizing the following model parameters:

$$\begin{array}{lll} a^* &=& 0.0625 \\ b^* &=& 0.0182 \\ m^* &=& 1.00 \\ I^* &=& 0.10 \\ \bar{C} &=& 0.35 \end{array} \qquad \begin{array}{lll} \overline{\Delta a} &=& 0.0063 \\ \overline{\Delta b} &=& 0.0018 \\ \overline{\Delta m} &=& 0 \\ A &=& -0.20 \ . \end{array}$$

The control design constants are

$$\begin{array}{lll} \alpha &=& 0.15 \\ \sigma &=& 1.00 \\ \mu &=& 2.00 \\ \epsilon &=& 0.0025 \ . \end{array}$$

The following uncertainty realizations are used in the simulation:

$$\Delta a(t) = \overline{\Delta a} \ , \quad \Delta b(t) = -\overline{\Delta b} \quad .$$

Figure 1(a) shows the evolution of the fraction of infected, $I(t)$, for the nominal parameter values, whereas Figure 1(b) shows the evolution of $I(t)$, for the given uncertainty realizations, both in the *absence of screening.*

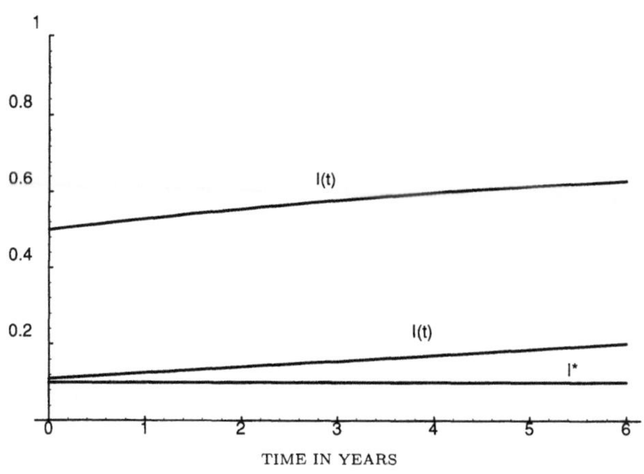

Figure 1(a) Infacted fraction: Nominal parameter values without screening

Figure 2 shows the evolution of $I(t)$ and the corresponding screening rate history for the given uncertainty realizations. The proposed screening scheme appears to be very efficacious. Not surprisingly, screening occurs at a relatively high rate at the outset until $I(t)$ reaches a "small" neighborhood of I^* and then proceeds at a low rate so as to maintain $I(t)$ "near" I^*.
Additional simulation results may be found in [14].

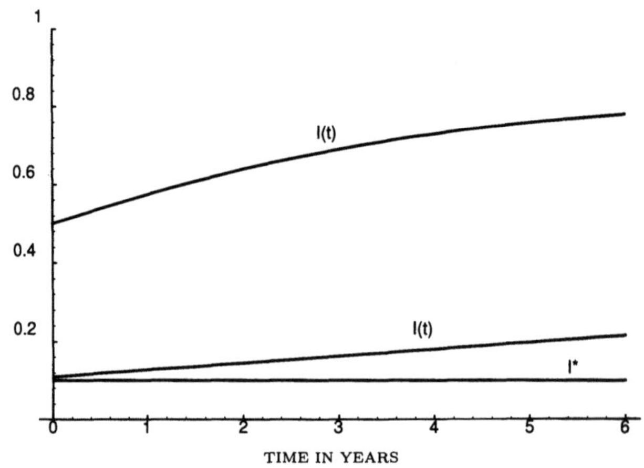

Figure 1(b) Infected fraction: Perturbed parameter values without screening

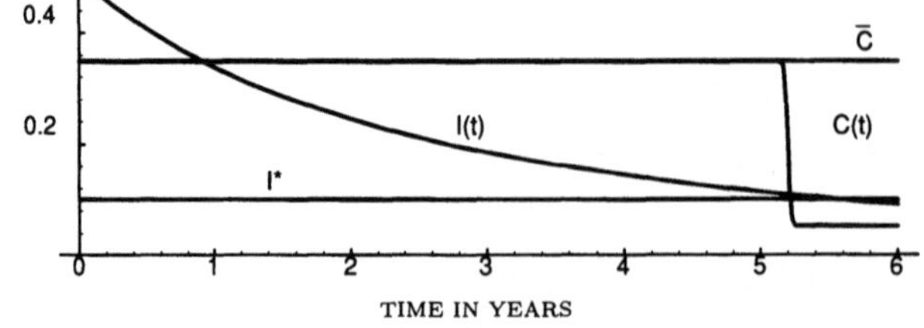

Figure 2 Infected fraction and screening rate: perturbed parameter values with screening

A Appendix

A1 Problem statement

We consider uncertain systems by

$$\dot{x}(t) = f(x(t)) + \sum_{i=1}^{l}[B_i u_i(t) + \Delta F_i(t, x(t), u_i(t))] \tag{A.1}$$

where $t \in \mathbf{R}$ is the time variable, $x(t) \in \mathbf{R}^{\mathbf{n}}$ is the state and $u_i(t) \in \mathbf{R}^{\mathbf{m_i}}$, $i = 1, 2, \ldots, l$, are control inputs. The continuous function f and the constant matrices B_i, $i = 1, 2, \ldots, l$, are known; they define the *nominal system*

$$\dot{x} = f(x(t)) + Bu(t), \tag{A.2}$$

where

$$u := \begin{bmatrix} u_1 \\ u_2 \\ \vdots \\ u_l \end{bmatrix}, \quad B := [B_1 B_2 \cdots B_l].$$

All the uncertainty and time-dependence in the system is represented by the terms ΔF_i which are assumed to be continuous functions.
Each control input u_i is subject to hard a constraint of the form

$$\|u_i(t)\| \le \bar{\rho}_i \tag{A.3}$$

where the bound $\bar{\rho}_i > 0$ is prescribed. We shall consider the control input u_i to be generated by a memoryless state feedback controller p_i, i.e.,

$$u_i(t) = p_i(x(t)). \tag{A.4}$$

The resulting *closed loop system* is described by

$$\dot{x} = F(t, x(t)) \tag{A.5}$$

with

$$F(t, x) := f(x) + \sum_{i=1}^{l}[B_i p_i(x) + \Delta F_i(t, x, p_i(x))]. \tag{A.6}$$

For any scalar $r \ge 0$, the *ball of radius* r is defined by

$$\mathcal{B}(r) := \{x \in \mathbf{R}^{\mathbf{n}} : \|\mathbf{x}\| \le \mathbf{r}\}.$$

Consider any scalar $\alpha > 0$ and any set $\mathcal{A} \subset \mathbf{R}^{\mathbf{n}}$ containing a neighborhood of the origin.

Definition A.1 System (A.5) is *uniformly exponentially convergent* to $\mathcal{B}(r)$ with rate α and *region of attraction* $\mathcal{A}$ if there exist a scalar $\beta \ge 0$ such that the following hold

(i) *Existence of solutions.* For each $t_0 \in \mathbf{R}$ and $x_0 \in \mathcal{A}$, there exists a solution $x(\cdot) : [t_0, t_1) \to \mathbf{R}^{\mathbf{n}}$, $\mathbf{t_0} < \mathbf{t_1}$, of (A.5) with $x(t_0) = x_0$.

(ii) *Indefinite extension of solutions.* Every solution $x(\cdot) : [t_0, t_1) \to \mathbf{R^n}$ of (A.5) with $x(t_0) \in \mathcal{A}$, has an extension $\bar{x}(\cdot) : [t_0, \infty) \to \mathbf{R^n}$, i.e., $\bar{x}(t) = x(t)$ for all $t \in [t_0, t_1)$ and $\bar{x}(\cdot)$ is a solution of (A.5).

(iii) *Uniform exponential convergence of solutions.* If $x(\cdot) : [t_0, \infty) \to \mathbf{R^n}$ is any solution of (A.5) with $x(t_0) \in \mathcal{A}$, then

$$\|x(t)\| \leq r + \beta \|x(t_0)\| exp[-\alpha(t - t_0)].$$

The problem we wish to consider is as follows.

Problem Statement. *Consider a system described by (A.1) subject to control constraints (A.3) and let $\alpha > 0$ and $r \geq 0$ be specified scalars. Find memoryless state feedback controllers p_i, $i = 1, 2, \ldots, l$, which render the closed loop system (A.5) uniformly exponentially convergent to $\mathcal{B}(r)$ with rate α.*

A2 Assumptions on uncertainty

The following assumption, which is sometimes referred to as a matching condition, is common in the literature on control of uncertain systems.
Assumption 1. For each $i = 1, 2, \ldots, l$, there is a function $e_i(\cdot)$ such that

$$\triangle F_i = B_i e_i. \tag{A.7}$$

Assumption 2. For each $i = 1, 2, \ldots, l$, there exist non-negative scalars k_{0i}, k_{2i}, with

$$k_{2i} < 1 \tag{A.8}$$

and a continuous non-decreasing nonnegative function k_{1i} such that

$$\|e_i(t, x, u_i)\| \leq k_{0i} + k_{1i}(\|x\|)\ x\| + k_{2i}\|u_i\| \tag{A.9}$$

for all $t \in \mathbf{R}$, $\mathbf{x(t)} \in \mathbf{R^n}$, $\mathbf{u_i(t)} \in \mathbf{R^{m_i}}$.
Assumption 3. For each $i = 1, 2, \ldots, l$,

$$\bar{\rho} > \frac{k_{0i}}{1 - k_{2i}}. \tag{A.10}$$

A3 Constrained control assuring exponential convergence

The proposed control meeting the requirement of the Problem Statement in A1 is

$$u_i = -\rho_i sat(\epsilon^{-1} B_i^T P x) - \tilde{\rho}_i sat(\tilde{\rho}_i^{-1} \gamma_i(\|x\|) B_i^T P x),\ i = 1, 2, \ldots, l, \tag{A.11}$$

where P satisfies
Assumption 4. There exist positive-definite symmetric matrices P and Q and a scalar $\sigma \geq 0$ which statisfy

$$2x^T P f(x) \leq -2\alpha x^T P x - x^T Q x + \sigma x^T P B B^T P x. \tag{A.12}$$

Remark *If the nominal system is linear and controllable, that is, the nominal system is*

$$\dot{x} = Ax + Bu$$

with (A, B) controllable, then (A.12) is met for each positive-definite symmetric Q and each σ and $\alpha > 0$, since there exist a positive-definite symmetric P which satisfies the Riccati equation

$$P(A+\alpha I) + (A+\alpha I)^T P - \delta PBB^T P + Q = 0. \tag{A.13}$$

Furthermore,

$$\rho_i := (1 - k_{2i}^{-1})k_{0i}, \tag{A.14}$$

γ_i is any continuous function which satisfies

$$\gamma_i(\|x\|) \geq \frac{\sigma}{2(1-k_{2i})} + \frac{lk_{1i}(\|x\|)^2}{2\mu(1-k_{2i})} \tag{A.15}$$

with

$$\mu := \lambda_{\min}(Q), \tag{A.16}$$

and the saturation function *sat* is given by

$$sat(y) := \begin{cases} y & \text{if } \|y\| \leq 1 \\ \|y\|^{-1}y & \text{if } \|y\| > 1 \end{cases},$$

and

$$\tilde{\rho}_i := \bar{\rho}_i - \rho_i. \tag{A.17}$$

The positive real scalar ϵ is chosen sufficiently small to satiesfy

$$\epsilon < \epsilon^* := \frac{\alpha c^{*2}}{k_0} \tag{A.18}$$

with

$$c^* := \min\{c_i : 1, 2, \ldots, l\} \tag{A.19}$$

where $c_i > 0$ satisfies

$$\lambda_i[\mu\sigma + lk_{1i}(\lambda c_i)^2]c_i \leq 2\mu(1-k_{2i})\tilde{\rho}_i \tag{A.20}$$

and

$$\lambda_i := \lambda_{\max}(B_i^T PB_i)^{\frac{1}{2}}, \quad \lambda := \lambda_{\min}(P)^{-\frac{1}{2}}.$$

Before introducing the main result, consider any real scalar $c \geq 0$ and define the Lyapunov ellipsoid

$$\mathcal{E}(c) := \{x \in \mathbf{R^n} : x^T Px \leq c^2\}.$$

Theorem 1 *Consider an uncertain system described by (A.1), satisfying assumptions 1-4 and subject to bounded control given by (A.11). Then the resulting closed loop system (A.5) is uniformly exponentially convergent to $\mathcal{B}(r_\epsilon)$ with rate α and region of attraction $\mathcal{A} = \mathcal{E}(c^*)$ where*

$$r_\epsilon = \left(\frac{\epsilon k_0}{\alpha\lambda_{\min}(P)}\right)^{\frac{1}{2}}, \quad k_0 = \sum_{i=1}^{l} k_{0i} \quad \textit{and} \quad \beta = \left(\frac{\lambda_{\max}(P)}{\lambda_{\min}(P)}\right)^{\frac{1}{2}}.$$

References

[1] Corless, M.; Leitmann, G.: *Bounded controllers for robust exponential convergence*, Journal of Optimization Theory and Applications, Vol. 76, pp. 1–12, 1993.

[2] Corless M.; Leitmann, G.: *Componentwise bounded controllers for robust exponential convergence*, Proceedings of the Variable Structure and Lyapunov Theory Workshop, Sept. 7–9, 1994, Benevento, Italy, pp. 64–69.

[3] Lee, C. S.; Leitmann, G.: *A bounded harvest strategy for an ecological system in the presence of uncertain disturbances*, Proceedings of the International Workshop on Intelligent Systems and Innovative Computations – The 6th Bellman Continuum, August 1–2, 1994, Hachioji, Tokyo, Japan, to appear in J. Computers and Mathematics.

[4] Leitmann, G.; Lee, C. S.: *Stabilization of a Competing Species System*, Proceedings of the EUROSIM Congress '95, September 11–15, Vienna, Austria.

[5] Cromer, T. L.: *Seasonal control for an endemic disease with seasonal fluctuations*, Theoretical Population Biology, Vol. 33, pp. 115–125, 1988.

[6] Hethcote, H. W.: *Asymptotic behavior in a deterministic epidemic model*, Bulletin of Mathematical Biology, Vol. 35, pp. 607–614, 1973.

[7] Lajmanovich, A.; Yorke, J. A.: *A deterministic model for gonorrhea in a nonhomogeneous population*, Mathematical Biosciences, Vol. 28, pp. 221–236, 1976.

[8] Hethcote, H. W.; Yorke, J. A.: *Gonorrhea Transmission Dynamics and Control*, Springer-Verlag, New York, 1985.

[9] Cooke, K. L.; Kaplan, J. L.: *A periodicity threshold theorem for epidemics and population growth*, Mathematical Biosciences, Vol. 31, pp. 87–104, 1976.

[10] Smith, H. L.: *On periodic solutions of a delay integral equation modeling epidemics*, Journal of Mathematical Biology, Vol. 4, pp. 69–80, 1977.

[11] Nussbaum, R. D.: *A periodicity threshold theorem for some nonlinear integral equations*, SIAM Journal of Mathematical Analysis, Vol. 9, pp. 356–376, 1978.

[12] Cromer, T. L.: *Asymptotically periodic solutions to Volterra integral equations in epidemic models*, Journal of Mathematical Analysis and Applications, Vol. 110, No. 2, pp. 483–494, 1985.

[13] Yorke, J. A.; Hethcote, H. W.; Nold, A.: *Dynamics and control of the transmission of gonorrhea*, Sex. Transm. Dis. Vol. 5, No. 2, pp. 51–56, 1978.

[14] Lee, C. S.; Leitmann, G.: *Control Strategies for an Endemic Disease in the Presence of Uncertainty*, in Recent Trends in Optimization Theory and Applications, edited by R.P. Agarwal, World Scientific, Singapore, 1994.

[15] Leitmann, G.: Guaranteed Ultimate Boundedness for a Class of Uncertain Linear Dynamical Systems, IEEE Transactions on Autom. Control, Vol. AC-23, No. 6, pp. 1109–1110, 1978.

International Series of Numerical Mathematics
Vol. 124, © 1998 Birkhäuser Verlag, Basel

Optimal Control of Sloshing Liquids

Harald Leonpacher* Dieter Kraft*

Abstract. This paper is concerned with optimal trajectory path planning for the motion of open, fluid filled containers moving in two dimensional space. The fluid and the mechanical facility that moves the container are subject to several constraints. The objective of the optimisation is a minimisation of time to transport the container from an initial position to its final destination. Optimisation criteria are investigated to control the movement of the fluid within the container. The system of ordinary differential equations and of partial differential equations, representing the dynamics of the models is solved numerically using a direct shooting method.

1. Introduction

Current demands of the global market compel manufacturing companies towards optimising their production. The demands of the market are greater flexibility, quicker response to market changes and a higher variety of products. One aspect of cost reduction is linked to the storage and distribution of goods. The development of automatic warehouses has minimised labour costs and increased the throughput of products.

The aim of this paper is to develop a strategy to store and extract open topped fluid filled containers in an automatised warehouse. During the whole process, the fluid is not allowed to slosh over. This problem is more difficult than the transportation of rigid parts, as in this case not only the motion of the platform, but also the motion of the fluid has to be controlled.

In the proposed case when a cart is being used to move an open fluid filled container, three major scientific computational techniques are required. The combination of modelling and simulation of the warehouse plant and control system, numerical simulation of the fluid in motion within the cart, and some form of optimisation technique are necessary to calculate a control of the cart that fulfils the objective function and reduces the sloshing behaviour of the fluid. To solve the optimisation problem a general 'hill climbing' algorithm will be used. Hill climbing describes the search for the fastest descent on the surface of states due to the control parameters and the dynamics of the system.

Therefore, the aim of the investigation is to integrate fluid dynamics analysis with machine dynamics within an optimisation strategy for application within automated warehouse technologies. The goal is to enable optimal transport paths to be found for liquid-filled open containers within a warehouse.

This problem is new and has not been treated in the literature before as an optimisation problem. Related references with respect to sloshing can be found in the survey[7].

*Fachhochschule München, Department of Mechanical Engineering, Dachauer Str. 98b, D-80335 München, Germany, leo@robot.fm.fh-muenchen.de

2. Model problem

At the Fachhochschule of Munich an automatic warehouse is currently being investigated in the laboratory of Computer Integrated Manufacturing and Manufacturing Management. The investigation concerning optimal control of sloshing liquids is performed on the basis of these mechanical facilities.

In order to simplify the large amount of complicated properties of the real world warehouse a small scale model system has been derived which represents the major properties of a warehouse. This model assumes that in a warehouse the main movements take place in a vertical plane. The container filled with liquid is assumed to be of rectangular shape. Hence, the fluid domain can be represented in two dimensions, while the third dimension is thought to be of infinite length. The general layout of this model system is illustrated in Figure(1). In this Figure, values are indicated with (c)

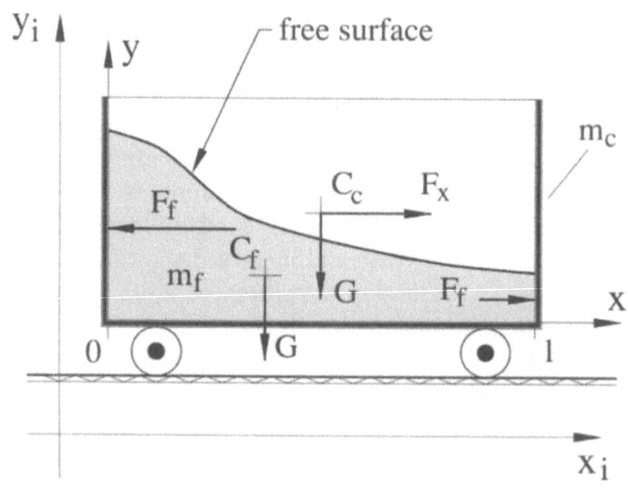

Figure 1: Model of the dynamic system.

and (f) corresponding to the cart and the fluid, respectively. The main movement of the fluid will occur while moving the container in the horizontal (x) direction. Thus the acceleration-vector of the cart and the liquid (F_x) is perpendicular to the force-vector (G) of the natural gravity field. The centres of gravity of the cart and the fluid are indicated with (C) and their masses with (m). Sloshing can only occur if a force orthogonal to gravity exists. The vertical movement in (y) direction results in a time dependent variation of (G). Investigations illustrated in this paper will only concern horizontal, one dimensional movement of the platform.

The cart is modelled using common mechanical laws of the interaction of masses, springs and damping devices. Their mathematical representation results in a system of ordinary differential equations (ODE). These equations are the transfer functions correlating the input data of the driving motors and the output motion of the cart.

Fluid flow is modelled and solved for numerically by using finite discretisation of the fluid domain. There are three different major discretisation strategies, finite volume-, finite element -, and finite difference techniques. Main properties of interest of the fluid domain are velocity, pressure and temperature at any given point in space and time. Free surface motion introduces additional information like surface tension and surface shape.

In our case of an accelerated open fluid container, temperature variation is not considered. On the one hand there is no temperature difference between the environment, the cart and the fluid (no heat source or sink q'''), and on the other hand temperature increase due to friction can be neglected. The motion of the viscous incompressible fluid is in our case assumed to be laminar. Surface tension is of minor influence and therefore has not been integrated in the model. The behaviour of the fluid is represented by the Navier–Stokes equations, a set of partial differential equations (PDE).

They consist of the conservation of mass

$$\frac{\partial u}{\partial x} + \frac{\partial v}{\partial y} = 0, \tag{1}$$

the conservation of linear momentum

$$\begin{aligned} \frac{\partial u}{\partial t} + u\frac{\partial u}{\partial x} + v\frac{\partial u}{\partial y} + \frac{1}{\rho}\frac{\partial p}{\partial x} &= \nu\left(\frac{\partial^2 u}{\partial x^2} + \frac{\partial^2 u}{\partial y^2}\right) + F(t,x,y) \\ \frac{\partial v}{\partial t} + u\frac{\partial v}{\partial x} + v\frac{\partial v}{\partial y} + \frac{1}{\rho}\frac{\partial p}{\partial y} &= \nu\left(\frac{\partial^2 v}{\partial x^2} + \frac{\partial^2 v}{\partial y^2}\right) + G(t,x,y), \end{aligned} \tag{2}$$

and the conservation of energy

$$\frac{\partial T}{\partial t} + u\frac{\partial T}{\partial x} + v\frac{\partial T}{\partial y} = \alpha\Delta T + q''' \tag{3}$$

in any finite small control volume of the fluid domain. The energy equation solves for the temperature distribution and has not been considered.

The discretisation of the fluid domain is modelled using the numerical finite difference code NaSt2D[3] developed at the Technical University Munich. This code solves the Navier–Stokes equations on a domain with free surface. The code has been investigated and further developed (see [6]) to allow special manipulations that are necessary for our problem.

These models of the fluid and the automatic warehouse incorporate all physical and mechanical interactions which influence the movement. In particular its trajectory shape, velocity, acceleration and further derivatives like jerk. The mechanical properties are based on Newton's law. The fluid properties are based on the Navier–Stokes equations.

3. Optimal control problem

The optimisation problem needs to be formulated mathematically, allowing computation of cost function and constraints via numerical algorithms. Only if the equations are linear in their solution can an algebraic single iteration be found. The model-equations for the mechanical system and the fluid flow are highly non-linear. Hence numerical iterative algorithms must be used to calculate an optimal solution. For this purpose a direct shooting method will be used which is more robust in comparison to collocation or multiple shooting methods. In addition, due to the complexity of the problem, direct shooting is easier to apply.

The formulation of the modelling problem illustrates the separation of two different models. The influences in the specific case of trajectory path planning for open fluid filled containers are separated into the effects within the fluid described by the velocity vector $\mathbf{v} = (u, v)$ in the coordinates $\mathbf{x} = (x, y)$. and those in the mechanical system providing the movement in the fixed coordinates $\mathbf{y} = (x_i, y_i)$ with $\mathbf{z} = (\mathbf{y}, \dot{\mathbf{y}})$. The optimisation procedure must combine those different models. The way this will be done is illustrated later on in conjunction with Figure(6).

The general optimal control problem is formulated as the minimisation of a real-valued objective function subject to the differential constraints (ODE's and PDE's), appropriate boundary conditions and algebraic state or control constraints by proper choice of control function $\mathbf{u}(t)$ and design parameters, e.g. the final time b.

In the case of optimal control of sloshing liquids, the optimal control problem (OCP) will be formulated as follows:

OCP

$$\min_{\mathbf{u}\in U} \quad \int_a^b \psi(t, \mathbf{z}(t), \mathbf{v}(\mathbf{x}, t), \mathbf{u}(t))dt \; + \; \phi(\mathbf{z}(a), \mathbf{z}(b), \mathbf{v}(\mathbf{x}, a), \mathbf{v}(\mathbf{x}, b))$$

subject to

$$\frac{\mathrm{d}}{\mathrm{d}t}\mathbf{z}(t) = \mathbf{f}(t, \mathbf{z}(t), \mathbf{u}(t)) \quad t \in [a, b], \tag{4}$$

$$\frac{\partial \mathbf{v}}{\partial t} + (\mathbf{v}\cdot\nabla)\mathbf{v} + \nabla p = \frac{1}{Re}\Delta\mathbf{v} + \mathbf{F} \quad t \in [a, b],\ \Omega_i \subset \Omega \tag{5}$$

$$\mathrm{div}(\mathbf{v}) = \mathbf{0} \quad t \in [a, b],\ \Omega_i \subset \Omega \tag{6}$$

$$\mathbf{g}(\mathbf{z}(t), \mathbf{u}(t), t) \geq \mathbf{0} \quad t \in [a, b], \tag{7}$$

$$\mathbf{h}(\mathbf{z}(a), \mathbf{z}(b), \mathbf{v}(\mathbf{x}, a), \mathbf{x}(a)) = \mathbf{0} \quad \mathbf{x} \in \Omega,\ \Omega \subset \mathcal{R}^2. \tag{8}$$

Within the mathematical representation of the optimal control problem equations (4-6) describe the differential constraints on the system. In particular, (4) represents the mechanical system using ordinary differential equations and (5, 6) represent the dimensionless Navier–Stokes equations modelling the fluid flow where Re is the Reynolds number.

The algebraic state or control constraints are given in equation (7). There are limitations on the control and on the state constraints due to the physical limitations of the motors driving the cart and the size of the warehouse or the small scale model:

$$\begin{aligned} \mathbf{u}_{\min} \;\leq\; \mathbf{u}(t) \;\leq\; \mathbf{u}_{\max}, &\qquad \forall\, t \in [a, b]. \\ \mathbf{z}_{\min} \;\leq\; \mathbf{z}(t) \;\leq\; \mathbf{z}_{\max}. & \end{aligned}$$

Additionally the limitation for the liquid not to slosh over is specified:

$$y(t)|_{x=0} \leq h, \quad y(t)|_{x=l} \leq h, \qquad \forall\, t \in [a, b].$$

The layout of the surface at (a) and the velocity and location of the cart at (a, b) need to be specified. Figure(1) illustrates the free surface within the container at some time-step $(t_i,\ t_i > a)$; the initial distribution of the fluid has a horizontal surface. This is specified within the boundary conditions:

$$\begin{aligned} y(a) &= y_a, \quad y(b) = y_b, \\ \mathbf{v}(\mathbf{x}, a) &= \mathbf{v}_{\mathbf{x},a}, \quad \mathbf{x}(a) = \mathbf{x}_a, \qquad \forall\, \mathbf{x} \in \Omega,\ \Omega \subset \mathcal{R}^2. \end{aligned}$$

In general Ω denotes a feasible domain for the fluid and Ω_i a sub-volume of the fluid filled region in which the Navier–Stokes equations are solved.

The optimal control problem is solved iteratively including the following steps:

1. control parameterisation
2. control approximation
3. solution of initial value problem
4. iterative procedure

This will result in a replacement of the optimal control problem by a nonlinear programming problem.

3.1 Control parameterisation

The infinite optimal control problem must be converted to a finite dimensional problem by introducing a finite set of control parameters $(\mathbf{x}_u)$ representing the infinite control function $\mathbf{u}(t)$ as described in the next paragraph. The resulting nonlinear programming problem NLP will be solved using "off the shelf" numerical software for finite dimensional optimisation like sequential quadratic programming (SQP).

NLP

$$\min_{\mathbf{x}_u \in \mathcal{R}^n} \quad f(\mathbf{x}_u)$$

$$\text{subject to} \quad \mathbf{g}(\mathbf{x}_u) \geq 0, \qquad \mathbf{h}(\mathbf{x}_u) = 0$$

With $\mathbf{u}(t)$ approximated by $(\mathbf{x}_u)$ the differential constraints are solved with initial values $\zeta(a), \nu(\mathbf{x}_u, a)$, where $\zeta(t),\ \nu(\mathbf{x}_u, t)$ are the approximation to $\mathbf{z}(t),\ \mathbf{v}(\mathbf{x}_u, t)$ respectively.

3.2 Control approximation

The control function $\mathbf{u}(t)$ will be defined differently for each iteration k and has to be approximated via $\tilde{\mathbf{u}}^k(t)$. The different steps are illustrated in Figure(2).

1. Interval partitioning: $t_i \in [a, b]$.
   Split the time interval $[a, b]$ in several sub-intervals, not necessarily equally spaced (dashed lines).
2. Choice of basis functions ppf.
   Choose an interpolation scheme e.g. piece-wise polynomial functions (ppf), or other linear or spline functions. In the given example four points are necessary to built a third order polynomial function (dotted line).

3. $\tilde{u}^k(t) \cong \mathbf{u}^k(t) : \{\text{ppf} \,|\, x^k_{u,i} = u^k(t_i)\}$
   Discretise the control using the points $x^k_{u,i}$ which represent the value of $\mathbf{u}^k$ at time t_i and interpolating with the basis functions in the intervals in between.

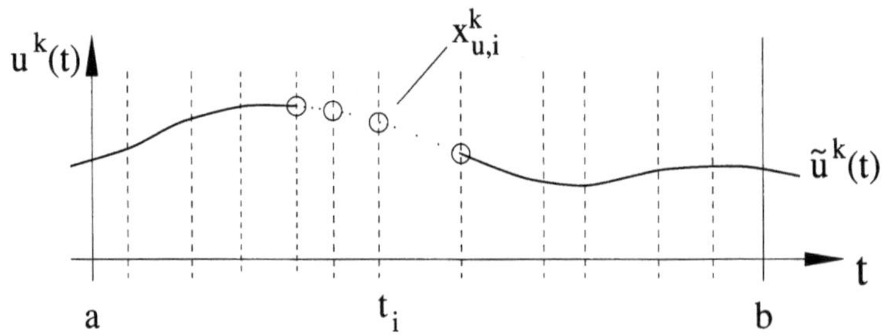

Figure 2: Approximation of $\mathbf{u}^k(t)$.

The result is a continuous function $\tilde{u}^k(t)$ as an approximation to $\mathbf{u}^k$.

3.3 Initial value problem

The direct shooting method is a calculation procedure that solves initial value problems as nucleus. Boundary value problems can be formulated as an iterative optimisation using and solving initial value problems. In our case the system of equations to be solved consists of a subsystem of first order differential equations, representing the mechanical system and the Navier–Stokes equations as representation for the fluid motion. The procedure of solving an initial value problem can be observed in Figure(3) and will be performed as follows:

1. Start the calculation with an initial value $\zeta_0 = y_0$.
   This value has been predefined within the boundary conditions.
2. Evaluation of the solution of the IVP using the control function $\tilde{u}^k(t)$,
   $$\zeta_b^k = \varphi(\zeta_0, \tilde{u}^k(t)),$$
   where k is the number of iteration.
3. The solution deviation is generally $d^k = \zeta_b^* - \zeta_b^k \neq 0$.
   With ζ_b^* being a single boundary element of $\mathbf{z}(t)$. Depending on the value of the state $\zeta_b^k = \zeta^k(b)$ a solution deviation d^k can be calculated which in the optimised case will be $d^* = 0$.

The constraints must be applied on the control function and the state function. It is very easy to limit the control during its evaluation. Constraint violation of the state will occur while proceeding towards an optimal solution within the iterative process. Point-wise verification as illustrated in Figure(4) will be used to value the constraint violation as follows:

1. Choose $\zeta_{\min}(t), \zeta_{\max}(t)$.
   The upper and lower bounds on the state are indicated with a grey shade. Generally they are time dependent. In Figure(4) these bounds are constant.

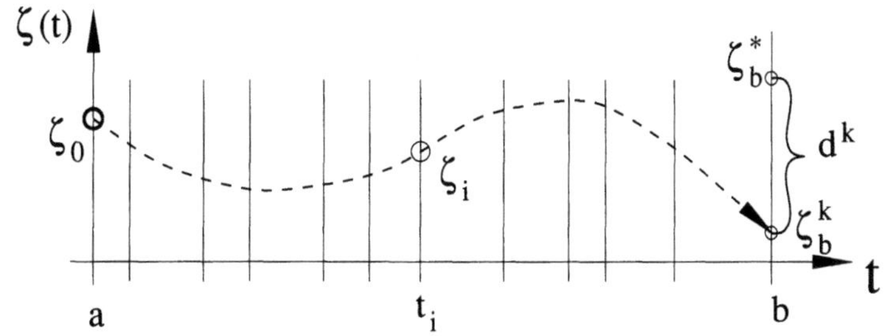

Figure 3: Initial value problem.

2. Set communication grid: $t_i \in [a, b],\ i = 0 \dots n$.
   The communication grid is illustrated with dotted lines. This grid can be different to the one used for the control discretisation.
3. Analyse function $e_{ub,i} \overset{!}{\geq} 0,\ \ \forall\, i = 0 \dots n$.
   The function will be analysed at all t_i. Figure(4) illustrates only the verification on the upper bound. The value of e_i is calculated by $e_{ub,i} = \zeta_{\max}(t_i) - \zeta(t_i)$.

A similar procedure can be formulated for the violation of the lower bound:

$$e_{lb}(t_i) \overset{!}{\geq} 0, \qquad e_{lb}(t_i) = \zeta(t_i) - \zeta_{\min}(t_i), \qquad t_i \in [a, b],\ i = 0 \dots n.$$

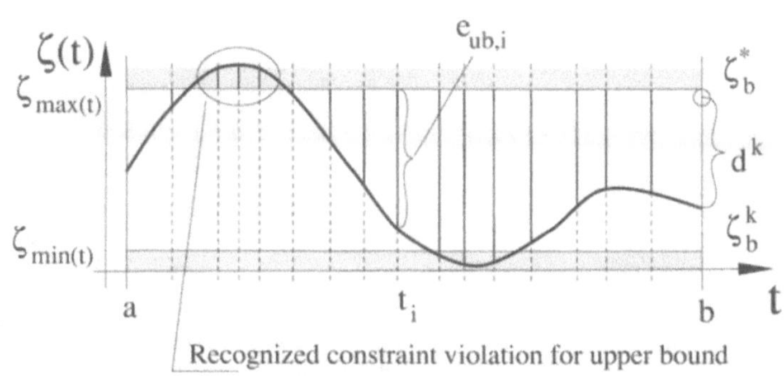

Figure 4: Pointwise verification of ζ.

By using the direct shooting procedure it is possible to calculate unique solutions based on the model system. A specific set of control parameters creates a specific solution deviation, constraint violations and a value for the objective function. The control parameters for the calculation of the fluid motion (proportional to **F**) are a function of the control parameters for the mechanical system **u**. Therefore it is not necessary to introduce a separate set of parameters.

3.4 Iterative procedure

The general idea of optimal control is the calculation of an optimal solution of an object function based on a system of non-linear differential equations limited by state constraints with time-dependent parameters. The problem of time-minimal movement of an open fluid-filled container is a problem of this class.

Optimal control problems can be solved using direct shooting or collocation methods by parameterisation as shown above. In comparison, indirect shooting or indirect collocation methods are transformed into a boundary value problem by applying the necessary conditions of optimality. These techniques need a reformulation of the problem creating additional differential equations for the Lagrange multipliers. They are often extended to multiple shooting methods to increase the stability of the algorithm. They need higher computational efforts.

Direct shooting in conjunction with sequential quadratic programming (SQP)[2] has been used by Kraft[4][5] for the calculation of an optimal trajectory for a six–degree–freedom robot. A method will be presented for solving the more complex problem of optimal control of sloshing liquids.

Further study has been undertaken by Barcley et al.[1] on the solution of large scale optimisation problems using SQP algorithms. These algorithms calculate a new direction and a new step-size due to gradient calculations on the virtual surface of the state. Based on sufficient conditions on the gradient vectors $\nabla f(\mathbf{x}_u)$, $\nabla \mathbf{g}(\mathbf{x}_u)$ and the tensors $\nabla^2 f(\mathbf{x}_u)$, $\nabla^2 \mathbf{g}(\mathbf{x}_u)$ (often called the Hessian, denoted by $\mathbf{H} = (h_{ij})$) an optimality criterion can be formulated. Figure(5) illustrates the computational approach of modelling and simulation of the process and subsequent optimisation.

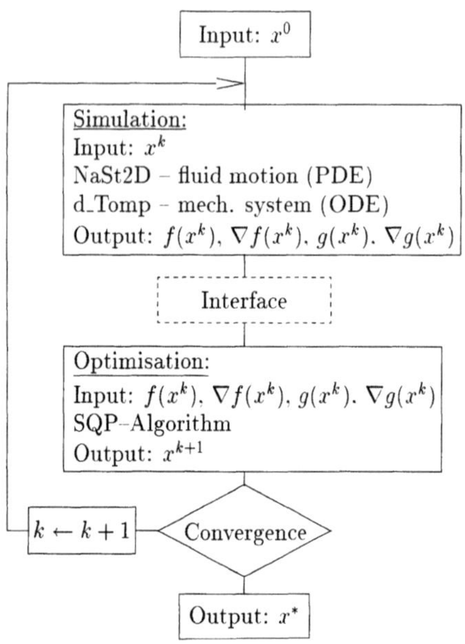

Figure 5: Program implementation.

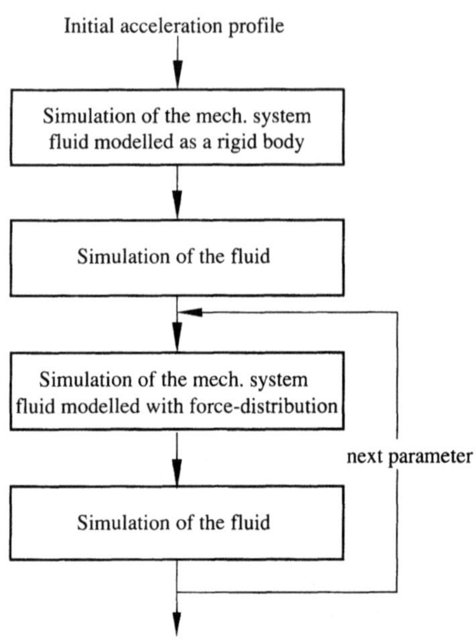

Figure 6: IVP solution evaluation.

The calculation of the response of the model system is illustrated in Figure(6).

1. A set of control parameters stimulates the model of the mechanical system. The fluid is modelled as a rigid body to evaluate an initial guess for the influence of the fluid motion. This set of parameters has been supplied initially by the user or by the optimisation module. The IVP solver calculates the response of the mechanical system. The control parameters will be altered due to the performance of the mechanical system.
2. The updated parameters will be used to determine the reaction of the sloshing liquid within the fluid model.
3. With the information of the sloshing behaviour of the fluid, the model of the mechanical system can be simulated again. This time the fluid is modelled with its full sloshing movement. The initial control parameters change once more.
4. The objective function and the sloshing limit can only be calculated while simulating the fluid motion. Therefore the final simulation of the overall system must involve the fluid dynamics. This calculation uses the new set of control parameters gained after step (3).

4. Results

There is still a deviation between reality and the results obtained using alternately calculation and simulation of the two different models as illustrated. This deviation decreases while increasing the number of alternating simulations. After four simulations, the mechanical system has influenced the fluid motion and vice versa. Wether this is sufficient must be investigated.

Within one iteration of the SQP algorithm, the first two steps (see inner loop in Figure(6)) could be eliminated due to the fact that the parameter variation for the calculation of gradients is very small. It can be assumed that a minimal alteration of one control parameter will not influence the behaviour of the fluid rapidly while simulating the mechanical system and vice versa.

A second, different approach solves the mechanical system within every time-step of the simulation of the fluid motion. Obviously, this algorithm will produce more precise results. Due to the simulation of the cart within every time step, the calculation of the iterative optimisation cycle will become very time consuming.

MATLAB routines have been developed by the first author to use the optimisation capabilities of its subroutines and m-files. A first insight into the motion of the container and the motion of the fluid within an optimised movement has been gained. This work has included performance studies of the optimisation due to initial conditions and parameter settings. Figure(7) illustrates an optimised acceleration profile and Figure(8) shows the corresponding amplitudes at the right and left boundary of the fluid domain. Note that the upper limit of the container has been reached by the liquid once.

Starting from the initial control distribution, marked as a dashed line, the solid line has been reached as an optimal control function. The upper and lower bounds are fulfilled only at specific points due to the communication grid. The value of 100 on the y-axis in Figure(7) is equivalent to a gravity field of $10.0[m/s^2]$. The modelled container is a cube with the dimensions of $10[cm]$ length in both X- and Y-direction.

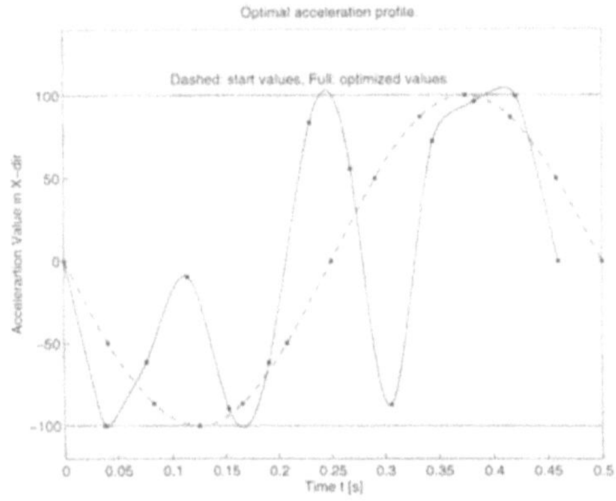

Figure 7: Acceleration profile.

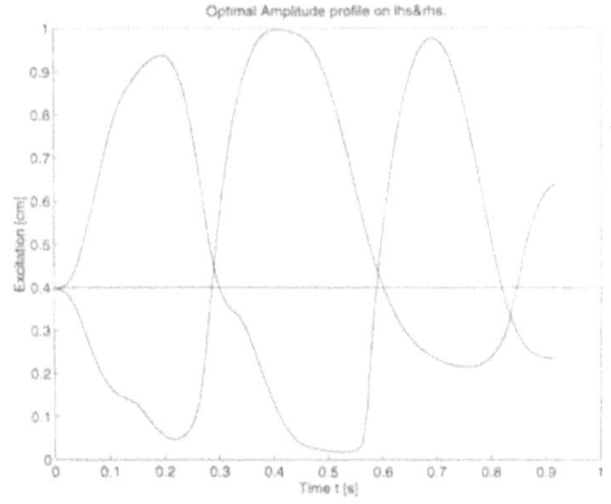

Figure 8: Amplitudes at boundaries.

The horizontal line in Figure(8) indicates the initial surface. The movement is time optimised with the aim to travel 30[cm] in positive horizontal direction. The mechanical system is, for reasons of simplicity, neglected.

References

[1] Barcley, A.; Gill, P. E.; Rosen, J. B.: *SQP methods and their application to numerical optimal control.* This volume.

[2] Gill, P. E.; Murray, W.; Wright, H. H.: *Large-scale SQP Methods and their Application in Trajectory Optimization.* In *Computational Optimal Control.* Edited by Bulirsch R., Kraft D., ISNM Vol. 115, Birkhäuser, Basel, 29–42, 1994.

[3] Griebel, M.; Dornseifer, T.; Neunhoeffer, T.: *Numerische Simulation in der Strömungsmechanik.* Vieweg, Braunschweig, Wiesbaden, 1995.

[4] Kraft D.: *TOMP – Fortran Modules for Optimal Control Calculations.* ACM Trans. Math. Softw., Vol. 20, No. 3, 262–281, 1994.

[5] Kraft D.: *Optimal Robot Path Planing.* In: Bestle D., Schiehlen W. (eds.): *IUTAM Symposium on Optimization of Mechanical Systems.*, NATO ASI Series 15, Springer, 261–280, 1995.

[6] Leonpacher H.: *The Problem of Sloshing Waves in a Rectangular Container as a Base to Illustrate Numerical Solution Strategies and Programs.* M.Sc. Dissertation, Liverpool, München 1995.

[7] Leonpacher H.: *Trajectory optimisation to reduce sloshing in open liquid filled containers.* M.Phil. Transfer Report, Liverpool, München 1996.

International Series of Numerical Mathematics
Vol. 124, © 1998 Birkhäuser Verlag, Basel

Free Surface Waves in a Wave Tank

Angela Pawell *

Abstract. We consider the motion of an inviscid and incompressible fluid in an closed or semi-infinite basin with free surface waves. An intrinsic feature of the problem is the formation of waves by a periodic motion of one part of the boundary of the basin. Sharp corners of the rectangular domain and between the free surface and the side walls of the basin are critical regions of the solution. The correct formulation and solution of the problem in these regions is necessary for the simulation of the free surface waves in a bounded domain. We formulate compatibility relations for the dynamic contact lines as necessary conditions for the existence of a solution to the free boundary problem for a potential flow. We derive a formulation and numerical solution in a compact way so that natural compatibility conditions can be satisfied at the interface between the walls of the basin and the free surface.

1. Introduction

During the last decades, a wide range of different approaches to the theory and numerical solution of free surface wave problems were developed. Nevertheless, this topic still comprises a series of open problems and challenging tasks. Intrinsic features of the problem are, e.g., the nonlinearity of the equations of motion and nonlinear free boundary conditions, the occurence of blow up effects for the norm of the solution to the Euler equations, or the interaction between fluids and rigid bodies or basins. For an overview concerning mathematical and numerical results in this field we refer to [3], [6]–[8]. In this paper we study free surface waves for an inviscid fluid in a closed domain. The special aim of this work is to simulate waves corresponding to experiments which were carried out at the ONR-Hinsdale Wave Facilities in Corvallis, Oregon. We will study free surface waves caused by a moving part of the boundary of the basin, the so-called wave-maker. In [9], [10] we derived a compact formulation of the free surface wave problem for a potential flow and a piston-like wave-maker motion. We proposed a numerical method and presented first numerical tests and a comparison with experimental data.

In this paper, we extend our results to more general types of wave-maker motions and boundary conditions admitting reflection of waves and the space and time dependent calculation of contact lines for a nonlinear wave propagation. In general, the free surface forms contact lines with the side walls of the basin as sharp edges and corners of the domain of fluid motion. The correct calculation of these contact lines as part of the solution intrinsicly determines the regularity of the solution. We shall derive formulas for the contact lines and compatibility relation as necessary conditions for the solvability of the free boundary problem.

*Brandenburgische Technische Universität Cottbus, Institut für Mathematik, PF 10 13 44, 03013 Cottbus, Germany, e-mail: pawell@math.tu-cottbus.de

I am grateful to Prof. Dr. R. B. Guenther from the Oregon State University for introducing me to this topic and for many helpful discussions and inspirations. Furthermore, I would like to thank Dr. K.-D. Krannich from the Technical University Cottbus for his support concerning the numerical solution and implementation of the problem.

2. Mathematical model

In this section we derive a mathematical model describing free surface waves of an inviscid and incompressible fluid in a closed domain. We denote by $v = (v_1, v_2, v_3)^T$ the velocity field, by p the pressure, by $\vec{g} = (0, 0, g)^T$ the gravity forces and by ρ the constant density of the incompressible fluid. The differential equations describing the motion of the fluid are given by the *conservation of momentum*

$$\rho\,\partial_t v + \rho\,(v \cdot \nabla)\,v = -\nabla p - \rho\,\vec{g} \tag{1}$$

and the *conservation of mass*

$$\nabla \cdot v = 0. \tag{2}$$

The horizontal components of the spatial vector will be denoted by $X \in \mathbb{R}^2$ and the vertical component by Y. The fluid is bounded from above by a free surface $Y = \eta(X, t)$. The rigid bottom of the basin is fixed and described by a smooth function $Y = h(X)$. The impermeability condition is given by

$$v \cdot \vec{n} = 0, \qquad \text{for } Y = h(X), \tag{3}$$

with the outer normal vector $\vec{n}$. Assuming atmospheric pressure outside the domain of fluid motion, we obtain a dynamic boundary condition at the free surface:

$$p = 0, \qquad \text{for } Y = \eta(X, t). \tag{4}$$

The derivation in time of $Y = \eta(X, t)$ yields the kinematic boundary condition:

$$\partial_t \eta + (\partial_{X_1} \eta)\,v_1 + (\partial_{X_2} \eta)\,v_2 = v_3, \qquad \text{for } Y = \eta(X, t). \tag{5}$$

System (1)–(5) describes free gravity waves in an open domain. Local solvability and uniqueness for this problem, augmented by a sufficiently smooth initial condition, was established by Shinbrot. He used a Lagrange formulation of the problem, power series expansions for the unknown functions v, p, η and a priori estimates which require analyticity of the solution [13], [14], [15]. The linearization of the problem using mixed Euler-Lagrange variables and the application of a limit analysis to the nonlinear problem enabled Tani to prove solvability of the free surface wave problem for the Euler equations under weaker regularity assumptions [21]. For related results in the viscous case we refer to [1], [2], [16]–[20] and the references there.

In the present paper we consider system (1)–(5) in a bounded domain. We will formulate and solve the free surface wave problem under the condition that the waves are caused by a moving wall of the basin. The formation of waves by the interaction of the fluid with the boundary of the domain requires an exact formulation of boundary and compatibility conditions. Naturally, the boundary conditions for the velocity at an impermeable boundary of the domain are given by

$$v \cdot \vec{n} = v_n(X, Y, t), \tag{6}$$

where v_n denotes the normal velocity of the boundary. Obviously, $v_n = 0$ for any fixed part of the boundary.

In the following we assume that the motion of the fluid is irrotational within a simply connected domain. This means that the velocity vector can be represented by the velocity potential φ with $\nabla\varphi = v$. Under this assumption, system (1)–(2) can be formulated in terms of the velocity potential

$$\varphi_t + \frac{1}{2}|\nabla\varphi|^2 + \frac{p}{\rho} + gY = 0, \qquad \text{in } \Omega, \tag{7}$$

$$\Delta\varphi = 0, \qquad \text{in } \Omega. \tag{8}$$

Let us assume now that the domain of fluid motion, Ω, is bounded by some fixed boundary parts, in particular a rigid bottom $\{Y = h(X)\}$, side walls $\{X_2 = 0 \vee X_2 = b\}$, $b > 0$ and $\{X_1 = H(X_2, Y)\}$ with a smooth function H and by a moving wall $\{X_1 = M(Y,t)\}$. Naturally, we require that $M(Y,t) < H(X_2, Y)$ for $t > 0$, $0 < X_2 < b$ and $h(x) < 0$ assuming that $\eta(X,t) = 0$ describes the still water level. The boundary of the domain is piecewise smooth and consists of the parts Γ_0 (fixed walls and bottom), the wave-maker Γ_w (moving wall), the boundary opposite to the wave-maker Γ_1 and the free surface Γ_s. A two-dimensional cross-section of the domain is sketched in Fig.1. Critical parts of the boundary are the edges and corners between the different parts of the boundary and the intersection between the free surface and the side walls. The domain of fluid motion $\Omega = \Omega(\eta, M)$ changes in time dependent on the position of the free surface and the prescribed motion of the wave-maker.

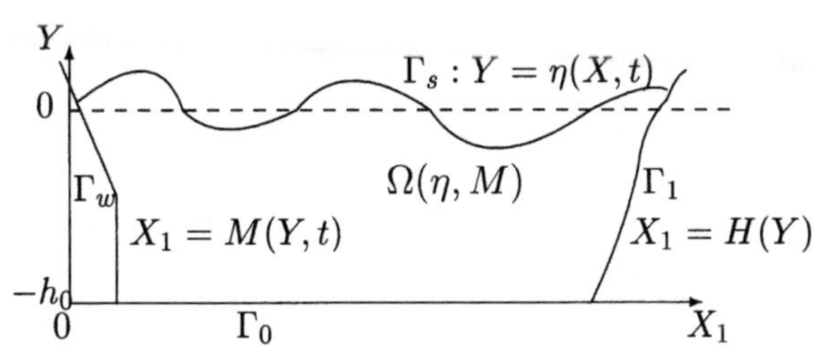

Fig. 1: Two-dimensional cross-section of the fluid domain

The following motions of the wave-maker are of special interest (see Fig. 2):

i) the piston type wave-maker,

$$X_1 = M(Y,t) = P_1 + a_0 \sin(\omega_0 t),$$

ii) the flap type wave-maker,

$$X_1 = M(Y,t) = P_1 + \frac{Y - P_2}{h_0} a_0 \sin(\omega_0 t),$$

iii) the joint type wave-maker,

$$X_1 = M(Y,t) = P_1 + a_0 \sin(\omega_0 t) + \max(Y - P_1 - h_1, 0)\frac{a_1}{h_0 - h_1} a_0 \sin(\omega_1 t),$$

where P_1, P_2 denote the X_1 and Y coordinate of the joint between the bottom and the wave-maker, ω_i $(i = 1,2)$ are the angle velocities of the wave-maker, a_i $(i = 1,2)$ are the amplitudes of the wave-maker strikes, h_0 is the depth of the fluid at the wave-maker (i.e. the difference between the still water level and P_2) and h_1 is the distance between the two wave-maker joints in case *iii)*.

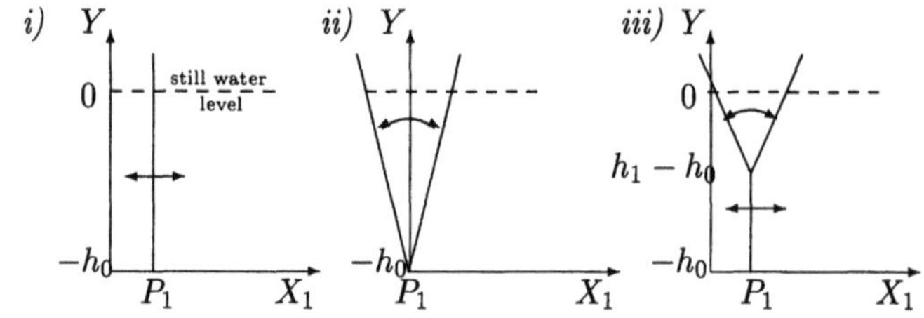

Fig. 2: Different types of wave-maker motion

3. Reduction to a non-local system of partial differential equations

We shall transform the free boundary problem (1)–(6) into a non-local system of partial differential equations for the velocity potential and the free surface. We denote by W the projection of the potential to the free surface, $W(X,t) = \varphi(X, \eta(X,t), t)$. From (7), (5) we obtain

$$\eta_t = -\nabla W \cdot \nabla\eta + \varphi_Y(1 + |\nabla\eta|^2), \qquad \text{for } (X,t) \in G, \tag{9}$$

$$W_t = -\frac{1}{2}|\nabla W|^2 + \frac{1}{2}\varphi_Y^2(1 + |\nabla\eta|^2) - g\eta, \qquad \text{for } (X,t) \in G, \tag{10}$$

with $\nabla = \left(\frac{\partial}{\partial X_i}\right)_i$ and

$$G = \{(X,t) \in \mathbb{R}^2 \times \mathbb{R}_+ \mid K_a(X_2,t) < X_1 < K_b(X_2,t), 0 < X_2 < b\}.$$

For the initial condition we suppose that the fluid is at rest at $t = 0$:

$$\eta(X,0) = 0, \qquad W(X,0) = 0.$$

Note that equations (9), (10) are given in the domain G which is bounded by the unknown functions K_a and K_b describing the X_1 coordinate of the contact points or lines between the free surface and the boundary parts Γ_w and Γ_1 respectively.

These equations are augmented by the continuity equation $\Delta\varphi = 0$, in $\Omega(\eta, M)$ with the boundary conditions:

$$\frac{\partial\varphi}{\partial n} = f(Y,t) = \begin{cases} -\dfrac{M_t(Y,t)}{\sqrt{1 + M_Y^2}}, & \text{at } \Gamma_w, \\ 0, & \text{at } \Gamma_0, \Gamma_1, \end{cases}$$

$$\varphi = W, \qquad \text{at } \Gamma_s.$$

We use the coordinate transformation $\Phi : \Omega(t) \longrightarrow \Omega(\eta, M)$,

$$X_1 = K_a(x_2, t) + \frac{x_1}{q(x,t)}, \quad X_2 = x_2, \quad Y = y, \quad q(x,t) = \frac{K_b(x_2, 0) - K_a(x_2, 0)}{K_b(x_2, t) - K_a(x_2, t)}$$

and the corresponding transformation matrix

$$Q(x,t) = \frac{\partial(x,y)}{\partial(X,Y)} = \begin{pmatrix} Q^x(x,t) & 0 \\ 0 & 1 \end{pmatrix}, \quad Q^x(x,t) = \begin{pmatrix} q(x_2,t) & \tilde{q}(x,t) \\ 0 & 1 \end{pmatrix}$$

to describe system (9), (10) in a fixed domain $B = \Phi^{-1}(G(t))$ (see Fig. 3). Note that the transformation matrix depends on the contact points or lines, i.e. it depends on the unknown position of the free surface.

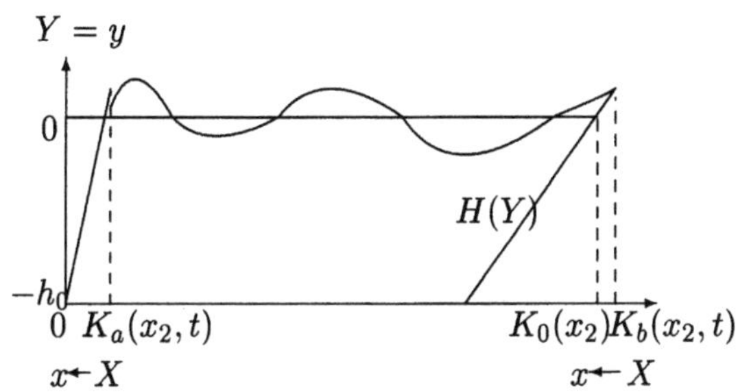

Fig. 3: Transformation of the horizontal variables to a fixed domain

Introducing the functions

$$s(x,t) := \eta(X,t), \qquad w(x,t) := W(X,t), \qquad u(x,y,t) := \varphi(X,Y,t),$$

we derive from (9),(10) a system of non-local, partial differential equations of first order for w and s in $B = \{(x,t) \in \mathbb{R}^2 \times \mathbb{R}_+ \mid 0 < x_2 < b, 0 < x_1 < K_0(x_2)\}$, $K_0(x_2) := K_b(x_2, 0) - K_a(x_2, 0)$:

$$s_t = u_y - (Q^x \nabla s \cdot Q^x \nabla w) + u_y \, |Q^x \nabla s|^2 + s_{x_1} \tilde{K}, \tag{11}$$

$$w_t = u_y \, s_t - (w_{x_1} - u_y \, s_{x_1}) \, \tilde{K} - \frac{1}{2}(u_y^2 + |Q^x \nabla w|^2 + u_y^2 \, |Q^x \nabla s|^2) + u_y \, (Q^x \nabla s \cdot Q^x \nabla w) - gs, \tag{12}$$

with the initial condition

$$s(x,0) = 0, \qquad w(x,0) = 0, \qquad x \in B. \tag{13}$$

The transformation coefficients $\tilde{K}$ and $\tilde{q}$ depend on the derivatives of the functions K_a and K_b and will be derived in the next section. The introduction of the projection w enables us to formulate the continuity equation with Dirichlet boundary conditions at one boundary. It is well known that this equation has a unique solution for given functions s and w. This means that the non-local term, u_y, is uniquely determined by the solution $u = u(s, w)$ to

$$Q\nabla(Q\nabla u) = 0, \qquad \text{in } \Omega(s), \tag{14}$$

with the boundary conditions

$$\begin{aligned} -\vec{n}\cdot Q^2\nabla u &= \frac{qM_t}{\sqrt{1+\tilde{q}^2+M_y^2}}, && \text{at } \Gamma_w,\\ -\vec{n}\cdot Q^2\nabla u &= 0, && \text{at } \Gamma_0, \Gamma_1,\\ u &= w, && \text{at } \Gamma_s. \end{aligned} \tag{15}$$

This reduction of system (7), (8) splits the problem into a linear and a nonlinear part. The linear operator, given by the equations (14)–(15), is formulated in the time dependent domain $\{(x, y, t)\,|\,(x, y) \in \Omega(t)\} \subset \mathbb{R}^3 \times R_+$, whereas the nonlinear operator, given by the equations (11)–(12), describes the velocity potential and the free surface in the fixed domain $\{(x, t) \mid x_1 \in (0, K(0)\} \subset \mathbb{R}^2 \times R_+$. Note that the regularity of the solution is determined by the partial differential equations in (7), (8). The transformation of the problem does not require any higher regularity.

4. Contact lines and compatibility conditions

In the last section we transformed the equation of motion and the free surface condition into a system of partial differential equations in a fixed domain. Nevertheless, the transformation matrix and the coefficients in system (11)–(12) depend on the contact lines between the free surface and the side walls of the basin. These contact lines are part of the solution of the problem. Natural relations between the free surface and the contact lines are given by

$$K_a(x_2, t) = M(s(0, x_2, t), t), \qquad K_b(x_2, t) = H(s(K_0(x_2), x_2, t), x_2),$$

$$\frac{\partial}{\partial t}K_a(x_2, t) = M_y(s(0, x_2, t), t)s_t(0, x_2, t) + M_t(s(0, x_2, t), t),$$

$$\frac{\partial}{\partial t}K_b(x_2, t) = H_y(s(K_0(x_2), x_2, t), t)s_t(K_0(x_2), x_2, t),$$

$$\frac{\partial}{\partial x_2}K_a(x_2, t) = M_y(s(0, x_2, t), t)s_{x_2}(0, x_2, t),$$

$$\frac{\partial}{\partial x_2}K_b(x_2, t) = H_{x_2}(s(K_0(x_2), x_2, t), t), x_2) + H_y(s(K_0(x_2), x_2, t), t)s_{x_2}(K_0(x_2), x_2, t) +$$
$$H_y(s(K_0(x_2), x_2, t), t)s_{x_1}(K_0(x_2), x_2, t)K_0'(x_2).$$

Consequently, we obtain

$$\tilde{K}(x, t) = \frac{q(x_2, t)}{K_0(x_2)}[(x_1 - K_0(x_2))\frac{\partial}{\partial t}K_a(x_2, t) - x_1\frac{\partial}{\partial t}K_b(x_2, t)].$$

Furthermore, we have

$$\tilde{q}(x,t) = \frac{K_0'(x_2)x_1}{K_0(x_2)} - q(x_2,t)[\frac{\partial}{\partial x_2}K_a(x_2,t) + \frac{x_1}{K_0(x_2)}(\frac{\partial}{\partial x_2}K_b(x_2,t) - \frac{\partial}{\partial x_2}K_a(x_2,t))].$$

From $s_{x_2}(0,t) = 0$, $\tilde{K}(0,t) = q(x_2,t)\frac{\partial}{\partial x_2}K_a(x_2,t)$ and (11), (15) we derive the relation

$$s_t(0,x_2,t) = u_y(0,x_2,s(0,x_2,t),t). \tag{16}$$

Similarly, we obtain compatibility conditions for the free surface and the potential along the contact lines at the wave-maker and the boundary opposite to the wave-maker:

$$w_t(0,x_2,t) = \frac{1}{2}\left(M_t(s(0,x_2,t),t) + M_y(s(0,x_2,t),t)u_y(0,x_2,s(0,x_2,t),t)\right)^2 + \frac{1}{2}u_y(0,x_2,s(0,x_2,t),t) - gs(0,x_2,t), \tag{17}$$

$$s_t(K_0(x_2),x_2,t) = u_y(K_0(x_2),x_2,s(K_0(x_2),x_2,t),t). \tag{18}$$

$$w_t(K_0(x_2),x_2,t) = \frac{1}{2}\left(M_y(s(K_0(x_2),x_2,t),t)u_y(K_0(x_2),x_2,s(K_0(x_2),x_2,t),t)\right)^2 + \frac{1}{2}u_y(K_0(x_2),x_2,s(K_0(x_2),x_2,t),t) - gs(K_0(x_2),x_2,t). \tag{19}$$

The relations (16)–(19) state consistency between the boundary conditions and the differential equations at the interfaces between the free boundary and the side walls of the basin. They are natural compatibility conditions for the solvability of problem (11)–(15).

5. Numerical tests

The initial boundary value problem (11)–(15) is a highly nonlinear system of partial differential equations. This system was formally split into a nonlinear and a linear part. However, these parts are coupled by the transformation coefficients which depend on the solution to the nonlinear system (11)–(13). We choose an implicit time discretization scheme for the numerical approximation of the problem in order to fulfill the compatibility conditions derived in the previous section in each time step. We solve the continuity equation (14)–(15) by a finite element method and couple this solution with the discretization of the nonlinear initial value problem (11)–(13). Discretizing the system first in space we obtain a system of ordinary differential equations which can be solved by backward differentiation formulas. To ensure stability of the numerical method, we use modified upwind schemes for the approximation of the right-hand side of (11)–(12). A more detailed discussion of the numerical method was published in [10].

We developed a program system based on the C++ programming language and using the advantages of object-oriented programming. The system consists of several independent parts, in particular a grid-generation tool for arbitrary domains in 2 and

3D, a finite element package for the solution of elliptical problems in 2 and 3D, a interface for the implementation of an ODE-solver, a class library containing application specified functions for different types of wave generation, reflection and absorption and corresponding transformation tools, a solver for an instationary system of equations of motion including the management of coupling of the different parts, as well as graphic supported pre- and post-processing tools. Numerical tests for the solution of the full nonlinear free surface wave problem in 3D are in work. Finally, we present some examples of two-dimensional instationary calculations. For a comparison with experimental data we refer to [10]. The following pictures show plots of the solution to the free surface wave problem for different types of wave-maker motions and different boundary conditions. The first example (Figures 4–5) was calculated in a rectangular domain with a joint-type wave-maker and reflecting boundary condition.

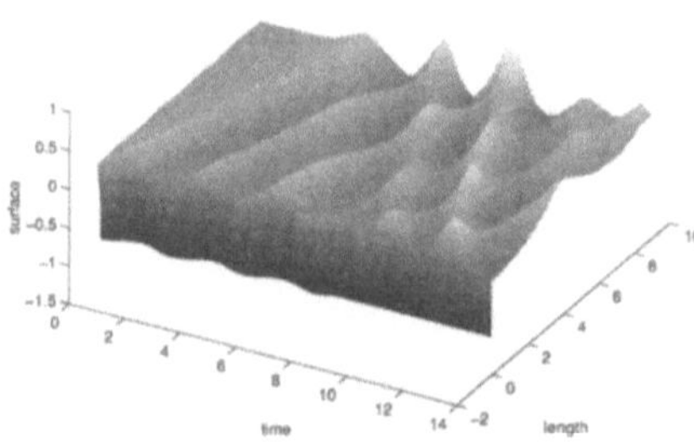

Fig. 4: Joint-type wave-maker motion and reflection at a vertical wall

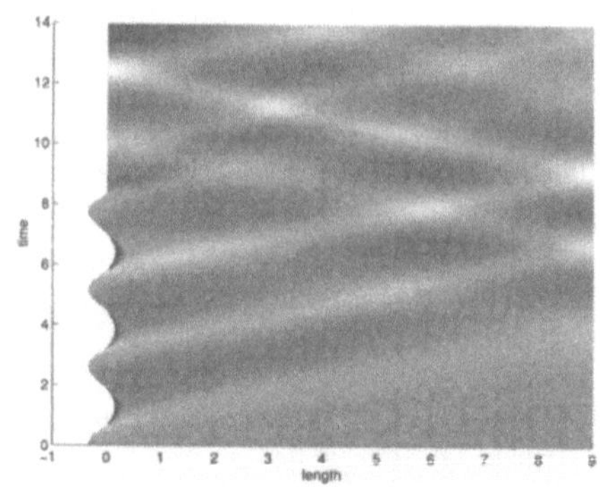

Fig. 5: 2-D plot corresponding to Figure 4

The second example (Figures 6–7) simulates a piston-type wave-maker and the reflection of waves at a slanted wall. A typical feature for this example is the time dependent X coordinate of the contact point between the free surface and the slanted wall. In this situation, waves of different amplitude and wave speed are formed and interact with each other.

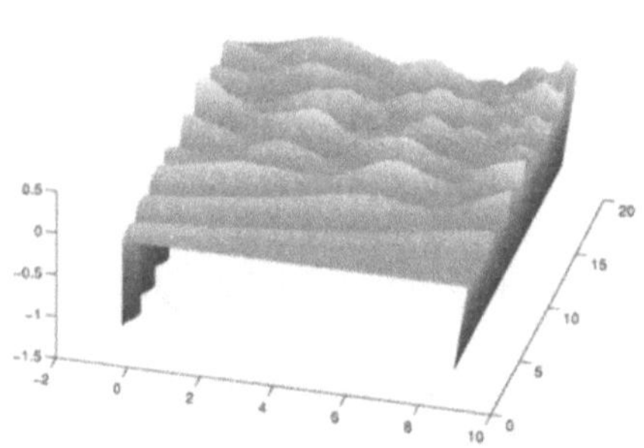

Fig. 6: Piston-type wave-maker motion and reflection at a slanted wall

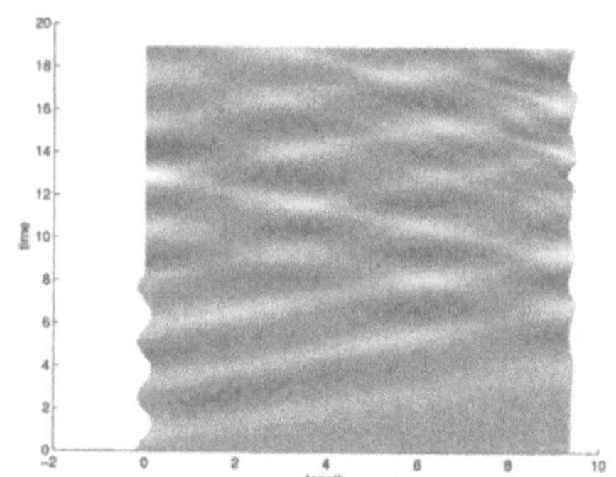

Fig. 7: 2-D plot corresponding to Figure 6

Finally, in Figures 8–9 we present some calculations of outgoing waves in an open basin caused by a flap-type wave-maker. The simulation of waves in an open domain requires an artificial reduction of the numerical problem to a bounded domain and the formulation of absorbing boundary conditions. In the presented context, this is realized by a extension of the domain for the continuity equation and the derivation of boundary conditions outside the original calculation domain from a travelling wave ansatz [10]. See also [4], [11], [12], [5], [23].

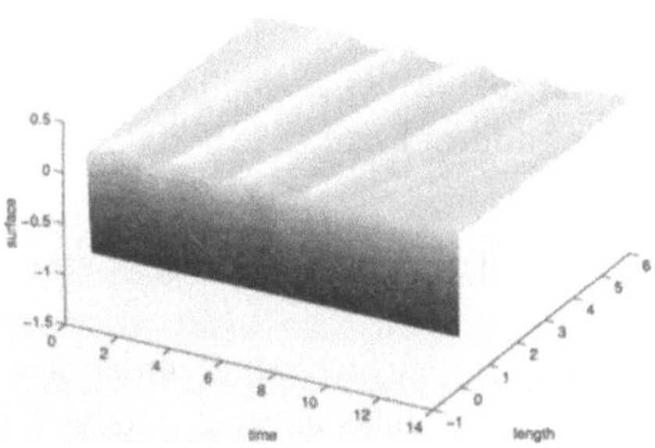

Fig. 8: Flap-type wave-maker motion and absorbing boundary conditions

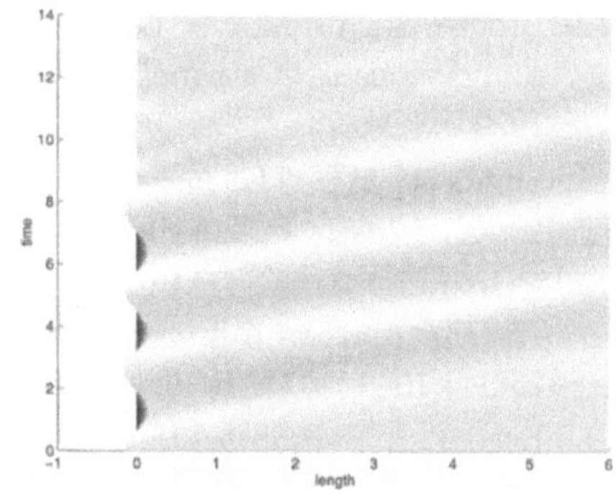

Fig. 9: 2-D plot corresponding to Figure 8

References

[1] Beale, J. T.: *The initial value problem for the Navier-Stokes equations with a free surface.* Communications on Pure and Applied Mathematics, XXXIV:359–392, 1981.

[2] Beale, J. T.: *Large-time regularity of viscous surface waves.* Arch. for Rat. Mech. and Anal., 84:307–352, 1984.

[3] Cohen, G. (editor): *Mathematical and numerical Aspects of wave propagation*, Philadelphia, 1995. SIAM.

[4] Kröner, D.: *Absorbing boundary conditions for the linearized Euler equations in 2-d.* Mathematics of Computation, 57(195):153–167, 1991.

[5] Higdon, R. L.: *Radiation boundary conditions for disipative waves* SIAM J. Numer. Anal. 31(1):64–100, 1994

[6] Lions, P.-L.: *Mathematical Topics in Fluid Mechanics, I incompressibe Models.* Oxford Lecture Series in Mathematics and its Applications. Clarendon Press, Oxford, 1996.

[7] Majda, A.: *Compressible Fluid Flow and systems of Conservation Laws in Several Space Variables*, volume 53 of Applied Mathematical Sciences. Springer Verlag, New York, Berlin, Heidelberg, Tokyo, 1984.

[8] Miloh, T. (editor): *Mathematical Approaches in Hydrodynamics*, Philadelphia, 1991. SIAM.

[9] Pawell, A.: *The motion of an inviscid and incompressible fluid in a non-smooth domain.* In Proceedings of the ICIAM 95, Applied Sciences – Especially Mechanics, ZAMM, 76(5):375–376. Akademie Verlag, Berlin, 1996.

[10] Pawell, A.; Guenther, R. B.: *A numerical solution to a free surface wave problem.* Top. Methods in Nonlin. Anal., 6(2), 1996.

[11] Romate, J. E.: *Absorbing boundary conditions for free surface waves.* Journal of computational Physics, 99:135–145, 1992.

[12] Broeze, J.; Romate, J. E.: *Absorbing boundary conditions for free surface waves, Simulations with a panel method.* Journal of computational Physics, 99:146–158, 1992.

[13] Shinbrot, M.: *The initial value problem for surface waves under gravity, I. The simplest case.* Indiana University Mathematics Journal, 25(3):281–300, 1976.

[14] Reeder, J.; Shinbrot, M.: *The initial value problem for surface waves under gravity, II. The simplest 3-dimensional case.* Indiana University Mathematics Journal, 25(11):1049–1071, 1976.

[15] Reeder, J.; Shinbrot, M.: *The initial value problem for surface waves under gravity, III. Uniformly analytic initial domains.* Journal of Mathematical Analysis and Applications, 67:340–391, 1979.

[16] Socolowsky, J.: *The solvability of a free boundary problem for the stationary Navier-Stokes equations with a dynamic contact line.* Nonlinear Analysis, Theory, Methods & Applications, 21(10):763–784, 1993.

[17] Socolowsky, J.: *Existence and uniqueness of the solution to a free boundary-value problem with thermocapillary convection in an unbounded domain.* Acta Applicandae Mathematicae, 37:181–194, 1994.

[18] Solonnikov, V. A.: *On some free boundary problems for the Navier-Stokes equations with moving contact points and lines.* Math.Ann., 302(4):743–772, 1995.

[19] Tanaka, N.; Tani, A.: *Large-time existence of compressible viscous and heat-conductive surface waves.* Sūrikaisekikenkyūsho Kōkyūroku, 824:138–150, 1993.

[20] Tanaka, N.: *Two-phase free boundary problem for viscous incompressible thermo-capillary convection.* Japan. J. Math., 21(1):1–42, 1995.

[21] Tani, A.: *Free boundary problems for the incompressible Euler equations.* Sūrikaisekikenkyūsho Kōkyūroku, 862:237–248, 1994.

[22] Radder, A. C.: *An explicit Hamiltonian formulation of surface waves in water of infinite depth.* J.Fluid.Mech, 237:435–455, 1992.

[23] Wagata, L.: *Absorbierende Randbedingungen für hyperbolische partielle Differentialgleichungen.* PhD thesis, Ludwig-Maximilians-Universität München, März 1982.

International Series of Numerical Mathematics
Vol. 124, © 1998 Birkhäuser Verlag, Basel

Efficient Convexification of Flight Path Optimization Problems

Gottfried Sachs* Rainer Mehlhorn* Michael Dinkelmann*

Abstract. A problem in the optimization of flight paths concerns the modelling of the thrust force direction. This problem is addressed by considering various mathematical modellings of thrust force directions in a unified approach. When modelling the thrust force direction which is basically related to the vehicle attitude as a linear or non-linear function of angle of attack, higher order optimality conditions are generally not met for interior thrust settings. A convexification technique is developed for efficiently computing a solution. There may be interior arcs as the limiting case of chattering arcs which concern both controls thrust setting and angle of attack of the original system. A numerical example of hypersonic range flight is presented.

Nomenclature

C_D drag coefficient	r_e radius of the Earth	α_T effective thrust inclination angle
C_L lift coefficient	S reference area	γ flight path angle
D drag	S switching function	δ_T throttle setting
g gravity acceleration	s range	ε_T thrust inclination angle (relative to α)
H Hamiltonian	T thrust	ζ control variable of modified system
h altitude	t time	λ adjoint variables vector
J performance criterion	t_f final time	ρ atmospheric density
L lift	u control variables vector	σ specific fuel consumption
M Mach number	V speed	Φ Mayer form of performance criterion
m mass	x state variables vector	ω_e angular velocity of the Earth
m_f fuel mass consumed	α angle of attack	

1. Introduction

In the optimization of flight paths of aircraft, a problem of a more basic nature exists which concerns the direction of the thrust force. Different types of modellings for the direction of the thrust force $\vec{T}$ are applied. These modellings which are illustrated in Fig. 1 may be described as:

$$\text{I. direction of } \vec{T} \text{ parallel to speed vector}: \quad \alpha_T = \alpha + \varepsilon_T = 0 \quad \text{where } \varepsilon_T = -\alpha \tag{1}$$

$$\text{II. direction of } \vec{T} \text{ fixed on vehicle}: \quad \alpha_T = \alpha + \varepsilon_T \quad \text{where } \varepsilon_T = \text{const} \tag{2}$$

$$\text{III. direction of } \vec{T} \text{ considered a control}: \quad \alpha_T = \alpha + \varepsilon_T \quad \text{where } \varepsilon_T : \text{arbitrary} \tag{3}$$

Modelling type I which is often applied in flight path optimization relates the direction of the thrust force to the speed vector the direction of which is tangential to the

*Institut für Flugmechanik und Flugregelung, Technische Universität München, D-85747 Garching, e-mail: sachs@lfm.mw.tu-muenchen.de

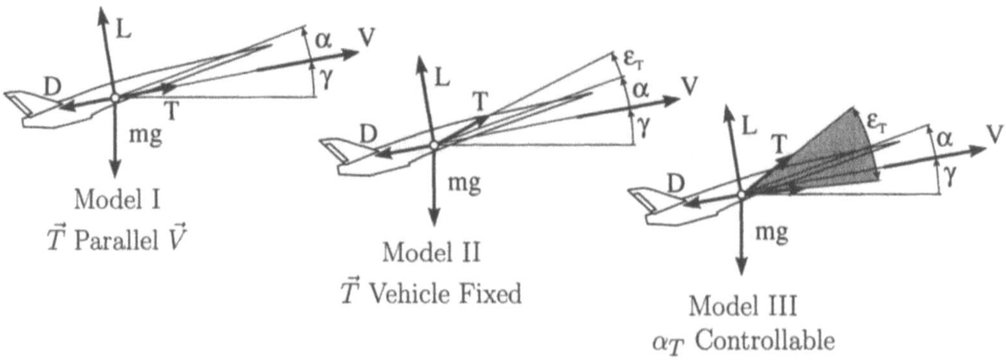

Figure 1: Modellings of thrust force direction

flight path. Relating the thrust force direction to the speed vector is in contrast with the vehicle-fixed nature of the engine as the producer of the thrust. This modelling implies that the thrust force direction would change its relation to the vehicle when the angle of attack is changed (Fig. 2). As a result, modelling I is basically a simplification which does not account for the vehicle-related nature of the thrust force.

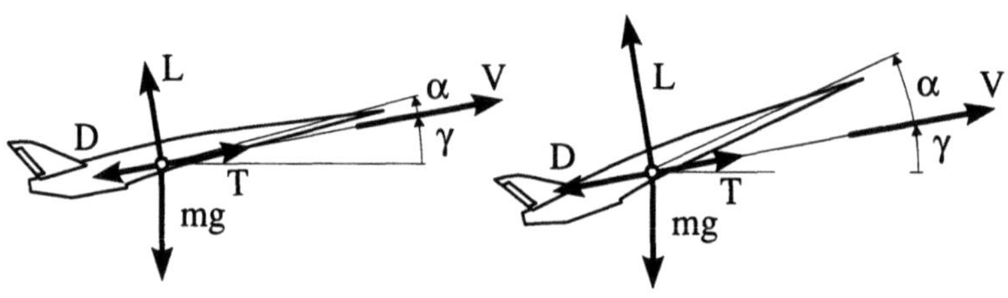

Figure 2: Change of thrust force direction with angle of attack for modelling I

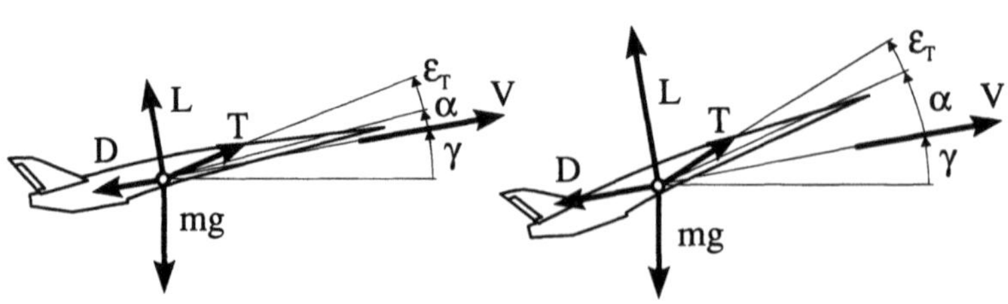

Figure 3: Change of thrust force direction with angle of attack for modelling II

Modelling type II relates the direction of the thrust force to the vehicle. Accordingly, the thrust force changes its direction relative to the speed vector with the angle of attack (Fig. 3). From Fig. 3 it also follows that the thrust force does not only have a component parallel to the speed vector but also perpendicular to it, and thus, parallel to the lift force. The thrust force component perpendicular to the speed vector can be used to support the lift force. Thus, less lift is required than in the case where the thrust force would not have such a component (i.e., when the thrust force would be parallel to the speed vector, modelling I). The lift supporting effect basically provides a performance advantage because the drag which is dependent on lift can be reduced when the thrust force direction takes on an appropriate value.

Modelling type III concerns a thrust force direction which is considered a control. Accordingly, it can be arbitrarily set independently of vehicle attitude or angle of attack. Such a control possibility, which requires a complex technical mechanization, has recently received great interest for enhancing the capabilities of high-performance aircraft (e.g., Refs. 1, 2).

The optimization problem addressed in this paper concerns modelling II. It is basically due to a general non-convexity of the hodograph in this case (Refs. 3, 4).

The purpose of this paper is to present an efficient approach for convexifizing the flight path optimization problem. Furthermore, the optimization problem is considered in a unified approach for various modellings of thrust force directions.

2. Flight dynamics modelling

A rather general flight dynamics modelling which can be used for speeds up to the hypersonic flight regime is applied. The equations of motion with reference to a spherical and rotating earth model can be expressed as

$$\dot{\boldsymbol{x}} = \boldsymbol{f}\,(\boldsymbol{x}, \boldsymbol{u}) \tag{4}$$

where $\boldsymbol{x}$ and $\boldsymbol{u}$ are the state and control variables vectors

$$\boldsymbol{x} = (V, \gamma, h, m)^{\mathrm{T}}, \qquad \boldsymbol{u} = (\alpha, \delta_T)^{\mathrm{T}} \tag{5}$$

and

$$\boldsymbol{f} = \begin{cases} \frac{T\cos\alpha_T - D}{m} - g\sin\gamma + \omega_e^2(r_e+h)\sin\gamma \\ \frac{T\sin\alpha_T + L}{mV} - \frac{g}{V}\cos\gamma + \frac{V}{r_e+h}\cos\gamma + \frac{\omega_e^2(r_e+h)}{V}\cos\gamma + 2\omega_e \\ V\sin\gamma \\ -\dot{m}_f \end{cases} \tag{6}$$

with

$$\alpha_T = \alpha + \varepsilon_T, \qquad g = g_0\left(\frac{r_e}{r_e+h}\right)^2$$

The aerodynamic forces can be written as

$$L = C_L(\alpha, M)\cdot(\rho/2)V^2S, \qquad D = C_D(\alpha, M)\cdot(\rho/2)V^2S \tag{7}$$

Thrust and fuel consumption are modelled as

$$T = \delta_T T_{\max}(h, M), \qquad \dot{m}_f = \dot{m}_{f,0}(h, M) + \delta_T\sigma(h, M)\cdot T_{\max}(h, M) \tag{8}$$

Control variables are the angle of attack and the throttle setting which are subject to the following constraints

$$\boldsymbol{u}_{\min} \le \boldsymbol{u} \le \boldsymbol{u}_{\max} \tag{9}$$

In addition, constraints for the state variables $\boldsymbol{x} = (V, \gamma, h, m)^{\mathrm{T}}$ may exist

$$\boldsymbol{x}_{\min} \le \boldsymbol{x} \le \boldsymbol{x}_{\max} \tag{10}$$

There are boundary and/or interior point conditions yielding additional expressions such as

$$\Psi_0(\boldsymbol{x}(0)) = 0, \qquad \Psi_f(\boldsymbol{x}(t_f)) = 0 \tag{11}$$

where t_f is considered the final time.

With the above flight dynamics model, a great variety of problems in the optimization of flight paths of aircraft can be dealt with. They include range and endurance cruise optimization like minimizing fuel per range or time. A performance criterion can be introduced for adequately covering the optimization goal. The performance criterion may be expressed in Mayer form as

$$J = \Phi(\boldsymbol{x}(t_f)) \tag{12}$$

3. Optimality conditions

Necessary optimality conditions can be derived by applying the Minimum Principle. For this purpose, the Hamiltonian is introduced

$$H(\boldsymbol{x}, \boldsymbol{u}, \boldsymbol{\lambda}) = \boldsymbol{\lambda}^{\mathrm{T}} \boldsymbol{f}(\boldsymbol{x}, \boldsymbol{u}) \tag{13}$$

where $\boldsymbol{\lambda} = (\lambda_V, \lambda_\gamma, \lambda_h, \lambda_m)^{\mathrm{T}}$ is the adjoint vector. The following relation holds

$$\dot{\boldsymbol{\lambda}} = -\partial H / \partial \boldsymbol{x} \tag{14}$$

The optimal control is such that the Hamiltonian is minimized. Thus, it can be determined from

$$\partial H / \partial \boldsymbol{u} = 0 \tag{15}$$

or by values on its bounds. Candidates for the optimal control are

$$\delta_T := \delta_{T,\min} \qquad \alpha := \arg\min_\alpha H(\alpha, \delta_{T,\min}) \tag{16}$$

$$\delta_T := \delta_{T,\max} \qquad \alpha := \arg\min_\alpha H(\alpha, \delta_{T,\max}) \tag{17}$$

$$\delta_T := \delta_{T,int} \in (\delta_{T,\min}, \delta_{T,\max}) \qquad \alpha := \arg\min_\alpha H(\alpha, \delta_{T,int}) \tag{18}$$

The optimality problem related to thrust force direction can be shown with the use of higher order conditions. For this purpose, the Legendre Clebsch condition is applied. It requires that

$$\det \begin{bmatrix} H_{\alpha\alpha} & H_{\alpha\delta} \\ H_{\delta\alpha} & H_{\delta\delta} \end{bmatrix} \geq 0, \qquad H_{\alpha\alpha} \geq 0, \qquad H_{\delta\delta} \geq 0 \tag{19}$$

for the controls interior to their admissible set.

4. Evaluation of optimality conditions for thrust force modellings I–III

An evaluation of optimality conditions is presented for the three modellings I-III of thrust force directions in a unified approach.

For this purpose, the thrust force direction described by the effective thrust inclination angle $\alpha_T = \alpha + \varepsilon_T$ is treated in a more general manner as an angle of attack function which may be non-linear

$$\alpha_T = \alpha_T(\alpha) \tag{20}$$

This relation is considered to include each thrust modelling I–III by introducing appropriate dependencies:

I. $\alpha_T(\alpha) = 0$

II. $\alpha_T(\alpha) = \alpha + \varepsilon_T$ where $\varepsilon_T = \text{const}$

III. $\alpha_T(\alpha)$: yielding the best contribution of thrust force direction to performance
ε_T can be computed using $\varepsilon_T := \alpha_T(\alpha) - \alpha$

By applying Eq. (20), the first order necessary conditions read

$$\begin{aligned} H_\alpha &= -\frac{\lambda_V}{m}\left[T\frac{\mathrm{d}\alpha_T}{\mathrm{d}\alpha}\sin\alpha_T + D_\alpha\right] + \frac{\lambda_\gamma}{mV}\left[T\frac{\mathrm{d}\alpha_T}{\mathrm{d}\alpha}\cos\alpha_T + L_\alpha\right] = 0 \\ H_\delta &= \frac{\lambda_V}{m}T_\delta\cos\alpha_T + \frac{\lambda_\gamma}{mV}T_\delta\sin\alpha_T - \lambda_m\frac{\partial\dot{m}_f}{\partial\delta_T} = 0 \end{aligned} \tag{21}$$

With the use of

$$\begin{aligned} H_{\alpha\alpha} &= -\frac{\lambda_V}{m}\left\{T\left[\frac{\mathrm{d}^2\alpha_T}{\mathrm{d}\alpha^2}\sin\alpha_T + \left(\frac{\mathrm{d}\alpha_T}{\mathrm{d}\alpha}\right)^2\cos\alpha_T\right] + D_{\alpha\alpha}\right\} \\ &\quad + \frac{\lambda_\gamma}{mV}T\left[\frac{\mathrm{d}^2\alpha_T}{\mathrm{d}\alpha^2}\cos\alpha_T - \left(\frac{\mathrm{d}\alpha_T}{\mathrm{d}\alpha}\right)^2\sin\alpha_T\right] \\ H_{\delta\delta} &= 0 \\ H_{\delta\alpha} &= H_{\alpha\delta} = \frac{T_\delta}{m}\frac{\mathrm{d}\alpha_T}{\mathrm{d}\alpha}\left[\frac{\lambda_\gamma}{V}\cos\alpha_T - \lambda_V\sin\alpha_T\right] \end{aligned} \tag{22}$$

the determinante in Eq. (19) reduces to

$$\det\begin{bmatrix} H_{\alpha\alpha} & H_{\alpha\delta} \\ H_{\delta\alpha} & H_{\delta\delta} \end{bmatrix} = -H_{\delta\alpha}^2 \tag{23}$$

This number can only be nonnegative if

$$H_{\delta\alpha} = 0 \tag{24}$$

Rewriting $H_{\delta\alpha}$ with the use of $H_\alpha = 0$ from Eq. (21) yields

$$H_{\delta\alpha} = \lambda_V \frac{T_\delta/m}{L_\alpha + T(\mathrm{d}\alpha_T/\mathrm{d}\alpha)\cos\alpha_T}\cdot\frac{\mathrm{d}\alpha_T}{\mathrm{d}\alpha}\left[D_\alpha\cos\alpha_T - L_\alpha\sin\alpha_T\right] = 0 \tag{25}$$

In the following we assume that $\lambda_V \neq 0$. An evaluation of relation (25) shows that in this case the Legendre Clebsch condition can only be met if

$$\frac{\mathrm{d}\alpha_T}{\mathrm{d}\alpha} = 0 \qquad \text{(equivalent to } \alpha_T = \text{const)} \tag{26}$$

or

$$\tan\alpha_T = \frac{D_\alpha}{L_\alpha} \tag{27}$$

As a result, a modelling showing a constant thrust inclination meets the optimality condition in mind. This result implies modelling I where $\alpha_T = 0$. From Eq. (26) it follows that any other constant thrust inclination modelling would also meet the above optimality condition.

Eq. (27) represents the case of optimal thrust direction control (modelling III) where the thrust vector angle can be arbitrarily set (e.g., Refs. 5, 6).

For any other α_T dependency on α, the Legendre Clebsch cannot be met. This implies the linear relationship as given by modelling II. In this case, the hodograph is not convex. For solving this problem, a convexification is required.

5. Convexification approach

A solution can be achieved for a relaxed problem the hodograph of which is the smallest convex hull of the original system. For constructing the relaxed system, a geometric approach is applied.

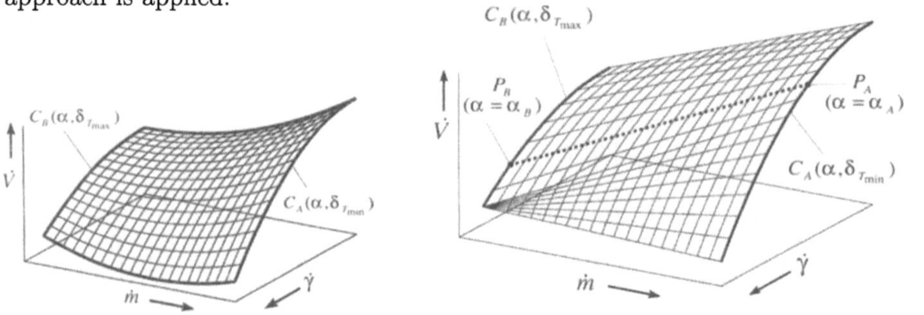

Figure 4: Hodograph for modelling II, original system (left) and relaxed system (right)

Reference is made to the hodograph of the original problem, Fig. 4 (left). From this hodograph, the boundary curves $C_A(\alpha, \delta_{T,\min})$ and $C_B(\alpha, \delta_{T,\max})$ are retained and used as boundaries for the hodograph of the relaxed system, Fig. 4 (right). The convexification is based on a geometric property which means that the tangents to curve C_A in $\boldsymbol{f}^*(\alpha_A, \delta_{T,\min})$ and to curve C_B in $\boldsymbol{f}^*(\alpha_B, \delta_{T,\max})$ and the vector $\boldsymbol{f}^*(\alpha_A, \delta_{T,\min}) \rightarrow \boldsymbol{f}^*(\alpha_B, \delta_{T,\max})$ must be in the same plane:

$$\det\left(\boldsymbol{f}^*_\alpha(\alpha_A, \delta_{T,\min}), \boldsymbol{f}^*_\alpha(\alpha_B, \delta_{T,\max}), \boldsymbol{f}^*(\alpha_A, \delta_{T,\min}) - \boldsymbol{f}^*(\alpha_B, \delta_{T,\max})\right) = 0 \tag{28}$$

where

$$\boldsymbol{f}^*(\alpha, \delta_T) := \boldsymbol{f}\,(\boldsymbol{x}; \alpha, \delta_T, \alpha_T(\alpha)) \tag{29}$$

The relation $H = \text{const}$ holds, if $\boldsymbol{f}^*(\alpha_A, \delta_{T,\min})$ and $\boldsymbol{f}^*(\alpha_B, \delta_{T,\max})$ result in a minimal value for H. With this algebraic condition, a reduction $\alpha_A = \alpha_A(\boldsymbol{x}, \alpha_B)$ is possible. Thus, a reduction from 3 to 2 control variables can be achieved. As a result, an efficiency improvement is possible for the computational process.

A new control ζ is introduced which linearly enters a modified Hamiltonian

$$H_{\text{mod}}(\boldsymbol{x}, \zeta, \alpha_B, \boldsymbol{\lambda}) := (1-\zeta)H(\boldsymbol{x}, \alpha_A(\boldsymbol{x}, \alpha_B), \delta_{T,\min}, \boldsymbol{\lambda}) + \zeta H(\boldsymbol{x}, \alpha_B, \delta_{T,\max}, \boldsymbol{\lambda})\,. \tag{30}$$

We define a new pair of systems of a relaxed problem

$$\dot{\boldsymbol{x}} := \partial H_{\text{mod}}/\partial \boldsymbol{\lambda}, \qquad \dot{\boldsymbol{\lambda}} := -\partial H_{\text{mod}}/\partial \boldsymbol{x} \tag{31}$$

With the switching function

$$S := \partial H_{\text{mod}}/\partial \zeta \tag{32}$$

candidates for the optimal control can be expressed as

$$\begin{array}{lll} \text{A)}\ \ S > 0: & \zeta := 0, & \text{boundary arc} \\ \text{B)}\ \ S < 0: & \zeta := 1, & \text{boundary arc} \\ \text{C)}\ \ S = 0: & \zeta := \zeta_{int} \in (0,1), & \text{interior arc} \end{array} \tag{33}$$

Case C) is of particular interest because it concerns an interior control. In case of a singular control, ζ_{int} can be determined in numerical computations with techniques appropriate for singular control problems (Refs. 6–8). A stabilization technique may be applied for overcoming numerical difficulties when constructing singular arc solutions (Ref. 9).

6. Numerical example

The numerical example concerns the optimization of a range flight of a hypersonic vehicle. For a given range (9000 km), fuel consumption is minimized. A performance criterion may be formulated as

$$J = \frac{m_f(t_f)}{s(t_f)} \tag{34}$$

where $m_f(t_f)$ is the fuel consumed and $s(t_f)$ is the range at the end of cruise

$$m_f(t_f) = m(0) - m(t_f), \qquad s(t_f) = \int_0^{t_f} V \cos\gamma dt \tag{35}$$

The boundary conditions of the cruise problem are presented in Table 1.

	h [m]	V [m/s]	γ [deg]	m [Mg]
Begin $(t = 0)$	500	150	0	244
End $(t = t_f)$	500	150	0	-

Table 1: Boundary conditions

A realistic modelling is applied for engine and aerodynamics characteristics which show complex dependencies on state and control variables. As an example, the aerodynamics characteristics of the vehicle are shown in Fig. 5. The vehicle is equipped with a turbo/ram jet engines combination which provides a speed capability up to about Mach 7. The transition from turbo to ram jet operation and vice versa takes place between Mach 3.0 and 3.5 as a specified process. The maximum Mach number is considered to be constrained.

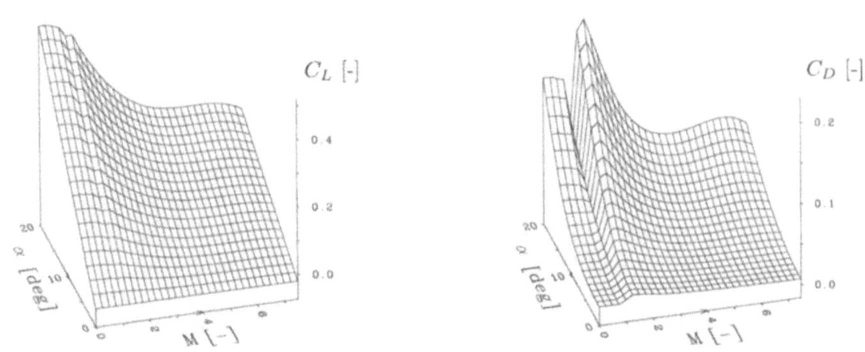

Figure 5: Lift and drag characteristics of a hypersonic vehicle

The hypersonic performance problem was chosen as an example because vehicles for this flight regime have a comparatively low lift-to-drag ratio which contributes to increasing the thrust inclination effects as represented by modellings II and III.

The optimal range trajectory for the vehicle with thrust direction modelling II is illustrated in Figs. 6 and 7 which show the histories of state and control variables. From Fig. 6 it follows that a hypersonic range trajectory consists of an accelerated climb, a cruise climb portion and a decelerated descent.

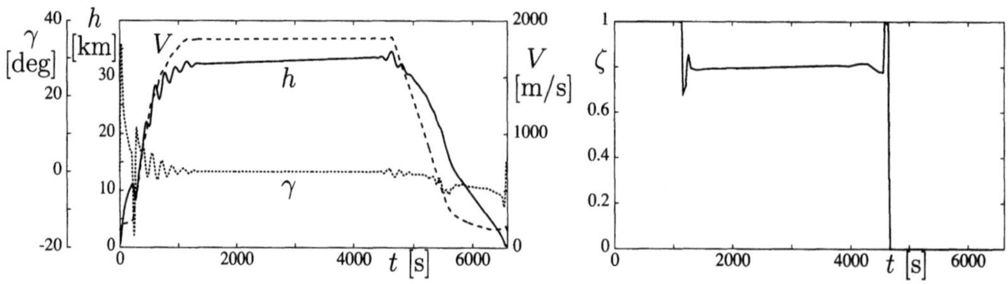

Figure 6: Optimal range trajectory of a hypersonic vehicle with modelling II (state and control variables), relaxed system

The control ζ introduced for treating the relaxed problem shows an interior arc for the cruise climb portion (Fig. 6). This portion describes a chattering behavior of the throttle setting of the original system (Fig.7). There is also a chattering in angle of attack between the values α_A and α_B for the original system (Fig.7).

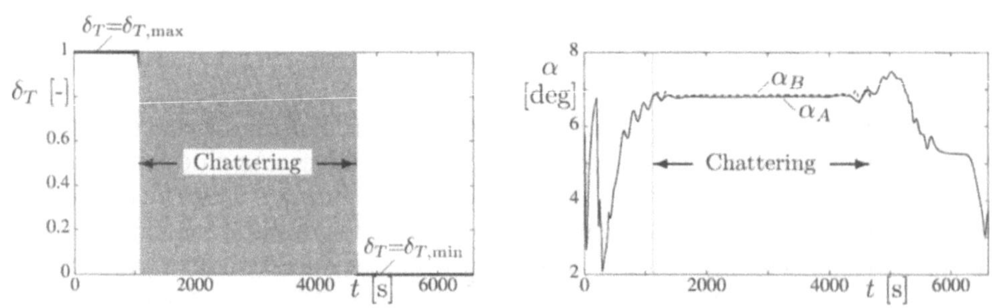

Figure 7: Optimal range trajectory of a hypersonic vehicle with modelling II (chattering control of thrust and angle of attack), original problem

For comparison, the modellings I and III were also considered for the same range flight optimization problem. The performance in each case may be summarized as:

- modelling I: $J = 67230$ kg
- modelling II: $J = 66539$ kg
- modelling III: $J = 66455$ kg
- modelling I_{mod} ($\alpha_T = 7.5$ deg): $J = 66517$ kg

The performance is the smallest in case of modelling I. This is because the thrust cannot provide a supporting effect on lift so that no reduction in drag can be achieved.

In case of modelling II, the performance is generally greater when compared with the results for modelling I. The thrust force can support the aerodynamic lift due to its inclination. As a consequence of this supporting effect, the drag can be decreased so that a reduction of the thrust as the propulsive force is possible and consequently a decrease in fuel consumption.

The achievable performance takes on the greatest values in case of modelling III. This is because the supporting effect of thrust to aerodynamic lift can be optimized by the separate control capability for the thrust force direction.

It may be of interest to note that a modification of modelling I with a constant non-zero inclination can show a performance similar to case II. This is indicated in the above results as modelling I_{mod} where $\alpha_T = 7.5$ deg.

An insight into the physical mechanism underlying the different performance values for modellings I–III can be provided when considering the supporting effect of thrust on lift in more detail. This is illustrated in Figs. 8 and 9 which show the time histories for the thrust directions as well as for the thrust component perpendicular to the speed vector and the aerodynamic lift for the optimized range trajectories in each case.

The time histories of thrust force directions (thrust angle of attack) are presented in Fig. 8 for each modelling I–III. Besides modelling I which implies a zero thrust angle of attack, a significant thrust inclination exists with the other cases.

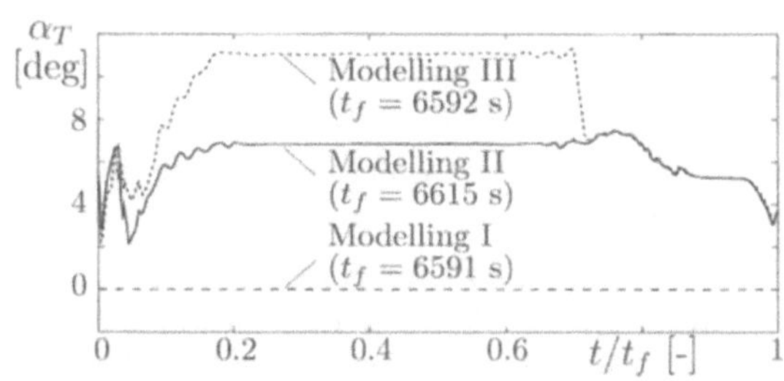

Figure 8: Time histories of thrust vector angle α_T for three thrust direction cases

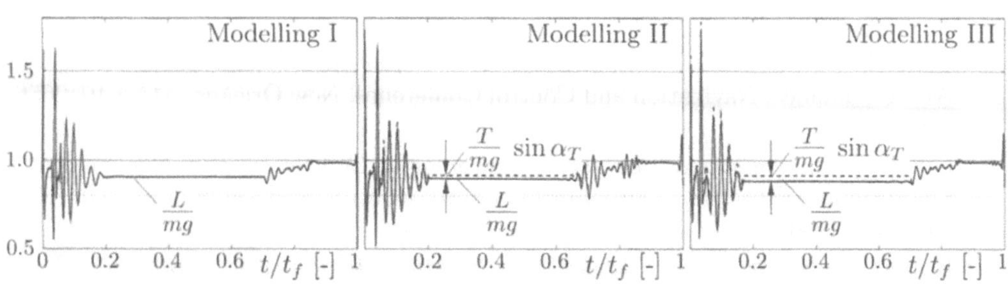

Figure 9: Time histories of aerodynamic lift and supporting thrust effect for modelling I (t_f=6591 s), modelling II (t_f=6615 s) and modelling III (t_f=6592 s)

Modelling I which leads to a lift history shown in Fig. 9 is used as a reference. There is no thrust component supporting the aerodynamic lift.

Fig. 9 shows the supporting effect of thrust for modelling II. Comparison with Fig. 9 reveals that less lift is required for the balance of forces perpendicular to the speed vector.

An effect basically similar holds for modelling III, Fig. 9. The possibility of controlling the thrust direction provides an optimal supporting effect of thrust to lift.

7. Conclusions

A problem in the optimization of flight paths concerns the modelling of the thrust force direction. Various modellings of thrust force directions are considered and a unified approach is presented. When modelling the thrust force direction which is related to the attitude of the vehicle as a linear or non-linear function of angle of attack, the Legendre-Clebsch condition may be generally not met for interior thrust settings. A solution based on the relaxation of the original problem can be developed. A convexification technique using geometric properties of the hodograph of the original system is developed for efficiently computing a solution. There may be interior arcs as the limiting case of chattering arcs which concern both controls thrust setting and angle of attack of the original system.

A hypersonic range trajectory problem is considered as a numerical example with a realistic modelling of engine and aerodynamics characteristics of a turbo/ram jet propelled vehicle.

References

[1] Friehmelt, H.: *Thrust Vectoring and Tailless Aircraft Design – Review and Outlook.* In AIAA Atmospheric Flight Mechanics Conference Proceedings, San Diego, AIAA-96-3412, pages 420–430, 1996.

[2] Berens, T.: *Thrust Vector Optimization for Hypersonic Vehicles in the Transonic Mach Number Regime.* In AIAA/DGLR 5th International Aerospace Planes and Hypersonics Technologies Conference, München, AIAA-93-5060, December 1993.

[3] Speyer, J. L.: *Periodic Optimal Flight.* Journal of Guidance, Control, and Dynamics, 19(4):745–755, July 1996.

[4] Schultz, R. L.; Zagalsky, N. R.: *Aircraft Performance Optimization.* Journal of Aircraft, 9(2):108–114, 1972.

[5] Bulirsch, R.; Chudej, K.: *Ascent Optimization of an Airbreathing Space Vehicle.* In AIAA Guidance, Navigation and Control Conference, New Orleans, AIAA-91-2656, August 1991.

[6] Oberle, H. J.: *Numerical Computation of Singular Control Functions in Trajectory Optimization Problems.* Journal of Guidance, Control, and Dynamics, 13(1):153–159, July 1990.

[7] Kumar, R.; Kelley, H. J.: *Singular Optimal Atmospheric Rocket Trajectories.* Journal of Guidance, 11(4):305–312, July 1988.

[8] Sachs, G.; Lesch. K.: *Periodic Maximum Range Cruise with Singular Control.* Journal of Guidance, Control, and Dynamics, 16(4):790–793, July 1993.

[9] Mehlhorn, R.; Lesch, K.; Sachs, G.: *A Technique for Improving Numerical Stability and Efficiency in Singular Control Problems.* In AIAA Guidance, Navigation and Control Conference Proceedings, pages 388–396, 1993.

International Series of Numerical Mathematics
Vol. 124, © 1998 Birkhäuser Verlag, Basel

Determining the Controllability Region for the Re-Entry of an Apollo-Type Spacecraft

Dietmar W. Tscharnuter *

Abstract. In closed and open control loops, particularly in optimal control problems, the controllability region of the perturbations is a factor of importance. One looks for those controls which guarantee that the system stays within the controllability region: small perturbations must be compensated by suitable controlling in such a way that the controllability region is not left. In general, optimal control problems are characterized by a given quality functional, which has to be minimized, and by boundary conditions. Especially in closed optimal loops the terminal condition has to be fulfilled either exactly or within a prescribed neighborhood of the terminal point, when talking about real-time controlling. For the special problem of the atmospheric re-entry of an Apollo-type capsula the aim of this paper is to determine the controllability region and to characterize it by a manifold. This manifold will be constructed by methods of the stability theory, since the controllability region can be interpreted as a generalized epsilon-tube, where the nominal orbit that can be determined beforehand is flight-path stable. Some numerical results are presented.

1. Introduction

For the re-entry manœuvre of Apollo-type spacecrafts the determination of the controllability region is one of the major, still unsolved problems. It belongs to the class of optimal control problems which can be treated by methods based on multipoint boundary value problem solvers (see, e.g., [8]). Furthermore, perturbations of the initial data characterizing the optimal trajectory ("fiducial trajectory") also demand an effective feedback control. Its application yields the desired neighboring extremals [6]. What is not known in the first place is whether or not the terminal conditions can be fulfilled. The task is to develop criteria, dependent on the chosen feedback control strategy, which guarantee that the trajectories will stay within the admissible domain of the state variables at the terminal epoch. This is a problem of stability theory, in particular, the determination of the *basin of attraction* for a given equilibrium point (see, e.g., [1], [2], [3], [4], [5], [10]). In our case the "equilibrium point" is the whole fiducial trajectory, if the perturbed neighboring extremals are described by relative coordinates. This procedure is outlined in Section 4.

Sections 2. and 3. deal with basic definitions and a simple application of LYAPUNOV's direct method to classify the stability behavior of equilibrium points. Some new ideas of how to proceed in the case of the re-entry problem and numerical results are discussed in Section 5. The last, Section 6., deals with new questions which have come up as a result of the numerical findings.

*Lehrstuhl für Höhere Mathematik und Numerische Mathematik — FORTWIHR, Technische Universität München, D-80290 München, Germany, email: tscharnu@mathematik.tu-muenchen.de

2. Basic definitions

There are several definitions of the rather intuitive notion of stability. In the sequel, stability in the sense of LYAPUNOV is used throughout.

We start out with the equation of motion $\dot{x} = f(t, x)$ for which a solution is assumed to exist and to be unique within the domain $\Omega = \{x : \|x\| < A\}$ for $t \geq 0$. Without loss of generality we may further assume $f(t, 0)$ to be identically zero for $t \geq 0$. Otherwise it is not difficult to find a suitable transformation which yields the desired result.

Definition 1 *A solution* $z = (z_1, \ldots, z_n)$ *of an equation of motion is called* stable, *if for every* $\epsilon > 0$ *there is a* $\delta = \delta(\epsilon)$, *so that any other solution* y *of the equation of motion, for which* $\|z - y\| < \delta$ *holds at the initial epoch* $t = t_0$, *satisfies the inequality* $\|z - y\| \leq \epsilon$ *for all* $t > t_0$.

In order to demonstrate the stability of an orbit, let me now introduce a particular type of function. Such functions can also be used for estimating the controllability region.

Definition 2 *The function* $V(t, x)$ *is called* positive definite, *if*

(1) $V(t, x)$ *is defined on* $R^+ \times \Omega \subset R^+ \times R^n$,

(2) $V(t, 0) = 0 \quad \forall t \geq 0$,

(3) $V(t, x)$ *dominates a certain continuous function* $W(x)$, $0 < W(x) \leq V(t, x)$ $\forall t \geq 0, \forall x \in \Omega - \{0\}$.

Definition 3 *If* $V(t, x)$ *is positive definite and, in addition,*

$$(4) \quad \frac{\partial V}{\partial x_i}(t, x) \in C^0(\Omega),\ i = 1, \ldots, n; \quad \text{hence } \dot{V}(t, x) = V_t + \nabla V \cdot f(t, x)$$

holds with

(5) $-\dot{V}(t, x)$ *being positive semi-definite*

then the function $V(t, x)$ *is called a* LYAPUNOV function.

3. Lyapunov's direct method

By means of a positive definite function satisfying the aforementioned conditions (4) and (5) it is possible to show the stability of the *equilibrium point* of a system of differential equations. $x_0 \in \Omega$ is called an equilibrium point, if $f(t, x_0) = 0\ \forall t \geq 0$. According to LYAPUNOV's theorem (see, e.g., [2], [10]), the origin is a stable equilibrium point if there exists a LYAPUNOV function in any neighborhood of the origin. Moreover, if $-\dot{V}(t, x)$ is positive definite, then the origin is even *asymptotically* stable.

It should be stressed that the existence of a LYAPUNOV function is only sufficient to guarantee the stability of the equilibrium, but not necessary and by no means unique. The big disadvantage of LYAPUNOV's stability theorem is the lack of a general method for constructing a LYAPUNOV function (see, e.g., [5]).

3.1 Basins of attraction

However, for autonomous systems, the situation is much more convenient. Since in this case the right-hand sides can be expanded into formal power series, a positive definite function $V(x)$ can be constructed, which contains information about the stability behavior on the one hand and, on the other hand, about the shape and extension of the so-called *basin of attraction*.

Definition 4 *The* basin of attraction $\mathcal{E}$ *consists of the set of all initial data, for which the respective orbits run into the equilibrium point if* $t \to \infty$.

$\mathcal{E}$ is open, connected, and invariant in the following sense: If a starting point, i.e., an initial value, belongs to the interior of the basin of attraction, then the complete trajectory for $t \to \pm\infty$ lies entirely in $\mathcal{E}$. The LYAPUNOV function $V(x)$ characterizing $\mathcal{E}$ becomes infinite if x approaches the boundary $\partial\mathcal{E}$.

Applying the ZUBOV transformation $V \to 1 - \exp(-V)$ we arrive again at a LYAPUNOV function which, however, approaches 1 if $x \to \partial\mathcal{E}$, and its domain can be extended over the whole R^n.

3.2 Construction of LYAPUNOV functions — an example

Let $x = (x_1, \ldots, x_n)$ be a n-vector, A a $n \times n$-matrix, and $(r_k(x))_{k=2,3,\ldots}$ a family of vector-valued functions homogeneous of order k, respectively. Then, for autonomous systems of the form

$$\dot{x} = Ax + \sum_{k\geq 2} r_k(x),$$

which are called *systems of differential equations with homogeneous right members* $r(x) = \sum_k r_k(x)$, the appropriate LYAPUNOV function is uniquely defined.

To verify this assertion, we represent $V(x)$ as a sum of functions $V_k(x)$ being homogeneous of order k, thus

$$V(x) = \sum_{k\geq 2} V_k(x).$$

The time derivative of $V(x)$ along the given trajectory has to be, by definition, negative definite:

$$\dot{V}(x) = \sum_{k\geq 2} \langle \nabla V_k(x), \dot{x} \rangle = -\sum_{k\geq 2} h_k(x), \quad h(x) := \sum h_k(x) > 0.$$

The $(h_k(x))_{k=2,3,\ldots}$ are arbitrary functions, so that the sum $\sum_{k\geq 2} h_k(x)$ is positive definite. We now combine terms of equal degree, and by comparison of the respective coefficients we end up with the following recursive system of equations:

$$\begin{array}{lll} k=2: & \langle \nabla V_2, Ax \rangle = -h_2(x) & \text{yields } V_2(x) \\ k=3: & \langle \nabla V_3, Ax \rangle = -h_3(x) - \langle \nabla V_2, r_2(x) \rangle & \text{yields } V_3(x) \\ \vdots & \quad\vdots \qquad\quad \vdots & \end{array}$$

The V_k are uniquely defined by this recursion; as a matter of fact, it is sufficient to prove that V_2 can be uniquely determined (see, e.g., [10]). The function

$$V^* := \sum_{k=2}^{N} V_k$$

can be used for estimating the basin of attraction $\mathcal{E}$. The following example (cf. [9]) illustrates the situation. Suppose we are given the following system of autonomous differential equations:

$$\begin{aligned} \dot{x} &= -y \\ \dot{y} &= -5x - 6y + xy^2 + 2y^3 \end{aligned} \tag{1}$$

It is obvious that the origin $(x, y) \equiv (0, 0)$ is a stationary solution, since the right-hand side consists of monomes only.

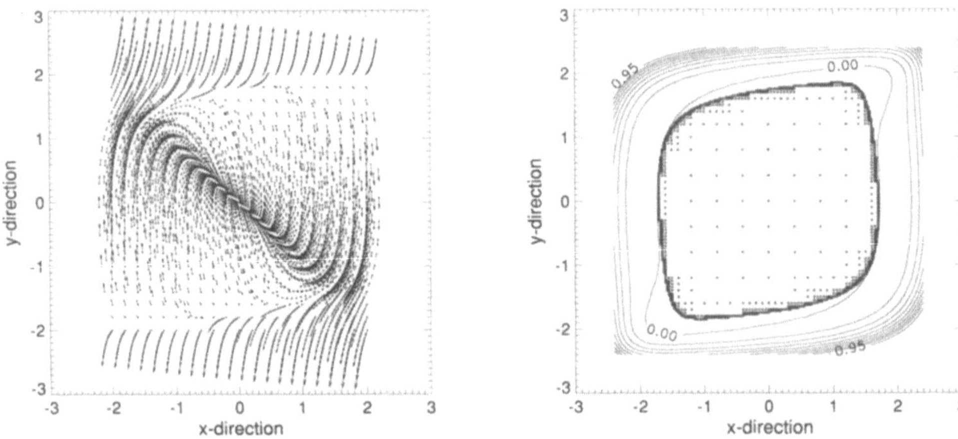

Figure 1: Solutions of Equations 1 ($n = 2$). Left panel: Trajectories and velocity field in the vicinity of the equilibrium point $(0, 0)$. Right panel: The basin of attraction determined numerically (dots) and analytically (contour lines).

Before estimating the basin of attraction $\mathcal{E}$ by construction of the appropriate LYAPUNOV function, as outlined above, it turned out to be a good idea to plot the trajectories for a set of initial values, equally spaced in the x- and y-direction. Figure 1 gives a nice impression of the qualitative behavior of the solutions. In the left panel the velocity field $(\dot{x}, \dot{y})$ in the phase space (R^2) is depicted. By an adaptive refinement of the numerical mesh it is possible to determine the boundary $\partial\mathcal{E}$ of the basin of attraction $\mathcal{E}$ numerically. The result is displayed in the right panel. Notice the sharply rising point concentration near $\partial\mathcal{E}$! The recursive method for constructing the LYAPUNOV function, i.e., $V(x, y) \approx V_N(x, y) := \sum_{k=2}^{N} V_k(x, y)$, say, with $N = 10$, turned out to be very effective for this example. As one can easily read off the right panel, the equicontour line for the zero level of the function $V_{10}(x, y) - V_9(x, y)$ can be regarded as a very good approximation of $\partial\mathcal{E}$. Figure 2 displays $V_{10}(x, y)$ as a wire frame surface.

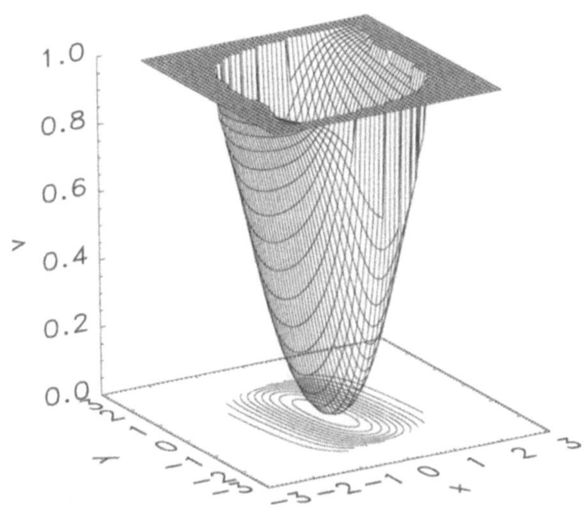

Figure 2: Stability of the equilibrium point $(0,0)$ of Equations 1. LYAPUNOV function after ZUBOV.

4. Stability and optimal control

We now move on to the investigation of the stability of the fiducial orbit pertaining to an optimum control problem. Here, the starting point is the following one: We are given state variables $x = (x_1, \ldots, x_n)$ which satisfy a certain system of equations of motion $\dot{x} = f(x,u)$, where u denotes the parameter pertinent to controlling. Furthermore, we assume the performance index as well as the initial and terminal conditions to be given.

The theory of optimum controlling then supplies us with *necessary* conditions, so that the problem can be formulated as a boundary value problem. Now, if one considers optimum control processes including perturbations, it is assumed that, for the perturbed "initial data", there is again an optimum solution. Since the perturbations are sufficiently small, this optimum solution will not leave a certain neighborhood of the fiducial orbit. This suggests linearization around the fiducial orbit.

However, in order to be able to investigate the stability behavior and to estimate the controllability region we first need to study the relative motion. If t_{f0} denotes the terminal epoch, we transform the time interval $[0, t_{f0}] \rightarrow [0,1]$ in the first place and can then write down the equations for the fiducial orbit $x_0(\tau), \tau \in [0,1]$ in the form

$$\begin{aligned}
\dot{x_0} &= f(x_0, \lambda_0)\, t_{f0} \\
\dot{\lambda_0} &= g(x_0, \lambda_0)\, t_{f0} \\
\dot{t_{f0}} &= 0.
\end{aligned}$$

Similarly, the equations of the relative motion read

$$\begin{aligned} \dot{\Delta x_0} &= f(x_0 + \Delta x, \lambda_0 + \Delta\lambda)\,(t_{f_0} + \Delta t_{f_0}) - f(x_0, \lambda_0)\,t_{f_0} \\ \dot{\Delta \lambda_0} &= g(x_0 + \Delta x, \lambda_0 + \Delta\lambda)\,(t_{f_0} + \Delta t_{f_0}) - g(x_0, \lambda_0)\,t_{f_0} \\ \dot{\Delta t_{f_0}} &= 0. \end{aligned} \tag{2}$$

The equations of the relative motion are nonlinear and, after all, non-autonomous. The origin $(\Delta x, \Delta\lambda, \Delta t_f) = (0, 0, 0)$ is the equilibrium point whose stability behavior has to be investigated. Notice that linearization of the right-hand side leads to the linear feedback law

$$\begin{pmatrix} \dot{\Delta x} \\ \dot{\Delta\lambda} \end{pmatrix} = J(t) \begin{pmatrix} \Delta x \\ \Delta\lambda \end{pmatrix}.$$

For studying the stability as such it is sufficient to consider only this variation equation. However, in doing so we are dealing with stability properties only of first order. We have to prove that, if $x \to 0$ the nonlinear part $r(x)$ approaches zero faster than the norm of x, i.e.,

$$\lim_{x \to 0} \frac{\|r(x)\|}{\|x\|} = 0$$

For estimating the controllability region, however, it is more convenient to rely upon the variation equation. In this case the controllability region is to be estimated with respect to the feedback control.

Definition 5 *The fiducial solution x_0 of Equations 2 is called* stable *if for every $\epsilon > 0$ there exist positive numbers $\delta(\epsilon)$, $\eta(\epsilon)$, so that for any other orbit x holds:*

$$\begin{aligned} \|x - x_0\| &< \epsilon \quad \text{if } t = t_f \quad \text{maximum deviation at the terminal epoch,} \\ \|x - x_0\| &< \delta \quad \text{if } t = t_0 \quad \text{maximum deviation at the initial epoch,} \\ \|x - x_0\| &< \eta \quad \text{if } t_0 < t < t_f \quad \text{(“}\eta\text{-tube”).} \end{aligned}$$

This definition of stability is compatible with the stability in the sense of LYAPUNOV. However, we consider a finite time interval and, in addition, we prescribe the maximum admissible deviation from the reference orbit at the terminal epoch t_f.

Because of the explicit time dependence of the relative motion, the LYAPUNOV function depends on space *and* time, according to the ansatz:

$$V(t,x) = \sum_{k \geq 2} V_k(t,x), \quad V_k = \sum_{i_1 + \ldots + i_n = k} c_{i_1 \ldots i_n}(t)\, x_1^{i_1} x_2^{i_2} \ldots x_n^{i_n}.$$

Since there is a non-vanishing contribution of the partial time derivative V_t to the total time derivative of V, we arrive at a *recursive system of differential equations* for the coefficients $c_{i_1 \ldots i_n}(t)$. Unfortunately, $V(t,x)$ is not uniquely defined, as the initial values can be chosen arbitrarily. This makes the estimation of the basin of attraction considerably more difficult. How to do that in detail has to be worked out yet.

5. Application to Apollo-type spacecrafts

As an example, I derived the equations describing the relative motion for the re-entry problem of an Apollo-type spacecraft and integrated them numerically for a variety of initial parameters. The results are displayed in Fig. 3 and 4. There are 4 state variables, v, γ, ξ, ζ, the velocity, flight path angle, the normalized altitude, and the distance on the earth's surface, respectively. These variables satisfy the equations of motion (cf. [6], [7], [8])

$$\begin{aligned}
\dot{v} &= -\frac{S}{2m}\rho v^2 c_D(u) - \frac{g\sin\gamma}{(1+\xi)^2} \\
\dot{\gamma} &= \frac{S}{2m}\rho v\, c_L(u) + \frac{v\cos\gamma}{R(1+\xi)} - \frac{g\cos\gamma}{v(1+\xi)^2} \\
\dot{\xi} &= \frac{v}{R}\sin\gamma \\
\dot{\zeta} &= \frac{v}{1+\xi}\cos\gamma
\end{aligned} \tag{3}$$

with prescribed initial and terminal conditions. The control variable $u < u_{\max}$ is the angle of attack of the wing-flap. R and g are the earth's radius and gravitational acceleration, respectively, $\rho = \rho(\xi)$ is the air density, S/m the area per mass of the spacecraft, $c_D(u) = c_{D_0} + c_{D_L}\cos(u)$ and $c_L(u) = c_{L_0}\sin(u)$ denote the (effective) drag and lift coefficients, respectively. The performance index is the so-called convective heating constraint

$$J(u) = \int_0^{t_f} 10\, v^3 \sqrt{\rho}\, dt.$$

As was mentioned before, the theory of optimal controlling furnishes us with necessary conditions, so that we are led to solve a boundary value problem. With λ_γ, λ_v being the Lagrange parameter and c_{L_0}, c_{D_L} the lift and drag coefficients, respectively, the optimal control then is given by

$$\tan u = -\frac{c_{L_0}}{c_{D_L}} \frac{\lambda_\gamma}{\lambda_v \cdot v}.$$

In order to estimate the controllability region in a first approximation, I have constructed the LYAPUNOV function with respect to the linearized equations of motion with the following ansatz:

$$\begin{aligned}
V &= \sum_{k=2}^{n} V_k(t,x), \quad x = (v,\gamma,\xi,\zeta) \\
V_k &= \sum_{i_1+\ldots+i_4=k} c^{(k)}_{i_1\ldots i_4}(t) v^{i_1}\gamma^{i_2}\xi^{i_3}\zeta^{i_4} \\
\dot{V} &= \langle \nabla V, Ax \rangle \equiv 0
\end{aligned}$$

This leads to a system of linear differential equations for the coefficients $c^{(k)}_{i_1\ldots i_4}(t)$. At the initial epoch, these coefficents are set equal 1.0. The integration has been carried out together with the integration of the reference orbit.

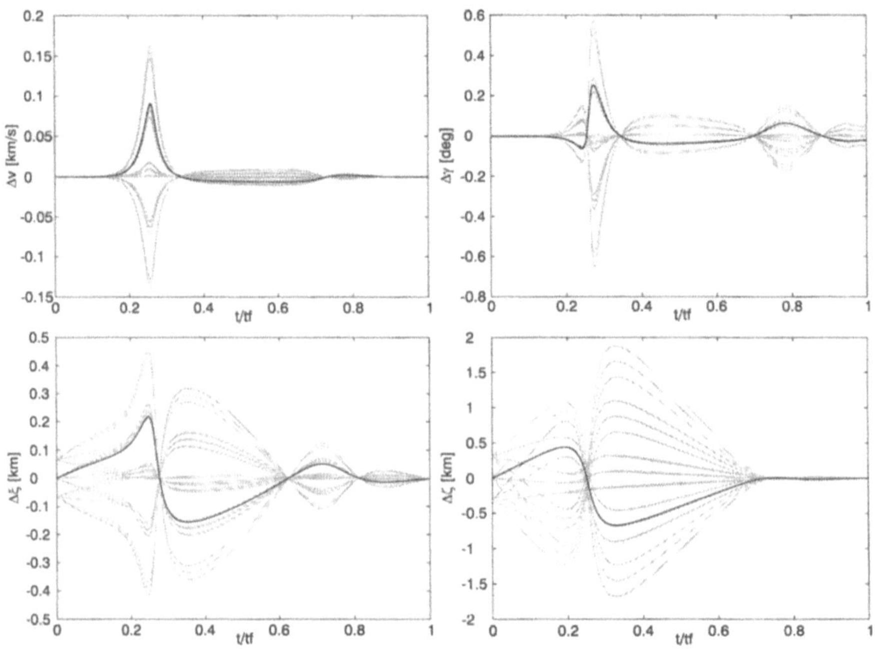

Figure 3: Solutions of Equations 3 (relative motions) for admissible initial data.

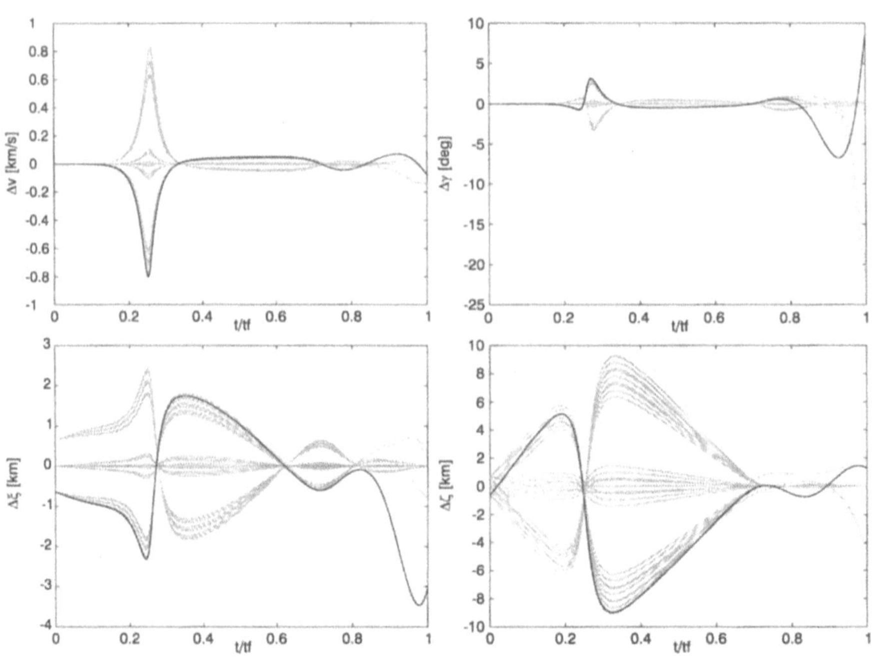

Figure 4: Some solutions of Equations 3 not matching the terminal conditions.

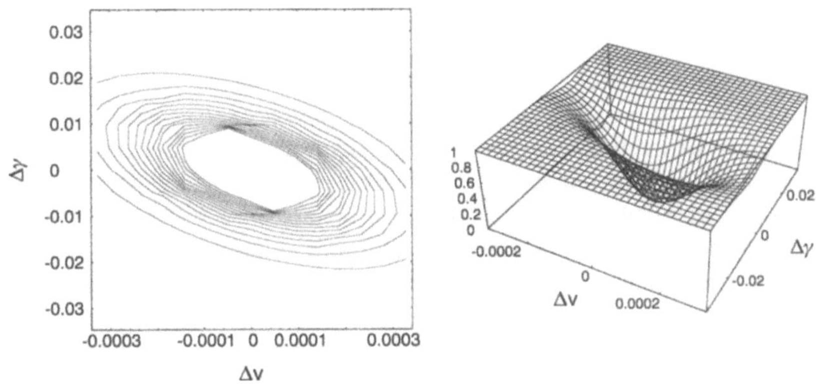

Figure 5: Estimation of the admissible domain for Δv and $\Delta\gamma$ ($\Delta\xi = \Delta\zeta = 0$) at the initial epoch t_0, visualized by equi-contour lines of the constructed LYAPUNOV surface before (left panel) and after (right panel) applying the ZUBOV transformation.

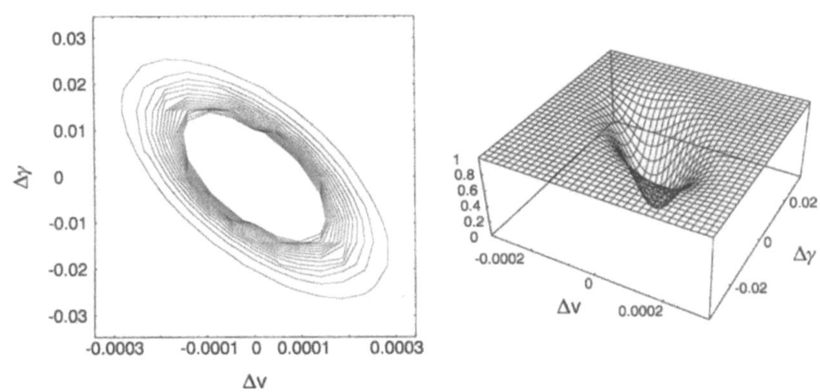

Figure 6: The same as in Fig. 5 at the terminal epoch t_f.

Figures 5 and 6 display the LYAPUNOV function for the initial epoch $t_0 = 0$ and the terminal epoch t_f, respectively. As the largest admissible deviations at t_f are given quantities, it looks promising to transform the function $V = \sum_{k=2}^{4} V_k$ in such a way that, for $t = t_f$, V takes on the value of approximately 0.99 at the boundary of the admissible domain, after the ZUBOV transformation has been applied:

$$
\begin{aligned}
V^*(x) &= \frac{-\log 0.01}{V(x^i_{\max})} V(x) = \sum_{k=2}^{4} \frac{-\log 0.01}{V(x^i_{\max})} V_k(x) \\
&= V_2\Big(\Big(\frac{-\log 0.01}{V(x^i_{\max})}\Big)^{1/2} x\Big) + V_3\Big(\Big(\frac{-\log 0.01}{V(x^i_{\max})}\Big)^{1/3} x\Big) + V_4\Big(\Big(\frac{-\log 0.01}{V(x^i_{\max})}\Big)^{1/4} x\Big),
\end{aligned}
$$

since V_k is homogeneous of degree k, i.e., $V_k(c\,x) = c^k\,V_k(x)$. The components of the vector $x^i_{\max}$ are all equal to zero, except the i-th one. Its value corresponds to the largest admissible deviation at $t = t_f$ for the respective component.

6. Open questions

The graphical representation of the individual state variables relative to the reference orbit, as is shown in Fig. 3 and 4, reveals an interesting, general property of the solutions: At certain, well-defined epochs the relative motions are confined to very small neighborhoods around the reference orbit. This is particularly true for the velocity and the flight path angle, being the most sensitive variables for which the admissible size of the neighborhood becomes extremely small and collapses almost to a point. By increasing the perturbation of the initial data to such an extent that the terminal condition is no longer satisfied, we arrive sooner or later at the situation where the orbits miss at least one of these node points, thus becoming inadmissible. This finding immediately raises the following questions:

(1) Do such nodes indeed exist?
(2) If yes, what about their classification?
(3) Are they of any significance for the stability, controllability, or feedback control?

References

[1] Bellman, R.: *Stability Theory of Differential Equations,* McGraw-Hill, New York, Toronto, London, 1953.

[2] Cesari, L.: *Asymptotic Behavior and Stability Problems in Ordinary Differential Equations.* Ergebnisse der Mathematik und ihrer Grenzgebiete, Springer-Verlag, Berlin, Göttingen, Heidelberg, 1959.

[3] Hahn, W.: *Stability of Motion.* Die Grundlagen der mathematischen Wissenschaften in Einzeldarstellungen, Band 138, Springer-Verlag, Berlin, Heidelberg, New York, 1967.

[4] La Salle, J.; Lefschetz, S.: *Die Stabilitätstheorie von Lyapunow — Die direkte Methode mit Anwendungen.* B.I. Hochschultaschenbücher, Mannheim, 1967.

[5] Parks, P. C.; Hahn, V.: *Stabilitätstheorie.* Hochschultext, Springer-Verlag, Berlin, Heidelberg, New York, 1981.

[6] Pesch, H. J.: *Real-time Computation of Feedback Controls for Constrained Optimal Control Problems, Part 1: Neighboring Extremals; Part 2: A Correction Method Based on Multiple Shooting,* Optimal Control Applications and Methods 10, 129–145; 147–171, 1989.

[7] Scharmack, D. K.: *An Initial Value Method for Trajectory Optimization Problems,* in: Advances in Control Systems Vol. 5 (ed. C.T. Leondes), Academic Press, New York, 1967.

[8] Stoer, J., Bulirsch, R.: *Introduction to Numerical Analysis,* 2nd edition, Springer-Verlag, New York, 1993.

[9] Tscharnuter, D. W.: *Charakterisierung des Einzugsbereichs der Ruhelage eines Differentialgleichungssystems,* Diplomarbeit TU Wien, 1994.

[10] Zubov, V. I.: *Methods of A.M. Lyapunov and their Application,* (ed. L.F. Boron), P. Noordhoff Ltd., Groningen, The Netherlands, 1964.

International Series of Numerical Mathematics

International Series of Numerical Mathematics *is open to all aspects of numerical mathematics. Some of the topics of particular interest include free boundary value problems for differential equations, phase transitions, problems of optimal control and optimization, other nonlinear phenomena in analysis, nonlinear partial differential equations, efficient solution methods, bifurcation problems and approximation theory. When possible, the topic of each volume is discussed from three different angles, namely those of mathematical modeling, mathematical analysis, and numerical case studies.*

The latest volumes:

ISNM 123 **C. Bandle, W.N. Everitt, L. Losonczi, W. Walter** (Eds): General Inequalities 7. 7th International Conference, Oberwolfach, November 13–18, 1995, 1997 (ISBN 3-7643-5722-3)

ISNM 122 **A.M. Khludnev, J. Sokolowski**: Modelling and Control in Solid Mechanics, 1997 (ISBN 3-7643-5238-8)

ISNM 121 **R. Jeltsch, M. Mansour (Eds)**: Stability Theory. Hurwitz Centenary Conference, Centro Stefano Franscini, Ascona, 1995, 1996 (ISBN 3-7643-5474-7)

ISNM 120 **W. Hackbusch:** Integral Equations: Theory and Numerical Treatment, 1995 (ISBN 3-7643-2871-1)

ISNM 119 **R. Zahar (Ed.):** Approximation and Computation: A Festschrift in Honor of Walter Gautschi, 1995 (ISBN 0-8176-3753-2)

ISNM 118 **W. Desch, F. Kappel, K. Kunisch (Eds):** Control and Estimation of Distributed Parameter Systems: Nonlinear Phenomena. International Conference on Control and Estimation of Distributed Parameter Systems, Vorau, July 18-24, 1993, 1994 (ISBN 3-7643-5098-9)

ISNM 117 **R.E. Bank, R. Bulirsch, H. Gajewski, K. Merten (Eds):** Mathematical Modelling and Simulation of Electrical Circuits and Semiconductor Devices, 1994 (ISBN 3-7643-5053-9)

ISNM 116 **P.W. Hemker, P. Wesseling (Eds):** Multigrid Methods IV. Proceedings of the Fourth European Multigrid Conference, Amsterdam, July 6-9, 1993 (ISBN 3-7643-5030-X)

MIX
Papier aus verantwortungsvollen Quellen
Paper from responsible sources
FSC® C105338

If you have any concerns about our products,
you can contact us on
ProductSafety@springernature.com

In case Publisher is established outside the EU,
the EU authorized representative is:
Springer Nature Customer Service Center GmbH
Europaplatz 3, 69115 Heidelberg, Germany

Printed by Libri Plureos GmbH
in Hamburg, Germany